621.381 3wks
DER

00

D1685946

Centre for Applied Sciences Library
City and Islington College
311-321 Goswell Road
London EC1V 7DD
Tel: 020 7520 7471
Email: scilib@candi.ac.uk

CITY AND ISLINGTON
COLLEGE

This book is due for return on or before the date stamped below. You may renew by telephone. Please quote the barcode number or your student number. This item may not be renewed if required by another user.

Fine : 10p per day

THREE WEEK LOAN

D04911

DO4911

Practical Interfacing in the Laboratory

Using a PC for Instrumentation,
Data Analysis, and Control

This text describes in practical terms how to use a desk-top computer to monitor and control laboratory experiments. The author clearly explains how to design electronic circuits and write computer programs to sense, analyze, and display real-world quantities, including displacement, temperature, force, sound, light, and biomedical potentials. The book includes numerous laboratory exercises and appendices that provide practical information on microcomputer architecture and interfacing, including complete circuit diagrams and component lists. Topics include analog amplification and signal processing, digital-to-analog and analog-to-digital conversion, electronic sensors and actuators, digital and analog interfacing circuits, programming, and data analysis and control. Only a very basic knowledge of electronics is assumed, making it ideal for college-level laboratory courses and for practicing engineers and scientists.

Stephen E. Derenzo is Professor-in-Residence in the Department of Electrical Engineering and Computer Sciences at UC Berkeley and a Senior Scientist at the Lawrence Berkeley National Laboratory. He has been teaching courses on electronic circuits, electronic transducers, and microcomputer interfacing for over 15 years and this book was developed from those courses. He has authored and co-authored over 150 technical publications, was awarded the 1992 Annual Merit Award and the 2001 Radiation Instrumentation Outstanding Achievement Award of the Nuclear and Plasma Sciences Society of the IEEE, and is a Fellow of the IEEE.

Practical Interfacing in the Laboratory

Using a PC for Instrumentation,
Data Analysis, and Control

Stephen E. Derenzo

University of California, Berkeley, California

CAMBRIDGE
UNIVERSITY PRESS

PUBLISHED BY THE PRESS SYNDICATE OF THE UNIVERSITY OF CAMBRIDGE
The Pitt Building, Trumpington Street, Cambridge, United Kingdom

CAMBRIDGE UNIVERSITY PRESS
The Edinburgh Building, Cambridge CB2 2RU, UK
40 West 20th Street, New York, NY 10011-4211, USA
477 Williamstown Road, Port Melbourne, VIC 3207, Australia
Ruiz de Alarcón 13, 28014 Madrid, Spain
Dock House, The Waterfront, Cape Town 8001, South Africa

http://www.cambridge.org

This edition © Cambridge University Press 2003

This book is in copyright. Subject to statutory exception
and to the provisions of relevant collective licensing agreements,
no reproduction of any part may take place without
the written permission of Cambridge University Press.

First published as *Interfacing: A laboratory approach using the microcomputer for instrumentation, data analysis and control* by Prentice-Hall, Englewood Cliffs, NJ, 1990.
First published by Cambridge University Press 2003

Printed in the United Kingdom at the University Press, Cambridge

Typefaces Times 10.5/14 pt, Helvetica Neue, and Arial   *System* LaTeX 2_ε [TB]

A catalogue record for this book is available from the British Library

Library of Congress Cataloguing in Publication data

Derenzo, Stephen E.
Practical interfacing in the laboratory : using a PC for instrumentation, data analysis, and control / Stephen E. Derenzo.
        p.   cm.
Rev. ed. of: Interfacing, c1990.
Includes bibliographical references and index.
ISBN 0-521-81527-4
1. Computer interfaces.   2. Microcomputers.   3. Automatic data collection systems.
4. Computer interfaces – Laboratory manuals.   5. Microcomputers – Laboratory manuals.
6. Automatic data collection systems – Laboratory manuals.   I. Derenzo, Stephen E.
Interfacing.
TK7887.5 .D42   2002
620′.0028′4 – dc21   2001052859

ISBN 0 521 81527 4 hardback

Dedicated to:

My mother, Alice, and my father, Stanley
for their lifelong support and encouragement

My brother, David
for our good times together

My wife, Carol
for being my partner, wife, friend, and adviser

My children, Jennifer and Julia
for reminding me of the joy of youth

My students and teaching associates
who, over the years, helped improve the
laboratory exercises and pointed out my errors

Contents

3   Analog ↔ digital conversion and sampling   153

4   Sensors and actuators   226

5   Data analysis and control   360

Preface

This text describes in practical terms how to use the microcomputer to sense real-world quantities such as temperature, force, sound, light, etc., to analyze the data rapidly, to display the results, or to use the results to perform a control function. It was written for practicing engineers and scientists, and as a textbook for laboratory courses in electronic transducers and microcomputer interfacing.

Our approach takes full advantage of the availability of relatively low-cost microcomputers that are powerful enough to support high-speed parallel input/output (I/O) ports, data-acquisition circuit boards, graphical operating systems, high-level programming languages, and fast double-precision calculations. This book shows in practical terms the range of problems in data acquisition, analysis, display, and control that can be tackled in a cost-effective manner without delving into the bus protocol or native language of a particular microprocessor.

The book contains five chapters, covering digital tools, analog tools, conversion between analog and digital signals, sensors and actuators, and data analysis and control. The 27 laboratory exercises can be used either in a college-level laboratory course or as working examples for practicing engineers and scientists who wish to apply sensor, low-level amplification, and microcomputer principles in their work in a practical and immediate way.

This material was developed for two one-semester laboratory courses in the Electrical Engineering and Computer Science Department at the University of California in Berkeley, EECS 145L: "Electronic Transducer Laboratory" and EECS 145M: "Microcomputer Interfacing Laboratory." The purpose of these two courses is to provide upper-level undergraduate students with the tools needed to sense and control "real-world" quantities, such as temperature and force, as well as to display the results of "real-time" analyses, such as least-squares fitting, the Student's t test, fast Fourier transforms, digital filtering, etc. **It is assumed that the students have had some exposure to elementary analog and digital electronics, differential calculus and linear algebra, and the C programming language.**

Over the years, we have used several different microcomputer systems in the laboratory, and the laboratory exercises were designed to be as machine-independent as possible. Special instructions (such as Appendices E and F) were provided for the particular counter/timer, parallel I/O port, and data-acquisition board that were used. A recent advance is software support in the form of C-callable drivers that make it relatively easy to perform single-word and block-data acquisitions and transfers in the Windows NT environment.

The C programming language was chosen because it is available for almost all microcomputers and is well-suited to data acquisition, analysis, and control. It provides word and byte I/O, bit manipulation, powerful conditional branching and data structures, a wide choice of accuracy and bit length for integer and floating point numbers, and high-speed execution.

Chapter 1, "Digital tools," briefly describes the overall organization of the microcomputer, binary and 2's complement number systems, and the digital components needed to perform data acquisition and control, such as digital timers, latches, registers, tri-state buffers, and parallel I/O ports. It goes on to describe the digital and control aspects of several data-acquisition procedures, and discusses the level of handshaking needed for various applications.

Laboratory Exercise 1 introduces the Windows NT operating system, the C compiler/editor, and the many ways that binary bit patterns can be interpreted as numerical quantities. Laboratory Exercise 2 provides examples using the microprocessor timer to measure human reaction times, and Laboratory Exercise 3 introduces the parallel I/O ports, reading switches, and controlling lights.

Chapter 2, "Analog tools," covers commonly used op-amp circuits, the instrumentation amplifier used for low-level differential amplification of sensor signals, noise sources, and the analog signal processing that can be used to enhance the signal-to-noise ratio. It goes on to describe a class B power amplifier that can be used to drive actuators.

Laboratory Exercises 4 and 5 explore op-amp circuits, instrumentation amplifiers, differential amplification, and noise sources, including electromagnetic interference. Laboratory Exercise 6 explores analog signal processing using the op amp, including active high-pass, low-pass, and notch filters.

Chapter 3, "Analog ↔ digital conversion," covers the data-conversion components needed to perform data acquisition and control, such as digital-to-analog (D/A) and analog-to-digital (A/D) converters, the sample-and-hold amplifier, and the comparator. It describes the commonly used methods for data sampling and introduces the notion of frequency aliasing resulting from inadequate sampling. (Considerations of aliasing in the Fourier domain are deferred to Chapter 5.) Chapter 3 lists and describes several commercially available circuit boards.

Laboratory Exercise 7 uses a commercial analog I/O board to provide an overview of both digital-to-analog and analog-to-digital conversion for those students who will not be doing Laboratory Exercises 8 and 9. The conversion between analog and digital is explored in Laboratory Exercises 8 and 9, using D/A and A/D integrated circuit chips. Laboratory Exercise 8 involves interfacing a D/A converter to a parallel input port and waveform generation. Laboratory Exercise 9 involves interfacing an A/D converter to a parallel output port, using a hardware "strobe" and "ready for data" and "data available" handshaking protocol. Laboratory Exercise 10 uses a commercial data-acquisition board for the periodic sampling of waveforms and demonstrates the concept of frequency aliasing in the time domain.

Chapter 4, "Sensors and actuators," covers the sensors (the first element in many data-acquisition systems), the real-world quantities that they sense, the nature of the signals (and the noise) that they produce, and actuators (essential in any control system).

Laboratory Exercises 11–14 explore the basic electronic transducers used to measure position, temperature, strain, force, and light. The thermoelectric heat pump is explored in Laboratory Exercise 15. Laboratory Exercise 16 investigates the ac and dc electrical properties of bare metal and Ag(AgCl) electrodes. Laboratory Exercises 17–19 explore physiological signals from the heart, skeletal muscles, and eyes.

Chapter 5, "Data analysis and control," covers data analysis, including statistical analysis; Student's t test; least-squares and Chi-squared fitting; continuous, discrete, and fast Fourier transforms, and some algorithms used for the control of real-world quantities.

Laboratory Exercise 20 explores analog-to-digital conversion for the storage of analog signals, digital-to-analog conversion for the analog recovery of those signals, and least-squares fitting for determining the accuracy of signal recovery. Laboratory Exercise 21 involves the sampling of sine, square, and triangle waves and the computation of their fast Fourier transforms (FFT). These techniques are applied in Laboratory Exercise 22 to the sampling and FFT of the human voice. Laboratory Exercise 23 compares analog to real-time digital filtering and Laboratory Exercise 24 demonstrates how the microcomputer can measure the impulse response of a linear, time-invariant system and use FFT techniques to determine the digital filter that can compensate for signal distortion caused by the system, provided that the frequency response of the system meets certain requirements. Laboratory Exercise 25 provides experience with analog temperature sensing and control. Laboratory Exercise 26 provides experience with computer-based digital temperature sensing and control using an electrical resistance oven and several algorithms. Laboratory Exercise 27 is similar to Laboratory Exercise 26, except that a thermoelectric heat pump is used with both the ability to heat and cool actively. An essential component is the LM12 power op amp.

In several laboratory exercises, a number of related circuits are built and examined. The *equipment* lists at the beginning of these exercises include all the parts needed for the students to build all the circuits before coming to the laboratory. As laboratory time is usually very limited, this approach works better than providing only the minimum number of parts needed and having the students dismantle one circuit during the laboratory period before they can build the next.

Each chapter is provided with problems derived from those used in midterm and final examinations.

Defined terms appear in the index followed by the word (definition) and the page number where they are first used. On that page, the term appears in bold face in the text that defines it.

Appendix A provides some physical and electronic units and constants for the problems at the end of the chapters, and Appendix B discusses issues of error propagation, and electrical shielding and grounds. Appendix C summarizes some hints useful in C programming. Appendix D provides C code listings and flow charts of some numerical methods, including the fast Fourier transform, nonlinear function minimization (used to fit curves to data), numerical integration using adaptive quadrature, and function inversion using both Newton's method and quadratic approximation. A program to compute the probability of exceeding Student's t is given as an example.

Appendix E describes the hardware and software needed to use the Data Translation DT3010 PCI plug-in board, and Appendix F describes how to use HP VEE to record waveforms on a digital oscilloscope. Appendix G discusses some potential electrical hazards and methods used to prevent them. Appendix H lists standard resistor and capacitor values and provides resistor color codes. Appendix I lists the ASCII character codes. Last is a glossary defining the technical terms used in the book.

Guide for the instructor

Although the entire book would serve for a full-year course, it is also possible to cover portions of the material in separate one-semester courses, as we do at Berkeley.

A one-semester course on *digital interfacing, data analysis, and control* would include Chapters 1, 3, and 5, and Laboratory Exercises 1–3, 8–10, 20–24, and 26 or 27.

A one-semester course on *sensors, low-level amplification, and analog signal processing* would include Chapters 2 and 4, and Laboratory Exercises 4–6 and 11–19. Portions of Chapter 5 and Laboratory Exercise 25 would provide an introduction to analog control.

A one-semester course on *bioengineering* would include Chapters 2, 4, and 5, and selections from Laboratory Exercises 2, 4–7, 11–19, and 20–22, depending on course emphasis.

A solutions set is available for this book – contact solutions@cambridge.org for details.

Acknowledgments

I am indebted to Kenneth Krieg, who was the cofounder of EECS 145M "Microcomputer Interfacing Laboratory" and, as teaching associate over a period of several years, made important contributions to most of the laboratory exercises. I also thank the numerous teaching assistants and students who contributed to the improvement of the laboratory exercises.

Special thanks to Professor Ted Lewis for contributions to Chapter 4, derived from his course EECS 145A, "Sensors, actuators, and electrodes," and to Dr Thomas Budinger for contributions to Chapter 5, derived from his course EECS 145B, "Computer applications in biology and medicine." Some of the laboratory exercises were derived from EECS 182, "Biological signals and transducers," developed by Professors Ted Lewis and Ed Keller at Berkeley during the 1970s, and to them I am grateful. I also thank John Cahoon, Matt Ho, and William Moses for discussions of circuit design, Ronald Huesman and Gerald Lynch for discussions of statistical analysis and fitting, and to Orin Dahl for discussions of pseudo-random number generators.

1 Digital tools

1.1 Introduction

In the past few years, enormous advances have been made in the cost, power, and ease of use of microcomputers and associated analog and digital circuits. It is now possible, with a relatively small expenditure, to purchase a microcomputer system that will take data, quickly analyze them, and display the results or control a process. This has been made possible by the development of technology that can fabricate millions of transistors, diodes, resistors, capacitors, and conductors on a single silicon **integrated circuit chip**.

Normally, the microcomputer is equipped with a number of standard items: the microprocessor chip and associated circuits, random-access memory chips, removable floppy and cartridge disk drives, magnetic hard disk drives, optical disk drives, keyboards, video display screens, serial interfaces, printers, and $x-y$ entry devices such as the mouse, trackball, joystick, bitpad, and touch-sensitive display screen. However, data acquisition and control require additional components, such as digital and analog input/output (I/O) ports, and counters/timers. Analog input ports contain analog multiplexers, sample-and-hold (S/H) amplifiers, and analog-to-digital (A/D) converters. Analog output ports contain digital-to-analog (D/A) converters.

Even for designs requiring only a microprocessor and a few additional circuits, there are considerable advantages to using the resources of the microcomputer during the development stage. These include program code editors and compilers, an operating system for the storage and manipulation of code and data files, and ample random-access memory.

In this chapter, we discuss digital interfacing concepts used in microcomputer-based data-acquisition and control systems (Figure 1.1), including parallel and serial input/output ports, handshaking, and digital counters/timers. Analog tools (amplification and filtering) are treated in Chapter 2, digital-to-analog and analog-to-digital conversion and sampling in Chapter 3, and sensors and actuators in Chapter 4.

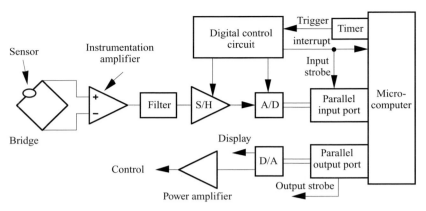

Figure 1.1 A microcomputer system interfaced to sensors and associated analog circuits for data acquisition, analysis, and control.

1.2   The microcomputer

In selecting a system for data acquisition and control, the **microcomputer** itself is a crucial component (Figure 1.2). The microcomputer is sufficiently small to fit on a laboratory bench (or desktop) and yet contains the following components:

1. The **microprocessor** is an integrated circuit that reads program instructions from memory and uses them to determine the sequence of actions that it performs. It is connected to memory and peripheral circuits by an address bus, a data bus, and control lines.

   These actions include reading data and instructions from memory, performing calculations, executing different instructions depending on the outcome of a calculation, printing data, and transferring data to and from peripheral devices such as hard disks. Microprocessors vary greatly in their speed and data-handling capability.

2. **Random-access memory (RAM)** usually consists of high-speed semiconductor memory chips that are used to store and retrieve program instructions and data. The highest data-acquisition speeds are achieved when external data are read directly into RAM, so the size of the RAM places a limit on the number of data values that can be sampled rapidly.

3. Common user interface devices are the keyboard, video display screen, printer, mouse, joystick, and trackball. Some systems provide voice input and synthesized speech output. The IEEE-1284 interface standard includes the standard parallel printer (SPP) port as well as other enhancements. The universal serial bus (USB) is the current standard for keyboards and pointing devices. For higher

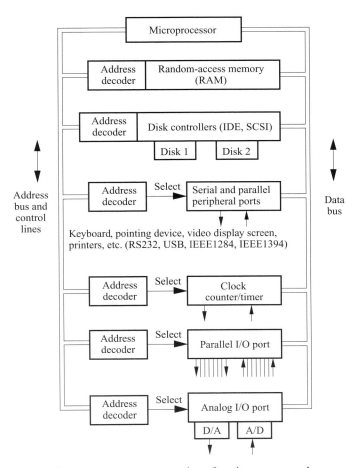

Figure 1.2 The microcomputer consists of a microprocessor that communicates with memory and input/output devices by address and data buses.

speed transfers (external hard drives, digital camcorders, HDTV), the IEEE-1394 standard (FireWire or i.Link) has recently been introduced.

4. **Magnetic disk memory** is used for the long-term storage of programs and data, and consists of one or more flat circular plates coated with a magnetic surface. Magnetic disk capacities range from 500 kbytes to 2 Mbytes for small removable floppy disks and from 1 to 20 Gbytes or more for hard disks. Access time consists of a fixed delay of tens of milliseconds (for the read/write head to locate the desired track) and a transfer time of typically 1 μs per 16-bit word.

5. **Optical disk memory** includes the CD-ROM and the DVD-ROM disks. The **CD-ROM** (compact disk-read-only memory) and DVD-ROM (digital versatile disk) drives use optical storage and retrieval technology that was developed for the music and entertainment industry. The capacity of the CD-ROM is over 600 Mbytes and about ten times larger for the DVD-ROM. Both are 12 cm in

diameter. Microcomputers and workstations are commonly shipped with a CD-ROM containing a back-up copy of the system software and on-line documentation, eliminating many floppy disks and thousands of pages of paper. CD-W (write once) and CD-RW (rewritable) and DVD-RAM (random-access) technology allows information to be written onto these disks.

6. The operating system permits the user to manipulate program and data files and supports a high-level compiled programming language (FORTRAN, Pascal, C, compiled BASIC, etc.).

7. A **compiler's** function is to translate a high-level language into microprocessor code that is able to:
    (i) perform numerical computations and conditional branching,
    (ii) communicate directly with a data-acquisition and control board or parallel I/O port (see below),
    (iii) read and write files to the disk.
    Additional useful features include:
    (i) a full range of scientific functions (sine, cosine, exp, log, etc.), and the ability to compute using floating-point representation, which can handle very small and very large numbers (for example, 80-bit extended precision can handle numbers from $\pm 10^{-4,932}$ to $\pm 10^{+4,932}$ with a precision of 19 decimal digits);
    (ii) the ability to write functions in assembly code for greater speed during data acquisition (some compilers permit intermixed assembly code and higher-level code);
    (iii) a built-in **editor** that displays lines causing compilation errors and permits immediate correction;
    (iv) a single command that compiles all changed program modules, links all necessary modules, and runs the result.

8. An analog input/output port (also called a data-acquisition and control circuit), with the required speed and number of A/D and D/A conversion circuits.

9. A parallel input/output port with sufficient speed, if item 8 is not available. In this case, it becomes necessary to design and build a data-acquisition circuit for connection to the parallel I/O port (this is demonstrated in Laboratory Exercise 9).

10. A counter/timer that can determine elapsed times to an accuracy of typically 1 μs, count input pulses, or produce output pulses of any desired width and period with an accuracy of typically 1 μs.

The microprocessor communicates with the other components of the microcomputer by an address bus, a data bus, and a number of control lines (Figure 1.2). The **address bus** allows the microprocessor to select particular components individually. Each component has a unique assigned **address** whether it is a RAM location, an I/O port register, or other peripheral circuit. An **address decoder** produces a **select** pulse whenever the assigned address appears on the address bus. For example, a 16-Mbit RAM chip has an internal address decoder with 24 input lines and 16 million select lines, one for each memory bit that can be selected. The **data bus** is used to transmit data words to and from the microprocessor and its associated circuits.

Note: In some systems, memory locations and external devices are distinguished from each other by a special control bit. In others, a large block of memory address space is reserved for external devices.

Since many devices are attached to the data and address buses and at any instant only one can be sending data, control lines are used to indicate when the bus is busy, when a sending device requests use of the bus, when use is granted, etc. These details are beyond the scope of this book and are mentioned to outline the organization of the microcomputer.

Laboratory Exercise 1 is designed to familiarize the reader with the particular editor and compiler that will be used for the rest of the exercises as well as review 2's complement, hexadecimal, real, and integer interpretations of binary numbers.

1.3   Number systems

1.3.1   Binary number representations

Binary numbers can be interpreted in a variety of ways. Table 1.1 shows the interpretation of 8-bit binary patterns as unsigned decimal, hexadecimal, Gray, and 2's complement numbers. The 16-bit and 32-bit numbers are logical extensions.

A/D converters and counters/timers produce binary bit patterns that are to be interpreted as unsigned numbers. The binary sequence runs continuously from all bits = 0 to all bits = 1 and the leftmost bit is the most significant bit (**MSB**).

Angle and position encoders usually produce **Gray code** that runs from all bits = 0 to all bits = 1, but the binary sequence is not continuous because it has the special property that advancing from one number to the next involves changing the state of only one bit. Gray code is described further in the following section.

Binary numbers can also be represented in **hexadecimal** form (base 16) for efficient notation. Note that each 8-bit byte can be represented as two hexadecimal digits. Octal (base 8) is less frequently used.

Binary bit patterns can also be interpreted as signed numbers, to include negative numbers (<0) as well as 0 and positive numbers (>0). Some computers use signed binary representation, where the leftmost bit represents the sign. While this representation is closer to that of the printed page, it is seldom used in computers because arithmetic operations take longer due to the need to process the sign bit.

Most microcomputers use **2's complement representation** to deal more efficiently with negative and positive numbers. In 2's complement representation, the sign of a number is changed by complementing (reversing) all its bits and then adding one. This is called the **2's complement operation**. By using this operation, the subtraction process

Table 1.1 *Interpretations of 8-bit binary numbers*

Binary	Unsigned decimal	Hexadecimal	Gray	2's complement
0000 0000	0	00	0	0
0000 0001	1	01	1	1
0000 0010	2	02	3	2
0000 0011	3	03	2	3
0000 0100	4	04	7	4
0000 0101	5	05	6	5
0000 0110	6	06	4	6
0000 0111	7	07	5	7
0000 1000	8	08	15	8
0000 1001	9	09	14	9
0000 1010	10	0A	12	10
0000 1011	11	0B	13	11
0000 1100	12	0C	8	12
0000 1101	13	0D	9	13
0000 1110	14	0E	11	14
0000 1111	15	0F	10	15
0001 0000	16	10	31	16
. . .	. . .	. . .	. . .	. . .
0111 1110	126	7E	65	126
0111 1111	127	7F	64	127
1000 0000	128	80	192	−128
1000 0001	129	81	193	−127
. . .	. . .	. . .	. . .	. . .
1111 1110	254	FE	129	−2
1111 1111	255	FF	128	−1

$a - b$ can be performed by adding a to the 2's complement of b. For an 8-bit number, 2 is represented as binary 0000 0010 (hexadecimal 02) and −2 is represented as binary 1111 1110 (hexadecimal FE). For example, $5 - 2 = 3$ in 2's complement arithmetic is:

5	0000 0101	simply add, but ignore
−2	1111 1110	the most significant carry bit
3	0000 0011	

Note that in 2's complement notation, positive numbers have their MSB = 0 and negative numbers have their MSB = 1.

Warning: sign extension

As demonstrated in Laboratory Exercise 1, if the MSB of a number is zero, then conversion from 8 to 16 bits or from 16 to 32 bits occurs as expected. However, if the

Table 1.2 *Typical variable types, storage, and ranges of values*

Type	No. bits	Decimal digits	Range
Char	8		-128 to $+127$
Unsigned char*	8		0 to 255
Short	16		$-32,768$ to $+32,767$
Unsigned integer	16		0 to 65,535
Int and long	32		$-2,147,483,648$ to $2,147,483,647$
Unsigned long*	32		0 to 4,294,967,295
Float	32	7	$\pm 1.2 \times 10^{-38}$ to $\pm 3.4 \times 10^{+38}$
Double	64	14	$\pm 2.3 \times 10^{-308}$ to $\pm 1.7 \times 10^{+308}$
Extended*	80	19	$\pm 1.7 \times 10^{-4932}$ to $\pm 1.1 \times 10^{+4932}$

*Standard in ANSI C, but not available on all C or Pascal compilers.

MSB is one, then the leftmost additional bits of the longer number will be filled with ones (**sign extension**). In this way, the converted number will have the same numerical value in 2's complement representation. For example, when transferred from char to int, 35 becomes 0035 and 8A becomes FF8A. Thus, if unsigned numbers are read from a counter/timer or A/D converter in blocks of eight bits, some precautions are necessary before they can be packed into 16- or 32-bit numbers. There are two approaches:

1. Mask the left half of the number with zeros (see Appendix C).
2. Declare all relevant variables to be "unsigned."

Table 1.2 shows the typical internal representations available on microcomputers. They are also explored in Laboratory Exercise 1. Each program variable is declared to be one of these types. The float, double, and extended have 8-, 11-, and 15-bit exponents and 23, 52, and 63 bits of precision, which correspond to 7, 15, and 19 decimal digits of precision, respectively.

1.3.2   Gray code

Gray code is used extensively in external devices such as digital position encoders because the transition from any number to the next involves a change of only one bit (Table 1.3). If binary code were used, erroneous values could result when more than one bit changed, since it is not possible to guarantee that all bits change simultaneously from one number to the next.

The exclusive-OR circuits shown in Figure 1.3 convert numbers from Gray code to binary code and from binary code to Gray code. See the following section for a review of the AND, inclusive-OR, and exclusive-OR logic circuits. It will be noted on the left-hand side of Figure 1.3 that bit 1, for example, cannot be determined until bit 2 is known, and bit 2 cannot be determined until bit 3 is known, etc. Thus the output is valid only after N gate propagation times. A "valid data" signal can be derived by connecting

Table 1.3 *Binary and Gray codes and their decimal equivalents*

Decimal	Binary	Gray	Decimal	Binary	Gray
0	00000	00000	16	10000	11000
1	00001	00001	17	10001	11001
2	00010	00011	18	10010	11011
3	00011	00010	19	10011	11010
4	00100	00110	20	10100	11110
5	00101	00111	21	10101	11111
6	00110	00101	22	10110	11101
7	00111	00100	23	10111	11100
8	01000	01100	24	11000	10100
9	01001	01101	25	11001	10101
10	01010	01111	26	11010	10111
11	01011	01110	27	11011	10110
12	01100	01010	28	11100	10010
13	01101	01011	29	11101	10011
14	01110	01001	30	11110	10001
15	01111	01000	31	11111	10000

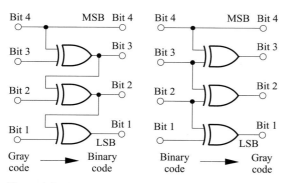

Figure 1.3 Circuits used to convert Gray code to binary and binary code to Gray code. Four bits are shown. The logic elements shown perform the exclusive OR, which has an output logic state that equals one only if the input logic states differ.

all input bits to an inclusive-OR circuit that is used as the input to a pulse generator. The output is read at the trailing edge of the pulse. Alternatively, a table lookup from computer memory or read-only memory (ROM) can be used to convert between Gray and binary codes.

1.4   Digital building blocks

This section describes the fundamental building blocks used to connect to a multiple-output bus, sample and store a logic state at a well-defined time, generate pulses,

Table 1.4 *Logic voltage ranges for TTL and ECL circuit families*

	TTL (V)	ECL (V)
Power supplies	0, +5 (±5%)	0, −5.2 (±5%)
Allowed "0" input range	−0.5 to +0.8	−5.0 to −1.4
Ambiguous input range	+0.8 to +2.0	−1.4 to −1.1
Allowed "1" input range	+2.0 to +5.5	−1.1 to +0.0
Nominal logic "0" output	+0.2	−1.75
Nominal logic "1" output	+3.2	−0.90
Allowed "0" output range	+0.0 to +0.4	−1.85 to −1.65
Ambiguous output range	+0.4 to +2.4	−1.65 to −0.96
Allowed "1" output range	+2.4 to +5.0	−0.96 to −0.81
Typical pulse risetime (10–90%)	10 ns*	1.5 ns†

*Low-power Schottky TTL.
†ECL 10,000.

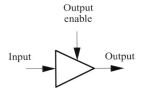

Figure 1.4 Tri-state buffer (see Table 1.5 for the function table and Figure 1.5 for a typical timing diagram).

and perform logical tests (AND, OR, etc.) of logic states. Table 1.4 lists the ranges of external voltages for the two most commonly used families of logic circuits, TTL (transistor–transistor logic) and ECL (emitter-coupled logic).

Note 1: To allow for voltage drop along conductors, the requirements for output are more stringent than for input.

Note 2: For both TTL and ECL, a logic 1 is always more positive in voltage than a logic 0.

1.4.1   Tri-state buffer

The **tri-state buffer** has three output states: asserted high, asserted low, and high impedance. In the high-impedance state, the output neither loads nor drives any circuit connected to it. This device has the usual logic input, but also has an additional enable input that determines whether the output follows the input or is put in the high-impedance state. The tri-state buffer is an essential component when several different outputs must be connected to form a common bus. See Figure 1.4 for the circuit schematic, Figure 1.5 for a typical timing diagram, and Table 1.5 for the function table.

Table 1.5 *Function table for tri-state buffer*

Input	Output enable	Tri-state output
H	L	H
L	L	L
X*	H	High impedance

*X = don't care.

Table 1.6 *Function table for edge-triggered* D-*type flip-flop*

Data D	Clock C	Flip-Flop output Q
H	↑[†]	H
L	↑	L
X*	H or ↓[§] or L	Previous state

*X = don't care.
[†]↑ = low-to-high edge.
[§]↓ = high-to-low edge.

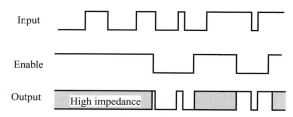

Figure 1.5 Typical timing diagram for the tri-state buffer (see Table 1.5 for the function table).

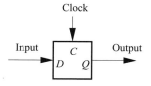

Figure 1.6 Edge-triggered D-type flip-flop (see Table 1.6 for the function table and Figure 1.7 for a typical timing diagram).

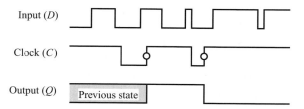

Figure 1.7 Typical timing diagram for the edge-triggered D-type flip-flop (see Table 1.6 for the function table).

Table 1.7 *Function table for transparent latch*

Data D	Gate G	Latch output Q
H	H	H
L	H	L
X*	L	Previous state

*X = don't care.

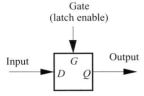

Figure 1.8 Transparent latch (see Figure 1.9 for a typical timing diagram and Table 1.7 for the function table).

1.4.2 Edge-triggered D-type flip-flop

The basic element in the parallel output port is the edge-triggered **D-type flip-flop**, which differs somewhat from the simple flip-flop that can be switched between two logic states. The edge-triggered D-type flip-flop has two inputs, a data input (D) and a clock input (C) (Figure 1.6). The output (Q) is set equal to the logic state of the input (D) during the clock (C) low-to-high edge. At all other times, the state of Q does not change even if D changes. See Table 1.6 for the function table and Figure 1.7 for a typical timing diagram.

Frequently, the outputs have tri-state buffers (see Figures 1.4 and 1.5) so that several outputs can be connected. The 74LS374 tri-state octal D-type edge-triggered flip-flop is such an example. The state of Q is only asserted at the output line when the "output-enable" line is asserted. When the output-enable line is not asserted, the output is in a high-impedance state that neither drives nor loads any other circuit connected to the output. Whenever two or more outputs are connected to a common line called a **bus**, they must all have tri-state outputs.

1.4.3 Transparent latch

The transparent latch (Figure 1.8) is similar to the edge-triggered D-type flip-flop, except that the output is equal to the input the entire time that the latch enable is asserted. See Table 1.7 for the function table and Figure 1.9 for a typical timing diagram.

Frequently, the outputs have tri-state buffers (see Figures 1.4 and 1.5) so that several outputs can be connected. The 74LS373 tri-state octal D-type transparent latch is such an example.

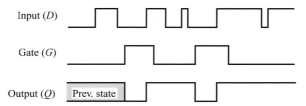

Figure 1.9 Typical timing diagram for transparent latch.

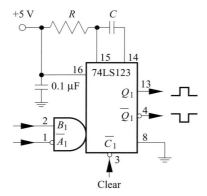

Figure 1.10 74LS123 dual retriggerable one-shot. Pin numbers correspond to section 1. Pulse width is determined by the external resistor and capacitor values.

1.4.4 One-shot

The one-shot produces output pulses of fixed width, where the width is determined by the values of an external resistor R and capacitor C. Figure 1.10 shows one of two sections of the 74LS123 dual retriggerable one-shot and its external components, and Table 1.8 shows the function table. The pulse width W can be estimated using the equation:

$$W = 0.37R(C + 22 \text{ pF})$$

The actual pulse width may differ by typically 20% due to component variations. To produce a more precise pulse width, it is common to use a variable resistor that is adjusted while observing the pulses on an oscilloscope.

The retriggerable one-shot has the property that if a new trigger is received while an output pulse is in progress, the output pulse is extended from that time by an amount W. The non-retriggerable one-shot ignores input triggers while an output pulse is in progress.

1.4.5 AND, OR, exclusive-OR gates

Integrated circuits are readily available to perform standard logic operations such as AND, OR, and exclusive OR (Figure 1.11). (The OR is also called the inclusive OR

Table 1.8 *Function table for 74LS123 retriggerable one-shot*

Clear C	Input A	Input B	Output Q	Output $\bar{Q}$
L	X	X	L	H
X*	H	X	L	H
X	X	L	L	H
H	L	↑	⊓	⊔
H	↓§	H	⊓	⊔
↑†	L	H	⊓	⊔

*X = don't care.

†↑ = Low-to-high transition.

§↓ = High-to-low transition.

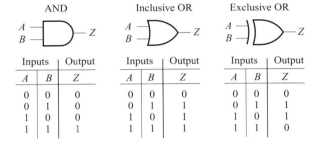

Figure 1.11   AND, inclusive-OR, and exclusive-OR logic gates.

to distinguish it from the exclusive OR.) The AND circuit is used to detect when two logic levels are both high, the inclusive-OR circuit is used to detect when either of two logic levels is high, and the exclusive-OR circuit is used to detect when two logic levels differ. When a circle is shown at the output, the output is complemented ($\bar{Z}$ rather than Z) and the device is called a NAND or NOR gate.

1.4.6   Set/reset latch

This circuit has two digital inputs that allow the output to be set or reset. The TTL 74LS279 contains four set/reset latches and each has the logic table shown in Table 1.9. It is used to convert pulses to stable logic levels. Specifically, if both inputs are initially H (their inactive level), an L pulse on $\bar{S}$ will set Q to H and an L pulse on $\bar{R}$ will reset Q to L.

1.5   Digital counters/timers

The digital counter/timer is a circuit that can count pulses that occur at arbitrary times, measure time by counting clock pulses, or produce pulses uniformly spaced in time. Normally, when executing a program, the microcomputer must perform other tasks that

Table 1.9 *Function table for 74LS279 quad set/reset latch*

Input $\bar{S}$	Input $\bar{R}$	Output Q
L	L	H*
L	H	H
H	L	L
H	H	Previous value

*May not persist when both inputs are set H.

make it impossible for the program to keep track of absolute time. Moreover, execution speed depends on the clock frequency of the particular computer used. A variety of integrated-circuit chips have been developed that can constantly keep track of time (and even of the date) while the microcomputer is busy with other tasks (or even turned off). For example, most microcomputers have battery-powered circuits that keep track of the date and time (to the nearest second) and this information is recorded whenever a disk file is created or changed. For data acquisition involving periodic sampling, a more precise, dedicated clock circuit is needed that can be read by the program or produce a series of external pulses evenly spaced in time. The use of a hardware counter/timer is explored in Laboratory Exercise 2, where human reaction time is measured. Sections 1.5.1–1.5.3 describe typical applications of digital counter/timer circuits and two of the more popular digital timer chips, the 8253 and the 9513.

1.5.1   Applications of digital counters/timers

Measuring the duration of a pulse
The counter is set to an initial value of 0 and to count up clock pulses when gated on. The pulse whose duration is to be measured is used as the gate pulse. The pulse duration is given by $T_w = N/f_c$, where N is the final value in the counter and f_c is the clock frequency.

Measuring the time difference between two events
The first event sets a logic level and the second event resets the logic level. The duration of the resulting pulse is then measured using the method just described. One well-known application is the timing of Olympic races to an accuracy of 1 ms.

Generating a pulse of precise duration
The counter is loaded with a number N and set to count down once per clock pulse. The output is high during counting and low after zero is reached. The duration of the pulse is given by $T_w = N/f_c$, where f_c is the clock frequency.

Measuring an average pulse frequency
One counter is used to produce a pulse of precise duration, using the method just described. A second counter counts pulses when gated on by the first counter. If the

pulse duration is T_w and the count in the second counter is M, then the average pulse frequency is given by $f_p = M/T_w$.

Producing pulses uniformly spaced in time

The counter is loaded with a number N and set to count down once per clock pulse. When its contents reaches zero, it produces an external pulse, reloads from a load register, and then resumes counting. The frequency of the resulting pulses is given by $f_p = f_{\text{clock}}/N$.

1.5.2   The 8253 programmable interval timer

This integrated-circuit chip (manufactured by Intel and others) has three 16-bit down counters that can be used to count clock pulses and can be written and read under program control. It has a number of functions that permit it to act as a pulse generator, a digital one-shot, or a digital square-wave generator. These functions are selected by writing to a control register. At typical 1-MHz clock rates, there are two issues:

1. It is not possible to read a rapidly changing accumulator directly, and it is necessary to latch the accumulator into a buffer (temporary storage) register. When the latch command is given, circuits on the chip transfer the contents of the specified counter to a buffer register that can be read later. If the counter value is in the process of changing, the circuits wait for the value to become stable before latching. Note that the read command reads the buffer register (not the counter itself) and, as a result, the value read is the counter value at the instant the latch command was given, not the counter value at the time of the read (Figure 1.12).

2. A 16-bit accumulator will overflow in 16 ms or less, and for counting longer periods, it is necessary to hardwire two accumulators in sequence. Since the two accumulators

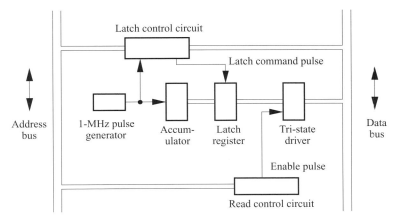

Figure 1.12 Circuits for accumulating 1-MHz pulses, for transferring a valid accumulator value to a latch register under computer control, and for reading the value into memory.

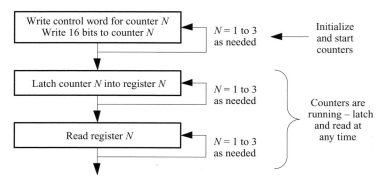

Figure 1.13 Sequence of operations for initializing, loading, latching, and reading the 8253 counter/timer.

must be latched by different instructions, an ambiguity arises whenever the faster accumulator passes through zero. The chip does not have circuits to handle this problem and the simplest solution is to reread the slower accumulator. (See the following warning about cascading counter/timer chips.)

The initialization, loading, latching, and reading sequence is shown in Figure 1.13.

1.5.3   The AM9513 system timing controller

This integrated-circuit chip (manufactured by Advanced Micro Devices) has many more features than the 8253 and requires more program steps to initialize (Figure 1.14). It has five independent 16-bit counters, a 1-MHz clock, and on-chip subscalers to permit divide-by-N counting for slower rates and longer time ranges.

Under program control, the input of any counter can be connected to any subscaler, the overflow output of another counter, or to an external input line. Similarly, the overflow output of any counter can be connected to the input of another counter or to an external output line.

The five counters can be latched in any combination with a single instruction. However, whenever two or more counters are cascaded, the ambiguity mentioned before is still present (although less frequent), and still requires rereading the slower counter. This interval timer is used in the Metra Byte parallel I/O board and the National Instruments analog data-acquisition board (among others).

Warning: cascading counter/timer chips

When two counter/timer chips are cascaded, the more rapidly moving counter (say, counter 1) receives input pulses from the system clock and increments until it reaches FFFF (for a 16-bit counter). As it transitions to 0000, it sends a carry pulse that increments the less rapidly moving counter (counter 2; see Figure 1.15).

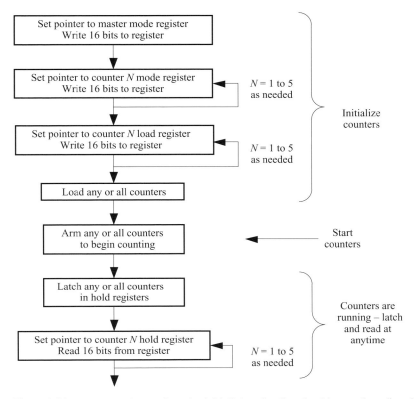

Figure 1.14 Sequence of operations for initializing, loading, latching, and reading the 9513 counter/timer. The counters may be latched and read repeatedly at anytime.

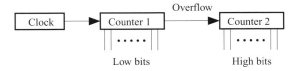

Figure 1.15 Timer consisting of a clock and two cascaded counters.

Unfortunately, it is not possible to guarantee that these two events happen simultaneously, and erroneous timing information can be read occasionally. For example, suppose that counter 2 reads A73D and counter 1 reads FFFF (Table 1.10). After counter 1 receives its next clock pulse, we would hope that counter 2 and counter 1 would change simultaneously and read A73E 0000, but if one changes before the other, and we latch the counters during this very brief period, we will get either A73E FFFF or A73D 0000. Both are in error by over 65,000 counts! Moreover, it is not possible to latch the outputs of two different counter circuits simultaneously, even if they could change simultaneously.

Table 1.10 *Correct and incorrect values of cascaded counters*

	Counter 1	Counter 2
Correct simultaneous increment of counters 1 and 2 (perfect timing)	FFFF	A73D
	0000	A73E
Incorrect: 1 increments before 2 (65,536 too low) →	FFFF	A73D
	0000	A73D
	0000	A73E
Incorrect: 2 increments before 1 (65,536 too high) →	FFFF	A73D
	FFFF	A73E
	0000	A73E

The general solution to this problem is to use the following steps:
1. Latch and read both counters.
2. If counter 1 is "near" 0000, go back to step 1.

The nearness condition is determined by the range of counter 1 values during which counter 2 could be changing.

For the 9513 in 1-MHz increment mode and "simultaneous" counter latching, counter 1 will increment at most 2 μs before its overflow pulse can increment counter 2. In this case the nearness condition in step 2 is "if counter 1 < 2."

For the 8253, where the counters decrement at 1 MHz and are latched by separate program statements that are separated by typically 20 μs, counter 1 would be latched immediately before counter 2, and a safe nearness condition in step 2 would be "if counter 1 > FF00."

1.6   Parallel and serial input/output ports

The **parallel input/output (I/O) port** allows the microcomputer to communicate directly with logic voltages in the external digital world and handles most of the problems of control and synchronization with the microprocessor address and data buses. The most convenient is the bi-directional port, which has separate input and output lines, simplifying the connection to external devices. Since all bits are transferred in parallel (at the same time), it is generally faster than the serial port. Laboratory Exercise 3 involves reading switches and writing to lights using a parallel port. Moreover, A/D and D/A converters naturally deal with parallel digital information and can be interfaced directly to a parallel I/O port, as demonstrated in Laboratory Exercises 8 and 9.

The parallel I/O port usually includes the following addressable internal registers:
1. **Data registers** that hold input data until the program can read them and hold output data while needed by the external circuit.

2. A **control register** that allows the program to write ones and zeros to control the mode of operation of the port or the logic state of external lines. These lines are typically used to notify an external circuit that the program: (i) has new output, (ii) has read new input, or (iii) is ready to accept new input.

3. A **status register** that can be read by the program to determine the status of the data register or the logic state of external lines. Various bits would be set when an external circuit: (i) has asserted and latched new data on the input port (and are ready to be read by the program), (ii) has read the contents of the output port, or (iii) is ready to read new data from the output port.

Some commercial parallel I/O ports have only data registers and no handshaking registers, but it is possible to assign some of the data bits to be used by the program to communicate with external circuits. This is described in the following sections on parallel input and output handshaking.

Most parallel I/O ports have both input and output data lines, whose functions cannot be changed (the bi-directional port), while others permit each data bit to be either input or output, as specified by the contents of a special control register.

The **serial I/O port** also has addresses for setting up the communication protocol and then can transfer data serially in time using only one input and one output line. The advantage over the parallel port is that existing circuits (specifically telephone communication lines) can transmit serial data over long distances, even to other continents. For connection to nearby peripheral devices the older RS232 serial port is being replaced by much faster USB and IEEE 1394 serial ports. (See Section 1.9 for more details.)

1.6.1   Handshaking considerations

Handshaking consists of the communication procedures used to ensure that both sender and receiver are ready for data transmission, that the sender tells the receiver when data are ready, and the receiver tells the sender that the data have been taken.

Parallel data can be correctly read only when all bits are stable. Handshaking is essential and allows the sender to signal the receiver when new data are ready and stable.

The following steps describe how handshaking can be implemented using two handshaking lines "ready for data" and "data available" in addition to the data lines. They can be used for data transmission in either direction between any combination of computers and external circuits. Between transactions, "ready for data" and "data available" are FALSE.

1. When the receiver is ready for new data, it sets "ready for data" TRUE.
2. The sender detects "ready for data" TRUE.
3. (a) If the receiver initiates the data transfers, the sender asserts the requested data as

soon as possible after step 2. ("Asserts" means setting voltages on the data lines that correspond to the 0s and 1s of the data.)

(b) If the sender initiates the data transfers, the sender can assert data anytime after step 2.

4. After the voltages on the data lines have settled, the sender sets "data available" TRUE. (It is essential that the data are valid *before* "data available" is set true.)

5. The receiver detects "data available" TRUE and reads the data.

6. The receiver sets "ready for data" FALSE. (At this point the receiver is not ready for data because it needs to do something with the data it just read.)

7. The sender detects "ready for data" FALSE and sets "data available" FALSE. (At this point the sender is relieved of the responsibility of asserting the data.)

Handshaking is required for both serial and parallel data when a series of data values must be transmitted faithfully and the receiver or the sender have an unpredictable response time or are transferring data at unpredictable times (asynchronous communication).

Handshaking is generally not necessary when the sender continually produces data (such as temperature measurements), and the receiver can tolerate occasional erroneous values that arise during bit changes. To avoid this problem, digital position encoders (see Chapter 4) use Gray code which has the property that neighboring values differ by only one bit (see Table 1.1).

Design tip

If a circuit asserts digital data for a brief period and the computer may not be able to read the data promptly:

1. Connect each output line of the circuit to a transparent latch.
2. When the circuit has data, it can store the data on the latch until it can be read by the computer.
3. Handshaking is needed so that the external circuit can inform the program that new data are available and the program can inform the external circuit that the data have been taken.

1.6.2   The parallel output port

The **parallel output port** reads a number from computer memory and converts the bit pattern to logic voltage levels on wires in the world "outside" the computer. Additional control lines and status registers may also be provided so that: (i) the external circuit can tell the computer program that it is ready to receive output data, (ii) the computer program can tell the external circuit that it has data in its internal registers, and (iii) the external circuit can tell the computer program that the output data have been taken.

The basic element in the parallel output port is the **register**, a circuit able to sample, store, and output digital data on command. This is usually achieved by using a set

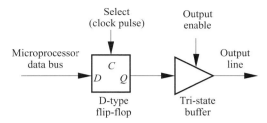

Figure 1.16 Edge-triggered D-type flip-flop as used for microcomputer output. A program write command provides the select pulse and the external circuit provides the output enable.

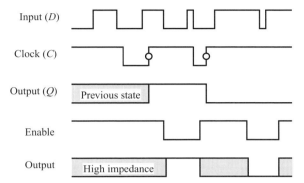

Figure 1.17 Typical timing diagram for edge-triggered D-type flip-flop shown in Figure 1.16. Output Q is set equal to the data D only on a low-to-high edge of clock C. Output Q is asserted on the output lines only when enable is low.

of edge-triggered D-type flip-flops followed by tri-state drivers, components that were introduced earlier in this chapter (Figures 1.4 and 1.6). This circuit has eight data inputs (D), a clock input (C), and an output enable input (Figure 1.16).

The output (Q) is set equal to the logic state of the input (D) only when the clock (C) goes from low to high. At all other times, the state of Q never changes (even if D changes). The state of Q is only asserted at the output line when the "output-enable" line is low. When the output-enable line is high, the output is in a high-impedance state that neither drives nor loads any other circuit connected to the output. Whenever two or more outputs are connected to a common line called a **bus**, they must all have tri-state outputs. See Figure 1.17 for a typical timing diagram.

The parallel output port consists of a set of registers that can be addressed and loaded from memory under program control (Figure 1.18). Each bit is controlled by the circuit shown in Figure 1.16. Whenever the output-enable line is asserted, the contents of the registers are asserted on the external output lines.

Almost all C compilers written for the IBM PC (and compatible microcomputers) provide output statements for 8-bit bytes and 16-bit words. For example, the Turbo C compiler (Borland International, Inc.) permits a byte "d" to be written to the parallel

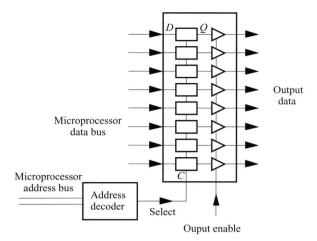

Figure 1.18 Parallel output port using an octal edge-triggered D-type flip-flop, address decoder, and tri-state output buffer. When addressed by the microprocessor, data from memory are transferred to the registers and are available at the inputs of the tri-state buffers until a new byte is written. The data appear at the output lines whenever the external circuit provides an output enable.

output port at address "a" by using the statement "outportb(a,d);". This function reads a byte d from memory, writes it to the parallel output port at address a, and the port latches the data to its internal registers by the following steps:

1. The function reads the address a and the byte d from memory.
2. The function writes a to the address bus and d to the data bus.
3. The parallel output port recognizes its address, reads the byte on the data bus, and clocks the data on its internal registers (usually D-type flip-flops).

The data are then available to the outside world whenever the output-enable line is asserted. By constantly asserting the output-enable line (connecting it high or low, depending on the circuit), the output port can be placed in a "transparent" mode so that a byte written to the port appears immediately on the output lines and does not change until a new byte is written.

In parallel output port handshaking, the computer is the sender and the external circuit is the receiver. For handshaking between a single parallel output port, the output enable can be constantly asserted so that the tri-state drivers are always in transparent mode. Starting with "output data available" = FALSE and "ready for output data" = TRUE, the steps are as follows (see Figure 1.19 for the numbered sequence of control and data flow, and Figure 1.20 for the timing diagram).

1. When the external circuit is ready for new data, it sets "ready for output data" TRUE.
2. The program waits until "ready for output data" becomes TRUE.
3. The program writes data to the output port.
4. The program sets "output data available" to TRUE. (In most cases the delay involved in executing this step is sufficient for the data written in step 3 to stabilize.)

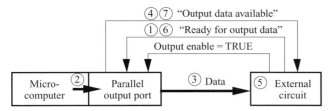

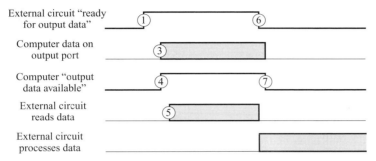

Figure 1.19 Handshaking between a parallel output port and a single external circuit. The sequence of numbered steps is described in the text.

Figure 1.20 Timing diagram for handshaking between a parallel output port and a single external circuit. The sequence of numbered steps is described in the text.

5. The external circuit detects "output data available" TRUE and reads the data.
6. The external circuit sets "ready for output data" FALSE, and starts to process the data.
7. The computer waits until "ready for output data" becomes FALSE and sets "output data available" FALSE (until this step, the computer cannot write new data to the output port).

In the case where the computer initiates the data transfers, steps 1 and 2 can occur any time before step 3. In the case where the external circuit initiates the data transfers, step 3 occurs as soon as possible after step 2.

In summary, the output port signals that new data are available by writing to the "output data available" control register and must maintain the data on its output lines until the external circuit signals by setting the "output data taken" status register.

Note that many commercial parallel output ports do not provide handshaking. For these, handshaking can be accomplished by assigning one input data line to carry the signal "output data available" and one output data line to carry the signal "output data taken." This consequently reduces the number of lines available for data transmission.

For the case where the role of the computer is to provide the most recent value of a number continuously to a single external circuit (such as in a control application), handshaking is not needed, steps 1, 2, and 5 are omitted, and the procedure is reduced

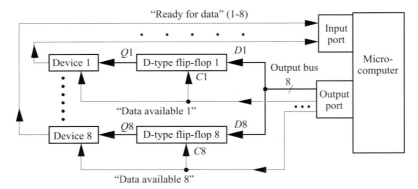

Figure 1.21 External output bus used for one sender and many receivers. The output-enable lines control tri-state drivers whose outputs are the port output lines. The external circuit enables only one output at one time and the other outputs are kept in a high-impedance state.

to the following steps:

3. The program writes new data to the internal registers of the output port at any time and they appear immediately on the output lines.
4. The external circuit reads the new data as needed.

The **external output bus** allows a single output port to be selectively connected to a number of devices using a common data bus (Figure 1.21).

For this system, the handshaking steps are:

1. The computer sets all eight "data available"/select lines FALSE.
2. The computer checks that device n (the one we want to write to) has its "ready for data" TRUE.
3. The computer writes the data bits to the inputs of all tri-state drivers.
4. The computer asserts "data available"/select output line n TRUE (or the computer asserts "data available" and the device selects an octal tri-state driver).
5. Device n detects its "data available" TRUE, reads the data, and sets its "ready for data" FALSE.
6. The computer detects the "ready for data" FALSE and sets "data available"/select FALSE.
7. After device n has processed the data, it sets its "ready for data" TRUE.

1.6.3   The parallel input port

The **parallel input port** converts a bit pattern of logic voltage levels in the world "outside" the computer to a number in computer memory. Additional control lines and status registers may also be provided so that: (i) the computer program can notify the external circuit that it is ready to receive data, (ii) the external circuit can notify the program that it has written data to the port's internal registers, and (iii) the computer program can notify the external circuit that the data have been read.

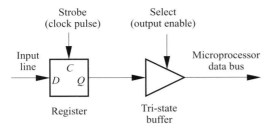

Figure 1.22 Edge-triggered D-type flip-flop as used for microcomputer input. The external circuit provides the strobe edge, and a program read command provides the select pulse. See Figure 1.23 for the timing diagram.

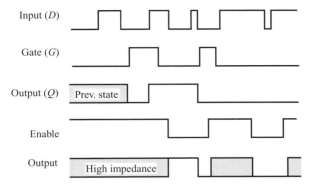

Figure 1.23 Typical timing diagram for the transparent latch and tri-state buffer (Figures 1.8 and 1.4) as used for microcomputer input. Q is set equal to the data D only when gate G is high. Q appears on the output of the tri-state buffer only when "select" is low.

As in the output port, the basic element in the parallel input port is the register, a circuit able to sample, store, and output digital data on command. As described in earlier sections, this is usually implemented with an edge-triggered D-type flip-flop followed by a tri-state buffer (Figure 1.22). When an external device asserts a low-to-high transition on the **strobe** line, the register output Q is set equal to the input D. Until the next high-to-low transition, the output Q does not change (even if D changes). When an output enable occurs, then the output is set equal to Q. In the absence of an output enable, the buffer output is in a high-impedance state and neither drives nor loads the microprocessor data bus.

A variation of this circuit is the **latch**, which makes the output Q equal to the input D only while the gate G is high (transparent mode; Figure 1.8). When G is low, the output Q does not change (even if D changes). See Figure 1.23 for a typical timing diagram.

The parallel input port (Figure 1.24) consists of a set of registers that stores whatever data are on the external input lines at the instant the input strobe line is pulsed. The microcomputer program can address and read these registers at a later time.

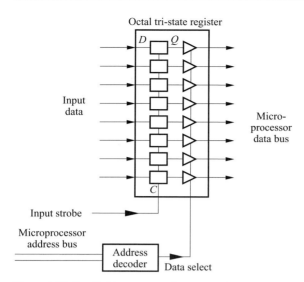

Figure 1.24 Parallel input port using an octal tri-state edge-triggered D-type flip-flop and address decoder.

Almost all C compilers written for the IBM PC (and compatible microcomputers) provide input statements for 8-bit bytes and 16-bit words. For example, the Turbo C compiler (Borland International, Inc.), permits a byte "d" to be read via the parallel port at address "a" by using the statement "d = inportb(a);". After data have been latched onto the internal registers of the parallel input port, the function and the port transfer the data to memory by the following steps:

1. The function reads the address a from memory.
2. The function writes the address a to the address bus.
3. The port recognizes its address and transfers the data from its internal registers to the data bus.
4. The function reads byte d from the data bus and stores it in memory.

Note that the number read by the program corresponds to the data on the input lines when the strobe line was pulsed, not when the program executed the read statement.

For full handshaking, the following steps are typical (see Figures 1.25 and 1.26 for the numbered sequence of control and data flow):

1. When the program is ready for new data, it sets "ready for input data" TRUE.
2. The external circuit detects "ready for input data" TRUE.
3. The external circuit detects that "input data taken" is FALSE and (when ready) asserts data on the port's input lines.
4. The external circuit pulses the input strobe line, which latches data on the port's internal registers. (This step is necessary for edge-triggered flip-flops, but can be omitted when using latches in transparent mode.)
5. The external circuit sets the status register "input data available" to TRUE.

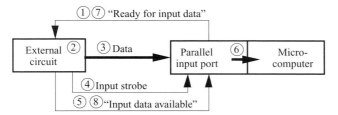

Figure 1.25 Parallel input port handshaking. The sequence of numbered steps is described in the text.

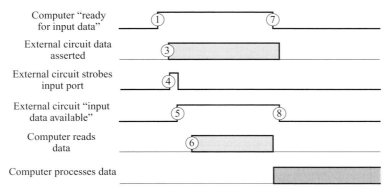

Figure 1.26 Timing diagram for parallel input port handshaking.

6. The program detects "input data available" and reads the data.
7. The program sets "ready for input data" to FALSE.
8. The external circuit detects "ready for input data" FALSE and sets "input data available" to FALSE.

For the example of the parallel input port circuit shown in Figure 1.24, the computer is the receiver, the external circuit is the sender.

In some parallel input ports, the input strobe line is used not only to latch the new data onto the port's input registers but also to set a register that can be read by the program to signal "input data available." This register must be reset by the input port when the program reads the new data. This provides for full handshaking, and combines steps 3 and 4, and steps 5 and 7 as follows:

1. The program sets a control register bit that sets "input data taken" FALSE (signals "ready for input data").
2. The external circuit detects that "input data taken" is FALSE and (when ready) asserts data on the port's input lines.
3. The external circuit pulses the input strobe line, which latches data on the port's internal registers, and sets "input data available" to TRUE.
4. The program detects "input data available" and reads the new data, which sets "input data available" to FALSE.
5. The program sets "input data taken" to TRUE.

For those parallel input ports that do not provide handshaking, one input data line can be assigned to carry the signal "input data available" and one output data line can be assigned to carry the signal "input data taken." This consequently reduces the number of lines available for data transmission.

Alternatively, if other tasks simultaneously require attention, the "input data available" signal is not used and the external circuit would pulse the microprocessor "interrupt" line at step 4, causing program execution to be temporarily paused, and a small service program reads the data registers and then transfers control back to the main program.

Commercial parallel input ports usually use the 74LS374 octal register, if a low-to-high edge strobe is desired (Figure 1.18), or the 74LS373 octal latch, if a control-level strobe (high = sample, low = hold) is desired (Figure 1.24).

The parallel input port is seldom operated in transparent mode because in this mode the program does not know when the external circuit may be changing the input bit pattern. If the new pattern involves a change of more than one bit, the external circuit cannot change them at precisely the same instant, so there is a brief period when the program could read erroneous data. When using handshaking, the program reads the input port only after the external circuit has signaled that the new data are stable.

The **external input bus** allows a number of parallel devices (such as A/D converters) to be read by a single parallel input port (Figure 1.27). Only one set of tri-state drivers are enabled at a time. Normally, the tri-state drivers are in a high-impedance state. In

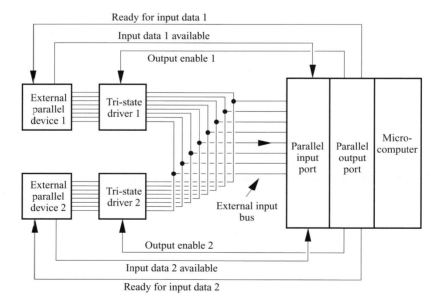

Figure 1.27 External input bus. The output-enable lines control tri-state drivers whose outputs are the port input lines. Only one tri-state driver is enabled at a time and the other tri-state driver outputs are kept in a high-impedance state.

this example we assume that the parallel input is operated in a transparent mode so that the values on the parallel input bus can be read by the computer at any time.

Design tip

To connect a single N-bit digital input port to $M \times N$-bit digital circuits:
1. Connect each output bit of each digital circuit to the input of a tri-state driver ($N \times M$ bits total).
2. Connect M corresponding tri-state outputs together for each bit to form a bus (N lines).
3. Connect the bus to the digital input.
4. Connect M lines from the computer's digital output port to the enable-input lines of the tri-state drivers.
5. The computer can then use M output lines to assert any of the M circuits selectively onto its input port.

1.7   Digital data-acquisition procedures

Periodic data acquisition from an external device usually requires:
1. A digital clock that provides a uniform sampling period, independent of variable software delays.
2. The ability to latch data from the external device onto registers that can also be read by the program. This external device can be an A/D converter, digital-position encoder, or any circuit that senses a physical quantity and converts it into a digital form.
3. A way for the program to determine that new data have been latched.

To perform these tasks, the parallel I/O ports and counters/timers are used in a variety of ways, as summarized in the sections that follow.

1.7.1   Software-trigger status-poll method

The simplest method uses the "software-trigger status-poll" to initiate sampling and to determine when the data were ready to be read into computer memory. **Polling** or **status polling** is the sequential testing of the status word of all relevant peripherals to determine whenever service is required. This method (shown in Figure 1.28) is used in Laboratory Exercise 9 to demonstrate interfacing an A/D to the parallel input port and periodic sampling. It consists of looping over steps 1 to 5:
1. The program generates a periodic series of external pulses to trigger the external acquisition circuit, usually by pulsing the "ready for input data" line on the parallel input port. Software loop delays are not accurate, and the best way for the program to produce periodic pulses with periods as short as 1 ms reliably is by reading the hardware timers in a loop. For shorter periods, the hardware-trigger method is recommended (described in the following section).

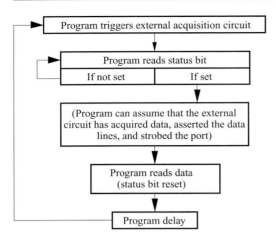

Figure 1.28 Flow chart for software-trigger status-poll method.

2. After the external circuit has new data, it asserts them on the input lines of the parallel port, pulses the strobe line to latch the data, and sets the "input data available" status bit.
3. Meanwhile, the program has been reading the "input data available" status bit in a continuous loop, waiting for it to be set.
4. When the status bit is set, the program detects it and reads the data. The program or the input port resets the status bit.
5. After the pre-determined delay, go back to step 1.

  This method is not recommended for accurate sampling time intervals shorter than 1 ms. It is used in the laboratory exercises for simplicity.

1.7.2   Hardware-trigger status-poll method

The hardware-trigger status-poll method (shown in Figure 1.29) provides trigger pulses that are uniformly spaced in time, but still uses software polling of the status bit, which is not a limitation if the program has no other tasks. The steps are:

1. The program initializes the hardware timer to load a number repetitively and put out a pulse every time the terminal count is reached. The counter is then armed to initiate counting.
2. Each pulse triggers the external circuit.
3. After the external circuit has new data, it asserts them on the input lines of the parallel port, pulses the strobe line to latch the data, and sets the "input data available" status bit.
4. Meanwhile, the program has been reading the "input data available" status bit in a continuous loop, waiting for it to be set.
5. When the status bit is set, the program detects it and reads the data. The program or the input port resets the status bit.
6. Go to step 2.

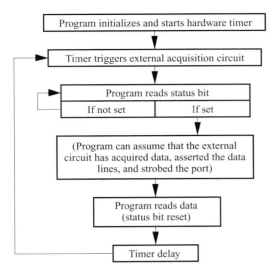

Figure 1.29 Flow chart for hardware-trigger status-poll method.

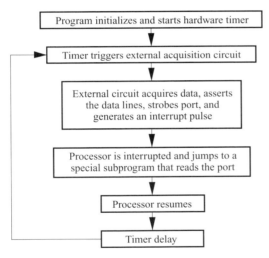

Figure 1.30 Flow chart for hardware-trigger hardware-interrupt method.

The advantage of this method is that the samples are taken at accurate time intervals. The disadvantages are: (1) the maximum rate is limited by the time required to service the interrupt, (2) the computer is not available for other tasks, and (3) system service interrupts can result in missing data samples.

1.7.3   Hardware-trigger hardware-interrupt method

The hardware-trigger hardware-interrupt method permits data analysis during data acquisition, because the reading of the parallel input port is initiated by an **interrupt**, which causes the program to pause whatever it is doing while a special jump occurs to an interrupt-service routine (Figure 1.30). When the routine has serviced the interrupt, it transfers back to the main program. The steps are:

1. The program initializes the hardware timer to load a number repetitively and put out a pulse every time the terminal count is reached. The counter is then armed to initiate the trigger pulses. An interrupt-service routine is initialized.
2. Each pulse triggers the external circuit.
3. After the external circuit has new data, it asserts them on the input lines of the parallel port, pulses the strobe line to latch the data, and generates an interrupt pulse. (Meanwhile, the computer has been performing other tasks, such as data analysis.)
4. The interrupt pulse causes: (i) the address of the current instruction and a number of registers to be saved in a "stack," and (ii) a jump to a special interrupt-service routine that reads the parallel input port, restores the registers, and then jumps back to resume execution. Note that this service routine could also do some digital signal processing and write the result to an output port.
5. Go to step 2.

The advantages of this method are: (1) the samples are taken at accurate time intervals, and (2) the computer is available for other tasks. The disadvantages are: (1) the maximum rate is limited by the time required to service the interrupt, and (2) system service interrupts can result in missing data samples.

1.7.4   Hardware-trigger direct-memory-access (DMA) method

The hardware-trigger DMA method permits the direct transfer of data to memory without interrupting program execution (Figure 1.31). This method is facilitated by a DMA controller circuit that can accept data from an external data-acquisition circuit, take command of the microprocessor bus, and write the data to a specific memory location. Most data-acquisition plug-in boards have timers and DMA circuits as well as software that makes it easy to use this high-speed transmission mode. The steps are:

1. The program tells the external data-acquisition circuit the number of samples to take, the sampling frequency, and where to store them in memory.

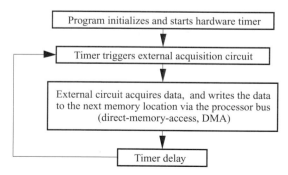

Figure 1.31 Flow chart for hardware-trigger direct-memory-access (DMA) method.

2. When the data-acquisition circuit has each new datum value, it sends it to the DMA controller, which stores the value into the next memory location. Note that it is also possible at this point to do digital signal processing and send the result to an output port, but this requires a custom DMA controller circuit to implement the processing algorithm.

3. Step 2 is repeated until the requested number of samples are taken.

4. The program is notified that the data are available in memory.

The advantages of this method are: (1) the samples are taken at accurate time intervals, (2) DMA is significantly faster than program acquisition, (3) the computer is available for other tasks, and (4) DMA transfers can occur during system interrupts.

1.7.5   Embedded processor method

The embedded processor method uses a data-acquisition card that contains a processor, parallel and analog I/O ports, and memory. The main computer downloads a data-acquisition program into the embedded processor and starts execution. This program can also perform complex digital signal processing tasks. When the program terminates, it can signal the main computer. Conversely, execution can be interrupted by the main computer. The embedded processor can efficiently handle specific data-acquisition and processing tasks because it can be dedicated to those tasks and is not burdened by a complex operating system. Graphical interface operating systems have evolved to be user friendly and in the process have become quite inefficient for real-time operations.

1.8   Switch debouncing

Switches are used to change logic levels manually, such as in keyboards or start buttons. After the switch is thrown, the moveable contact strikes the intended opposing contact and rebounds, so that the output is briefly in an open-circuit condition (Figure 1.32). This is called **contact bounce**. After many bounces, the switch reaches the intended

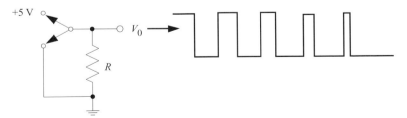

Figure 1.32 Single-pole double-throw switch used to connect load R either to +5 V or to ground. The output waveform shows the effect of contact bounce after the switch has been thrown from +5 V to the grounded position. The output V_0 is not stable for a period of tens of milliseconds.

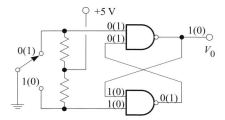

Figure 1.33 Cross-coupled NAND gates used to debounce a single-pole double-throw switch. Logic values are shown when the switch is up and (in parentheses) when the switch is down.

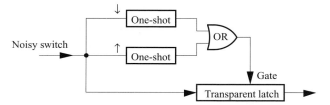

Figure 1.34 Switch debounce circuit for cases of extreme noise. The two one-shots detect any change in the input and put the transparent latch into hold mode for a pre-determined time. During this time all changes in the input are ignored.

state. This effect is demonstrated in Laboratory Exercise 3, where a mechanical switch is used to generate a parallel port input strobe. Methods for generating a clean logic pulse are called **debouncing**. The sections below describe the use of cross-coupled NAND gates, the one-shot, and software logic.

1.8.1   Cross-coupled NAND gates

This method uses a pair of cross-coupled NAND gates to produce an output that does not change even when the switch is in the open-circuit condition (Figure 1.33). When the switch momentarily touches the top contact, the output becomes high and remains high until the switch touches the bottom contact. When the switch momentarily touches the bottom contact, the output becomes low and remains low until the switch touches the top contact. When the switch is between the top and bottom contacts, the output is held at the value determined by the last contact. This circuit is available as an integrated circuit (for example, the TTL 74LS279 quad set/reset latch).

Note that this method does not work if the switch action is so violent that the switch rattles between the two contacts. In that case, the one-shot or software methods (described in the following sections) should be used.

1.8.2   One-shots

This solution uses a pair of edge-triggered one-shots that are OR'd together to produce a fixed-duration pulse after the input changes state in either direction (Figure 1.34).

The pulse width is chosen to be somewhat longer than the maximum settling time of the switch. This pulse then causes a transparent latch to be put in hold mode for a fixed period of time after any change in state. During this hold period changes in the input do not appear at the latch output. After the pulse, the latch returns to transparent mode. It is important that the various circuit delays allow the first state change to appear at the latch output slightly before the latch is put into hold mode.

1.8.3   Software debouncing

In this method, the switch output to the microcomputer is allowed to exhibit contact bounce, but the program detects the first contact, and then uses a delay to ignore subsequent data until the switch has had time to stabilize. The program can even adapt to a particular switch behavior by asking the user to toggle the switch several times, and then observing the maximum contact bounce duration. The program can then use the experimentally determined delay.

1.9   Digital interfacing standards

In previous sections, we have discussed the uses of parallel I/O ports for interfacing to a digital computer. This section describes several serial digital interfaces, such as RS-232 and RS-422, several parallel interfaces, such as IEEE 488 and SCSI, and the VME standard, which allows up to 21 controllers and many other circuits to communicate as a larger system. The following section describes some commercial circuits based on these standards. See Chapter 3 for a listing of some commercial analog I/O interface boards.

1.9.1   RS-232C

RS-232 is a serial, asynchronous interfacing standard. **Serial** means that the bits are transmitted one at a time and **asynchronous** means that data transmission can occur at any time. Because it has separate conductors for receive and transmit, it is called **full-duplex**. It is the oldest of the more commonly used interface standards and is used to interface microcomputers to modems, keyboards, video display terminals, serial printers, and other computers. The distances can be modest, such as local area networks that connect computers and printers within the same building, or they can be great, such as where a video display terminal or a local computer is connected to a distant computer using modems and long-distance telephone lines. A **video display terminal** consists of a display device (usually a cathode ray tube), a serial interface circuit, buffer memory sufficient to keep the display refreshed, and a keyboard. It does not have a microprocessor or additional memory. Video display terminals were used extensively in the 1970s and early 1980s, before the advent of low-cost desktop computers.

A serial interface circuit is required to convert parallel data from the microcomputer data bus to the serial RS-232C format. This circuit is usually called a **UART**, which stands for "universal asynchronous receiver/transmitter." One example is the National Semiconductor 8250 UART. Typically, a microcomputer would be equipped with several UARTs for communication with the keyboard, the video display, a printer, or another computer. In addition to handshaking signals (described below), the UART can generate signals when three different types of errors are detected in the serial bit input stream. The **parity error** results when the parity bit does not correspond to the number of ones in the 7-bit **ASCII code**. See Appendix I for ASCII character set codes. If odd parity is chosen, the number of ones in the eight data bits (including the parity bit) must be odd. Two other choices are even parity, and ignore parity (used when all eight bits represent data). The **overrun error** occurs when the UART receives a character before the microprocessor has read the previous character. To avoid this error, one of the UART handshaking lines is usually connected to an interrupt pin on the microprocessor for rapid response. The **framing error** occurs when the receiver has read the eight data bits, does not detect the "1" stop bit, but reads a "0" instead. This is usually caused by noise or by reading at the wrong clock rate.

Because RC-232 can communicate data (without handshaking) using only two signal wires and a common ground, it has been used for decades to connect video display terminals to remote computers using modems and pre-existing telephone lines. The primary limitation to long-distance data rates is the voice-grade telephone bandwidth of 300–3,300 Hz. Such telephone lines are not suited for the transmission of logic levels but work quite well at moderate data rates for the transmission of audio frequency tones. A device called a **modem** (which stands for modulator/demodulator) is used to convert logic levels to harmonic tones (modulation) for telephone line transmission and to convert those tones back to logic levels (demodulation). There are two types of modems, the older **acoustic modems** and the newer **direct coupling modems**.

The acoustic modem is used when direct electrical connection to the telephone wires is not possible, such as in some hotel rooms or in a telephone booth. When the modem is sending, it converts the binary data into a two-tone audio signal (1,200 and 2,100 Hz) that is acoustically coupled to the microphone in the telephone handset. The modem receives the two-tone signal from the speaker in the handset and converts it back into binary data. The method of sending logic zeros and ones as two different tones is called frequency shift keying (**FSK**) modulation. Data rates are limited to 300 baud (300 bits/s).

Direct coupling modems transmit and receive the tones directly on the telephone wires without using microphones and speakers. FSK is used at 300 baud and differential phase-shift keying (**DPSK**, which uses four phase changes to encode four two-bit data symbols), is used at 1,200 to 2,400 baud. Operation at 9,600 to 56,000 baud over long distances is possible using special techniques such as combined phase and amplitude encoding, data compression, adaptive compensation of line characteristics, and digital signal processing. Full handshaking cannot be used for distant connections because standard telephone connections do not provide enough conductors.

Telephone companies also offer special 56–144 kbaud service by transmitting digital information over existing conductors. Higher data rates are possible because the signal is regenerated during transmission with digital discriminators that have higher effective bandwidth and noise immunity than the standard 3,300 Hz bandwidth analog amplifiers. The cost of this service compares well with standard analog voice lines and often provides simultaneous voice and data transmission.

For local communication that does not involve the telephone service, RS-232 logic levels are sent directly over dedicated wires and modems are not needed. RS-232 at 9,600 baud is commonly used between microcomputers and their keyboards, video display devices, and dot matrix printers. For connections between computers and printers within a building (e.g. local area networks), operation at 56 kbaud is common.

The RS-232 interfacing standard specifies particular cable plugs and signal levels. A 25-pin D male connector plug (DB25P) is used on the ends of the cable and plugged into a 25-pin female socket (DB25S) mounted on the case. Either circuit can be sender or receiver. The receiver should interpret signals between +3 and +25 V as "HI," and signals between −3 and −25 V as "LO." The sender should deliver the corresponding signals as voltage levels between +5 and +15 V, and −5 and −15 V, respectively. Standard TTL outputs cannot be used reliably for line lengths greater than 0.5 m. Special **line driver** circuits (the Motorola MC1488 driver and MC1489 receiver, or the TI SN75188 driver and SN75189 receiver) can be used to convert to and from TTL logic levels ("HI" > 2.0 V, "LO" < 0.8 V). At 20 kbaud the recommended maximum line length is 17 m.

The RS-232 line numbers and functions are:

1. Protective ground (PC), usually connected to the cable shield and metal enclosures.
2. Transmitted data output (−TxD). "HI" represents a logic "0" and "LO" represents a logic "1."
3. Received data input (−RxD). "HI" represents a logic "0" and "LO" represents a logic "1."
4. Request to send (RTS) is an output that is set "HI" when the circuit wishes to send data. It must remain "HI" during transmission.
5. Clear to send (CTS) is an input that the other circuit sets "HI" when it is ready to accept data. It must remain "HI" during transmission.
6. Data set ready (DSR) is an input used to detect the "data terminal ready" (line 20) status of the other circuit. On hardwired modems that can operate in either voice (human speech) or data mode, this line is used to set the mode.
7. Signal ground (SG) for the signal transmitted from line 2.
8. Carrier detect (CD) input, used to indicate that the other circuit is "on line" and that the link is usable. Usually activated by the other circuit when the "request to send" is granted. On modems, this signal is often used to control an indicator light.

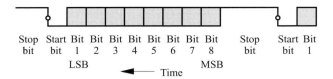

Stop   Start   Bit   Bit   Bit   Bit   Bit   Bit   Bit   Bit   Stop   Start   Bit
bit   bit   1   2   3   4   5   6   7   8   bit   bit   1
      LSB   MSB
                ◄——— Time

Figure 1.35 RS-232 serial data transmission. The stop bit is a "1" and can be of any length. The start bit is a "0." The receiving circuit synchronizes on the stop/start bit edge and reads the next eight bits as data.

20. Data terminal ready (DTR) is an output that is "HI" when the device is in the "on line" mode and "LO" when the device is "off line." For printers and video display terminals, this selection is usually made manually.

22. Ring indicator (RI) is an input sent by a modem to devices that are able to answer incoming calls automatically.

Lines 9, 11, 18, and 25 have been used in older circuits for a current loop interface that is not part of the RS-232 standard. The other line numbers are not assigned.

The data bits on lines 2 and 3 are transmitted in groups of eight, usually preceded by a "0" start bit and a "1" stop bit of arbitrary length (Figure 1.35). The receiving circuit synchronizes on the edge between the stop bit and the start bit of the next byte. It then latches the next eight bits, which are assembled and stored as an 8-bit byte. The clocks on the sender and receiver need only be accurate to a few percent and the sender can send the next byte at any time after the stop bit.

The RS-232 interfacing standard allows for a wide variety of handshaking protocols, depending on the requirements of the two circuits being interfaced. A common handshaking sequence follows:

1. Initially, DTR is set "HI," the control lines RTS and CTS are set "LO," and the data lines –TxD and –RxD are set "LO," which is their idle state.
2. The transmitter sets RTS "HI" to signal the receiver that data are available.
3. The receiver detects the RTS signal and sets CTS "HI" to signal the transmitter that it is ready to accept characters.
4. The transmitter transmits a start bit "HI," eight data bits, and a stop bit "LO," which can be of any length.
5. The receiver detects the "LO" to "HI" transition of the leading edge of the start bit, stores the next eight bits as data, and waits for the next start bit.
6. Steps 4 and 5 are repeated until all characters are transmitted.
7. RTS and CTS are set "LO" until more data need to be transmitted.

For minimal RS-232 interfacing (no handshaking), connections are made (Figure 1.36) that make each device think that the other is always ready to accept data. Either device can send data at any time on output line 2, and both are continuously waiting for a start bit on input line 3. These connections are:

• The RTS output (line 4) is connected to the CTS input (line 5), so that it appears that the other circuit is always ready to accept data.

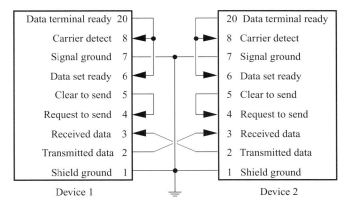

Figure 1.36 Connections used in simple RS-232 interfacing that make each device think that the other is always ready to accept data.

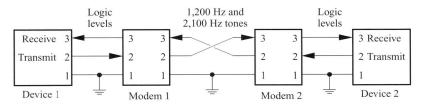

Figure 1.37 Connections between devices such as a computer and a video display terminal using modems and tones sent on the telephone system.

- The DTR output (line 20) is connected to the DSR (line 6) and CD (line 8) inputs.
- The data input and output (lines 2 and 3) are cross-connected.
- The grounds of both circuits (lines 1 and 7) are connected together.

An RS-232 cable for connecting devices directly has lines 2 and 3 crossed (as in Figure 1.36) and is called a **null modem**.

When connecting a computer or a video display terminal to a modem, a normal modem cable is used that does not cross lines 2 and 3. When using modems, the receive/transmit lines are crossed in the telephone system (Figure 1.37).

1.9.2   RS-422 and RS-423

The RS-422 interfacing standard is a variant of the RS-232 standard that uses balanced differential drivers and receivers on twisted pairs of wires rather than single wires with a common ground. This interface performs in the presence of ± 3 V common-mode noise, and has less cross-talk to other circuits, since the voltage and current in each twisted pair sum to zero. The recommended maximum line length is 1,200 m and the maximum rate is 10 Mbaud (1 Mbyte/s). Proper line termination is essential for good performance, since reflections at the ends of the conductors are not common-mode and appear as delayed differential signals.

When used with standard telephone lines (only two nonground conductors), RS-422 permits only one-way communication (this is called **half-duplex**). By using handshaking and a tri-state differential driver, the talker informs the listener when it has finished transmitting, disconnects its differential driver from the conductor pair, and becomes a listener. When either partner has something to transmit, or has been asked to respond by the other partner, it connects its differential driver to the conductor pair and becomes a talker.

The RS-423 interfacing standard is a variant of RS-422 that can be used when only a single nonground conductor is available. In this standard, one of the differential driver/receiver conductors is connected to a common ground conductor. Because the ground return conductor is shared by several differential drivers, performance is not as high as the RS-422 standard. The recommended maximum line length is 1,200 m and the maximum rate is 100 kbaud (10 kbyte/s).

1.9.3   RS-485

The RS-485 interfacing standard is a variant of RS-422 that allows up to 32 devices to communicate over long distances using a single pair of nonground conductors as a half-duplex bus or two pairs as a full-duplex bus. It uses differential drivers and receivers like the RS-422, but in addition defines a device address code that is incorporated in the transmitted data. A talker can use this address code to communicate with a particular listener. The other devices ignore bus communications until they are specifically addressed. The recommended maximum line length is 1,200 m and the maximum rate is 100 kbaud (10 kbyte/s). The maximum rate is lower than the RS-422 standard because termination is poorer when many devices are connected to a single pair of conductors.

1.9.4   IEEE-488

Developed in 1965 by the Hewlett Packard Corporation as the HP-IB (Hewlett Packard interface bus), this bus was adopted as a general standard by the Institute of Electrical and Electronic Engineers in 1975 as IEEE-488. It is also commonly referred to as GPIB (general purpose interface bus) and is an available feature on hundreds of peripheral devices and instruments, including printers, plotters, digital thermometers, function generators, oscilloscopes, and digital multimeters.

This bus has eight control lines, (five management lines and three handshaking lines), and eight data lines. The cable has 24 conductors, eight of which are ground. All lines operate at TTL voltage levels with negative logic. A voltage less than 0.8 V is a logic 1 and a voltage greater than 2.5 V is a logic 0. Each end of the cable has a connector with both male and female plug edges, so that many devices can be connected to the bus in either a linear or star configuration, or in a combination of the two. Moreover, devices need only one plug mounted on their cases.

A maximum of 15 devices can be connected by a single bus, and these can be either controllers or normal devices. Only one controller can be active at any one time and any other controllers must act as normal devices. Usually a controller is a microcomputer equipped with a special GPIB interface circuit. The controller specifies a maximum of 14 devices for listening and only one (possibly itself) for talking. Each device is assigned an address from 0 to 30, usually by manually setting hardware dual-in-line (DIP) switches. In some applications, where one device is always the talker and the other devices are always listeners, a controller is not needed.

Data transmission requires full handshaking between the talker and all listeners, so the transmission rate is determined by the slowest device. Typical data-transmission clock rates are 300–500 kHz, and 1 MHz is the maximum permitted by the standard. The cable length is limited to 20 or 2 m times the number of devices.

The three handshaking lines are:

NRFD (not ready for data): A negative logic wired "OR" that can be asserted low (TRUE) by any device on the line. When this signal is high (FALSE), all devices are ready for data.

DAV (data valid): A negative logic signal is low (TRUE) when data have been asserted on the bus and are ready for reading.

NDAC (not data accepted): A negative logic wired "OR" that can be asserted low (TRUE) by any device on the line. When this signal is high (FALSE), all devices have accepted the data.

The connections for these handshaking lines are shown in Figure 1.38.

The five management lines are:

ATN (attention): A negative logic signal that is low (TRUE) when the eight data bits represent a command or a device address. This signal is high (FALSE) during data transmission.

IFC (interface clear): This line is asserted low (TRUE) by the active controller to clear the bus and put all talkers and listeners in an idle state.

REN (remote enable): When this line is asserted, bus devices can be programmed by the active talker.

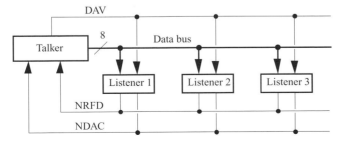

Figure 1.38 Data and handshaking lines used in the IEEE-488 interface standard. DAV = "data valid;" NRFD = "not ready for data;" NDAC = "not data accepted."

EOI (end or identify): Signals that the last byte of a multi-byte sequence is being transmitted. When asserted with **ATN**, initiates parallel polling, which allows the active controller to check the status of up to eight devices at a time.

SRQ (service request): A negative logic wired "OR" that can be asserted low (TRUE) by any device on the line. The active controller can then perform a serial polling loop to determine which device requires servicing and what is required.

The GPIB handshaking procedure for data transfer is:

1. The active controller specifies a device to talk (possibly itself) and devices to listen.
2. The talker sets **DAV** high (FALSE). All listeners set both **NRFD** and **NDAC** low (TRUE).
3. The talker places the first data byte on the eight data lines and waits for the data to settle.
4. As soon as each listener is able to take data, it asserts **NRFD** high (FALSE). When all listeners have done this, **NRFD** will go high.
5. The controller detects that **NRFD** is high and then asserts **DAV** low (TRUE) to inform the listeners that the data may be read.
6. When each listener detects that **DAV** is low (TRUE), it asserts **NRFD** low, and reads the data byte, and sets **NDAC** high (FALSE).
7. When all listeners have asserted their **NDAC** line high (FALSE), the **NDAC** line will become high.
8. When the controller detects that the **NDAC** line is high, it sets **DAV** high.
9. The individual listeners detect that **DAV** is high and assert **NDAC** line low (TRUE).
10. Repeat steps 3–9 until the last byte is read. The talker uses the EOI management line to indicate the last byte.

Note that in terms of the parallel port handshaking signals, NRFD corresponds to "ready for data," NDAC corresponds to "data taken," and DAV corresponds to "data available."

1.9.5   VME-BUS

VME-BUS is a computer architecture developed by Motorola, Mostek, and Signetics. The name was coined (Versa Module Eurocard) and submitted to IEEE as a public domain standard, which means that anyone can make and sell VME products royalty-free. It became available in 1981 and is widely used in industry for interfacing processors and data-acquisition and control circuits. The design is flexible and microprocessor independent. Standard interfaces are less costly than custom interfaces for small numbers of systems, and are quicker to develop.

The general features of the VME-BUS are:

- The address bus is 32 bits wide, allowing addressing of up to 4 Gbyte of memory.
- Data bus widths of 8, 16, 24, or 32 bits may be used.
- Data transfers are asynchronous, non-multiplexed, and not governed by a central

clock. Data transfer speeds can vary from 0 to 40 Mbytes/s, determined by the slowest participant.

• Up to seven levels of interrupt can be used.
• There can be from 1 to 21 processing "masters." (The VME-BUS is a multiprocessing bus.)
• A "master" must acquire the bus from the system controller before it can transfer data.
• There can be many "slaves" that respond to requests by "masters."
  The VME-BUS has a system module, which performs the following functions:
• The bus timer performs a reset if a data transfer takes too long. This feature is used to prevent lockups due to hardware errors.
• An arbiter monitors bus requests from masters and grants control of the data transfer bus, one controller at a time.
• A 16-MHz system clock is available to all modules. (Bus timing is determined by slowest partner.)
  VME-BUS user modules can be masters or slaves. A master can initiate data transfer bus cycles as follows (examples: microcomputers, peripherals with DMA controllers):

1. The master handshakes with the arbiter – when the arbiter is not busy, the master requests the data transfer bus.
2. When the bus is granted, the master selects another master or slave and performs handshaking.
3. After data are transferred, the master relinquishes the bus.
   A slave, when selected by a master, participates in a bus cycle (examples: memory, I/O, sensor, actuator modules).
   Some of the mechanical specifications of the VME-BUS are:

• The crate contains power supplies (5, −12, +12 V) and slots for 10–20 circuit cards.
• The single height card measures 16 × 10 cm.
• The double height card measures 23.3 × 10 cm.
• The connectors are DIN 603-2 with 96 pins (3 wide × 32 high) each.
• The single height card has 96 pins, and the double height card has two connectors with a total of 192 pins.

1.9.6   Small computer standard interface (SCSI)

The SCSI interface is an interface standard for connection to external hard disks and CD-ROM disk drives. It is also used for high-speed communications between processors in high-end systems such as those manufactured by Silicon Graphics, Inc.

1.9.7   Universal serial bus (USB)

Most modern computers are equipped with the universal serial bus (USB) for direct connection to keyboards, pointing devices, scanners, printers. The maximum data rate is 12 Mbits/s.

1.9.8   IEEE-1394

The IEEE-1394 interface standard is called FireWire by Apple, Inc., and i.Link by Sony, Inc. It is about 30 times faster than USB and intended for hard disks and digital camcorders.

1.9.9   ISDN, ADSL, and cable internet connections

The earliest high-speed internet connection was the integrated services digital network (ISDN), which provided both voice and data communications on a single leased telephone line. More recent technology includes asynchronous digital subscriber line (ASDL) and bi-directional cable. Both provide high-speed internet access at a monthly cost that many home-computer users can afford.

1.9.10   Intercomparsion of digital interfacing standards

Table 1.11 summarizes and compares the clock rates, number of parallel lines, and transmission distances of the digital interfacing standards described in the preceding sections.

1.10   Problems

1.1 You are given a data-acquisition system consisting of:
* A circuit that senses temperature and digitizes the signal whenever its "take data" input goes from low to high. At 0 °C it produces an output of 0, and at 128 °C it produces an output of 2,048. When the digitized output is ready, it produces a 12-bit data word (corresponding to the temperature), and its "data available" output goes from low to high. It brings "data available" low whenever "take data" is brought low.
* A parallel I/O port with 16 input lines and 16 output lines. Your program can use "get(&value)" and "put(value)" to read from and write to the digital port. A word written to the output lines is continuously asserted until a new word is written.

Your design goals are
* to measure the temperature under program control, and
* to read the digitized values of temperature into computer memory for later analysis.

Do the following:
(a) Sketch a block diagram of all components and essential interconnections, and label each.

Table 1.11 *Intercomparison of digital interfacing options*

Interfacing standard	Max. data clock rate (kHz)	Number of parallel lines	Transmission distance (m)
Parallel I/O port	1,000	8, 16, or 32	10
RS-232C (local)	20	1	17
RS-232C (with modem)	56	1	Unlimited*
RS-422[†]	10,000	1	1,300
RS-423	100	1	1,300
RS-485[†]	100	1	1,300
IEEE-488	500	8	20
VME	40,000	8, 16, 24, or 32	
PC serial	115	1	
PC parallel	200	8	
SCSI	500	8	7
USB 1.1	12,000	1	
USB 2	480,000	1	
IEEE-1394 (FireWire)	400,000	1	
ATA/66	66,000	16	
ISDN	128	1	
ADSL	1,000	1	5,000
Cable[§]	10,000	1	§

*Transmission to any part of the Earth possible via conventional telephone lines.
†Requires two nonground conductors per signal line.
§Transmission to any part of the Earth using cable and fiber optic lines. Maximum data rate assumes many users per cable.

(b) Give the sequence of events (hardware and software) that takes place every time this system makes one temperature measurement and stores the result in computer memory.

(c) Draw a timing diagram for the signals and data described in (b).

1.2 A colleague has just designed a digital data-acquisition system using a microcomputer, a digital input port with edge-triggered D-type flip-flop input registers, and the following handshaking protocol:

1. When the program is ready for data, it sets "ready for input data" TRUE.

2. When the external circuit detects "ready for input data" TRUE, it pulses the clock input of the D-type flip-flops.

3. The external circuit asserts data on the input registers of the port (the D-type flip-flops) and makes "input data available" TRUE.

4. The program detects "input data available" TRUE and reads the input port registers (the output of the D-type flip-flops).

5. The program sets "ready for input data" FALSE, processes the data, and then returns to step 1.

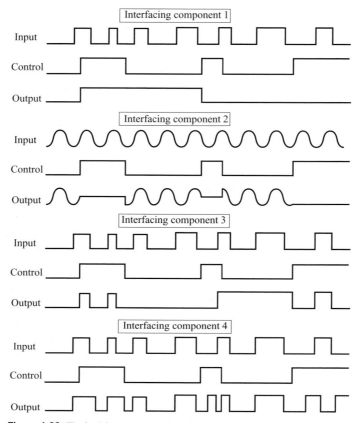

Figure 1.39 Typical input, control, and output signals for four interfacing components.

Your colleague complains that his design does not work, and that the values read during step 4 have nothing to do with the digital input data. After carefully examining his steps, you find that two serious errors were made. What are these errors, and how would you fix them?

1.3 Chapter 1 describes four interfacing components that have one input signal, one control input, and one output signal. Timing diagrams for these four components as they would occur in typical use are shown in Figure 1.39.

For each interfacing component listed below in column one, enter its number in column two:

Name	Interfacing component number
Transparent latch	
Tri-state driver	
Edge-triggered flip-flop	
Sample-and-hold amplifier	

1.4 Show how you would convert Gray code numbers to unsigned binary using software only. Prepare a flow chart or a list of program steps.

1.5 Given an 8-bit binary number where all bits are one, what is the numerical value when it is interpreted as:

(a) 2's complement integer,

(b) an unsigned integer,

(c) hexadecimal.

Given a 16-bit binary number where all bits are one, what is the numerical value when it is interpreted as:

(d) 2's complement integer,

(e) an unsigned integer,

(f) hexadecimal.

1.6 In 2's complement arithmetic, the sign of a number is changed ($a \rightarrow -a$) or ($-a \rightarrow a$) by taking the complement of each bit and adding 1. Using the 8-bit binary and 2's complement representations given in Table 1.1, show explicitly that:

(a) The number 0 and the 2's complement of 0 have the same bit pattern (this is expected, since $0 = -0$).

(b) The number -127 and the 2's complement of 127 have the same bit pattern (this is expected, since $-127 = -127$).

(c) The number -128 and the 2's complement of -128 have the same bit pattern (this is not expected, since $-128 \neq -(-128)$).

1.7 Using the 8-bit binary and 2's complement representations given in Table 1.1, show explicitly that for the numbers $1, -1, 16$, and 127, taking the 2's complement twice results in the original number: $a = (-(-a))$.

1.8 After each of the four interfacing standards listed below, write only the lettered characteristics that apply:

1. RS-232   _____

2. RS-422   _____

3. IEEE-488 (GPIB) _____

4. VME   _____

(a) Data transmitted serially (i.e. one bit at a time).

(b) Data transmitted in parallel (i.e. many bits transmitted simultaneously on as many wires).

(c) Can be implemented using existing telephone lines.

(d) Allows up to 21 controllers (e.g. microprocessors) to interface with each other and with the circuits that they control.

(e) Permits data transfer at or above 1,000 byte/s.

(f) Permits data transfer at or above 1 Mbyte/s.

(g) Permits data transfer at or above 10 Mbyte/s.

(h) Serial data transfer using differential ± 3 V signals on a twisted pair of wires.

1.9 Design a computer system for reading *eight* 16-bit digital counters with *one* 16-bit parallel input port.

The design objectives are:
- The eight counters are to be latched as simultaneously as possible (within 0.1 μs) and read sequentially by the 16-bit parallel input port.
- This operation is to be performed once per second with a timing accuracy of 1 μs.
- Counter number 1 is to count the elapsed time in seconds.

The components provided are:
- A digital timing circuit that produces a 1-μs wide active low pulse with a period of 1.000,000 s ± 1 μs.
- Eight 16-bit counters with pulse input, clear, latch command, output-enable, and 16 output lines (Figure 1.40). A low-to-high transition on the latch command transfers the contents of the internal counter to a set of 16 internal registers. When the output-enable is high, the internal registers are asserted on the 16 output lines. When the output-enable is low, the 16 output lines are in a high-impedance state. (This design allows the contents of a rapidly changing counter to be transferred to storage registers at a precise instant, and the storage registers read at a later time.)

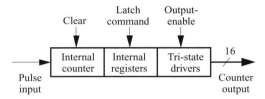

Figure 1.40 16-bit counter with internal registers and tri-state drivers.

- A 16-bit parallel output port operated in transparent mode. The write operation takes 2 μs.
- A 16-bit parallel input port that you decide to operate in transparent mode by letting its "strobe" input float HI. The read operation takes 2 μs.

Do the following:
(a) Draw a block diagram of all components and essential interconnections. Label all components, control lines, and data lines.
(b) Draw a timing diagram that shows the data-acquisition sequence for counter 1 and counter 8.
(c) Draw a flow chart (or a list of what the steps do) for your microcomputer control program.
(d) For your design, compute how long it takes to transfer the contents of all eight counters into memory.
(e) Indicate which best describes this method of data acquisition: (i) software-trigger, status-poll; (ii) hardware-trigger, status-poll; or (iii) hardware-trigger, direct memory access.

1.10 A microcomputer has a parallel I/O port, and eight external digital devices are connected to the parallel output port to form a parallel output bus. The I/O port has 16 bits of input, 16 bits of output, and all external devices have eight bits of input.

Do the following

(a) Sketch a block diagram showing and labeling all essential components and interconnections. (You only need to sketch two of the external devices – use dots to represent the other six.)

(b) Describe all the steps necessary for the microcomputer to write data to one specific external device and not to the others, using full handshaking.

1.11 Design a system for measuring human reaction times, using a parallel I/O port, a mechanical push button for user input, and a high-power, bright light emitting diode (LED) for the visual prompt. This is similar to Laboratory Exercise 2 but avoids the random delays in refreshing the display screen (60 Hz) and the use of a slow (9,600 baud) serial port to detect the pressing of the return key. Assume the following:

• The first press of the button alerts the computer that the user is ready.
• The computer program then waits a random time from 5 to 10 s and then lights the LED.
• The user presses the button a second time to respond to the LED prompt.
• The maximum rating of the LED is 0.1 A at 0.6 V (forward biased diode). Assume that the LED comes to full brightness within 1 μs after the power is applied.
• You have a power amplifier (voltage gain = 1) to drive the LED.
• The computer has a digital I/O port with eight bits of input and eight bits of output ("0" = 0 V, "1" = 5 V).
• Your program can read a 1-kHz system timer
• Ignore system interrupts that would cause your program to pause.

Do the following:

(a) Sketch a block diagram of your system, showing and labeling all essential components, connections, and signals. (You may draw the power amplifier with a single input, a single output, and a common ground.)

(b) List the steps (hardware and software) that must take place for a single measurement of the reaction time.

1.12 When using a push button for computer input, two problems arise. Firstly, contact bounce could cause the computer to detect a single button push as a series of signals. Secondly, system interrupts can last longer than the shortest button push the user could make (even with contact bounce). This would cause the program to miss short button pushes entirely. Show how you would use a set/reset latch interfaced to a digital I/O port to prevent these problems.

1.13 Design a system for measuring human reaction times, using the following components:
   - a microcomputer with a 32-bit parallel I/O port,
   - a mechanical push button for user input,
   - a set/reset latch,
   - an LED (light emitting diode),
   - a 32-bit counter that counts pulses from an internal 1-MHz clock.

   This is similar to Problem 1.11 but uses a dedicated 32-bit timer to avoid errors due to software delays in reading the system timer and uses the set/reset latch solution from Problem 1.12 to prevent errors due to contact bounce and system interrupts.

   Assume the following:
   - The first press of the button alerts the computer that the user is ready.
   - The computer program then waits a random time from 5 to 10 s and then lights the LED.
   - The user presses the button a second time to respond to the LED prompt.
   - The maximum rating of the LED is 0.1 A at 0.6 V (forward biased diode). Assume that the LED comes to full brightness within 1 μs after the power is applied.
   - You have a power amplifier (voltage gain = 1) to drive the LED.
   - The counter has two control lines:
     a pulse on the "reset" line sets the counter to zero and starts counting,
     a pulse on the "stop" line stops the counter.
   - The program command get(&value) will read the contents of the counter.
   Do the following:
   (a) Sketch a block diagram of your system, showing and labeling all essential components, connections, and signals. (You may draw the power amplifier with a single input, a single output, and a common ground.)
   (b) List the steps (hardware and software) that must take place for a single measurement of the reaction time.

1.14 Assuming that you have the system described in Problem 1.13, how would you use the system and statistical analysis to determine whether racecar drivers or jet fighter pilots have the faster reaction time? List all the steps that you need to accomplish to make a valid determination.

1.15 You have been chosen to design a microcomputer system for timing the swimming events in the summer Olympic games.
   - There are 12 swimmers and the pool has 12 lanes. Each swimmer starts at the one end of the pool and, at the sound of a gunshot, jumps in and swims to the opposite end of the pool in their own lane.
   - When they reach the opposite end of the pool, the swimmers make contact with a switch (called a "touch plate") mounted on the wall of the pool. When the switch is touched, the contacts stay closed until manually reset.

- The athletic event is started by the starter's pistol, which closes an electrical contact when the trigger is pulled.
- Your computer system detects the contact closure and immediately sends a pre-recorded gunshot sound to 12 speakers, each located behind a swimmer (this gives each swimmer a fair start and also avoids using chemical explosives).
- There is an external timing circuit mounted near each touch plate (Figure 1.41). Each has a 24-bit counter that is set to zero by a "start" input pulse, increases by one every 100 μs, and is stopped by a "stop" input pulse. The start and stop input lines float high when disconnected and can be brought low by connecting to ground.

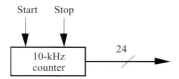

Figure 1.41 Digital timing circuit.

- Your microcomputer has three 16-bit input ports, two 16-bit output ports, and **NO** analog I/O. The input port lines float high when disconnected and can be brought low by connecting to ground.
- The gunshot sound is in a digital file and you have a software function that sends the digital data to one of the output ports at the correct speed.
- You have an external 12-bit D/A converter and a power amplifier, and any digital circuits described in Chapters 1 or 3.

The requirements for your design are:

- The system must record the time for every swimmer to an accuracy of 100 μs even if several swimmers touch their plates in the same 100 μs.
- The lane numbers and time for each swimmer (in units of seconds) are to be written to the computer display screen and to a file as soon as possible after the swimmer finishes.

Do the following:

(a) Sketch your design, showing and labeling all essential components and lines. (You only need to show two touch plate switches, timing circuits and speakers.)

(b) Describe the events (hardware and software) that must take place from the start of the race to after the last swimmer finishes.

1.11   Additional reading

Michael Andrews, *Programming Microprocessor Interfaces for Control and Instrumentation*, Prentice Hall, Englewood Cliffs, NJ, 1982.

Bruce A. Artwick, *Microcomputer Interfacing*, Prentice Hall, Englewood Cliffs, NJ, 1980.

David M. Auslander and Paul Sagues, *Microprocessors for Measurement and Control*, Osborne/McGraw-Hill, Berkeley, CA, 1981.

George C. Barney, *Intelligent Instrumentation*, Prentice Hall, Englewood Cliffs, NJ, 1988.

S. J. Cahill, *Designing Microprocessor-Based Digital Circuitry*, Prentice Hall, Englewood Cliffs, NJ, 1985.

Joe Campbell, *Crafting C Tools for the IBM PCs*, Prentice Hall, Englewood Cliffs, NJ, 1986.

IEEE Standard 488-1981, *IEEE Standard Digital Interface for Programmable Instrumentation*, The Institute of Electrical and Electronics Engineers, New York, NY, 1981.

IEEE-488 Control Data Acquisition and Analysis for Your Computer, National Instruments, Austin, TX, 1994.

B. W. Kernighan and D. M. Ritchie, *The C Programming Language*, Prentice Hall, Englewood Cliffs, NJ, 1988. This book defines ANSI C, which is not compatible with and supersedes the previous "Standard C."

Harold S. Stone, *Microcomputer Interfacing*, Addison-Wesley, Reading, MA, 1982.

Willis J. Tompkins and John G. Webster, *Interfacing Sensors to the IBM PC*, Prentice Hall, Englewood Cliffs, NJ, 1988.

Laboratory Exercise 1

Introduction to C programming

Purpose

To use a compiler/editor to write and compile a simple C program. To investigate the 2's complement unsigned, and hexadecimal representation of 8-, 16-, and 32-bit numbers, to use the printf function for printed output, and the scanf function for interactive program control and data entry.

Equipment

- IBM-compatible Pentium microcomputer with Windows NT operating system and Microsoft Visual C++ compiler
- Printer (shared with other laboratory stations)

Additional reading

Appendix C: C programming hints

Section 1.3 Number systems

B. W. Kernighan and D. M. Ritchie, *The C Programming Language*, Prentice Hall, Englewood Cliffs, NJ, 1988.

Procedure

1. Microsoft Visual C++ programming environment

You will need a separate project file and folder for each laboratory exercise that requires programming. Rather than creating your own project folder from scratch, copy a "starter project" folder into your file storage area. The laboratory assistant or teaching associate will tell you where to find it on your local network. This folder contains a starter project file with all the library modules that your project will require, and C code files you may need, such as InitAll and fft.c.

Rename the starter project folder and the .dsw project file in it, but be sure to keep the .dsw extension. Start the Microsoft Visual C++ developer studio by double-clicking the shortcut on the desktop. Open the project by clicking on the "File: Open Workspace" command and select your project .dsw file. "File: Close Workspace" will close the project you are currently working on. Note that only one project can be open at a time.

To create a text window to enter your C code, click on "File: New Text File." Save the file as a ".c" file (C file), not as a ".cpp" file (C++ file). Note that some of the C library functions used with the data-acquisition board in other laboratory exercises will not work with C++ files.

To insert a code file into a project file do step 1 and one of steps 2(a), (b), or (c).

1. Open the workspace associated with the project (see instructions above). The project should be listed on the left-hand side of the screen. Use the "File View" selection to see which files are currently linked to the project.

2. (a) Open the code file and right click on the code file window. Select "Insert File Into Project," *OR*

   (b) right click on the project name and click on "Add Files to Project" to select the desired code files by name, *OR*

   (c) click on the "project" in the menu bar and select "Add to Project" to select the desired code files by name.

A project can have only one file containing the main() code. To delete a code file from a project, go to the workspace window under "file view" mode, select the file you want to remove, and press "delete."

After you have added your main() and other code files to your project, compile your program by using the "Build" pull-down menu. Select "Build" under the "Build" menu to compile all C codes and link all Object codes. Select "Compile" under the "Build" menu to compile only new or changed code files. Select "Execute <run-time file name>" under the "Build" menu to run your program. "Build," "Compile," and "Execute" also have short-cut buttons on one of the menus.

Notes to the instructor and system manager for the laboratory computers:

1. Provide a desktop shortcut linked to the Microsoft Visual C++ developer studio for quick access by the students.

2. Provide a starting project folder and tell the students how to access the folder over the network. Before starting the programming assignment for each laboratory exercise, the students are to copy the starting project folder into their file storage area. They can then rename the folder and project file as appropriate, and create and add their code files.

3. To create a project folder from scratch, click on the "File: New" command and select the "Win 32 console application" in the "Project" section. Be sure to select the desired folder and enter the starter project name before clicking "OK."

4. Right click on the starter project name in the Workspace window (under the "File View" mode) and select "Settings...." Go to the "Links" section of the settings window and find the "Object/Library modules" section.

   Link the following Object/Library modules to the starting project: advapi32.lib, comdlg32.lib, gdi32.lib, graph32.lib, kernel32.lib, odbc32.lib, odbccp32.lib, oldaapi32.lib, ole32.lib, oleaut32.lib, olmem32.lib, shell32.lib, user32.lib, uuid.lib, and winspool.lib.

5. Place any additional code files for the students' use (such as InitAll and fft.c) in the starting project folder.

2.   Program

Use the editor to write the following C program, which prompts the user for a starting integer n1 and a final integer n2, and then prints out various bit-pattern representations of the numbers from n1 to n2:

```c
#include <stdio.h> /* needed to use printf, scanf */
main ()
{
char c;
int i,n1,n2,number;
long k;
float f;
while(1)
    {
    printf("enter first number: ");
    scanf("%d", &n1);
    printf("%d\n", n1);
    printf("enter last number: ");
    scanf("%d", &n2);
    printf("%d\n", n2);
    for (number=n1; number <= n2; number++)
        {
        c = number;
        i = number;
        k = number;
        f = number;
```

```
                printf("char:(dec)=\t%12d (uns dec)=%12u (hex)=%12x\n",c,c,c);
                printf("int:(dec)=\t%12d (uns dec)=%12u (hex)=%12x\n",i,-3,i);
                printf("long:(dec)=\t%12ld (uns dec)=%12lu (hex)=%12lx\n",k,k,k);
                     /* note - this is %12 followed by the letters ld, not 121
        followed by d */
                printf("f: (float)=\t %12.3f\n\n\n", f);
                }/* end for */
            }/* end while */
        }/* end main */
```

3.   Numerical representation

Run the program and investigate numbers in the ranges from 0 to 5, from 125 to 130, and from −130 to −125. Note that 0 and small positive integers have the same representation, whereas a negative number such as −129 has six different representations, depending on the variable type and the printf format type.

4.   Packing 8-bit bytes into 16-bit integers

Write a function pack(a,b) to pack two 8-bit "char" variables a and b into a 16-bit "int" variable. (*Hint*: Use the left shift operator ≪.) Enter a and b from the keyboard as hex input and print the packed variable as a decimal number.

Check your pack function for the following cases:

pack(0,80) = 128   pack(0,7F) = 127
pack(1,0) = 256   pack(0,FF) = 255
pack(80,00) = −32,768   pack(7F,FF) = 32,767
pack(FF,01) = −255   pack(FF,00) = −256

Laboratory report

1.   Setup

Sketch a simple block diagram of the major components used in this exercise (e.g. keyboard).

2.   Data summary

2.1 Tabulate the ten responses for each of the numbers you printed with your program. It is acceptable to paste a concise table of numbers from your printer output into your laboratory report. Include column headings so that the reader knows what the numbers are.

2.2 Tabulate the input and output numbers you used to test your pack(a,b) function.

3.   Discussion

3.1 Discuss the reasons why 0 and small positive integers have the same representation, whereas a negative number such as -129 has six different representations, depending on the variable type and the printf format type. Use examples to support your discussion.

3.2 Draw a simple flow diagram of your function pack(a,b).

4.   Questions

4.1 Do Problem 1.5.
4.2 Do Problem 1.6.
4.3 Do Problem 1.7.

5.   Laboratory data sheets

Attach printouts of your program code and raw output.

Laboratory Exercise 2

Measuring event times

Purpose

To write and use a C program that measures human response times. To use the Student's t test to determine whether the difference of two means is statistically significant.

Equipment

- IBM-compatible Pentium microcomputer with Windows NT operating system and Microsoft Visual C++ compiler
- Printer (shared with other laboratory stations)

Background

1.   Timer function

The Pentium processor board has a dedicated counter that increments (adds one) at a rate of 1 kHz. The counter is reset to zero when the computer is started. It is read with the following C function:

time1 = GetTickCount();

where the variable time1 is declared with:

DWORD   time1;   /* 32-bit storage */

2.   "Beep" function

The computer board can produce an audible tone using the function Beep (frequency, duration), where "frequency" is in hertz and "duration" is in milliseconds. "Frequency" = 550 is a good value. This function takes over control of the computer for the duration of the sound. Since the human reaction time is measured from the start of the sound and the computer can only detect a keystroke after the sound has finished, we want "duration" to be as short as possible, yet long enough so that it can be clearly heard. "Duration" = 50 is a good value. With that value, the shortest possible response time you can record will be 50 ms. Such a value is faster than any reasonable human response time and is usually the result of pressing the response key (and "pre-loading" the scanf buffer) *before* the visible or audible prompt.

Note that both GetTickCount and Beep need the header file <windows.h>.

3.   Student's *t*

Given two sets of values a_i, $i = 1$ to m_a, and b_i, $i = 1$ to m_b, measured under different experimental conditions, the averages $\bar{a}$ and $\bar{b}$ are given by:

$$\bar{a} = \frac{1}{m_a} \sum_{i=1}^{m_a} a_i \quad \text{and} \quad \bar{b} = \frac{1}{m_b} \sum_{i=1}^{m_b} b_i$$

The rms deviations σ_a and σ_b are given by

$$\sigma_a = \sqrt{\frac{1}{m_a - 1} \sum_{i=1}^{m_a} (a_i - \bar{a})^2} \quad \text{and} \quad \sigma_b = \sqrt{\frac{1}{m_b - 1} \sum_{i=1}^{m_b} (b_i - \bar{b})^2}$$

The standard error of the means $\sigma_{\bar{a}}$ and $\sigma_{\bar{b}}$ are given by:

$$\sigma_{\bar{a}} = \sqrt{\sigma_a^2 / m_a} \qquad \sigma_{\bar{b}} = \sqrt{\sigma_b^2 / m_b}$$

Use Table 5.2, pp. 369–370, to determine the probability that the difference $d = \bar{a} - \bar{b}$ could have arisen by chance. The number of degrees of freedom is given by $n_f = m_a + m_b - 2$, and the value of Student's t is given by:

$$t = \frac{d}{\sigma_d} = \frac{\bar{a} - \bar{b}}{\sqrt{\sigma_{\bar{a}}^2 + \sigma_{\bar{b}}^2}}$$

If the difference between $\bar{a}$ and $\bar{b}$ is small compared to the rms deviations σ_a and σ_b, then $|t|$ will be small, the probability of getting a value of $|t|$ of that size or larger by chance will be large, and we will not be able to conclude that the difference is statistically significant.

But if the difference between $\bar{a}$ and $\bar{b}$ is large compared to the rms deviations σ_a and σ_b, then $|t|$ will be large, and we should be able to conclude with some confidence that the different experimental conditions *caused* the averages $\bar{a}$ and $\bar{b}$ to differ. Table 5.2 allows us to state that confidence with a probability number.

One way of looking at the $|t|$ distribution is by first assuming that the null hypothesis is correct and the conditions a and b do not cause any difference in the data. Under this assumption, Student's t distribution will center at zero, with small tails extending to large positive and negative values of t. If the actual experiments produce large values of $|t|$, then the null hypothesis can be rejected. As a rule of thumb, if the probability of exceeding $|t|$ is below 0.1%, the null hypothesis can be rejected. Even so, if the null hypothesis is correct, a $|t|$ value large enough for $P(>|t|) = 0.1\%$ will occur once in 1,000 experiments!

A note on experiment size. If the difference in the experimental conditions actually causes a difference in the measured quantity, then making m_a and m_b sufficiently large will make $|t|$ large. If, however, the different conditions do not affect the measured quantity, then t will average 0 and $|t|$ will average ≈ 1, no matter how large m_a and m_b are.

Additional reading

Section 5.2 The Gaussian-error distribution
Section 5.3 Student's t test
For more complete treatments, see:

Lyman Ott and William Mendenhall, *Understanding Statistics*, Duxbury Press, Boston, MA, 1985.
George. W. Snedicor, *Statistical Methods*, Iowa State University Press, Ames, IA, 1965.

Procedure

1.   Program

1. Include needed header files:

```
#include <dos.h>
#include <windows.h>
#include <math.h>
```

Define variables:

```
DWORD   time1, time2, time3;
char dummy;
double dtime;
```

2. (a) Delay 5 s:

```
time1 = GetTickCount();
time2 = time1;
while (time2 < (time1 + 5000) )
    time2 = GetTickCount();
```

   (b) Random delay between 5 and 14.9 s:

```
time1 = GetTickCount();
time2 = time1;
while (time2 < (time1 + 5000 + 100*(time1%100)) )
    time2 = GetTickCount();
```

   Note: (time1%100) is a pseudo-random number between 0 and 99.

3. (a) Generate a prompting character on the display screen:

```
printf("a");      /* visible prompt– write character to screen*/
```

   Note: The electron beam scans the phosphor on the inside of the display screen about 60 times per second, so it may take as long as 17 ms for a character to appear.

   (b) Generate a prompting tone. The following function produces a 550-Hz tone with a duration of 50 ms, and then returns control to the program:

```
Beep(550, 50);
```

4. Pause until the return key is pressed. This can be done by reading a dummy variable from the keyboard with the following statement:

```
scanf("%c", &dummy);          /* pause until return key is pressed */
```

   Note: The keyboard interface is a 9,600-baud (9,600 bit/s) serial port and there are about 10 bits per character (an 8-bit byte plus start and stop bits). So keyboard I/O takes a minimum of 1 ms per character.

5. Read the tick counter after the human response delay and convert the delay to seconds.

```
time3 = GetTickCount();
dtime = 0.001 * (time3 - time2);
```

6. Print the time response on the screen and write the results to a file for subsequent printing.

7. Loop back to step 2 to try again.

8. After 10 tries, compute and print the 10 individual reaction times, the mean, the standard deviation, and the standard error of the mean.

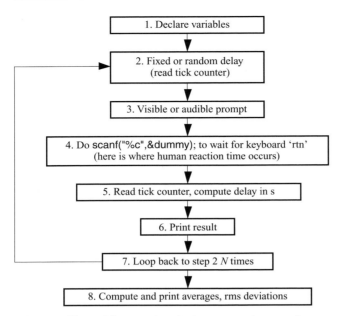

Laboratory Figure 2.1 Flow chart for human reaction experiments.

2.   Program flow chart

Your program should have the flow chart shown in Laboratory Figure 2.1.

3.   Learning curve

Using a fixed delay (see under Program, point 2(a)) and a visible prompt (Program, point 3(a)), each laboratory partner measures several series of 10 response times. Note how the averages change as the task is learned. A response time of a few milliseconds is probably due to pressing the return key *before* the prompt. Such data should be used only for determining the system response time.

4.   Response to a visible prompt with fixed delay

After the training provided in the previous section, each laboratory partner measures 10 response times. Here the delay between running the program and the first event (the prompting character) is fixed, so the subject might be able to anticipate it, using an "internal" clock in the mind.

5.   Response to a visible prompt with a random delay

Change the program at step 2 to make the delay random (replace the code in 2(a) with 2(b)). Here we will use the two least significant digits of the tick counter to generate a random delay from a minimum of 5 s to a maximum of 14.9 s.

With a random delay, it should be harder to anticipate the prompting character.

6.   Response to an audible prompt with random delay

Repeat the previous section using an audible prompt (Program, point 3(b)) rather than a visible prompt (Program, point 3(a)).

7.   Computer response time

Determine the minimum delay that you can measure using the visible prompt and the audible prompt. (*Hint*: "preload" the scanf by pressing the return key during the delay loop and before the prompt.)

Laboratory report

1.   Setup

Sketch a simple block diagram of the major components used in this exercise (e.g. keyboard).

2.   Procedure and data summary

Tabulate the response times measured for procedure section 3 (learning curve, difference between two laboratory partners), section 4 (visible prompt, fixed delay), section 5 (visible prompt, random delay), section 6 (audible prompt, random delay), and section 7 (computer response time for both visible and audible prompts).

3.   Analysis

3.1 Average delays and uncertainties. For each procedure section, i.e. 4 (visible prompt, fixed delay), 5 (visible prompt, random delay), 6 (audible prompt, random delay), and 7 (computer response), use the equations in background section 2 to compute and tabulate the average delay, the rms deviation, and the standard error of the mean. In one of these sections, you should have data for two different subjects. Tabulate the results. If you do not have time in the laboratory, perform the analysis at "home."

3.2 Student's *t* test (visible – fixed versus random). This compares the reaction time for a visible prompt between fixed and random delays. Use the equations in background section 2 to compare the difference between the results of procedure sections 4 and 5. Compute and tabulate the following:

- $d = \bar{a} - \bar{b}$,
- σ_d,
- the number of degrees of freedom $m_f = m_a + m_b - 2$,
- Student's t,
- the probability P of exceeding $|t|$ by chance – use Table 5.2 to determine (by interpolation, if necessary) the value of P.

If the probability $P(\geq |t|)$ is less than 1%, it is likely that the difference between the average response times was not due to chance.

3.3 Student's t test (random – visible versus audible). This compares the reaction time for a random delay between visible and audible prompts. Repeat the preceding analysis section 3.2 for the difference between the results of procedure sections 4 and 5.

3.4 Student's t test (different laboratory partners). Repeat the preceding analysis section 3.2 for the difference between two laboratory partners in any one of procedure sections 3, 4, 5, or 6.

4.   Discussion and conclusions

4.1 Discuss the subject's perceptions and any improvement in reaction time during the learning process for a fixed delay (procedure sections 3 and 4).

4.2 Discuss the difference in the reaction times of the two laboratory partners. Using your Student's t analysis, was one individual significantly faster?

4.3 Discuss the subject's perceptions in reacting to a random delay (procedure section 5) compared with a fixed delay (procedure section 4). Using your Student's t analysis, was one reaction time significantly faster?

4.4 Discuss the subject's perceptions in reacting to an audible prompt (procedure section 6) compared with a visible prompt (procedure section 5). Using your Student's t analysis, was one reaction time significantly faster?

4.5 Discuss how the ability of an experiment to determine whether a measurable quantity is actually different under two different experimental conditions is determined by the design of the experiment. Consider the accuracy of each measurement, the number of observations, control over confusing variables, using subjects as their own controls, etc.

5.   Questions

5.1 Write a C program (or draw a flow chart) that would sound an audible tone periodically with a period of 10 s, without accumulating a systematic error. In other words, the Nth tone would occur at a time $10N$ s as accurately as the Laboratory Exercise 2 equipment allows.

5.2 Suppose that you are convinced that a measured quantity really depends on some experimental variable, but after doing a preliminary experiment, you find that the Student's t value is too low to prove your case. What could you do to pursue the issue further?

5.3 What system delays did you expect for the visible and audible prompts? Did they agree with your observations in procedure section 7?

6.   Program and laboratory data sheets

6.1 Include printouts of your program code, data, and raw output.

6.2 Include your handwritten data sheets (or a copy), which should consist of a log of the procedures you used, any special circumstances, and the measurements you recorded manually.

Laboratory Exercise 3

Digital interfacing: switches and lights

Purpose

To write and test a C program that uses a microcomputer parallel interface to read simple external devices such as switches and to turn on light-emitting diodes, and to gain familiarity with handshake lines and digital interfacing protocols.

Equipment

- IBM-compatible Pentium microcomputer with Windows NT operating system and Microsoft Visual C++ compiler
- Printer (shared with other laboratory stations)
- Data Translation DT3010 interfacing board
- +5-V power supply
- Digital multimeter
- Superstrip circuit board
- One 10-μF 25-V electrolytic capacitor (put between power and ground at circuit board binding posts) (green post = ground)
- Three 0.1-μF CK-05 bypass capacitors (put between power and ground on all integrated circuits)
- Eight 330-Ω resistors
- One DIP unit of eight switches
- One 74LS244 octal buffer
- One 74LS374 octal edge-triggered flip-flop
- Eight light-emitting diodes
- One 74LS47 BCD to seven-segment decoder (negative logic, open collector outputs – output low corresponds to lit segment)
- One seven-segment LED display (common anode for open collector inputs)

- *OR* one integrated seven-segment LED display/decoder in place of the individual decoder and display listed above

Background

1.   Parallel I/O ports

To communicate with digital devices in the outside world, a microcomputer must have the ability to transfer a byte (8 bits) or a word (16 or 32 bits) between its random-access memory and external wires. We will be using a plug-in interfacing board (the Data Translation DT3010) with associated C-code function libraries that were specially written to work with the microprocessor, the interfacing board, and the Windows NT operating system.

Since the Data Translation DT3010 board we will be using has no handshaking capability, we will use an octal edge-triggered flip-flop integrated circuit to buffer data at the input, and we will use one of the input data lines to allow for simple handshaking with your program.

2.   Software

Your program should start out like this:

```
#include <windows.h>
#include <stdio.h>
#include "DAboard.h"
int main()
{
    unsigned int val;
    InitAll();          /* necessary to initialize the DT3010 data
acquisition board */
```

The "DAboard.c" file must be included in the compiling of the project.

3.   Reading from the DT3010 parallel input port

The DT3010 has a binary I/O device with two 8-bit parallel ports that can be configured for either input or output. In this course we will be using pins 89–96 for input. A high input voltage (5 V) reads as a binary one and a low input voltage (0 V) reads as a binary zero. The C code function for binary input is:

```
olDaGetSingleValue(hDin, &val, 0, 1.0);
```

4. Writing to the DT3010 parallel output port

The C code function for binary output is:

olDaPutSingleValue(hDout, val, 0, 1.0);

Since "Strobe" and "Input Data Available" float high when disconnected, the easiest way for you to put a pulse on these lines is to disconnect them briefly from ground and then connect them to ground again.

Note that the purpose of "Strobe" is to control the 74LS374 flip-flops, whereas the purpose of "Input Data Available" is to allow the external circuit to communicate to the computer program. The data will be held in the registers for your C program to read as long as "Strobe" is low.

Note that at power up, all the output lines are initially high. When a word is written to the output lines, they will take the corresponding level (5 V for a one, 0 V for a zero) and hold the voltages until a new word is written.

In summary, here are the steps in time sequence:

1. The program initiates some data-acquisition command (via timer pulses, attention pulse, keyboard prompt, etc.).
2. The program loops, waiting for "Input Data Available" bit to go high.
3. The external device asserts data on the parallel input port lines, and then pulses the "Strobe" line low–high–low. This freezes the data on the 74LS374 input lines onto the output lines.
4. The external device then pulses "Input Data Available" low–high–low.
5. The program detects the high state on "Input Data Available" and reads the latches (transfers data from latches to memory).

Additional reading

Appendix E Summary of data translation DT3010 PCI plug-in board

Procedure

1. Circuit

1. Before connecting the power supply to your circuit, monitor the output with the digital multimeter and adjust the output to 5.0 V. Use the ohmmeter to determine which binding post is connected to the metal plate (should be green). Connect this binding post to your power-supply ground and all your circuit grounds. Connect another binding post to +5 V and the appropriate points of your circuit.

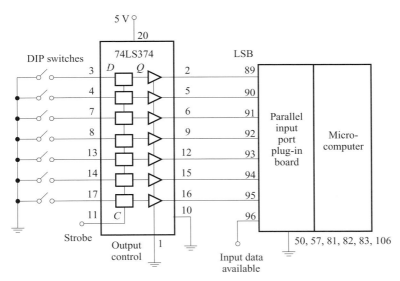

Laboratory Figure 3.1 Circuit diagram for switches and octal edge-triggered flip-flops connected to the parallel input port. "Strobe" and "Input Data Available" are normally grounded and briefly disconnected by the user to perform their functions.

2. As shown in Laboratory Figure 3.1, connect the seven DIP switches between ground and the input pins of the 74LS374 octal edge-triggered flip-flop. Connect the seven outputs of the 74LS374 to input bits 0–6 of the binary input port (pins 89–95). The 74LS374 has internal "pull up" resistors so that an open line is "high" and a grounded line is "low."

3. Ground the output control line (pin 1) so that the 74LS374 tri-state outputs are always active.

4. As shown in Laboratory Figures 3.2 and 3.3, connect output bits 0–6 of the binary output port (pins 97–103) to the inputs of the 74LS244 octal buffer. Connect the outputs to the current limiting resistors and LEDs as shown. Ground pin 1 to enable buffers 1–4 and ground pin 19 to enable buffers 5–8. Connect pin 10 to ground and pin 20 to +5 V. Connect a 0.1-μF capacitor between pin 20 and ground.

5. Connect the DT3010 grounds (pins 50, 57, 81, 82, 83, 106) to your external power-supply ground.

Note 1: For all laboratory exercises, connect a 10-μF electrolytic capacitor between each power-supply voltage (+5 V, +12 V, −12 V) and ground at the binding posts of your circuit board. Observe capacitor polarity! Electrolytics can explode when connected backwards! These capacitors help stabilize the supply voltage levels at low frequencies (such as 60 Hz) but are not as effective in reducing spikes caused by fast (<1 μs) circuit-switching transients. To reduce these fast spikes, connect 0.1-μF capacitors between power and ground at all integrated circuits.

Note 2: Use your 5-V power supply for the 74LS244 buffer. Never connect anything to the microcomputer power-supply lines. One mistake could short the microcomputer power supply, causing expensive damage.

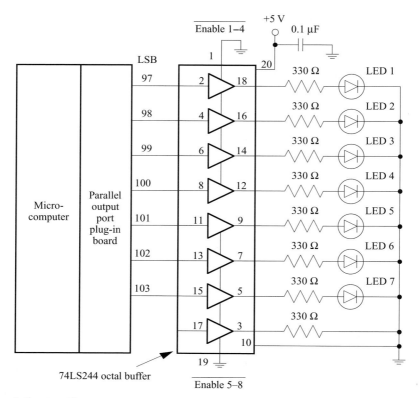

Laboratory Figure 3.2 Schematic for parallel output port, octal buffer, current-limiting resistors, and LEDs. Ground pins 1 and 19 to keep the tri-state outputs active.

74LS244 pinout

Enable 1–4	1	20	+5 V
Input 1	2	19	Enable 5–8
Output 8	3	18	Output 1
Input 2	4	17	Input 8
Output 7	5	16	Output 2
Input 3	6	15	Input 7
Output 6	7	14	Output 3
Input 4	8	13	Input 6
Output 5	9	12	Output 4
Ground	10	11	Input 5

Top view

Laboratory Figure 3.3 Schematic for parallel output port, octal buffer, current-limiting resistors, and LEDs.

2.   Program

Write a C program that does the following:

1. Ask the user (using printf) to (1) set the seven switches, (2) press the enter key, (3) briefly disconnect the "Strobe" line from ground, and (4) briefly disconnect the "Input Data Available" line from ground.

2. Wait for the enter key:

   scanf("%c", &dummy);

3. Wait until the status bit is pulsed:

   val = 0;
   while (val < 128)
       olDaGetSingleValue(hDin, &val, 0, 1.0);

4. Mask the seven switch values:

   val = val & 0x7F;

5. Write the value to the lowest seven bits of the output port:

   olDaPutSingleValue(hDout, val, 0, 1.0);

6. Write the value to the terminal screen.
7. Loop back to step 1.

3. Reading from switches and writing to lights

Run the program of procedure section 2 as follows:
1. Set a bit pattern on the switches, and press the return key.
2. Briefly disconnect the "Strobe" line from ground (1–2 s will suffice). When the "Strobe" line goes high, the octal flip-flop outputs will take on the value of the external switches. When the "Strobe" line goes back low, the input port registers will retain those values. This is important, since your program may not read them immediately.

   (*Briefly disconnect from ground* means to pull the wire from its grounded connection, allow it to float high (takes less than a second), and then to ground it again.)
3. Pulse "Input Data Available" line by briefly disconnecting it from ground.
4. The program detects the first pulse on the "Input Data Available" bit and reads the data. Contact bounce may produce subsequent pulses (see Figure 1.31), but they will be over long before the user can complete step 1 and the program can get back to step 4. In a higher-speed data-acquisition situation, however, this contact bounce can cause trouble, and requires a "debouncing" circuit, such as that in Figure 1.32.
5. The program writes the data to the lights and the terminal screen.

   Repeat the procedure, varying the bit pattern set on the switches. Verify that the pattern of LED lights agrees.

4. Testing the functions of "Strobe" and "Input Data Available"

Use the previous program but modify the procedure as follows:
1. Set a bit pattern in the switches and press the return key.

2. Briefly unground the "Strobe" line, which transfers the switch values to the 74LS374 outputs.
3. Change the bit pattern in the switches.
4. Briefly unground "Input Data Available," which causes your program to read the port.
5. Record both the pattern set in the switches and the pattern of lit LEDs. Note whether the value read by your program was the value on the switches during the "Strobe" line pulse (set in step 2) or during the read statement (set in step 4).
6. Briefly unground "Strobe" line.
7. Briefly unground "Input Data Available," which causes your program to read the port.
8. Record both the pattern set in the switches and the pattern of lit LEDs. Note whether the value read by your program was the value set in step 2 or step 4.

5.   From switches to timer to lights

Change the program of procedure section 2 as follows:
1. Start by writing the seven switch values to the lights, but then decrease the value of the number by one every 0.5 s. Use the GetTickCount function as you did in Laboratory Exercise 2 to keep track of elapsed time.
2. Write the decreasing value to the lights in a tight loop. This will constantly display the value of the number.
3. The program should stop when the number reaches 0.

Run the program with varying switch settings and record the time taken to count from 127 to 0.

6.   Seven-segment decoder driver

Connect the three most slowly varying bits that you used in procedure section 5 to the input of a seven-segment decoder driver and connect the output to a seven-segment LED display (Laboratory Figure 3.4).

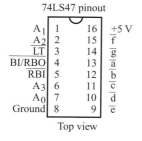

Laboratory Figure 3.4 BCD to seven-segment decoder. Four lines of input data A_0 to A_4 control the lighting of seven LED segments to produce a number from 0 to 9. $\overline{LT}$ (lamp test) turns all segments on when low. $\overline{BI/RBO}$ and $\overline{RBI}$ are used to suppress leading and/or trailing zeros.

Repeat procedure section 5, observing the changes in the seven-segment display. This demonstrates the conversion of binary to octal using a single chip.

Laboratory report

1.   Setup

Draw a simple block diagram of your experimental setup, showing all components on your circuit board as well as the connections to the timer and I/O ports.

2.   Data summary

Summarize your observations from procedure sections 3, 4, 5, and 6 above.

3.   Discussion and conclusions

3.1 Describe the operation of the hardware and your software in procedure section 3.

3.2 Discuss how procedure section 4 differs from procedure section 3.

3.3 Describe your observations in procedure section 5.

3.4 Describe your observations in procedure section 6.

3.5 Describe a computer program to read four switches, put the bits in a binary number ranging from 0000 to 1001, and light up the various segments of a seven-segment LED array to display the equivalent decimal digit. (Use only software commands and a parallel I/O port, not the seven-segment decoder circuit used in procedure section 6.)

3.6 Give two specific data-acquisition examples (not in this laboratory exercise) where a strobe pulse is needed before valid digital input data can be read by a computer.

4.   Questions

4.1 What role did the edge-triggered flip-flops play in this exercise?

4.2 How would you modify the program and circuit to be able to detect the closing of a switch and to display the number of minutes (max 99) and seconds continuously since the switch was closed to an accuracy of 0.001 s on seven seven-segment displays?

4.3 Do Chapter 1, Problem 1.11. In this problem you will use the techniques you learned in Laboratory Exercise 3 to improve the timing accuracy of Laboratory Exercise 2.

4.4 Do Problem 1.12

5.   Program and laboratory data sheets

5.1 Include a printout of your program code.

5.2 Include your handwritten data sheets (or a copy), which should consist of a log of the procedures you used, any special circumstances, and the measurements you recorded manually.

2 Analog tools

2.1 Introduction

This chapter describes a number of analog tools and techniques for processing sensor signals so that they can be digitized and read by the microcomputer, including: (i) the operational amplifier and useful circuits that can be used for low-level amplification, summation, and rectification; (ii) instrumentation and isolation amplifiers; (iii) noise sources and some of the factors that determine the signal-to-noise ratio; and (iv) analog filtering circuits. It also describes the power amplifier that allows the microcomputer to drive an actuator such as a motor, light, or heating element.

At this point it may be useful to review the fundamental differences between the analog signals that are treated in Chapter 2 and the digital signals that were treated in Chapter 1.

Analog signals:
- Single conductor (plus ground return)
- Continuous range of voltage levels
- Produced by most electronic sensors (temperature, pressure, light level, sound, nerve and muscle activity, etc.)

Serial digital signals:
- Single conductor (plus ground return) or single light beam
- Only two logic levels
- Digital bits sent single-file down the wire or fiber
- Good for transmission over phone wires, coaxial cable, and optical fiber

Parallel digital signals:
- One conductor per bit (plus ground return)
- Only two logic levels
- Generally used for short distances where many parallel conductors can be provided
- Good for high transmission rates

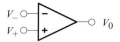

Figure 2.1 Operational amplifier. $V_0 = A(V_+ - V_-)$, where A is the open-loop gain.

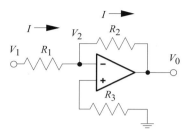

Figure 2.2 Inverting amplifier. $G = V_0/V_1 \approx -R_2/R_1$.

2.2   Operational-amplifier circuits

One of the most useful building blocks in analog circuit design is the operational amplifier, or op amp, which is available as a low-cost integrated circuit (Figure 2.1).

The **ideal op amp** has the following characteristics:
1. differential amplification with infinite gain,
2. infinite input impedance (no current enters the two inputs V_- and V_+),
3. zero output impedance (output current has no limit).

The **realistic op amp** differs in the following important ways:
1. Inputs V_+ and V_- produce an output $V_0 = A(V_+ - V_-)$, where the open-loop gain A is finite and decreases as 1/frequency. The product of gain and frequency is typically 10^5–10^8 Hz.
2. The input impedance is finite, typically 10^6 Ω for bipolar and 10^{12} Ω for FET input.
3. When V_- and V_+ are shorted to ground, V_0 may exhibit an offset potential due to unbalanced internal resistances and currents.
4. When the input terminals V_- and V_+ are connected to the same potential through equal resistors, V_0 may not be zero due to unequal external bias currents.
5. At 10 kHz in an amplifier circuit with a voltage gain of 100 (see the following sections), the output impedance is typically 10 Ω, and is approximately proportional to the gain–frequency product.
6. The maximum output current is typically 20 mA.

2.2.1   Inverting amplifier

The inverting amplifier will be the first example of a circuit that uses the op amp with negative feedback (Figure 2.2). As we shall see in Laboratory Exercise 4, the negative feedback establishes a fixed and well-defined **voltage gain** $G = V_0/V_1$ (the ratio of the output to the input voltage) over a wide range of frequencies. G is also called the **closed-loop gain**.

We compute the closed-loop gain G using the ideal op-amp defining relationships given in the last section. The output voltage is given by $V_0 = -AV_2$. Since the op amp has very high input impedance, no current flows into its input, and the current I flowing between terminals V_0 and V_1 flows through both resistors. Using Ohm's law, we have:

$$I = (V_1 - V_2)/R_1 = (V_2 - V_0)/R_2$$

Eliminating fractions:

$$V_1 R_2 - V_2 R_2 = V_2 R_1 - V_0 R_1$$

Substituting V_2 with $-V_0/A$, we have:

$$V_1 R_2 + (R_1 + R_2)V_0/A = -V_0 R_1$$

Collecting terms in V_0 and V_1:

$$V_1 R_2 = -V_0[R_1 + (R_1 + R_2)/A]$$
$$G = \frac{V_0}{V_1} = \frac{-R_2}{R_1 + (R_1 + R_2)/A} \approx -\frac{R_2}{R_1}$$

Note that $V_2 = -V_0/A$, and is therefore a very small voltage. It is commonly said that point V_2 is at a "virtual ground." Conversely, measurements of V_0 and V_2 for a sine-wave input can be used to compute the open-loop gain $A = -V_0/V_2$ as a function of frequency. As demonstrated in Laboratory Exercise 4, V_2 increases in magnitude at high frequency due to the decrease in open-loop gain A.

To the extent that bias currents $I_B{}^+$ and $I_B{}^-$ are equal, the resistive paths to ground from the $+$ and $-$ inputs should be equal for minimum offset error. This is justified by the following factors: (1) the two input circuits are close together on the same piece of silicon, (2) they were fabricated together, and (3) they are nearly at the same temperature. In the usual case where the source resistance feeding V_1 is small compared with R_1, and the output load resistance is small compared with R_2, this means that $R_3 = R_1 || R_2 = R_1 R_2/(R_1 + R_2)$.

The virtual short rule

If the op amp is in a negative feedback circuit, the output will adjust itself to keep both V_+ and V_- terminals at nearly the same potential. We can see that this is so if two conditions are met: (i) the output is not saturated ($|V_0| < 10\,\text{V}$), and (ii) the open-loop gain A is large ($>10^6$). From the op-amp equation, we immediately have $V_+ - V_- = V_0/A < 10\,\mu\text{V}$ and V_+ and V_- differ by less than $10\,\mu\text{V}$.

Virtual short rule

If an op amp is in a negative feedback circuit, if its output is not saturated, and if the open-loop gain is high, then negative feedback acts to keep the + and − terminals at the same potential.

Assuming that $A \gg 1$, we can use the virtual short rule to compute the gain of the inverting amplifier. Firstly we note that the V_- terminal cannot receive or produce current, so any current through R_1 must pass through R_2 and:

$$I_{R1} = (V_1 - V_2)/R_1 = I_{R12} = (V_2 - V_0)/R_2$$

The virtual short rule says $V_2 = 0$, so:

$$V_1/R_1 = -V_0/R_2$$

and the voltage gain is given by:

$$G = V_0/V_1 = -R_2/R_1$$

Output saturation

Op-amp output saturation results whenever the output would be driven to a potential beyond the power supply voltages. When this happens, the output seems to be "nailed" at about -10 or $+10$ V for all input signals. This condition is usually caused by a circuit error that prevents negative feedback.

2.2.2   Noninverting amplifier

It is also possible to arrange the feedback to create a noninverting amplifier with high input impedance (Figure 2.3).

From Ohm's law and the op-amp equation, we have:

$$I = (V_0 - V_2)/R_2 = V_2/R_1$$
$$V_0 = A(V_1 - V_2) \qquad V_2 = V_1 - V_0/A$$

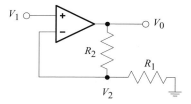

Figure 2.3 Noninverting amplifier. $G = V_0/V_1 \approx (R_1 + R_2)/R_1$.

Eliminating V_2 and solving for V_1 in terms of V_0, we have:

$$V_0 R_1 - V_1 R_1 + V_0 R_1/A = V_1 R_2 - V_0 R_2/A$$
$$V_0[R_1 + (R_1 + R_2)/A] = V_1(R_1 + R_2)$$

The gain is given by:

$$G = \frac{V_0}{V_1} = \frac{R_1 + R_2}{R_1 + (R_1 + R_2)/A} \approx \frac{R_1 + R_2}{R_1}$$

At sufficiently high frequencies, A is not large and $G \approx A$. For minimum offset error, the resistive paths to ground from the V_+ and V_- inputs should be equal. In the usual case where the V_1 source resistance is small compared to the op-amp input impedance, and the output load resistance is small compared with R_2, this means that $R_1 || R_2 = R(\text{source})$.

Assuming that $A \gg 1$, we can use the virtual short rule to compute the gain of the noninverting amplifier. The op amp produces whatever output voltage V_0 is necessary to maintain $V_2 = V_1$. We thus have $V_0/(R_1 + R_2) = V_1/R_1$ and the voltage gain is given by $G = V_0/V_1 = (R_1 + R_2)/R_1$.

2.2.3   Differential amplifier

By combining these circuits, we have a differential amplifier with a fixed gain over a range of frequencies (Figure 2.4). The input impedance is determined by the value of resistors R_1 and R_3, which must be much higher than the source impedance. If this condition is not met, the amplifier circuit will reduce the potential produced by the source and degrade accuracy.

For general values of R_1, R_2, R_3, and R_4 the output is given by:

$$V_0 = V_2 \frac{(R_1 + R_2)R_4}{(R_3 + R_4)R_1} - V_1 \frac{R_2}{R_1}$$

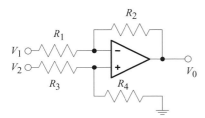

Figure 2.4 Differential amplifier.

For the special case where $R_1 = R_3$, and $R_2 = R_4$, we have:

$$V_0 = \frac{R_2}{R_1}(V_2 - V_1)$$

For minimum offset error, the resistive paths to ground from the $+$ and $-$ inputs should be equal. In the usual case where the source resistances feeding V_1 and V_2 are small compared with R_1 and R_3, and the output load resistance is small compared with R_2, this means that $R_1||R_2 = R_3||R_4$.

Design tip

If a sensor has an unwanted temperature sensitivity, consider using a second sensor that is at the same temperature but does not see the signal. Differential amplification can then be used to extract the signal.

2.2.4   Voltage follower

A special case of the noninverting amplifier is the **voltage follower**, or unity-gain buffer, which is very useful for amplifying small currents from sensors having sufficient voltage but a high output impedance (Figure 2.5).

The op-amp equation and voltage-divider equations are:

$$V_0 = A(V_1 - V_0) \quad \text{and} \quad V_0 + A V_0 = A V_1$$

Solving, we have:

$$V_0 = \frac{V_1 A}{1 + A} \approx V_1$$

Note that the voltage follower is a special case of the noninverting amplifier for infinite R_1. For minimum offset error, $R = R(\text{source})$.

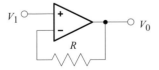

Figure 2.5 Unity-gain buffer. $V_0 = V_1$.

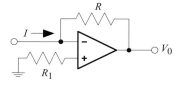

Figure 2.6 Current-to-voltage converter. $V_0 = -IR$. R_1 is chosen to minimize offset error.

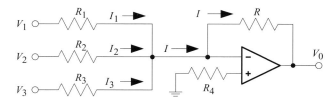

Figure 2.7 Summing amplifier. For equal input resistors R_1, R_2, and R_3, the output voltage is the sum of the input voltages. R_4 is chosen to minimize offset error.

Design tip

A resistive voltage divider is accurate only when the current passing through its resistors is much larger than the current passing through the load. If this is not the case, consider using a voltage follower between the output of the voltage divider and the load.

2.2.5   Current-to-voltage converter

By setting the input resistor R_1 in Figure 2.2 to zero, we have the current-to-voltage converter (Figure 2.6). In this case, the negative feedback through R cancels the input current I and produces an output voltage $V_0 = -IR$. It is important that the impedance of the input current source not be too large, or op-amp leakage currents will cause a large offset voltage or even output saturation.

2.2.6   Summing amplifier

The previous principle can be used to build a voltage-summing amplifier, shown in Figure 2.7. The current reaching the virtual ground at the negative op-amp terminal is $I = I_1 + I_2 + I_3 = V_1/R_1 + V_2/R_2 + V_3/R_3$, and the output voltage V_0 is given by:

$$V_0 = -IR = -(I_1 + I_2 + I_3)R = -\left(\frac{V_1}{R_1} + \frac{V_2}{R_2} + \frac{V_3}{R_3}\right)R$$

This assumes that the negative terminal of the op amp is a good virtual ground, which requires a high open-loop gain over the frequencies of interest. In the case where

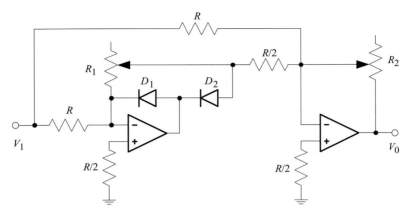

Figure 2.8 Full-wave rectifier circuit. Ideally, $R_1 = R_2 = R$, and $V_0 = |V_1|$. See Figures 2.9 and 2.10 for $V_1 < 0$ and $V_1 > 0$ equivalent circuits.

$R_1 = R_2 = R_3 = R$, we have a voltage-summing amplifier $V_0 = -(V_1 + V_2 + V_3)$. In the case where $V_1 = V_2 = V_3 = V$, we have a current-summing amplifier, and the current through each input leg is determined by the corresponding resistor value. The latter is used in the digital-to-analog converter (Chapter 3).

Design tip

Combine sensor (and other) voltages in any proportion by using a summing amplifier. The proportion of each voltage is determined by the corresponding series resistor.

2.2.7   Full-wave rectifier

The **full-wave rectifier** is an op-amp circuit whose output is equal to the absolute value of the input. High open-loop gain permits operation for very small input signals. It is used to rectify a waveform to determine the envelope of a carrier (such as in amplitude demodulation) or the average peak-to-peak amplitude of a noisy signal (such as in processing the electromyogram in Laboratory Exercise 18). The op-amp circuit shown in Figure 2.8 is commonly used for this purpose.

The circuit can be analyzed as two equivalent circuits. For the case $V_1 < 0$ (Figure 2.9), diode D_1 conducts and effectively removes the first op amp from the circuit by making its output a virtual ground. The current into the second op amp is V_1/R and its output is $V_0 = -V_1(R_2/R)$. Ideally, $R_1 = R_2$ and $V_0 = -V_1$ when $V_1 < 0$.

For the case $V_1 > 0$ (Figure 2.10), diode D_2 conducts and the first op amp becomes an inverting amplifier with an output given by:

$$V_2 = -V_1 R_1/R$$

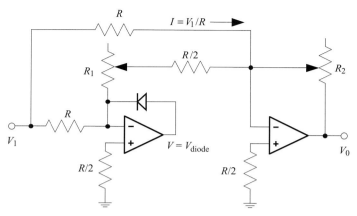

Figure 2.9 Equivalent full-wave rectifier circuit for $V_1 < 0$.

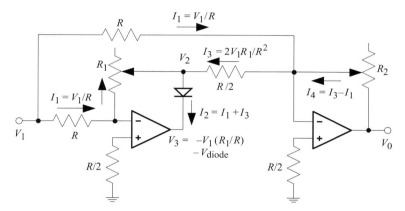

Figure 2.10 Equivalent full-wave rectifier circuit for $V_1 > 0$.

The current through the $R/2$ resistor is given by:

$$I_3 = V_2/(R/2) = 2V_1 R_1/R^2$$

The output of the second op amp is given by:

$$V_0 = R_2 I_4 = R_2(I_3 - I_1) = V_1\left(\frac{R_2}{R}\right)\left(\frac{2R_1}{R} - 1\right)$$

Ideally, $R = R_1 = R_2$ and $V_0 = V_1$ when $V_1 > 0$.

To adjust resistors R_1 and R_2, use a sine wave as a test input. If alternate lobes of the rectified waveform have different amplitudes, the gain of the first stage should be adjusted by varying its feedback resistor R_1. If the input and output have different magnitudes, the gain of the second stage should be adjusted by varying its feedback resistor R_2. Ideally, proper operation occurs when R_1 and R_2 are equal to R.

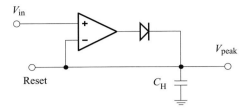

Figure 2.11 Peak detector op-amp circuit.

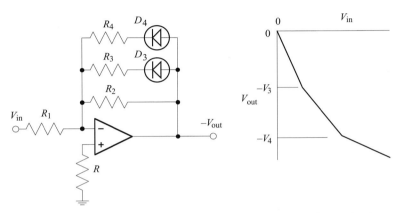

Figure 2.12 Curve shaper amplifier circuit. Gain decreases with increasing input voltage.

2.2.8   Peak detector

The **peak-detector circuit** is an op-amp circuit whose output is equal to the largest value that appears at the input (Figure 2.11). When $V_{in} > V_{peak}$, the diode is forward biased and negative feedback acts to keep $V_{peak} = V_{in}$. When $V_{in} < V_{peak}$, the diode disconnects the holding capacitor from the op-amp output and V_{peak} remains at the highest value. The holding capacitor is reset by grounding the reset input.

2.2.9   Curve shaper amplifiers

The **curve shaper amplifier circuit** is an op-amp circuit that can provide a nearly arbitrary relationship between the input voltage and the output voltage. Figure 2.12 shows a curve shaper that uses Zener diodes to produce a decreasing gain with increasing input voltage. Figure 2.13 shows a curve shaper that uses Zener diodes to produce an increasing gain with increasing input voltage. The critical points where the gain changes are controlled by Zener diodes D_i with corresponding Zener voltages V_i.

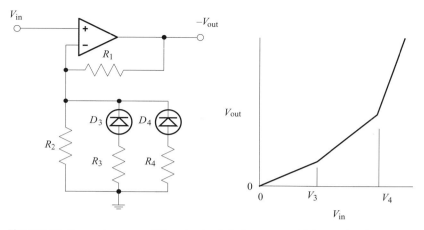

Figure 2.13 Curve shaper amplifier circuit. Gain increases with increasing input voltage.

2.3 Op-amp characteristics

2.3.1 Input and output offset voltages

For an ideal op amp, when the two inputs are both connected to ground ($V_- = V_+ = 0$ V), the output V_0 should be zero. However, differences in internal leakage currents can create a nonzero output called the **total output offset voltage** V_{TOO}. This offset voltage has a contribution from a voltage offset V_{RTI} (RTI = with respect to input) generated by the input circuits, and a voltage offset V_{RTO} (RTO = with respect to output) generated by the output circuits. Since V_{RTI} is amplified by the circuit gain G, we have:

$$V_{TOO} = G V_{RTI} + V_{RTO}$$

By measuring V_{TOO} for low and high values of G, it is possible to solve for V_{RTI} and V_{RTO}. Note that V_{RTI} cannot be measured directly. These offset voltages are affected by several factors: power-supply variations, temperature, and unequal resistance paths.

Temperature variations
The internal offset voltages and bias currents are all generally functions of temperature, so that it is necessary to refer to the data sheets to estimate the variation in input and output offset over the anticipated temperature range.

Unequal resistance paths
Even if both op-amp input terminals have equal bias currents, an offset voltage can develop due to unequal external resistance paths. This effect can be minimized by

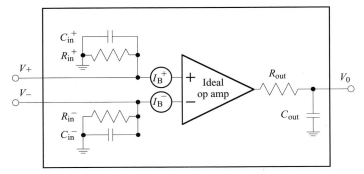

Figure 2.14 Realistic op amp shown as an ideal op amp combined with input leakage currents, input resistances, input capacitances, output resistance, and output capacitance. An output offset voltage is produced by unbalanced input leakage currents and/or unequal input resistance paths to ground.

choosing external resistors as described in the text for Figures 2.2–2.10. Conversely, the op-amp bias currents can be measured by intentionally making the external resistance paths unequal. Suppose that when both inputs are grounded, the output voltage is V_0. Now if we ground one of the inputs through an external resistor R (say, 1 MΩ), any change in V_0 can be related to the bias current I_B of that input and the closed-loop gain G:

$$I_B = \frac{\Delta V_0}{RG}$$

Figure 2.14 shows an equivalent circuit for the realistic op amp, using the ideal op amp, input resistors and capacitors, input leakage current sources, and an output resistance and capacitance. The input resistance and capacitance should be provided in the data sheets and describe the frequency-dependent input impedance.

Caution: The input bias currents of many op amps can cause large output offset voltages or even saturation if the external input resistors are too large. For example, if one input is grounded and the other is connected only by a capacitor so that its dc voltage is allowed to "float," then a leakage current of only 10 pA acting on an input impedance of 10^{12} Ω will try to develop 10 V at the input!

2.3.2   Op-amp dynamic response

The primary dynamic characteristics of the operational amplifier are given below.

The **slewing rate** is the maximum rate of output change (in volts per millisecond) for a large input step change. Under these conditions negative feedback in the various gain stages can fail when capacitors require more current through circuit impedances than the power-supply voltage can provide. The circuit will then be current limited and the capacitor voltages will change linearly with time until they approach their final voltage levels and the circuit comes out of saturation. Note that for small signals the current requirements are small, the amplifier has a linear R–C response, and the slewing rate is not limited by the circuit. These features are shown in Figure 2.15.

Table 2.1 *Some commercially available operational amplifiers*

Model	LF356	AD OP-07A	AD OP-37E	ADLH003
Manufacturer	National semi.	Analog dev.	Analog dev.	Analog dev.
Gain = 1 frequency	3 MHz	300 kHz	10 MHz	100 MHz
Gain at 1 Hz	2×10^5	3×10^5	10^6	
Input impedance	$10^{12}\ \Omega$	5 MΩ	5 MΩ	$10^{11}\ \Omega$
Input offset current	<3 pA	<2 nA	<50 nA	
Input bias current	<50 pA	<2 nA	<60 nA	<150 pA
Input offset voltage	<1 mV	<60 μV	<20 μV	<10 mV
Common-mode rejection	100 dB	126 dB	126 dB	
Maximum slew rate	12 V/μs	0.17 V/μs	11 V/μs	1000 V/s
Input noise voltage (nV/$\sqrt{\text{Hz}}$) (at 1 kHz)	12	10	4	18

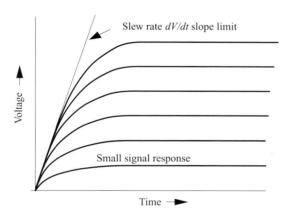

Figure 2.15 Response of amplifier to small and large signals, showing limited slewing rate.

The **unity-gain frequency** is the frequency at which the open-loop gain is equal to one.

The **gain at 1 Hz** is the open-loop gain at a frequency of 1 Hz. To provide stability at low frequencies, op-amp manufacturers often provide a small amount of internal negative feedback to limit the gain at very low frequencies (say, below 10 Hz). For this reason, the gain at 1 Hz is usually numerically less than the unity-gain frequency.

The **gain–bandwidth product** is the product of the open-loop gain and the frequency at high frequencies where the open-loop gain is falling as 1/frequency. Usually the gain–bandwidth product is numerically equal to the unity-gain frequency.

Table 2.1 shows some commercially available monolithic (single integrated-circuit) operational amplifiers. Note the wide range of offset currents and maximum slewing rates.

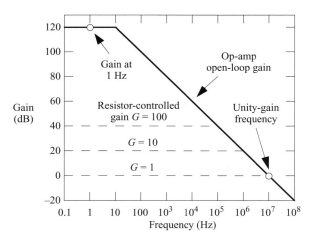

Figure 2.16 Bode plot of gain versus frequency, where G is the closed-loop gain.

2.3.3   Dynamic response with negative feedback

By using negative feedback, as shown in the previous circuits, it is possible to design an amplifier that has a constant gain over a wide range of frequencies (Figure 2.16).

The **gain–bandwidth product** of an amplifier is the product of the closed-loop gain and frequency band over which the gain is approximately constant (usually from 0 Hz to the upper 3-dB corner frequency). Since an op amp can be modeled as a single-pole low-pass filter, the effective bandwidth is larger than the upper 3-dB corner frequency (see the following section.) As the closed-loop gain is increased, the corner frequency formed by the intersection with the open-loop gain decreases in proportion and the gain–bandwidth product remains nearly constant.

The **settling time** of an amplifier is the time required for the output to remain within a specified error band of its final value after a step input. The settling time is limited by the closed-loop bandwidth of the op-amp circuit and should be independent of input step size, provided that the output is sufficiently small so that it is not limited by the op-amp slewing rate. The 0.1% settling time for an ideal linear 6.02-dB/octave amplifier is 6.9 exponential time constants, or $6.9/\omega_c = 1.10/f_c$, where ω_c is the 3.01-dB corner frequency in radians per second and f_c is the corner frequency in hertz.

The **risetime** of an amplifier is the time required for the output to rise from 10 to 90% of its final change after a step input. This requires only about two exponential time constants and for quantitative work is not as important a specification as the settling time.

2.3.4   Relationship between *RC* time constant, risetime, and bandwidth

The limited bandwidth of an amplifier acts as a single-pole low-pass filter with an RC time constant $\tau = RC$. (In Figure 2.14, the R and C are modeled by R_{out} and C_{out}.)

After a step-function input, the output rises from an initial value V_1 to a final value V_2 as:

$$V(t) = V_1 + (V_2 - V_1)\left(1 - e^{-t/\tau}\right)$$

The 10–90% risetime is given by $[\ln(0.9) - \ln(0.1)]\tau = 2.2\tau$ and the -3-dB corner frequency is given by $f_c = 1/(2\pi\tau)$.

For a circuit with a voltage gain $G(f)$ as a function of frequency f, and a gain G_0 in the passband, the bandwidth Δf is given by:

$$\Delta f = \int G^2(f)\, df / G_0{}^2$$

For a low-pass one-pole filter with unity gain in the passband ($G_0 = 1$):

$$\Delta f = \int_0^\infty G^2(f)\, df = \int_0^\infty \frac{df}{1 + (f/f_c)^2} = f_c \frac{\pi/2}{\sin(\pi/2)} = f_c(\pi/2)$$

So for a low-pass single-pole filter with $\tau = 1\,\mu s$, the risetime is $2.2\,\mu s$, $f_c = 159\,kHz$, and the bandwidth is $\Delta f = (\pi/2)f_c = 1/(4\tau) = 250\,kHz$. Note that the gain of the single-pole filter has a tail that extends into higher frequencies than f_c, and the effective bandwidth Δf is larger than f_c. For a filter with a larger number of poles, the corner falls off more sharply, and f_c and Δf are more nearly equal.

2.4   Instrumentation and isolation amplifiers

2.4.1   Instrumentation amplifiers

The **instrumentation amplifier** has all the properties of the op amp: (1) differential amplification, (2) high input impedance, (3) low output impedance, (4) low common-mode gain; but it has additional important properties: (i) the differential gain can be accurately set by choosing resistor values, and (ii) the differential gain is constant over a wide frequency band. Table 2.2 lists the properties of several of the amplifiers discussed in the preceding section and compares them to the instrumentation amplifier.

Table 2.2 *Comparison of amplification circuits*

	Op amp	Invert. amplifier	Noninvert. amplifier	Diff. amplifier	Instr. amplifier
High Z_{in}	Yes	No	Yes	No	Yes
Differential input	Yes	No	No	Yes	Yes
Defined gain over a frequency band	No	Yes	Yes	Yes	Yes

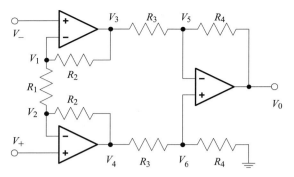

Figure 2.17 Classical instrumentation amplifier design using three op amps. The differential gain is given by $G_\pm = (R_4/R_3)(1 + 2R_2/R_1)$.

While many of the previously discussed op-amp circuits have constant gain, one must choose between high input impedance Z_{in} (e.g. the noninverting amplifier in Figure 2.3) and differential amplification (e.g. the differential amplifier in Figure 2.4). The instrumentation amplifier provides both properties.

The instrumentation amplifier is usually constructed using three op amps: the first two provide high input impedance and differential amplification and the third provides a buffered output referenced to ground (Figure 2.17). In this way, the primary deficiency of the single op-amp differential amplifier (Figure 2.4) is overcome (i.e. its low input impedance is overcome). It is good practice to get most of the voltage gain from the first stage to reduce the effect of noise from the third op amp. Note that the first stage has good common-mode rejection and can handle large common-mode signals even at high gain (see the examples that follow).

Instrumentation amplifiers are available in integrated-circuit chips as well as hybrid and discrete forms.

An instrumentation amplifier has two inputs, V_+ and V_-, with a common ground. A single output is produced, V_0, also referenced to the common ground.

The output depends on the sum and difference of the two inputs:

$$V_0 = G_c V_c + G_\pm (V_+ - V_-) \qquad V_c = (V_+ + V_-)/2$$

The **common-mode gain** G_c is given by:

$$G_c = dV_0/dV_c$$

The **differential gain** $G_\pm$ is given by:

$$G_\pm = dV_0/d(V_+ - V_-)$$

Normally, $G_\pm$ is the useful signal gain and G_c should be as small as possible. Their ratio is defined as the common-mode rejection ratio (CMRR):

$$\text{"CMRR"} = G_\pm/G_c \quad \text{(typically } 10^3\text{--}10^7\text{)}$$

Converting to decibels, we have the common-mode rejection (CMR):

"CMR" $= 20 \log_{10}(G_{\pm}/G_c)$ (typically 60–140)

For the circuit in Figure 2.17, we have $G_c = 0$ and:

$$G_{\pm} = \frac{V_0}{V_+ - V_-} = \frac{R_4}{R_3}\left(1 + \frac{2R_2}{R_1}\right) \tag{2.1}$$

By varying the resistor values, typically gains from 10 to 1,000 can be selected.

At high frequencies, the differential gain decreases due to the limited gain–bandwidth of the internal amplifiers. On the other hand, at high frequencies the common-mode gain increases due to the effect of stray capacitance. For both of these reasons, the CMRR drops at high frequencies.

As described for the op amp, the instrumentation amplifier has a total output offset voltage $V_{TOO} = G V_{RTI} + V_{RTO}$. V_{RTI} is the output offset voltage with respect to the input, and V_{RTO} is the output offset voltage with respect to the output.

The instrumentation amplifier, its gain as a function of frequency, and its common-mode rejection ratio are studied in Laboratory Exercise 5.

EXAMPLE 2.1

For the instrumentation amplifier in Figure 2.17, derive the common-mode gain.

Step 1: In pure common mode, we have $V_- = V_+$.

Step 2: By the virtual short rule, we have $V_- = V_2 = V_+ = V_1$.

Step 3: Since no current flows through R_1, no current flows through either R_2, and we have $V_- = V_2 = V_4 = V_+ = V_1 = V_3$.

Step 4: The first-stage common-mode gain is:

$$G_c = \frac{V_3 + V_4}{V_- + V_+} = 1$$

The second-stage common-mode gain is zero and the overall common-mode gain is zero.

EXAMPLE 2.2

For the instrumentation amplifier in Figure 2.17, derive the differential gain (Eq. (2.1)).

Step 1: The first two op amps have infinite open-loop gains so that the negative feedback current sets $V_1 = V_-$ and $V_2 = V_+$.

Step 2: The same current flows through all three feedback resistors (the upper R_2, R_1, and the lower R_2), so:

$$\frac{V_2 - V_1}{R_1} = \frac{V_4 - V_3}{R_1 + 2R_2}$$

Thus, the differential gain of the first stage is given by:

$$\frac{V_4 - V_3}{V_+ - V_-} = \frac{R_1 + 2R_2}{R_1}$$

Step 3: No current enters the third op-amp inputs, so:

$$\frac{V_0 - V_5}{R_4} = \frac{V_0 - V_3}{R_3 + R_4} \quad \text{and} \quad \frac{V_6}{V_4} = \frac{R_4}{R_3 + R_4}$$

Step 4: Assume that the third op amp has infinite open-loop gain, so that $V_5 = V_6$. The equations in step 3 can be combined to give:

$$\frac{V_0}{R_4} - \frac{V_0}{R_3 + R_4} = \frac{V_5}{R_4} - \frac{V_3}{R_3 + R_4} = \frac{V_4 - V_3}{R_3 + R_4}$$

which can be simplified to give:

$$\frac{V_0 R_3}{R_4} = V_4 - V_3$$

and the second-stage gain is given by:

$$\frac{V_0}{V_4 - V_3} = \frac{R_4}{R_3}$$

Step 5: Combining, we have the overall gains:

$$\boxed{G_\pm = \frac{V_0}{V_+ - V_-} = \frac{R_4}{R_3}\left(\frac{R_1 + 2R_2}{R_1}\right)} \qquad \boxed{G_c = \frac{V_0}{(V_+ + V_-)/2} = 0}$$

EXAMPLE 2.3

For the instrumentation amplifier in Figure 2.17, derive the maximum common-mode input potential as limited by output saturation.

Step 1: Solving the equations in Example 2.2 for V_3 and V_4, we have:

$$V_4 + V_3 = V_+ + V_-$$
$$V_4 - V_3 = (V_+ - V_-)(R_1 + 2R_2)/R_1$$
$$V_4 = (V_+ + V_-)/2 + (V_+ - V_-)(R_1 + 2R_2)/(2R_1)$$
$$V_3 = (V_+ + V_-)/2 - (V_+ - V_-)(R_1 + 2R_2)/(2R_1)$$

Step 2: Since the maximum values of $|V_3|$ and $|V_4|$ are limited by output saturation at V_{sat}, the common-mode voltage:

$$V_c = (V_+ + V_-)/2$$

is limited by:

$$|V_c| < V_{\text{sat}} - |V_+ - V_-|(R_1 + 2R_2)/(2R_1)$$

The maximum common-mode voltage can approach the maximum output voltage of the op amps minus one-half of whatever output is being produced by the amplification of the differential signal. Typically, V_{sat} is about 2 V below the supply voltages, or 10 V for ± 12 V supplies.

EXAMPLE 2.4

If the first-stage differential gain $(V_4 - V_3)/(V_+ - V_-) = 100$ and the second-stage differential gain $V_0/(V_4 - V_3) = 5$, for the instrumentation amplifier circuit shown in Figure 2.17, what are the voltages V_3, V_4, and V_0 for $V_+ = 1.010$ V and $V_- = 1.000$ V?

Using the equations derived in Example 2.3:

$$V_3 = (V_+ + V_-)/2 - (50)(V_+ - V_-) = 0.505 \text{ V}$$
$$V_4 = (V_+ + V_-)/2 + (50)(V_+ - V_-) = 1.505 \text{ V}$$
$$V_0 = 5.00 \text{ V}$$

Design tip

If the circuit or sensor providing a signal to an op amp or instrumentation amplifier has an impedance so high that leakage currents can cause an unwanted output voltage shift or even saturation, consider two options:
1. Add a comparable current path to the other differential input. This assumes that while leakage currents cannot be predicted, they are similar for both inputs.
2. Use the output offset adjustment. This requires manual adjustment for each circuit.

2.4.2   Isolation amplifier

The **isolation amplifier** is similar to the instrumentation amplifier, having a fixed differential gain over a wide range of frequencies, high input impedance, and low output impedance, but it also has an input circuit that is isolated from the output circuit and its power supply. This isolation is designed so that relatively large dc or 60-Hz voltages applied to the output circuit do not appear on the input circuit. Of course, the output depends on the input just as it does for an instrumentation amplifier. Two commonly used methods of isolation are electromagnetic isolation and optical isolation.

Electromagnetic isolation involves modulating a high-frequency carrier with the amplified signal and coupling the signal to the output circuit using an air-core transformer. The transformer efficiently passes the modulated carrier but blocks (isolates) at low frequencies, such as at dc and 60 Hz, where dangerous currents can occur. This is different than the iron-core transformers that are designed to efficiently transform low-frequency (50–60 Hz) ac power. A simplified schematic is shown in Figure 2.18.

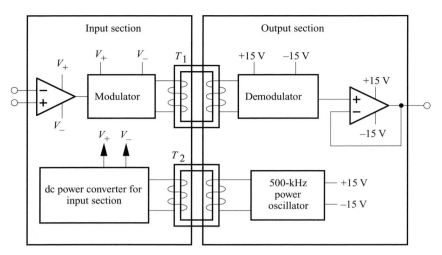

Figure 2.18 Simplified schematic of transformer-coupled isolation amplifier.

The signal is amplified differentially in the input section and then modulated. Transformer T_1 carries the modulated signal from the input amplifier to the output section, where it is demodulated and further amplified. Transformer T_2 carries high-frequency power to the input section where it is converted into direct current for the input amplifier and modulator. By using this design, isolated power is provided to the input section without using batteries.

Applications of the electromagnetic isolation include: (1) isolating a subject with heart-monitoring electrodes from the 110-V power in the recording electronics, and (2) isolating sensors that are "floating" at dangerous voltages from personnel who are operating associated test equipment.

Optical isolation involves modulating the output of a solid-state photoemitter with the amplified signal and then detecting the light with a photodetector in the output circuit. The advantage of this approach is the complete elimination of electrical paths between the input and output sections, and the disadvantage is the need for battery power for the input section.

Applications of the optical isolation include: (1) transmitting signals through electromagnetically noisy regions without interference, and (2) safety isolation as described above.

2.5   Noise sources

- **Noise** is any unwanted component in the signal of interest, and can be due to external interference or generated in the circuit itself. It can be repetitive or random.
- **White noise** is random and has equal noise power in each (linear) frequency interval.

- **Pink noise** is random and has equal noise power in each frequency octave (i.e. each factor of two in frequency).
- **Interference noise** is picked up from other circuits and is usually repetitive.

In the following sections, random noise is described in terms of the **rms** (the root mean square, or the square root of the average of the squares of the random fluctuations). See Chapter 5 for a discussion of random processes and the rms deviation.

2.5.1   Johnson noise

Johnson noise is a white-noise voltage developed across a resistor due to the thermal agitation of the charge carriers (electrons) within the resistor. Although there is a vast number of electrons moving in all directions within the resistor, their motion is random and from instant to instant, there are more electrons moving in one direction than another. Since these random motions never cease, the imbalance varies endlessly and results in a fluctuating noise voltage across the resistor. The average voltage is zero, and the rms voltage is given by:

$$V_{rms} = \sqrt{4kTR\Delta f}$$

where k (Boltzmann's constant) $= 1.380 \times 10^{-23}$ JK $= 1.380 \times 10^{-23}$ V^{2}s$/\Omega/$K, R is the resistance in ohms; T is the temperature in degrees kelvin; and Δf is the bandwidth in hertz. For a circuit with a constant gain G_0 between frequencies f_1 and f_2, and zero gain at all other frequencies, $\Delta f = f_2 - f_1$. If the circuit has a gain that varies with frequency, but has a nominal value G_0 over some frequency range, then the bandwidth is given by:

$$\Delta f = G_0^{-2} \int_0^\infty G^2(f) \, df$$

At 300 K (close to room temperature), the Johnson noise in resistor R at bandwidth Δf can be expressed as:

$$V_{rms} = D_J\sqrt{R\Delta f}, \quad \text{where } D_J = 1.287 \times 10^{-10} \text{ V}/\sqrt{\Omega}/\sqrt{\text{Hz}}$$

For $\Delta f = 10^6$ Hz and $R = 10^6 \, \Omega$, $V_{rms} = 129 \, \mu$V. For $\Delta f = 10^9$ Hz and $R = 50 \, \Omega$, $V_{rms} = 28.8 \, \mu$V.

2.5.2   Shot noise

Shot noise arises because an electrical current is not a smooth flow of charge but the motion of a finite number of charge carriers (electrons) per unit time. These electrons are moving independently, so that the actual number passing any point per unit time varies randomly. As discussed in Section 5.5 (the chi-squared statistic), these statistical fluctuations result in an rms (root mean square) deviation from the mean that is equal to the square root of their number. For large currents (>1 mA) the number of electrons per microsecond is very large (6.242×10^9), but in many cases, the signal consists of a much smaller current, in the nanoamp or picoamp range. Amplifying such a small current also amplifies the shot noise.

For an average current I_{ave}, the number of electrons passing in time T is $N = IT/q$, $I_{ave} = qN/T$, and the shot-noise fluctuations are described by:

$$I_{rms} = \frac{q\sqrt{N}}{T} = \frac{q}{T}\sqrt{\frac{IT}{q}} = \sqrt{\frac{qI}{T}}$$

Converting to the bandwidth $\Delta f = f_2 - f_1$, we have:

$$I_{rms} = \sqrt{2qI\Delta f}$$

where $q = 1.602 \times 10^{-19}$ C (charge of the electron). Note that the effective bandwidth of a rectangular pulse of width T is given by $\Delta f = 1/(2T)$.

For $I = 1\,\mu A$ and $\Delta f = 10^6$ Hz, $I_{rms} = 0.566$ nA. For $I = 1$ pA, $I_{rms} = 0.566$ pA, a significant fraction of the average current.

The voltage noise V_{rms} across a resistor R due to the shot noise of a current I through it is given by:

$$V_{rms} = R\sqrt{2qI\Delta f} = \sqrt{2qVR\Delta f}$$

2.5.3   Amplifier noise

Noise in an amplifier is a combination of Johnson and shot noises from the various elements of the circuit, both of which are proportional to the square root of the bandwidth. As a result, the noise specifications for the input and output sections of an amplifier are of the form $V_{rms} = D\sqrt{\Delta f}$, where D is a constant, and Δf is the bandwidth. Typical values are $D_1 = 5$ nV/$\sqrt{Hz}$ with respect to the input and $D_0 = 50$ nV/$\sqrt{Hz}$ with respect to the output (measured when the input is grounded). Since these input and output noise sources are uncorrelated, we combine them as the square root of the sum of the squares. For closed-loop gain G, the overall noise at the output is:

$$V_{rms} = \sqrt{\Delta f[(D_1 G)^2 + (D_0)^2]}$$

Measurement tip

To determine the amplifier noise contributions D_1 with respect to the input and D_0 with respect to the output at a bandwidth Δf, do the following:

1. Set the amplifier gain to unity and measure the rms of the output noise V_1.
2. Set the amplifier gain to a large value G and measure the rms of the output noise V_G.

3. Solve the two equations for the two unknowns D_0 and D_1:

$$D_0 = \sqrt{\frac{G^2 V_1^2 - V_G^2}{\Delta f(G^2 - 1)}} \qquad D_1 = \sqrt{\frac{V_G^2 - V_1^2}{\Delta f(G^2 - 1)}}$$

2.5.4   Electrical interference

In our modern world, electricity controls and powers nearly everything. As a result, there are wires in the walls of every building, carrying hundreds or thousands of amperes of 60 Hz. A 1-m long unshielded wire can pick up 100 mV of 60 Hz from these sources. Note that the actual amplitude (and phase) of the induced voltage depends critically on conductor geometry. By using two wires close together, it is possible to detect very small signals using *differential amplification* because the 60 Hz induced on the two wires has very nearly the same amplitude and phase. The use of a pair of shielded cables reduces the pickup still further.

Electrical interference can also arise from high-frequency communication sources (radio and television) and from high-speed switching (digital electronics, computers). This can be greatly reduced by a conductive shield placed around signal lines and circuits. Effectiveness is reduced if the conductive shield is also used to carry current back to the power supply.

Good measures against electrical interference are:
1. Amplify the signal as close to the sensor as possible. Once the signal amplitude is well above any interference, longer wires can be used for connection to later amplification stages. The amplifier close to the sensor is often called a pre-amplifier.
2. Shield the sensor/pre-amplifier with a metal enclosure to reduce electromagnetic interference on the signal lines. Use shielded cables to carry power and signals between enclosures.
3. If interference is still a problem, arrange a second wire close to the signal-carrying wire so that it picks up essentially the same interference but is not connected to the signal source. Differential amplification will then cancel most of the interference.
4. If the signal and the unwanted noise contain significantly different frequency content, frequency filtering can help remove the noise while retaining as much of the signal as possible. Maximization of the signal-to-noise ratio is often used as a guide. Any loss in signal can be recovered by amplification.

2.5.5   Inadequate grounds

The circuit ground is generally a common reference point through which currents are sent before being returned to the power supply. Digital circuits, especially, have large current transients that can generate brief potentials due to the resistance and inductance of the "ground" wires. For this reason, it is important to keep the analog and digital grounds separate (see Appendix A for circuit diagrams). Remember that whereas digital circuits have some noise immunity, *analog circuits have none*.

If after diligent application of differential amplification, shielding, and grounding, some unwanted noise persists, analog filtering can often be helpful (see Section 2.6).

2.6   Analog filtering

Analog filtering is used to reduce selected frequency components of the signal. It is useful whenever the signal of interest has a frequency content that is different than the frequency content of unwanted signals, electromagnetic pickup, or other noise. Frequency filtering should be used to reduce electromagnetic pickup *only after* proper shielding, grounding, and differential amplification have reduced it as much as is practical. The following are common applications of filtering:

1. Reducing high frequencies that contain a fundamental white noise (noise power per hertz is independent of frequency) such as Johnson or shot noise, or to reduce 60-Hz components in dc power voltages.
2. Reducing low frequencies that contain flicker ($1/f$) noise (noise power per hertz is inversely proportional to frequency).
3. Reducing low frequencies (usually less than 1 Hz) such as those caused by instabilities in electrodes or electronics (baseline restoration).
4. Rejecting input frequencies that are higher than one-half the sampling frequency of a data-acquisition system. (This point is discussed further in Chapters 3 and 5.)
5. Extracting a signal from a residual carrier wave after frequency demodulation or chopping.
6. Rejecting an unwanted waveform that is an unavoidable product of the transducer or some other component of the measuring system.
7. Using a notch filter to reduce 60-Hz electromagnetic pickup (after shielding, grounding, and differential amplification).

The general characteristics of analog filters are:

1. The **passband**, the range of frequencies that are passed unfiltered.
2. The variations in voltage gain in the passband.
3. The **stop band**, the range of frequencies that are rejected.
4. The incomplete rejection of frequencies in the stop band.
5. The **corner frequency**, where the amplitude has dropped by 3.01 dB (a factor of $2^{-1/2} = 0.707$) from the passband.
6. The **filter order**, which determines how rapidly the filter gain drops in the stop band. For a Butterworth low-pass filter of order n, the filter gain falls as f^{-n}, or n decades ($20n$ dB) in gain for each decade in frequency, or n powers of 2 ($6.02n$ dB) in gain for each octave in frequency.
7. The **phase shift** between the input and the output as a function of frequency.

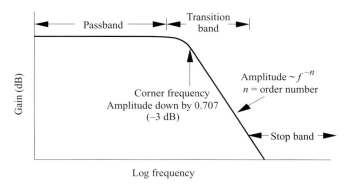

Figure 2.19 Characteristics of the low-pass filter.

Figure 2.20 General schematic for analog filter.

8. The **risetime**, the time required for the output to rise from 10 to 90% of its final change after a step input.
9. The **bandwidth**, the effective frequency response Δf, defined so that Δf times the square of the voltage gain in the passband (G_0^2) is equal to the total power:

$$\Delta f = \int_0^\infty G^2(f)\,df/G_0^2$$

The **low-pass filter** is designed to pass frequencies below a specified corner frequency and attenuate higher frequencies. It is most commonly used to suppress carrier waves, frequency aliasing, and Johnson noise. See Figure 2.19 for a schematic of frequency characteristics.

The **high-pass filter** is designed to pass high frequencies and attenuate low frequencies.

The **band-pass filter** has a passband between two stop bands. It is used whenever the signal of interest has a narrow frequency content compared with other unwanted signals. Since all filters have an upper corner frequency due to limited amplifier bandwidth, all "high-pass" filters are actually band-pass filters.

The **notch filter** uses frequency-dependent cancellation to block a particular frequency. Due to imperfect components, the cancellation is not perfect, and nearby frequencies are reduced to a lesser extent.

In general, the complex gain G of a filter (Figure 2.20), can be described as the filter output $V_0(t)$ for a unit cosine input $V_1(t) = \cos(\omega t)$, either in terms of the magnitude of the gain $|G|$ and a phase shift ϕ, or in terms of the real and imaginary components

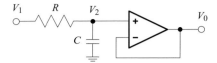

Figure 2.21 Passive low-pass filter with buffer output.

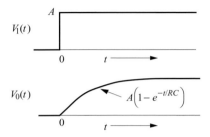

Figure 2.22 Step function response of passive low-pass filter.

of the complex gain:

$$V_0(t) = |G|\cos(\omega t + \phi) = |G|\cos(\omega[t + \Delta t])$$
$$= \text{Re}(G)\cos(\omega t) + \text{Im}(G)\sin(\omega t)$$

The phase shift ϕ is given by:

$$\tan(\phi) = \text{Im}(G)/\text{Re}(G)$$

The phase shift can also be thought of as a frequency-dependent delay Δt between an input and output harmonic:

$$\Delta t = \phi/\omega = \phi/(2\pi f)$$

In Laboratory Exercise 6, we explore the low-pass one-pole filter, the high-pass one-pole filter, the Butterworth low-pass two-pole filter, and the notch filter, which are described in the following sections.

2.6.1   Simple passive filters

As a review of the properties of filters, consider the simple low-pass passive filter in Figure 2.21. The buffer amplifier serves only to provide a means for measuring V_2 without drawing current.

Step function response of a passive low-pass filter

Initially, $V_1 = V_2 = V_0 = 0$ V. After $t = 0$, the step function sets $V_1 = A$, which causes a current $I = (V_1 - V_2)/R$ to flow through resistor R and charge the capacitor C (Figure 2.22).

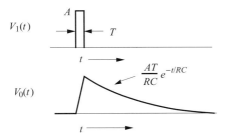

Figure 2.23 Impulse response of passive low-pass filter.

The rate of change in the voltage V_2 is given by:

$$dV_2/dt = (dQ/dt)/C = I/C = (A - V_2)/RC$$

The solution of this differential equation is:

$$V_0(t) = V_2(t) = A\left(1 - e^{-t/RC}\right)$$

Impulse response of a passive low-pass filter

The impulse input can be thought of as a brief step function of amplitude A, before and after which the amplitude is 0 (Figure 2.23). The duration of this step is assumed to be short ($T \ll RC$). During time T, a current $I = A/R$ will flow through the resistor R and places a charge $Q = IT = AT/R$ on the capacitor C, charging it to a voltage $V_2(T) = Q/C = AT/(RC)$. After the impulse, $V_1 = 0$ V, and the voltage on the capacitor will exponentially decrease with time constant RC. The ideal impulse response corresponds to the limit where $T \to 0$ while the product AT is kept constant.

Frequency response of a passive low-pass filter

The frequency and phase response can be derived from the voltage-divider equation for complex impedances:

$$V_2 = V_1 \left(\frac{1/j\omega C}{1/j\omega C + R} \right) = V_1 \left(\frac{1}{1 + j\omega RC} \right)$$

$$\left| \frac{V_2}{V_1} \right| = \left| \frac{1 - j\omega RC}{1 + (\omega RC)^2} \right| = \frac{\sqrt{1 + (\omega RC)^2}}{1 + (\omega RC)^2} = \frac{1}{\sqrt{1 + (\omega RC)^2}} = \frac{1}{\sqrt{1 + (2\pi f RC)^2}}$$

At the corner frequency $f_c = 1/(2\pi RC)$, $|V_2/V_1| = 1/\sqrt{2}$, or -3.01 dB. At low frequencies ($f \ll f_c$), $V_2 \approx V_1$. At high frequencies ($f \gg f_c$), $|V_2/V_1| \approx f_c/f$, which decreases as 6.02 dB per octave or 20 dB per decade.

Now we consider the simple high-pass passive filter in Figure 2.24.

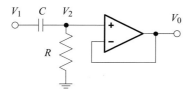

Figure 2.24 Passive high-pass filter with buffer output.

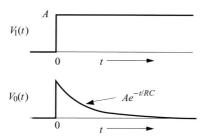

Figure 2.25 Step function response of passive high-pass filter.

Step function response of a passive high-pass filter

Initially, $V_1 = V_2 = V_0 = 0$ V. After $t = 0$, the step function sets $V_1 = V_2 = A$, which causes a current $I = -V_2/R$ to flow through resistor R and charge the capacitor C (Figure 2.25).

The rate of change in the voltage V_2 is given by:

$$dV_2/dt = (dQ/dt)/C = I/C = -V_2/RC$$

The solution of this differential equation is:

$$V_0(t) = V_2(t) = Ae^{-t/RC}$$

Impulse response of a passive high-pass filter

The impulse input can be thought of as a brief step function of amplitude A, before and after which the amplitude is 0 (Figure 2.26). The duration of the impulse is $T \ll RC$. The input step from 0 to A immediately produces an output step from 0 to A. During time T, a current $I = A/R$ will flow through the resistor R and places a charge $Q = IT$ on the capacitor C, charging it to a voltage $V(T) = A(1 - T/(RC))$. The input step from A to 0 causes the output $V_2(t)$ to step from $A(1 - T/(RC))$ to $-AT/(RC)$. The capacitor then discharges through resistor R, and $V_2(t)$ exponentially increases to 0 with time constant RC. The ideal impulse response corresponds to the limit where $T \to 0$ and the product AT is kept constant.

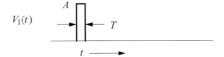

Figure 2.26 Impulse response of passive high-pass filter.

Frequency response of a passive high-pass filter

The frequency and phase response can be derived from the voltage-divider equation:

$$V_2 = V_1\left(\frac{R}{1/j\omega C + R}\right) = V_1\left(\frac{j\omega RC}{1 + j\omega RC}\right)\left(\frac{1 - j\omega RC}{1 - j\omega RC}\right)$$

$$= V_1\left[\frac{j\omega RC + (\omega RC)^2}{1 + (\omega RC)^2}\right]$$

$$\left|\frac{V_1}{V_2}\right| = \left|\frac{j\omega RC + (\omega RC)^2}{1 + (\omega RC)^2}\right| = \frac{\sqrt{(\omega RC)^2 + (\omega RC)^4}}{1 + (\omega RC)^2} = \frac{\omega RC\sqrt{1 + (\omega RC)^2}}{1 + (\omega RC)^2}$$

$$= \frac{\omega RC}{\sqrt{1 + (\omega RC)^2}} = \frac{2\pi f RC}{\sqrt{1 + (2\pi f RC)^2}}$$

At the corner frequency $f_c = 1/(2\pi RC)$, $|V_2/V_1| = 1\sqrt{2}$, or -3.01 dB. At high frequencies ($f \gg f_c$), $V_2 \approx V_1$. At low frequencies ($f \ll f_c$), $|V_2/V_1| \approx f/f_c$, which increases as 6.02 dB per octave or 20 dB per decade.

Square-wave response of passive high-pass and low-pass filters

From the preceding sections, the low-pass and high-pass filter response to a periodic series of square waves depends on the relative frequency f of the square-wave period and the corner frequency $f_c = 1/(2\pi RC)$ of the filter (Figure 2.27). Note that a low-pass filter acts as an integrator for high frequencies ($f \gg f_c$), where the output falls as $1/f$, and that a high-pass filter acts as a differentiator for low frequencies ($f \ll f_c$), where the output rises as f. Thus we see that integrating a waveform corresponds to multiplying its frequency spectrum by $1/f$ and that taking the derivative of a waveform corresponds to multiplying the frequency spectrum by f.

Design tips

1. To measure the area under a pulse of duration T, integrate the signal with a low-pass filter with corner frequency $f_c \ll 1/(2\pi T)$.

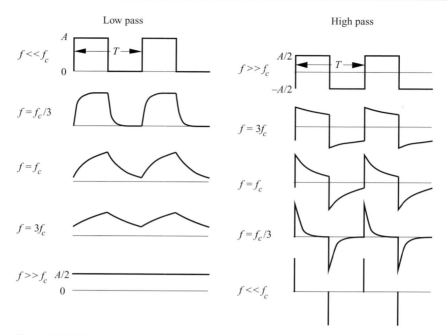

Figure 2.27 The output waveform of low- and high-pass filters for various values of the corner frequency $f_c = 1/(2\pi RC)$ relative to the square-wave frequency $f = 1/T$.

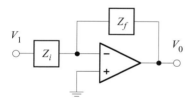

Figure 2.28 Generalized filter circuit with input and feedback blocks having complex impedance.

2. To take a moving average of a waveform $V(t)$ with an averaging window T, integrate the signal with a low-pass filter with corner frequency $f_c = 1/(2\pi T)$. Components with $f \ll f_c$ are unchanged. Components with $f \gg f_c$ will be integrated (summed).

3. To measure the rate-of-change dV/dt of a waveform $V(t)$, differentiate the signal with a high-pass filter (f_c = corner frequency). Components with $f \gg f_c$ are unchanged. Components with $f \ll f_c$ will be differentiated.

2.6.2   Computing the Bode plot of op-amp filters

Frequency filter circuits can also be constructed using the inverting op-amp voltage gain equations shown in Section 2.2 but with input and feedback blocks having complex impedance (Figure 2.28).

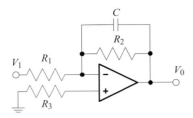

Figure 2.29 Low-pass one-pole filter.

The complex voltage gain G and phase ϕ are given by:

$$G = \frac{-Z_f}{Z_i} \qquad |G| = \sqrt{G_{\text{Re}}^2 + G_{\text{Im}}^2} \qquad \tan(\phi) = \frac{G_{\text{Im}}}{G_{\text{Re}}}$$

2.6.3   Low-pass one-pole filter

The expressions for the frequency-dependent gain $G = V_{\text{out}}/V_{\text{in}}$ of simple op-amp filters can often be derived by remembering that the voltage gain of an op amp with a negative feedback network is the ratio of the feedback impedance to the input impedance. Note that both of these impedances may be complex numbers, that the impedance of a resistor is R, and that the impedance of a capacitor is $1/(j\omega C)$, where $j = \sqrt{-1}$.

For example, the gain of the low-pass one-pole filter (Figure 2.29) is the ratio of the feedback impedance (the parallel combination of C and R_2) to the input impedance (the resistor R_1):

$$G = -\frac{Z_f}{Z_i} = -\frac{R_2 \| C}{R_1} = -\frac{1}{R_1(1/R_2 + j\omega C)}$$

$$= -\left(\frac{R_2}{R_1}\right)\left(\frac{1}{1 + j\omega R_2 C}\right) = -\frac{R_2}{R_1}\left[\frac{1 - j\omega R_2 C}{1 + (\omega R_2 C)^2}\right]$$

We assume that the open-loop gain of the op amp is very large for all the frequencies in the passband.

It is also possible to analyze filters in the complex frequency plane ($s = j\omega$ plane). The function G is also the amplitude of the complex function:

$$\frac{V_0}{V_1} = -\frac{R_2}{R_1}\left[\frac{1/(R_2 C)}{s + 1/(R_2 C)}\right]$$

along the real axis. In this form, the circuit has a pole at $s = -1/(R_2 C)$.

The **magnitude** of G is given by:

$$|G| = \sqrt{G_r^2 + G_i^2} \tag{2.2}$$

where G_i and G_r are the imaginary and real parts, respectively.

$$|G| = \frac{R_2/R_1}{\sqrt{1 + (\omega R_2 C)^2}} = \frac{R_2/R_1}{\sqrt{1 + (f/f_c)^2}} \qquad (2.3)$$

where $f_c = (2\pi R_2 C)^{-1}$ is the corner frequency.
   The phase ϕ of G is given by:

$$\tan \phi = -\omega R_2 C = -f/f_c$$

For frequencies well below the corner frequency ($f \ll f_c$) the gain is given by $|G| = R_2/R_1$, independent of frequency, and the phase shift is $-180°$, determined by the amplifier inversion. At the corner frequency ($f = f_c$), the gain has dropped by 3.01 dB (decibels) to $0.707 R_2/R_1$, and the phase shift is $-45° - 180° = -225°$. Somewhat above the corner frequency, the gain drops as $1/f$ (6.02 dB per octave or 20 dB per decade). For large frequencies ($f \gg f_c$), the phase shift is $-90° - 180° = -270°$. An octave is a factor of 2 in frequency, and a change of $20n$ dB corresponds to a factor of 10^n in voltage, or a factor of 10^{2n} in power.

2.6.4   High-pass one-pole filter

By placing a capacitor and resistor in series as the input impedance, we have the high-pass one-pole filter (Figure 2.30).
   We compute the gain as the (complex) ratio of the feedback impedance to the input impedance. The magnitude of the gain is the square root of the sum of the squares of the real and imaginary parts.

$$G = \frac{-R_2}{R_1 + 1/j\omega C} = -\omega R_2 C \left[\frac{j + \omega R_1 C}{1 + (\omega R_1 C)^2} \right]$$

$$|G| = \frac{\omega R_2 C}{\sqrt{1 + (\omega R_1 C)^2}} \qquad \tan \phi = \frac{1}{\omega R_1 C} \qquad (2.4)$$

At frequencies well below the corner frequency, $\omega_c = (R_1 C)^{-1}$, $|G| = \omega R_2 C$. At frequencies well above ω_c, $|G| = R_2/R_1$. However, all op amps have a limited gain–bandwidth product, which results in a decrease in gain at sufficiently high frequencies.

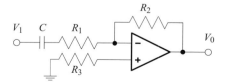

Figure 2.30 High-pass one-pole filter.

Strictly speaking, all active high-pass filters are actually band-pass filters (see Example 2.5 following).

EXAMPLE 2.5

Compute the gain of the high-pass one-pole filter, taking under consideration the finite gain–bandwidth product of the op amp.

Using the open-loop gain equation, we have:

$$V_0 = -AV_2$$

The voltage-divider relationship gives:

$$\frac{V_0 - V_1}{1/j\omega C + R_1 + R_2} = \frac{V_0 - V_2}{R_2} = \frac{V_0 + V_0/A}{R_2}$$

Simplifying, we have:

$$R_2(V_0 - V_1) = (1 + 1/A)(1/j\omega C + R_1 + R_2)$$
$$V_1 R_2 = V_0 R_2 - V_0(1 + 1/A)(1/j\omega C + R_1 + R_2)$$

Solving for the gain:

$$G = \frac{V_0}{V_1} = \frac{R_2}{R_2 - (1 + 1/A)(1/j\omega C + R_1 + R_2)}$$
$$= \frac{-R_2\omega C[R_1\omega C + (R_1 + R_2)\omega C/A + j(1 + 1/A)]}{[R_1\omega C + (R_1 + R_2)\omega C/A]^2 + (1 + 1/A)^2}$$

Computing the magnitude:

$$|G| = \left|\frac{V_0}{V_1}\right| = \frac{AR_2\omega C}{\sqrt{(\omega C)^2(AR_1 + R_1 + R_2)^2 + (1 + A)^2}} \tag{2.5}$$

At lower frequencies, where the open-loop gain $A \gg (R_1 + R_2)/R_1$, this reduces to Eq. (2.4), as expected. At high frequencies, where the open-loop gain A has fallen, so that $A \ll (\omega C)(R_1 + R_2) - 1$ and $A \ll (R_1 + R_2)/R_1$, Eq. (2.5) reduces to:

$$|G| = \left|\frac{V_0}{V_1}\right| \approx \frac{AR_2}{R_1 + R_2} = \frac{KR_2/\omega}{R_1 + R_2}$$

where $K = A\omega$ is the gain–bandwidth product of the op amp. The op amp introduces a pole at high frequencies, so that $|G|$ decreases linearly with increasing frequency.

2.6.5   Notch filter

The **notch filter** rejects a narrow band of frequencies and passes all others. It is particular useful in eliminating a specific frequency (such as 60 Hz) while retaining higher and lower frequencies. In Laboratory Exercise 6, we explore the properties of a notch filter designed to reject 60 Hz.

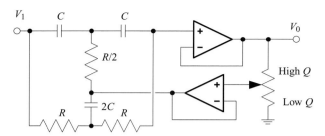

Figure 2.31 Notch filter. Notch frequency $f_n = (2\pi RC)^{-1}$.

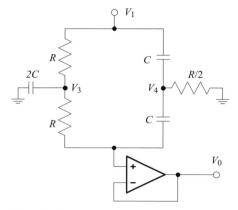

Figure 2.32 Simplified circuit of low-Q notch filter. See text for circuit analysis.

The circuit consists of two parallel T filters (Figure 2.31). The R–$2C$–R section is a low-pass filter with corner frequency $f_c = (4\pi RC)^{-1}$. The C–$R/2$–C section is a high-pass filter with $f_c = (\pi RC)^{-1}$. For ideal components, the phases of these two filters cancel perfectly at the notch frequency, $f_n = (2\pi RC)^{-1}$. In a practical circuit, it is possible to achieve notch depths of 30 dB, using 5% components, and 60 dB, using 1% components. At frequencies above 1 kHz, the capacitors become small, and stray capacitances degrade circuit performance.

The complex voltage gain of the simplified low-Q notch filter (Figure 2.32) is given by:

$$G = \frac{V_0}{V_1} = \frac{[1 - (\omega RC)^2][1 - (\omega RC)^2 - 4j(\omega RC)^2]}{1 + 14(\omega RC)^2 + (\omega RC)^4}$$

The phase is given by (Figure 2.33):

$$\phi = \tan^{-1}\left[\frac{4\omega RC}{1 - (\omega RC)^2}\right]$$

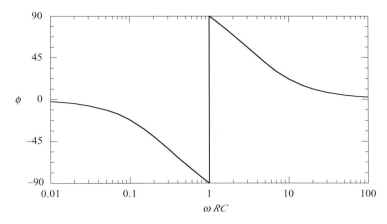

Figure 2.33 Bode phase plot of low-Q notch filter.

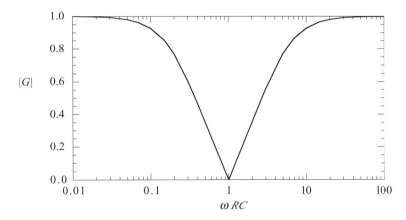

Figure 2.34 Bode gain plot of low-Q notch filter.

And the magnitude of the gain is given by (Figure 2.34):

$$|G| = \frac{|1 - (\omega RC)^2|}{\sqrt{1 + 14(\omega RC)^2 + (\omega RC)^4}}$$

A triangle or square wave has a fundamental frequency f_0 and higher harmonics at multiples nf_0 of that frequency. The effect of a notch filter with notch frequency $f_n = f_0$ is to remove the fundamental and leave the higher harmonics (Figure 2.35).

Design tip

Notch filters are useful for rejecting a narrow band of frequencies. However, to keep $R > 1{,}000\ \Omega$ and $C > 100\ \text{pF}$, $RC > 10^{-7}\ \text{s}$, so the maximum practical frequency is $1/(2\pi RC) \approx 1\ \text{MHz}$.

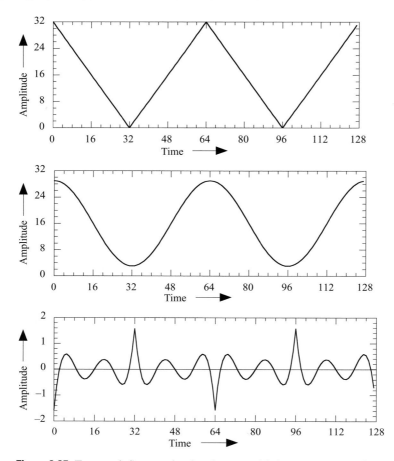

Figure 2.35 Top panel: Symmetric triangle wave with frequency one cycle per 64 samples. Center panel: Harmonic with one cycle per 64 samples that is removed by a notch filter. Lower panel: Higher harmonics of the triangle wave that are passed by the notch filter.

2.6.6 High-order low-pass filters

The basic RC filters discussed in Sections 2.6.2–2.6.4 are used primarily for noncritical applications where sharp frequency selection is not needed. The Butterworth, Bessel, transitional, and Chebyshev filters described in this section are designed for more critical requirements, such as a constant gain for all frequencies in the passband, a rapid falloff from the passband to the stop band, a low gain for frequencies in the stop band, and the ability to transmit a pulse with little change in shape.

The **Butterworth** filter has a flat frequency response below the corner frequency f_c, but responds poorly to transients because the phase–frequency relationship is nonlinear. It is commonly used for anti-aliasing in circuits that sample analog waveforms (described in Chapters 3 and 5) because its gain is maximally flat in the passband.

Table 2.3 *Effective bandwidth of Butterworth filters of order n and corner frequency* f_c

n	$\Delta f / f_c$
1	1.571
2	1.111
4	1.026
6	1.012
8	1.006

The nth order Butterworth filter with corner frequency f_c has a gain $|G|$ and phase ϕ given by:

$$|G| = \frac{1}{\sqrt{1 + (f/f_c)^{2n}}} \qquad \tan\left(\frac{\phi}{n}\right) = \frac{-f}{f_c}$$

Note that for a Butterworth filter of any order n, when $f = f_c$, then $|G| = 2^{-1/2} = 0.707$ (i.e. the gain has dropped 3.01 dB below the passband gain), and $\phi = \pi n/4$.

The effective bandwidth of the Butterworth low-pass filter is given by:

$$\Delta f = \int\limits_0^\infty G^2(f)\, df = \int\limits_0^\infty \frac{1}{1 + (f/f_c)^{2n}} = f_c \int\limits_0^\infty \frac{dx}{1 + x^{2n}} = f_c \left[\frac{\pi/2n}{\sin(\pi/2n)}\right]$$

In Table 2.3, $\Delta f / f_c$ is tabulated for several values of the filter order n. Note that for high filter orders, the filter cuts off sharply at the corner frequency f_c and the effective bandwidth Δf is approximately equal to f_c.

The **Bessel** filter has a phase shift that is proportional to frequency. Each Fourier component is shifted by the same amount of time, and the signal is transmitted without a change in shape but with a fixed delay. This filter is desired for transmitting pulses with minimum distortion.

The **transitional**, or **Paynter**, filter (also called the "Besselworth") has properties intermediate between those of either the Butterworth or Bessel filter.

The **Chebyshev** filter maximizes the sharpness of the frequency roll-off, but introduces ripples in the passband. This filter is actually a family of filters classified by the amplitude of the ripples (in decibels). Achieving the intended response requires accurate component values (typically 1–5%) and low-leakage capacitors. Inductors are rarely used as they are bulky and not very ideal.

The basic circuit realizations are the unity-gain Sallen–Key filter (Figures 2.36 and 2.37, Table 2.4) and the equal-component-value (or VCVS, voltage-controlled voltage source) Sallen–Key filter (Figures 2.38 and 2.39, Table 2.5). Each of these circuits provides two poles of low-pass or two poles of high-pass filtering. Higher-order filters use cascaded stages. The equal-component design has the advantage that each stage amplifies the signal as it reduces its bandwidth so that the rms amplitude is kept nearly

Table 2.4 *Unity-gain Sallen–Key low-pass and high-pass filters. Refer to Figures 2.36 and 2.37 for circuit diagrams*

Poles	Butterworth		Transitional		Bessel		Chebyshev (0.5 dB)	
	k_1	k_2	k_1	k_2	k_1	k_2	k_1	k_2
2	1.414	0.707	1.287	0.777	0.907	0.680	1.949	0.653
4	1.082	0.924	1.090	0.960	0.735	0.675	2.582	1.298
	2.613	0.383	2.206	0.472	1.012	0.390	6.233	0.180
6	1.035	0.966	1.060	1.001	0.635	0.610	3.592	1.921
	1.414	0.707	1.338	0.761	0.723	0.484	4.907	0.374
	3.863	0.259	2.721	0.340	1.073	0.256	13.40	0.079
8	1.019	0.981	1.051	1.017	0.567	0.554	4.665	2.547
	1.202	0.832	1.191	0.876	0.609	0.486	5.502	0.530
	1.800	0.556	1.613	0.615	0.726	0.359	8.237	0.171
	5.125	0.195	3.373	0.268	1.116	0.186	23.45	0.044

Source: Brian K. Jones, *Electronics for Experimentation and Research*, Prentice Hall, Englewood Cliffs, NJ, 1986. By permission of Prentice-Hall International (UK) Ltd, London.

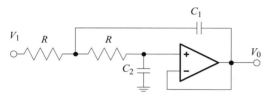

Figure 2.36 Unity-gain Sallen–Key low-pass two-pole filter. $RC_1\omega_c = k_1$ and $RC_2\omega_c = k_2$. Higher-order filters use cascaded stages. See Table 2.4 for values of k_1 and k_2.

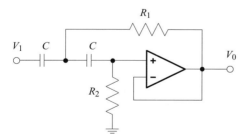

Figure 2.37 Unity-gain Sallen–Key high-pass two-pole filter. $R_1C\omega_c = 1/k_1$ and $R_2C\omega_c = 1/k_2$. Higher-order filters use cascaded stages. See Table 2.4 for values of k_1 and k_2.

constant. This allows larger signals to be used, which reduces the effect of op-amp noise.

EXAMPLE 2.6

Design a Butterworth high-pass four-pole filter with $f_0 = 10\,\text{kHz}$ ($\omega_c = 62.83\,\text{krad/s}$). From Table 2.4, the first stage has $k_1 = 1.082$ and $k_2 = 0.924$. Thus, $R_1C = 1/(k_1\omega_c) =$

Table 2.5 *Equal-component-value Sallen–Key low-pass and high-pass filters. Refer to Figures 2.38 and 2.39 for circuit diagrams*

Poles	Butterworth		Transitional		Bessel		Chebyshev (0.5 dB)	
	k_3	G	k_3	G	k_3	G	k_3	G
2	1.000	1.586	1.000	1.446	0.785	1.268	1.129	1.842
4	1.000	1.152	1.023	1.123	0.704	1.084	1.831	1.582
	1.000	2.235	0.977	2.035	0.628	1.759	1.060	2.660
6	1.000	1.068	1.030	1.056	0.622	1.040	1.332	2.627
	1.000	1.586	1.009	1.492	0.591	1.364	1.355	2.448
	1.000	2.483	0.962	2.293	0.524	2.023	1.029	2.846
8	1.000	1.038	1.034	1.032	0.561	1.024	3.447	1.522
	1.000	1.337	1.021	1.284	0.544	1.213	1.708	2.379
	1.000	1.889	0.996	1.765	0.510	1.593	1.188	2.711
	1.000	2.610	0.951	2.436	0.455	2.184	1.017	2.913

Source: Brian K. Jones, *Electronics for Experimentation and Research*, Prentice Hall, Engelwood Cliffs, NJ, 1986. By permission of Prentice-Hall International (UK) Ltd, London.

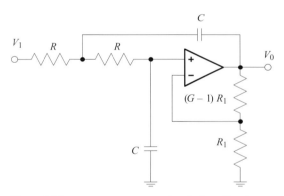

Figure 2.38 Equal-component-value Sallen–Key low-pass two-pole filter. $RC\omega_c = k_3$ and R_1 is chosen for convenience. Higher-order filters use cascaded stages. See Table 2.5 for values of gain G and k_3.

1.471×10^{-5} and $R_2C = 1/(k_2\omega_c) = 1.722 \times 10^{-5}$. Choosing $C = 1,000$ pF, we have $R_1 = 14.71$ kΩ and $R_2 = 17.22$ kΩ. Similarly, the second stage has $k_1 = 2.613$ and $k_2 = 0.383$. Choosing $C = 1,000$ pF, we have $R_1 = 6.10$ kΩ and $R_2 = 41.6$ kΩ.

The op-amp feedback establishes $V_3 = V_0$:

$$I_1 = \frac{V_1 - V_2}{R_1} \qquad I_2 = \frac{V_2 - V_0}{R_2} = V_0 j\omega C_2 \qquad I_3 = (V_2 - V_0)j\omega C_1$$

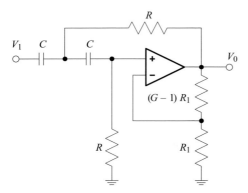

Figure 2.39 Equal-component-value Sallen–Key high-pass two-pole filter. $RC\omega_c = 1/k_3$ and R_1 is chosen for convenience. Higher-order filters use cascaded stages. See Table 2.5 for values of gain G and k_3.

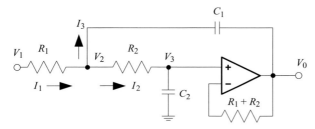

Figure 2.40 Diagram for analysis of unity-gain Sallen–Key, two-pole low-pass filter.

EXAMPLE 2.7

Derive the voltage-response function for the low-pass filter in Figure 2.40.

Since no current flows into the op-amp inputs, we have $I_1 = I_2 + I_3$:

$$I_1 = (V_1 - V_2)/R_1 = I_2 + I_3 = V_0 j\omega C_2 + (V_2 - V_0) j\omega C_1$$

Solving for V_1:

$$V_1/R_1 = V_2/R_1 + j\omega V_0 C_2 + j\omega V_2 C_1 - j\omega V_0 C_1$$

Using the equation for I_2:

$$I_2 = (V_2 - V_0)/R_2 = V_0 j\omega C_2$$
$$V_2 = V_0(1 + j\omega R_2 C_2)$$
$$\frac{V_1}{R_1} = \left(\frac{V_0}{R_1}\right)(1 + j\omega R_2 C_2) + j\omega V_0 C_2 + V_0 j\omega C_1(1 + j\omega R_2 C_2) - j\omega V_0 C_1$$
$$V_1 = V_0[1 - \omega^2 R_1 R_2 C_1 C_2 + j\omega C_2(R_1 + R_2)]$$

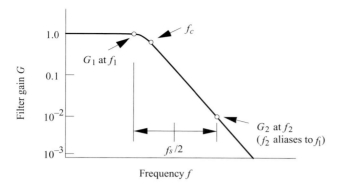

Figure 2.41 Gain versus frequency plot for a Butterworth filter.

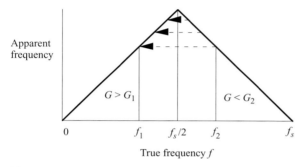

Figure 2.42 Aliasing of frequencies above $f_s/2$ to apparent frequencies below $f_s/2$.

Substituting $s = j\omega$, we then have the s-space amplitude:

$$\frac{V_0}{V_1} = \frac{1}{s^2 R_1 R_2 C_1 C_2 + s(R_1 + R_2)C_2 + 1}$$

which has poles defined by the characteristic equation

$$s^2 - 2s\omega_c \cos\theta + \omega_c^2 = s^2 + \omega_c s/Q + \omega_c^2 = 0$$

where ω_c is the corner frequency, θ is the polar angle of the pole on the s plane, and Q is the fractional energy loss per cycle. The overall gain of a series of low-pass filters is the product of the V_0/V_1 terms.

The low-pass Butterworth filter is of special significance because it is used as an anti-aliasing filter to pass amplitudes accurately in the passband and effectively eliminate frequencies above one-half the sampling frequency.

EXAMPLE 2.8

Design an nth order Butterworth filter with a gain $>G_1$ for frequencies below f_1 (the passband) and a gain $<G_2$ for frequencies above f_2 that could alias below f_1 (the stop band) (Figures 2.41 and 2.42). Frequencies between f_1 and f_2 are in the transition band.

Table 2.6 *Frequency ratio f/f_c for a Butterworth low-pass filter with n poles as a function of filter gain*

$\|G\| =$	0.999	0.990	0.900	0.707	0.010	0.001	0.000,1
$(n=2)f/f_c =$	0.212	0.377	0.696	1.000	10.000	31.623	100.000
$(n=4)f/f_c =$	0.460	0.614	0.834	1.000	3.162	5.623	10.000
$(n=6)f/f_c =$	0.596	0.723	0.886	1.000	2.154	3.162	4.642
$(n=8)f/f_c =$	0.678	0.784	0.913	1.000	1.778	2.371	3.162
$(n=10)f/f_c =$	0.733	0.823	0.930	1.000	1.585	1.995	2.512
$(n=12)f/f_c =$	0.772	0.850	0.941	1.000	1.468	1.778	2.154

For a sampling frequency f_s, f_2 aliases to $f_s - f_2$. Since we want f_2 to alias to f_1, we have $f_s = f_1 + f_2$. Given f_1, G_1, and G_2, the design procedure is:
1. Choose a filter order n.
2. Compute $f_1/f_c = \sqrt[2n]{G_1^{-2} - 1} = \exp[\ln(G_1^{-2} - 1)/(2n)]$ or use Table 2.6.
3. Compute f_c.
4. Compute $f_2/f_c = \sqrt[2n]{G_2^{-2} - 1}$ or use Table 2.6.
5. Compute $f_s = f_1 + f_2$.
6. If f_s is too high, get a faster sampling system or increase n and go back to step 2.

2.6.7   High-order high-pass filters

The Butterworth, Bessel, transitional, and Chebyshev high-pass filters are analogous to the low-pass filters of the previous section. The unity-gain Sallen–Key high-pass filter is shown in Figure 2.37 and the equal-component-value Sallen–Key high-pass filter is shown in Figure 2.39.

As an example, consider a unity-gain Sallen–Key, Butterworth high-pass four-pole filter with $f_0 = 10\,\text{kHz}$ ($\omega_0 = 62.83\,\text{krad/s}$). From Table 2.4, the first stage has $k_1 = 1.082$ and $k_2 = 0.924$. Thus, $R_1 C = 1/(k_1\omega_0) = 1.471 \times 10^{-5}$ and $R_2 C = 1/(k_2\omega_0) = 1.722 \times 10^{-5}$. Choosing $C = 1,000\,\text{pF}$, we have $R_1 = 14.71\,\text{k}\Omega$ and $R_2 = 17.22\,\text{k}\Omega$. Similarly, the second stage has $k_1 = 2.613$ and $k_2 = 0.383$. Choosing $C = 1,000\,\text{pF}$, we have $R_1 = 6.10\,\text{k}\Omega$ and $R_2 = 41.55\,\text{k}\Omega$.

Design tip

To suppress electromagnetic interference from external sources:
1. Mount your circuits in conductive boxes. Use coaxial cables to carry signals between boxes. Remember that external electric fields cannot penetrate a conductive shield.
2. If your sensor and its external wiring cannot be shielded but it is possible to block the signal, use your sensor to record the signal + interference and use a second identical sensor to

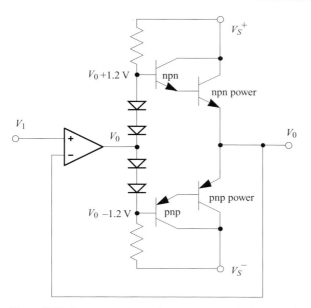

Figure 2.43 Op-amp and complementary cascaded emitter follower used for efficient bipolar current amplification. See the LM12 data sheets for a detailed circuit implementation.

record only the interference. Differential amplification can then be used to extract the signal.

3. If the interference is in a different frequency band than the signal, use analog filtering to reduce the interference (notch filters are useful against narrow-band interference <1 MHz).

Design tip

Analog filters are designed for single inputs. Therefore, if a differential signal must be amplified and filtered, firstly amplify with an instrumentation or differential amplifier, then filter.

2.7   The power amplifier

As we shall see in Chapter 3, the microcomputer can generate analog voltages by using a digital-to-analog converter, but the currents available are too small to operate many actuators. The digital-to-analog converter, the op amp, and the instrumentation amplifier can only provide a few tens of milliamperes at most, while many actuators, such as motors and heating elements require many amperes. Basically, what is required is current amplification, using power transistors.

Figure 2.43 shows a current amplifier using an op-amp, two *npn*, and two *pnp* power transistors in a complementary cascaded emitter-follower (Darlington) configuration.

The op-amp negative feedback compensates for the base-emitter voltage drop and $V_0 \approx V_1$ from $V_1 = V_S{}^+ - 1.2\,\mathrm{V}$ to $V_S{}^- + 1.2\,\mathrm{V}$. This circuit draws $<1\,\mu\mathrm{A}$ and, using large power transistors with heat sinks, can produce an output current of several amperes. The four diodes serve to bias the transistors and reduce the **crossover distortion**, that occurs when the input waveform crosses zero. Without the diodes the op-amp output would have to swing $2.4\,\mathrm{V}$ to turn one Darlington off and bring the other into conduction. During this swing, the output voltage cannot follow the input voltage. See the LM12 data sheets for the implementation of this circuit in an 80-W op amp.

2.8   Problems

2.1 In the circuit shown in Figure 2.44, assume:
- the op-amp open-loop gain $A = 10^6\,\mathrm{Hz}/f$ for $f > 10\,\mathrm{Hz}$,
- op-amp input currents are zero,
- output offset can be neglected,
- the wave generator output V_1 is 1 V peak-to-peak at all frequencies,
- the wave generator has zero output impedance.

Do the following:
(a) Derive expressions for V_0, V_3, and V_2 as a function of frequency f and open-loop gain A.
(b) Evaluate the above expressions at $f = 10\,\mathrm{Hz}$, $100\,\mathrm{Hz}$, $1\,\mathrm{kHz}$, $10\,\mathrm{kHz}$, $100\,\mathrm{kHz}$, and 1 MHz.

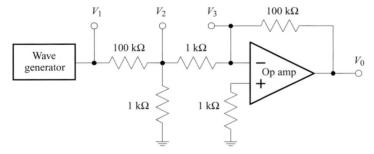

Figure 2.44 Inverting op-amp circuit driven by a wave generator and voltage divider.

2.2 A project requires an op-amp circuit with a gain of 100 from dc to $100\,\mathrm{kHz}$ (Figure 2.45). Assume that an op amp with field-effect transistor inputs is used, similar to the LF356 that you used in Laboratory Exercise 4.
(a) What is the value of R in your design?
(b) For an open-loop gain A that varies inversely with frequency f as $A = B/f$, derive an expression for the closed-loop gain G as a function of the constant B and the frequency f.
(c) For $B = 10^7\,\mathrm{Hz}$, sketch the Bode plot (closed-loop dB gain versus frequency).

(d) At what frequency does the closed-loop gain G equal unity?

(e) What are typical input and output impedances of this circuit at 1 kHz and 100 kHz? (*Hint*: look at the LF156/LF356 data sheets.)

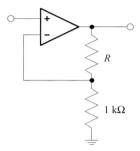

Figure 2.45 Noninverting amplifier op-amp circuit.

2.3 Analyze the op-amp circuit shown in Figure 2.46 (assume infinite open-loop gain):

(a) What are the currents flowing through each of the three input resistors?

(b) What is the current flowing through the op-amp feedback resistor?

(c) What is V_0 in terms of the quantities R, V_1, V_2, and V_3?

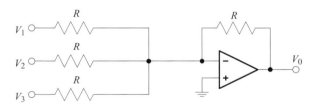

Figure 2.46 Current summing op-amp circuit.

2.4 You have an application that requires an instrumentation amplifier with the following requirements:

• differential voltage gain 1,000 from dc to 10 kHz

• common-mode rejection > 100 dB

(a) What is the required minimum gain–bandwidth product?

(b) What is the maximum common-mode gain?

(c) With the two inputs grounded, the output offset is adjusted for zero output voltage by equalizing the leakage currents of the two inputs. Then the two inputs are connected to ground through external resistors of 1 and 2 MΩ. What is the maximum leakage current specification of the instrumentation amplifier that will guarantee that the output offset will be below 1 mV?

2.5 For the differential amplifier circuit shown in Figure 2.4, and assuming that the open-loop gain A is infinite, do the following:

(a) Derive the expression for the output V_0 as a function of the four variables R_1, R_2, R_3, and R_4. Your result should be of the form $V_0 = aV_2 - bV_1$.

(b) Derive expressions for the differential gain $G_\pm$ and the common-mode gain G_c in terms of R_1, R_2, R_3, R_4, using the expression from (a) and the following:

$$V_0 = aV_2 - bV_1 = (a+b)(V_2 - V_1)/2 + (a-b)(V_2 + V_1)/2$$
$$= G_\pm(V_2 - V_1) + G_c V_c, \text{ where } V_c = (V_1 + V_2)/2$$

(c) Using your expressions from (b), derive an expression for the CMRR in terms of R_1, R_2, R_3, and R_4.

(d) Under what conditions does $G_c = 0$?

(e) When $G_c = 0$ is satisfied, what does your expression for $G_\pm$ reduce to?

(f) For $G_\pm = 1$ and $G_\pm = 1,000$, first set $R_2/R_1 = R_4/R_3 = G_\pm$ and then find the percentage variation in R_3 that causes CMR ≈ 120 dB. Note that $G_\pm$ changes very little. Comment on the resistor accuracy required for a good CMRR at the two differential gains.

Note: $G_\pm$ is primarily determined by R_2/R_1 and R_3 (or R_4) can be used to "fine tune" G_c.

2.6 For the differential amplifier circuit shown in Figure 2.4, and finite open-loop gain A, do the following:

(a) Derive an expression for the closed-loop differential gain $V_0/(V_2 - V_1)$ as a function of R_1, R_2, R_3, R_4, and open-loop gain A.

(b) For $R_1 = R_3 = 1\,\text{k}\Omega$, $R_2 = R_4 = 9\,\text{k}\Omega$, and open-loop gain–bandwidth product $AB = 10^7$ Hz, sketch the Bode amplitude plot (decibel gain versus log frequency).

2.7 You want to evaluate the LM363A instrumentation amplifier for an audio biofeedback project:

(a) For the 16-pin package, show external connections and components for gains of 100 and 250. *Note*: Finding data sheets for this integrated circuit is part of the exercise.

(b) What would you have to add to be able to adjust: (i) the input bias current, (ii) the input offset voltage, and (iii) the output offset voltage? (Look at the data sheets and sketch the circuits.)

(c) Devise a set of procedures for adjusting the input and output offset voltages so that when the inputs are connected to ground through equal 1-kΩ resistors, the output voltage is nearly zero, both at a gain of 1 and at a gain of 1,000. (Do not worry about adjusting the input bias current.)

(d) After performing the procedures in (c), at a gain of 1,000, how large could the output voltage be if one of the 1-kΩ resistors were replaced with a 10-kΩ resistor? (*Hint*: Look at the data sheets for the maximum input bias current over the full temperature range.)

(e) How would you measure the small signal gain for frequencies from dc to 1 MHz? Prepare a diagram and a list of procedures.

(f) How would you measure the common-mode rejection for the same frequencies? Prepare a diagram and a list of procedures.

(g) Analyze the following data, assuming zero output when both inputs are connected to ground through equal-value resistors. Compute the differential gain, the common-mode gain, and the common-mode rejection ratio at 10 kHz.

Case I: The differential input is a 10-kHz, 1-mV p–p sine wave. The output is a 10-kHz, 100-mV sine wave.

Case II: The common-mode input is a 10-kHz, 5-V p–p sine wave connected to both inputs through equal resistors. The output is a 10-kHz, 2.5-mV p–p sine wave.

(h) In the frequency range from 10 Hz to 10 kHz and with a gain of 100, what is the input noise in $nV/\sqrt{Hz}$? (*Hint*: Look at the data sheets.) What value resistor has the same Johnson noise in $nV/\sqrt{Hz}$?

(i) For a gain of 100 and both inputs connected to ground, estimate the output noise (in microvolts) in the full bandwidth of the amplifier. What value input resistor would produce an equivalent amount of output noise?

(j) Look at the data sheets for the AD625A and the LM363A and prepare a table that compares the following quantities:
- input offset voltage and its temperature dependence;
- input bias current and its temperature dependence;
- bandwidth for gains of 10, 100, and 1,000;
- common-mode rejection for gains of 10, 100, and 1,000;
- settling time;
- maximum slew rate.

2.8 You are given an instrumentation amplifier with a gain that is adjustable from 1 to 1,000. At a gain of 1, the bandwidth is 10^6 Hz. At a gain of 1,000, the bandwidth is 10^4 Hz.

(a) Both input terminals are connected to ground with 5-MΩ resistors. If the input leakage currents on the two inputs are 0.5 and 1.5 nA, what is the resulting output offset voltage at a gain of 1,000? At a gain of 1?

(b) What is the output noise in the 10^4-Hz bandwidth at a gain of 1,000 due only to the room-temperature Johnson noise in the 5-MΩ resistors? (*Hint*: If two uncorrelated noise sources are added, the rms noises combine as the square root of the sum of the squares.)

(c) When the inputs are connected directly to ground, the output-voltage noise with a gain of 1,000 is 1 mV rms in the 10^4-Hz bandwidth. When the gain is reduced to 1, the output-voltage noise is 0.1 mV rms in the 10^6-Hz bandwidth. What is the amplifier noise with respect to the input (D_1) and the output (D_0)? (Express the noise in units of $nV/\sqrt{Hz}$.)

2.9 Design a system for the filtering of EEG (brain-wave) data, given that:
  • the EEG signal amplitude is 50 μV and is in the 0.5- to 30-Hz frequency band,
  • the EMG background amplitude from the head muscles is 100 μV and is in the 100-Hz to 3-kHz band,
  • the EM interference is 10 mV at 60 Hz,
  • you wish to minimize the number of circuit components.
  (a) What type of filtering would you use to see the EEG signal undistorted (variations in gain less than 1%) while reducing all other backgrounds to 1 μV? (Do not work it out in detail, just give the number of poles, corner frequency, etc.)
  (b) Sketch the response of the filter circuit, showing $|V_{out}/V_{in}|$ from 0.1 Hz to 10 kHz, marking the values at 0.5, 30, 60, 100, and 3,000 Hz.

2.10 Design a high-pass one-pole filter using the op-amp circuit shown in Figure 2.24. The op-amp specifications are:
  • infinite input impedance, no input leakage currents;
  • above 10 Hz, the open-loop gain varies as 1/frequency and reaches unity gain at 10^7 Hz.
  The high-pass filter circuit specifications are:
  • lower −3-dB corner frequency at 100 Hz,
  • gain = 10 in the passband.
  (a) Sketch the circuit, using $C = 1.59$ μF, and show values for all resistors.
  (b) What is the 3-dB upper corner frequency? (*Hint*: Do not forget the limited gain–bandwidth product of the op amp.)
  (c) Give typical values for the input and output impedances of the filter circuit at 10 kHz.
  (d) Sketch the Bode plot (decibel gain versus frequency) from 1 to 10^8 Hz.

2.11 Figure 2.24 shows a high-pass one-pole filter op-amp circuit, which has a lower 3-dB corner frequency $\omega_c = (R_1 C)^{-1}$. Using Eq. (2.5), which describes the gain for all frequencies by including the op-amp open-loop gain A, compute the upper 3-dB corner frequency. Assume that $A = K/\omega$, where K is a constant gain–bandwidth product. Your answer should be a function of K, C, R_1, and R_2.

2.12 Design a Butterworth high-pass four-pole filter using two cascaded stages of the unity-gain Sallen–Key circuit shown in Figure 2.36 and the filter parameters in Table 2.3. The design constraints are that the −3-dB corner frequency is $f_c = 1$ kHz, and $C = 0.1$ μF.
  (a) Determine the component values R_1 and R_2 for each filter section.
  (b) Sketch the Bode amplitude plot, assuming that the op amp has infinite open-loop gain at all frequencies.
  (c) What happens when you apply a 100-Hz square wave? Sketch the approximate resulting waveform (amplitude versus time) and explain. (*Hint*: Think in terms of the time domain, rather than the frequency domain.)

2.13 Design a Butterworth high-pass four-pole filter using two cascaded stages of the equal-component-value Sallen–Key circuit shown in Figure 2.37 and the filter parameters in Table 2.4.

(a) For a 3-dB corner frequency f_c of 1 kHz and $C = 0.1\,\mu\text{F}$, sketch the filter circuit and determine R_1 and R_2 for each filter section.

(b) Sketch the Bode plot assuming that the op amp has an infinite open-loop gain at all frequencies. (*Hint*: See Problem 2.14 below.)

(c) What happens when you apply a 100-Hz square wave? Sketch the approximate resulting waveform (amplitude versus time) and explain. (*Hint*: Think in terms of the time domain, rather than the frequency domain.)

2.14 Design a Butterworth filter that passes frequencies from 0 Hz to 1 kHz with an accuracy of 0.1 dB and rejects frequencies above 10 kHz by a factor of 100 dB.

The nth order Butterworth filter has a gain magnitude $|G|$ and phase shift ϕ given by:

$$|G| = \frac{1}{\sqrt{1 + (f/f_c)^{2n}}} \qquad \tan\left(\frac{\phi}{n}\right) = \frac{f}{f_c}$$

(a) What is the minimum order n and the corresponding corner frequency f_c that will satisfy the requirements?

(b) What are the phase shifts at 100 Hz and 1 kHz?

(c) What are the time delays at 100 Hz and 1 kHz associated with those phase shifts?

(d) What can you say about the ability to preserve the shape of a 100-Hz square wave? Consider both the effects of the filter on amplitude and phase. A symmetric square wave of frequency f_0 and amplitude ±1 may be represented by the Fourier series:

$$V(t) = \left(\frac{4}{\pi}\right) \sum_{n=1,3,5,\ldots}^{\infty} \left(\frac{1}{n}\right) \cos(2\pi n f_0 t)$$

(e) The Bessel filter has a phase shift that is proportional to frequency. How would this filter preserve the shape of the 100-Hz square wave?

2.15 A power amplifier with a gain $V_0 = G V_1$ can be described by the equivalent circuit shown in Figure 2.47.

(a) What is V_0 in terms of V_{in}?

(b) What is V_{in} in terms of V_S?

(c) What is V_0 in terms of V_S?

(d) You want to use this circuit to amplify 1-mV signals from a magnetic tape head (output impedance 1 MΩ) and drive a speaker (input impedance 8 Ω) at 10 V amplitude. What are the requirements on R_{in} and R_{out} so that V_{in} is within 1% of V_S and V_0 is within 1% of $G V_{in}$?

(e) Comment on the design requirements for R_{in} and R_{out} necessary for specific applications.

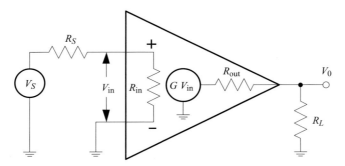

Figure 2.47 Power amplifier equivalent circuit.

2.16 The classic instrumentation amplifier circuit is shown in Figure 2.13. Assume:
- $R_1 = 100\,\Omega$, $R_2 = 5\,\mathrm{k}\Omega$, $R_3 = 1\,\mathrm{k}\Omega$, $R_4 = 10\,\mathrm{k}\Omega$;
- input $V_+(t) = 1.0\,\mathrm{V} + 1.0\,\mathrm{mV}\sin(\omega t)$;
- input $V_-(t) = 1.0\,\mathrm{V} - 1.0\,\mathrm{mV}\sin(\omega t)$;
- $f = 2\pi\omega = 1\,\mathrm{kHz}$;
- power supply voltages are -10 and $+10\,\mathrm{V}$.

Answer the following:
(a) What is $V_3(t)$?
(b) What is $V_4(t)$?
(c) What is $V_4(t) - V_3(t)$?
(d) What is $V_0(t)$?

2.17 A new instrumentation amplifier circuit has been proposed, as shown in Figure 2.48.

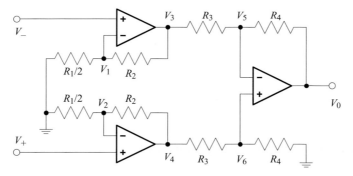

Figure 2.48 Alternative instrumentation amplifier circuit.

Assume:
- $R_1/2 = 50\,\Omega$, $R_2 = 5\,\mathrm{k}\Omega$, $R_3 = 1\,\mathrm{k}\Omega$, $R_4 = 10\,\mathrm{k}\Omega$;
- input $V_+(t) = 1.0\,\mathrm{V} + 1.0\,\mathrm{mV}\sin(\omega t)$;
- input $V_-(t) = 1.0\,\mathrm{V} - 1.0\,\mathrm{mV}\sin(\omega t)$;
- $f = 2\pi\omega = 1\,\mathrm{kHz}$;
- power supply voltages are -10 and $+10\,\mathrm{V}$.

Answer the following:
(a) What is $V_3(t)$?
(b) What is $V_4(t)$?
(c) What is $V_4(t) - V_3(t)$?
(d) What is $V_0(t)$?
(e) Is this circuit design better than the one in Problem 2.16? Explain your answer.

2.18 You have been given the assignment of designing an amplifier and filtering circuit that meets the following requirements:
  • differential input;
  • operational temperature range 10–30 °C;
  • differential gain 10^6 between 1 and 1,000 Hz, with an accuracy of 30%;
  • differential gain <1 for frequencies $>10,000$ Hz;
  • common-mode gain $<10^{-2}$ for all frequencies.
  Assume:
  • Since you cannot reliably get a differential gain $>10^4$ from a single instrumentation amplifier, your circuit will need additional amplification.
  • To avoid saturation, all amplifier outputs must be between -10 and $+10$ V.
  • The input offset voltage of the first instrumentation amplifier varies by 1 mV over the range from 10 to 30 °C, but the direction and magnitude of variation cannot be predicted because it differs from part to part (assume all other offset voltages are much less important and can be neglected).
  • It is not possible to measure the temperature of the circuit.
  Do the following:
  (a) Draw a block diagram of your design, showing all necessary components.
  (b) Sketch the differential gain versus frequency for your circuit from 0.01 Hz to 10 kHz.
  (c) What is the requirement for the common-mode rejection ratio of the instrumentation amplifier in your circuit?
  (d) If a 1-kΩ resistor is connected to one input and the other input is grounded, approximately how much Johnson noise does the resistor contribute to the output of the circuit?
  (e) If both inputs are connected to ground through 1-kΩ resistors, approximately how much Johnson noise do the resistors contribute to the output of the circuit?

2.19 Design a circuit that uses Johnson noise in a resistor to measure absolute temperature. The design requirements are:
  • output voltage proportional to absolute temperature over the range from 100 to 1,000 K, with output $= 0.100$ V at 100 K, 1.00 V at 1,000 K;
  • the output varies slowly, responding to temperature change frequencies <1 Hz.
  Your circuit consists of the following elements:
  • A 1-MΩ resistor with an accurate resistance from 100 to 1,000 K.

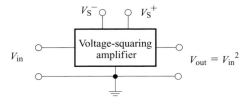

Figure 2.49 Voltage-squaring amplifier.

- Two wires (twisted pair) from the ends of the 1-MΩ resistor to the input of an instrumentation amplifier (below). The wires pick up 60-Hz electromagnetic interference with a common-mode voltage of ±100 and ±1 mV differential.
- An instrumentation amplifier circuit with a gain of 426 and a bandwidth of 1 MHz. To simplify the problem, ignore leakage currents.
- A circuit whose output (in volts) is equal to the square of the input (in volts) (Figure 2.49).
- Any additional filtering or amplification needed, using circuits covered in this book. For filters you only need to specify type, order n, and corner frequency f_c. For amplifiers, specify gain.

Hint: The Johnson noise on a 1-MΩ resistor in a 1-MHz bandwidth is given by:

$$V_{rms} = \sqrt{4kTR\Delta f} = 7.43 \, \mu V \sqrt{T}$$

For $T = 100$ and 1000 K, $V_{rms} = 74.3$ and 235 μV, respectively.

Do the following:
(a) Sketch a block diagram of your circuit in enough detail so that a skilled technician can build it and understand how it meets the design objectives.
(b) Sketch the differential input waveform to the instrumentation amplifier when the resistor is at 1,000 K (for this item and those following, label both voltage and time axes).
(c) Sketch the common-mode input waveform to the instrumentation amplifier when the resistor is at 1,000 K.
(d) Sketch the output waveform of the instrumentation amplifier when the resistor is at 1,000 K.
(e) Sketch the input waveform to the voltage-squaring amplifier when the resistor is at 1,000 K.
(f) Sketch the output waveform of the voltage-squaring amplifier when the resistor is at 1,000 K.
(g) Sketch the output waveform of your entire circuit while the resistor temperature is changed from 1,000 to 500 K in 0.1 s.

2.20 At room temperature (300 K) the AD625 instrumentation amplifier has an input noise:

$$V_{\mathrm{rms}} = 4\,\mathrm{nV}\sqrt{\Delta f}/\sqrt{\mathrm{Hz}}$$

What value resistor has Johnson noise equal to the above?

2.21 You have been given an instrumentation amplifier and asked to measure and characterize its differential and common-mode gains.

(a) How would you measure the common-mode gain and differential gain as a function of frequency?

You find that the differential gain can be modeled as an ideal instrumentation amplifier with a differential gain of 1,000 followed by a Butterworth low-pass filter of order one with a corner frequency of 1 kHz. You also find that the common-mode gain can be modeled as a Butterworth high-pass filter of order one with unity gain and a corner frequency of 10^6 Hz.

(b) Write an equation for the common-mode gain as a function of frequency and sketch the result from 10^6 Hz to 100 MHz.

(c) Write an equation for the differential gain as a function of frequency and sketch the result from 10 Hz to 100 MHz.

(d) Sketch the common-mode rejection ratio (CMRR) as a function of frequency and sketch the result from 10 Hz to 100 MHz.

2.9   Additional reading

Glenn M. Glasford, *Analog Electronic Circuits*, Prentice Hall, Englewood Cliffs, NJ, 1986.

John L. Hilburn, *Manual of Active Filter Design*, McGraw-Hill, New York, 1983.

Paul Horowitz and Winfield Hill, *The Art of Electronics*, 2nd edition, Cambridge University Press, New York, 1989.

Robert G. Irvine, *Operational Amplifier Characteristics and Applications*, Prentice Hall, Englewood Cliffs, NJ, 1987.

Brian K. Jones, *Electronics for Experimentation and Research*, Prentice Hall, Englewood Cliffs, NJ, 1986.

Daniel H. Sheingold, *Transducer Interfacing Handbook*, Analog Devices, Norwood, MA, 1981.

Sidney Soclof, *Application of Analog Integrated Circuits*, Prentice Hall, Englewood Cliffs, NJ, 1985.

Laboratory Exercise 4

Operational-amplifier circuits

Purpose

To construct several op-amp circuits using negative feedback for voltage and current amplification, and to measure their output offset, noise, and gain as a function of frequency. To gain familiarity with the properties and limitations of the operational amplifier (op amp) such as open-loop gain, leakage currents, the effects of heating, offset voltage, noise, and bandwidth.

Equipment

- Superstrip circuit board with ground plane and connections for ground, $+5$ V, ± 12 V
- Two 10-μF 25-V electrolytic capacitors (put between power and ground at circuit board binding posts)
- Six 0.1-μF CK-05 bypass capacitors (put between power and ground on all chips)
- Three LF356 op amps
- Oscilloscope
- Three 20-kΩ trimpots (op-amp offset adjust)
- Six 1-kΩ resistors
- Three 100-kΩ resistors
- One 10-MΩ resistor
- $+5$-V, ± 12-V power supplies
- Wave generator
- Heat gun (shared with other laboratory groups)
- Dial thermometer

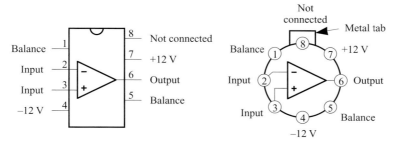

Laboratory Figure 4.1 LF356 pinout for dual inline package (DIP) and T0-5 metal can (top views).

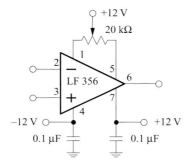

Laboratory Figure 4.2 LF356 external connections. The bypass capacitors at pins 4 and 7 are required for circuit stability and noise reduction. The trimpot between pins 1 and 5 is used to adjust output offset.

Background

The operational amplifier is one of the most important building blocks in analog-circuit design and is used both for amplification and filtering. Chapter 2 describes the fundamental properties of the op amp and the effect of negative feedback on gain, bandwidth, and input impedance.

The pin assignments for the LF356 op amp that you will be using is shown in Laboratory Figure 4.1. Pins 1, 4, 5, and 7 are to be connected as shown in Laboratory Figure 4.2.

Note: For all laboratory exercises, connect 10-μF electrolytic capacitors between each power-supply voltage ($+5$, $+12$, and -12 V) and ground at the binding posts of your circuit board. Observe capacitor polarity! Electrolytics can explode when connected backwards! These capacitors help stabilize the supply voltage levels at low frequencies (such as 60 Hz), but are not effective in reducing spikes caused by fast (<1 μs) circuit-switching transients. To reduce the fast spikes, connect 0.1-μF capacitors between power and ground at all integrated circuits.

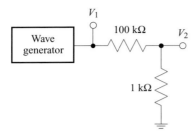

Laboratory Figure 4.3 Small-signal generator circuit.

Additional reading

Section 2.2 Operational amplifier circuits
Section 2.3 Op-amp characteristics

Procedure

1.   Small-signal generator

Construct the small-signal generator circuit shown in Laboratory Figure 4.3. It consists of a simple 100-kΩ/1-kΩ resistor divider, and produces an output voltage $V_2 = 0.01V_1$, provided that its output is connected to a circuit with input impedance $\gg 1$ kΩ. Measure the resistors accurately, since V_2 is small and difficult to measure directly. Instead, for the following sections, compute V_2 from your measured value of V_1 and the resistor values.

2.   Inverting amplifier with gains $= -100$ and -1

Construct the amplifier circuit shown in Laboratory Figure 4.4 on your superstrip breadboard with $R_1 = 1$ kΩ, but do not connect the 20-kΩ offset adjust trimpot yet. You will use $R_2 = 100$ kΩ for gain $= -100$ and $R_2 = 1$ kΩ for gain $= -1$. Laboratory Figure 4.1 shows the pinout diagram of the LF356 op amp, and Laboratory Figure 4.2 shows the connection for pins 1, 4, 5, and 7.

2.1 dc offset with $R_2 = 100$ kΩ. Connect the input V_2 to ground, and do the following:

1. *Unadjusted offset*: Record the output offset voltage V_0 (20-kΩ offset-adjust trimpot not connected).

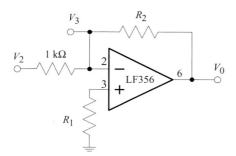

Laboratory Figure 4.4 Inverting amplifier with gain $G = V_0/V_2 = -R_2/1\,\text{k}\Omega$. See Laboratory Figure 4.1 for pinout. Connect pins 1, 4, 5, and 7 as shown in Laboratory Figure 4.2.

2. *Offset range*: Connect the 20-kΩ trimpot as shown in Laboratory Figure 4.2 and record V_0 for the two extreme values of the trimpot (full clockwise and full counter-clockwise).

   Note: Twenty turns of the adjusting screw will be required.

3. *Temperature effect*: Adjust the trimpot for $V_0 = 0$ V. Heat the op amp about $10\,°\text{C}$ (check with dial thermometer), and record V_0.
4. *Reproducibility*: Wait about 5 min, for the op amp to return to room temperature, and again record V_0.
5. *Leakage currents*: Adjust the trimpot for $V_0 = 0$ V. Temporarily change R_1 to $10\,\text{M}\Omega$ and record V_0. Then set $R_1 = 1\,\text{k}\Omega$.

   Note: In the preceding section you will have measured a total of six offset voltages.

2.2 Noise with $R_2 = 100\,\text{k}\Omega$. With $R_1 = 1\,\text{k}\Omega$ and V_2 connected to ground, adjust the trimpot for $V_0 = 0$ V. Connect V_0 to the oscilloscope input and record the amplitude of the output noise (extreme oscilloscope gain required). Distinguish between random "fuzz" and repeating waveforms such as 60-Hz interference. Repeat for $R_1 = 10\,\text{M}\Omega$.

   Note: The rms measurement feature of the digital oscilloscope combines random and periodic deviations from the average.

2.3 Small-signal gain with $R_2 = 100\,\text{k}\Omega$. Replace $R_1 = 1\,\text{k}\Omega$ and connect the input of the amplifier circuit (V_2) to the small-signal generator output (Laboratory Figure 4.3). Adjust the generator output for a 1-kHz sine wave with $V_1 = 1$ V peak-to-peak (p–p) as seen on your oscilloscope. Measure the p–p sine-wave amplitudes of V_1, V_2, and V_0 at 10 Hz, 100 Hz, 1 kHz, 10 kHz, 100 kHz, and 1 MHz. Note the following: (1) the wave generator amplitude V_1 will vary somewhat as the frequency is changed, and (2) the op-amp amplifier circuit has a 1-kΩ input impedance.

2.4 Open-loop gain with $R_2 = 100\,\text{k}\Omega$. Increase the generator output V_1 to 10 V p–p and measure V_0 and V_3 at 10 Hz, 100 Hz, 1 kHz, 10 kHz, 100 kHz, and 1 MHz (extreme oscilloscope gain required for V_3). In the analysis section you will compute the open loop gain A as V_0/V_3.

   2.5 dc offset with $R_2 = 1\,k\Omega$. Make the following measurements.

1. *Unadjusted offset*: Disconnect the 20-kΩ offset-adjust trimpot and record the output offset voltage V_0.
2. *Temperature effect*: Replace the 20-kΩ offset-adjust trimpot and adjust the trimpot for $V_0 = 0$ V. Heat the op amp about 10 °C (check with dial thermometer), and record V_0.

   Note: In the preceding section you will have measured two offset voltages.

3. Noninverting amplifier with gains = 101 and 1

Construct the circuit shown in Laboratory Figure 4.5, but do not connect the 20-kΩ trimpot yet. You will use $R_2 = 100\,k\Omega$ for gain = 101 and $R_2 = 0\,k\Omega$ for gain = 1. Laboratory Figure 4.1 shows the pinout diagram of the LF356 op amp, and Laboratory Figure 4.2 shows the connection for pins 1, 4, 5, and 7.

   3.1 dc offset with $R_2 = 100\,k\Omega$. Connect V_2 to ground with a resistor $R_1 = 1\,k\Omega$, and do the following:

1. *Unadjusted offset*: Record the output offset voltage V_0 (20-kΩ offset-adjust trimpot not connected).
2. *Offset range*: Connect the 20-kΩ trimpot, as shown in Laboratory Figure 4.2 and record V_0 for the two extreme values of the trimpot (full clockwise and full counterclockwise).

   Note: Twenty turns of the adjusting screw will be required.

3. *Temperature effect*: Adjust the trimpot for $V_0 = 0$ V. Heat the op amp about 10 °C (check with dial thermometer), and record V_0.
4. *Reproducibility*: Wait about 5 min for the op amp to return to room temperature, and again record V_0.
5. *Leakage currents*: Adjust the trimpot for $V_0 = 0$ V. Temporarily change R_1 to 10 MΩ and record V_0.

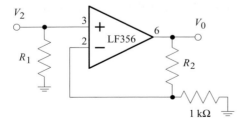

Laboratory Figure 4.5 Noninverting amplifier with gain $G = V_0/V_2 = (R_2 + 1\,k\Omega)/1\,k\Omega$.

3.2 Noise with $R_2 = 100\,\text{k}\Omega$. With $R_1 = 1\,\text{k}\Omega$ between V_2 and ground, record the output noise as you did in procedure section 2.2. Repeat for $R_1 = 10\,\text{M}\Omega$.

3.3 Small-signal gain with $R_2 = 100\,\text{k}\Omega$. Remove R_1 and connect the input of the amplifier circuit (V_2) to the small-signal generator output. Adjust the generator output for a 1-kHz sine wave with $V_1 = 1$ V peak-to-peak (p–p) as seen on your oscilloscope. Measure the p–p sine-wave amplitudes of V_1, V_2, and V_0 at 10 Hz, 100 Hz, 1 kHz, 10 kHz, 100 kHz, and 1 MHz.

3.4 dc offset with $R_2 = 0\,\text{k}\Omega$. Replace R_2 with a wire and make the following measurements:

1. *Unadjusted offset*: Disconnect the 20-kΩ offset-adjust trimpot and record the output offset voltage V_0.
2. *Temperature effect*: Replace the 20-kΩ offset-adjust trimpot and adjust the trimpot for $V_0 = 0$ V. Heat the op amp about 10 °C (check with dial thermometer), and record V_0.

Laboratory report

1.   Setup

To make your report a complete description of the laboratory exercise, include copies of the diagrams provided here. Be sure to indicate any variations from the suggested setup.

2.   Data summary

Tabulate your measurements and summarize your observations from procedure sections 2 and 3.

3.   Analysis

3.1 Output offset voltages. For the two amplifier circuits, tabulate the six measured output offset voltages in procedure sections 2.1 and 3.1; and the three output offset voltages in procedure sections 2.5 and 3.5. Using the measured unadjusted output offset voltages for gains $G = 1$ and $G = 100$, and $V_{\text{TOO}} = G V_{\text{RTI}} + V_{\text{RTO}}$, compute V_{RTI} (offset voltage with respect to input) and V_{RTO} (offset voltage with respect to output) for the op amp. Compute the change of input and output offset voltages with temperature (in microvolts per degrees celsius). Compute the leakage current using the change in input offset voltage that occurred when the 10-MΩ resistor was used.

3.2 Noise. From your measurements of closed-loop gain G, rms output noise V_{rms}, and bandwidth Δf (3-dB frequency) for the two circuits, compute the input noise figure for 1-kΩ input resistors:

$$D = \frac{V_{\text{rms}}}{G\sqrt{\Delta f}}$$

which is usually expressed in the units $nV/\sqrt{Hz}$.

Repeat for the noninverting and buffer amplifiers when the 10-MΩ resistor was used.

3.3 Bode amplitude plot. From the small-signal gain data you took for the two op-amp circuits (procedure sections 2.3 and 3.3), tabulate gain $G = V_0/V_2$ versus frequency. Either plot G versus frequency on log–log graph paper or plot decibels $(\text{dB}) = 20\log_{10}(|G|)$ versus frequency on semilog graph paper.

3.4 Open-loop gain. From your measurements of V_0 and V_3 in procedure section 2.4, tabulate and plot the open-loop gain of the LF356 as a function of frequency. For comparison, include open-loop gain data from the LF356 data sheet.

3.5 Summary table. Compare your measurements with the claims in the data sheets: input bias current, input and output voltage offset, adjustment range, temperature dependence, open-loop gain–frequency product. For definitions of these and other parameters, see Chapter 2.

4.   Discussion and conclusions

4.1 Based on your data and analysis of the two amplifier circuits, discuss how the characteristics of the real op amp differed from those of the ideal op amp.

4.2 Discuss how negative feedback is used in this laboratory exercise to establish a desired gain that is constant over a range of frequencies.

4.3 Compare the gain behavior of the two amplifier circuits at high frequencies.

4.4 Discuss the difficulties of using the circuit in Laboratory Figure 4.4 to measure the open-loop gain at low frequencies.

4.5 Discuss situations where each of the two amplifier circuits would best be used.

5.   Questions

5.1 When you were measuring the small-signal gain in procedure sections 2.3 and 3.3, why were V_2 and V_0 less for the inverting amplifier than for the noninverting amplifier?

5.2 Why did V_2 increase for the inverting amplifier at high frequencies? (*Hint*: See Problem 2.1.)

5.3 How well does your measured open-loop unity-gain frequency agree with the data sheet?

5.4 How accurately do you think you were able to measure the open-loop gain at 100 kHz? Are your data consistent with the data sheet?

5.5 Find the "Open Loop Frequency Response" in the LF356 op-amp data sheets. Prepare a table with three columns: (1) frequency = 100 Hz, 1 kHz, 10 kHz, 100 kHz,

1 MHz, and 10 MHz; (2) "open loop voltage gain;" and (3) the product of the two (i.e. gain–bandwidth product).

Note: Convert the open-loop voltage gain from decibel (as provided in the data sheets) to numerical gain.

6.   Laboratory data sheets

Include your handwritten data sheets (or a copy), which should consist of a log of the procedures that you used, any special circumstances, and the measurements you recorded manually.

Laboratory Exercise 5

Instrumentation amplifiers

Purpose

To gain familiarity with the instrumentation amplifier, to demonstrate differential amplification, to measure the output offset voltage with balanced and unbalanced inputs, the dc and ac common-mode rejection, and to measure the Bode plot of gain versus frequency. To measure the ability of differential amplification and coaxial cables to suppress electromagnetic interference.

Equipment

- Superstrip circuit board with ground plane and connections for ground, 5 V, ±12 V
- Two 10-μF, 25-V electrolytic capacitors (put between power and ground at circuit board binding posts)
- Two 0.1-μF, CK-05 bypass capacitors (put between power and ground on all chips)
- +5-V, ±12-V power supplies
- Wave generator
- Oscilloscope
- Two 1-m long coax cables
- Two 1-m insulated wires
- Two BNC-to-alligator clip adapters (POMONA 91836)
- One 100-Ω resistor
- Two 10-kΩ resistors
- Two 1-MΩ resistors
- One heat gun (shared with other laboratory groups)
- One dial thermometer
- One AD625 or LH0036 instrumentation amplifier
- For the AD625 instrumentation amplifier, use resistors below:
  - One 20-kΩ trimpot (offset adjust)

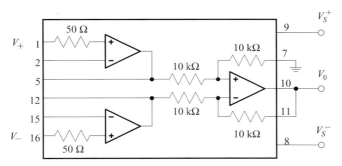

Laboratory Figure 5.1 Simplified circuit of the AD625 instrumentation amplifier.

- Two 20-kΩ resistors (R_F)
- One 390-Ω resistor (R_G for gain of 100)
- One 3.9-kΩ resistor (R_G for gain of 10)
- For the LH0036 instrumentation amplifier chip, use resistors below:
  - One 100-kΩ trimpot (offset adjust)
  - One 510-Ω resistor (R_G for gain of 100)
  - One 3.3-kΩ resistor (offset adjust)
  - One 5.1-kΩ resistor (R_G for gain of 10)
  - One 33-kΩ resistor (offset adjust)

Background

The instrumentation amplifier (shown in Laboratory Figure 5.1) is a very useful circuit element, especially when the signal to be amplified is the difference between two potentials (as from a bridge) and the amplifier input impedance must be very high.

Additional reading

Section 2.4 Instrumentation and isolation amplifiers
Section 2.5 Noise sources

Procedure

1. Circuit construction

Set up the circuit shown in Laboratory Figures 5.2 or 5.3. Set the instrumentation differential gain at about 100. For the AD625, use $R_F = 20\,\text{k}\Omega$ and $R_G = 390\,\Omega$. For

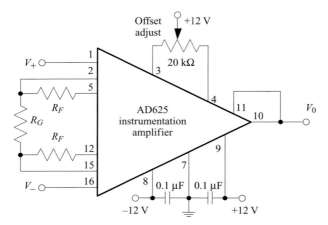

Laboratory Figure 5.2 AD625 instrumentation amplifier circuit. In the absence of common-mode gain, differential gain is given by $G_\pm = V_0/(V_+ - V_-) = 1 + 2R_F/R_G$.

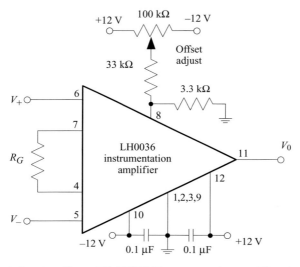

Laboratory Figure 5.3 LH0036 instrumentation amplifier circuit. In the absence of common-mode gain, differential gain is given by $G_\pm = V_0/(V_+ - V_-) = 1 + 50\,\text{k}\Omega/R_G$.

the LH0036, use $R_G = 510\,\Omega$ ($R_F = 25\,\text{k}\Omega$ resistors are internal). Set the power supply for $V_S^+ = +12\,\text{V}$ and $V_S^- = -12\,\text{V}$.

2.   Offset voltage

2.1 Unadjusted offset. With the offset-adjustment trimpot removed, and both inputs connected directly to ground ($V_- = V_+ = 0\,\text{V}$), measure the unadjusted output offset voltage V_0.

2.2 Offset-adjustment range. With both inputs connected directly to ground, install the trimpot and record the output offset voltage for the full trimpot range (full clockwise and full counterclockwise).

Note: This will require 20 turns.

2.3 Effect of temperature on offset. With both inputs connected directly to ground, firstly adjust the trimpot for $V_0 = 0$ V, and then heat the circuit about $10\,°C$ with the heat gun (monitor with dial thermometer), and record any change in V_0.

2.4 Effect of unequal input resistors and temperature on offset. With both inputs connected directly to ground, firstly adjust the trimpot for $V_0 = 0$ V. Then connect one of the inputs to ground through a 1-MΩ resistor and record V_0. Then heat the circuit about $10\,°C$ with the heat gun (monitor with dial thermometer), and record any change in V_0.

2.5 Effect of power-supply voltage on offset. With both inputs connected directly to ground, increase the $+12$-V power supply by 5% to 12.6 V and record any change in V_0. Make sure that you do not change the -12-V supply. (With some power supplies, one knob adjusts both voltage polarities and another knob adjusts the ratio.) After you have made your measurement, readjust the $+12$-V supply and adjust the offset-adjustment trimpot for zero output.

3.   Noise

With both inputs connected directly to ground, use the oscilloscope to observe the amplifier output noise V_0 (extreme oscilloscope gain required). Record the amplitude and nature of the noise. Distinguish between electromagnetic interference (fixed frequency content, amplitude depends on conductor geometry) and Johnson noise (wideband noise with random fuzzy waveform). A sketch will be helpful here.

With both inputs connected to ground through 1-MΩ resistors, record the amplitude and nature of the noise. Johnson noise in the 1-MΩ resistors will be amplified and combined with the amplifier output noise that occurred when the inputs were connected directly to ground.

4.   Common-mode gain

Set up the circuit shown in Laboratory Figure 5.4.

4.1 DC common-mode gain. With $V_1 = 0$ V, adjust the trimpot for $V_0 = 0$ V. Change V_1 to $+5$ V, and record V_0. Since the differential input is zero, we can determine the common-mode gain G_c as V_0/V_1.

4.2 AC common-mode gain. With V_1 equal to the maximum undistorted sine-wave amplitude (not to exceed 10 V peak-to-peak), measure V_0 at frequencies 10 Hz,

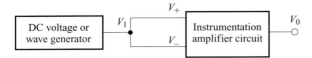

Laboratory Figure 5.4 Circuit for measuring dc and ac common-mode gain.

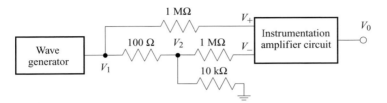

Laboratory Figure 5.5 Circuit for measuring the small-signal differential gain.

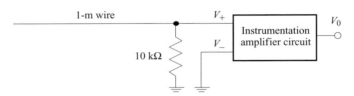

Laboratory Figure 5.6 Circuit for measurement of electromagnetic pickup with one input grounded. The pickup is large, and a gain of 10 is used.

100 Hz, 1 kHz, 10 kHz, 100 kHz, and 1 MHz. Since the differential input is zero, we can determine the common-mode gain G_c as V_0/V_1.

5.   Small-signal differential gain

Set up the circuit shown in Laboratory Figure 5.5 with a gain of 100.

Set the sine-wave generator for 1 kHz, and adjust the amplitude for 1 V peak-to-peak (p–p) at V_1 as seen on your oscilloscope. Record the amplitude at V_2. Measure V_0 and V_1 at 10 Hz, 100 Hz, 1 kHz, 10 kHz, 100 kHz, and 1 MHz. Because a large common-mode signal is present at the input, the output V_0 will be the sum of a differential term $G_\pm(V_1 - V_2)$ and a common-mode term $G_c(V_1 + V_2)/2 \approx G_c V_1$.

Assume that the differential input $V_1 - V_2 = V_1/101$.

6.   Electromagnetic pickup

6.1 Electromagnetic pickup with insulated wires. Reduce the differential gain to 10. If using the AD625, set $R_G = 3.9\,\text{k}\Omega$. If using the LH0036, set $R_G = 5.1\,\text{k}\Omega$. As shown in Laboratory Figure 5.6, attach a 1-m long insulated wire to one of the instrumentation amplifier inputs, connect the same input to ground with a 10-kΩ resistor, and ground the other input. Record the output amplitude and any prominent frequencies.

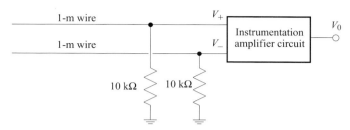

Laboratory Figure 5.7 Circuit for measurement of differential electromagnetic pickup. A gain of 100 is used.

As shown in Laboratory Figure 5.7, attach 1-m insulated wires to both inputs, and connect both inputs to ground with 10-kΩ resistors. Arrange the wires so that they run parallel to each other but do not touch. Again record the amplitude and prominent frequencies. Now twist the wires around each other without letting the conductors touch. Again record the amplitude and frequency.

6.2 Electromagnetic pickup with coaxial cables. Restore the differential gain to 100. With both 10-KΩ resistors in place, record for the following: (1) one differential input connected to a 1-m coaxial cable with grounded shield and the other differential input grounded; (2) both differential inputs connected to 1-m coaxial cables with grounded shields.

Laboratory report

1.   Setup

Sketch a simple block diagram of the major components used and their interconnections.

2.   Data summary

Summarize your data and observations from procedure sections 2 to 6.

3.   Analysis

3.1 Offset voltages. For procedure section 2, tabulate the five output offset voltages.

3.2 Power-supply rejection ratio. Compute the power-supply rejection ratio as $\Delta V_S/\Delta V_0$, where ΔV_S is the change in power-supply voltage and ΔV_0 is the corresponding change in output voltage. Compare with the data-sheet specification "offset referred to the input versus supply," which is $G_{\pm}(\Delta V_S/\Delta V_0)$.

3.3 Input leakage current. From your observations in procedure section 2.4, estimate the input leakage current (picoamperes) and its dependence on temperature (picoamperes per degree celsius). *Hint*: Calculate the leakage current as (input offset voltage)/(1 MΩ).

3.4 Noise coefficient. Use your measurements of amplifier noise with both inputs connected directly to ground (procedure section 3), closed-loop differential gain $G_{\pm}$, and bandwidth Δf to determine the noise coefficient of the instrumentation amplifier. Assume that the amplifier output noise V_{amp} is given by:

$$V_{\text{amp}} = K G_{\pm} \sqrt{\Delta f}$$

where K is the amplifier noise coefficient in $\text{nV}/\sqrt{\text{Hz}}$ with reference to input (RTI), and the bandwidth $\Delta f = 1.57 f_c$, where f_c is the 3-dB corner frequency in your Bode plot.

The measured output noise voltage with two 1-MΩ resistors connected between the inputs and ground includes: (1) the amplifier noise with both inputs connected directly, and (2) the amplified Johnson noise of the two resistors. As described in Chapter 2, the Johnson noise at room temperature is given by:

$$V_{\text{Johnson}} = D_J \sqrt{R \Delta f}, \qquad \text{where } D_J = 1.287 \times 10^{-10} \, \text{V}/\sqrt{\Omega}/\sqrt{\text{Hz}}$$

where Δf is the bandwidth of the instrumentation amplifier circuit in hertz and D_J is the Johnson noise coefficient. Compute V_{Johnson} for each 1-MΩ resistor, and combine these with the measured amplifier output noise with inputs grounded. Uncorrelated noise sources are combined as the square root of the sum of the squares of their rms amplitudes. Combining both amplifier and resistor noise

$$V_{\text{rms}} = \sqrt{V_{\text{amp}}^2 + 2 G_{\pm}^2 V_{\text{Johnson}}^2}$$

Compare this calculated value with the value of output noise you measured in procedure section 3 using the 1-MΩ resistors.

3.5 Common-mode and differential gains. Tabulate the common-mode gain $G_c = V_0/V_1$ (procedure section 4) as a function of frequency. From your data of procedure section 5:

$$V_0 = V_1 G_{\pm}/101 + V_1 G_c$$

compute and tabulate the differential gain $G_{\pm}$ as a function of frequency. (*Note*: We assume here that $V_1 - V_2 = V_1/101$.)

3.6 Common-mode rejection. Compute the common-mode rejection ratio CMRR $= G_{\pm}/G_c$ and plot as a function of frequency on semilog paper.

3.7 Bode amplitude plot. Plot the differential gain $G_{\pm}$ (in decibels) versus frequency. Plot either $G_{\pm}$ versus frequency on log–log graph paper or decibels (dB) $= 20 \log_{10}(|G_{\pm}|)$ versus frequency on semilog graph paper.

3.8 Data-sheet comparison. In a table, compare the following data-sheet values with your measurements:
- differential gain versus frequency,
- CMRR versus frequency,
- output offset voltage,
- output offset voltage change per temperature change,
- power-supply rejection ratio,
- amplifier noise coefficient (for the gain = 100 bandwidth).

3.9 60-Hz pickup. Using your data from procedure section 6, prepare a table of 60-Hz noise amplitude as follows:

Amplifier input	60-Hz amplitude
One 1-m straight wire on one input, other input grounded	
Two 1-m straight wires, one on each input	
Two twisted 1-m wires, one on each input	
One coax with alligator adapter on one input, other input grounded	
Two coaxes with alligator adapters, one on each input	

4.   Discussion and conclusions

4.1 Describe output offset voltages for typical gains and temperatures and discuss how they can be corrected.

4.2 Discuss the effect of input leakage currents with both equal and unequal resistance paths to ground, and how they can be measured.

4.3 Discuss typical output noise levels and how they vary with bandwidth, gain, and external resistance.

4.4 Discuss what you have learned about amplifying small differential signals in the presence of large common-mode signals and electromagnetic pickup.

4.5 Discuss the advantages of the instrumentation amplifier over a single op amp.

4.6 Discuss situations where differential amplification is necessary.

5.   Questions

5.1 Did the output change when one of the power supplies was changed?

5.2 In procedure section 3, did the observed Johnson noise agree with the formula given above?

5.3 Which type of signal lead permits the observation of the smallest signals, twisted pair, or coaxial cable?

6.   Laboratory data sheets

Include your handwritten data sheets (or a copy), which should consist of a log of the procedures that you used, any special circumstances, and the measurements you recorded manually.

Laboratory Exercise 6

Analog filtering

Purpose

To construct active high-pass, low-pass, and notch filter circuits using the LF356 op-amp integrated-circuit chip. To measure the filtering characteristics of these circuits (Bode amplitude and phase plots).

Equipment

- Superstrip circuit board with ground plane and connections for ground, 5 V, ±12 V
- Three 10-μF 25-V electrolytic capacitors (put between power and ground at circuit board binding posts)
- Ten 0.1-μF CK-05 bypass capacitors (put between power and ground on all chips)
- Five LF356 op amps
- +5-V, ±12-V power supplies
- Wave generator
- Oscilloscope
- Eight 10-kΩ resistors
- One 100-kΩ resistor (notch filter)
- One 1-MΩ resistor (notch filter)
- One 1-MΩ resistor (5% for notch filter)
- Two 2-MΩ resistors (5% for notch filter)
- One 0.01-μF capacitor (low-pass Butterworth filter)
- One 1000-pF capacitor (low-pass Butterworth filter)
- Two 0.015-μF capacitors (high- and low-pass one-pole filters)
- One 0.022-μF capacitor (low-pass Butterworth filter)
- Two 150-pF capacitors (notch filter)
- Two 1200-pF capacitors (5% for notch filter)
- One 2700-pF capacitor (5% for notch filter)

Background

Analog filters can be described in terms of their output $V_0(t)$ for a cosine input $V_1(t) = A \cos(\omega t)$, either in terms of the magnitude of the gain $|G|$ and a phase shift ϕ, or in terms of the real and imaginary components of the complex gain:

$$V_0(t) = |G|A \cos(\omega t + \phi) = |G|A \cos[\omega(t + \Delta t)]$$
$$= \text{Re}(G)A \cos(\omega t) + \text{Im}(G)A \sin(\omega t)$$

The magnitude of the voltage gain $|G|$ is given by:

$$|G| = |V_0|/A$$

The phase shift ϕ is given by:

$$\tan(\phi) = \text{Im}(G)/\text{Re}(G)$$

The phase shift can also be thought of as a frequency-dependent delay Δt between the input and output waveforms:

$$\Delta t = \phi/\omega = \phi/(2\pi f)$$

Note: It is always possible to shift the time scale so that any sinusoidal input has the form $V_1(t) = A \cos(\omega t)$.

1.   Low-pass one-pole filter

A low-pass one-pole filter is shown in Laboratory Figure 6.1. This filter consists of a passive low-pass filter followed by a buffer amplifier.

The magnitude of the voltage gain is given by:

$$|G| = \frac{1}{\sqrt{1 + (\omega RC)^2}} = \frac{1}{\sqrt{1 + (f/f_c)^2}}$$

and the phase shift is given by $\phi = \tan^{-1}(-\omega RC) = \tan^{-1}(-f/f_c)$.

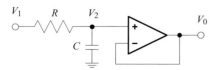

Laboratory Figure 6.1 Low-pass one-pole filter.

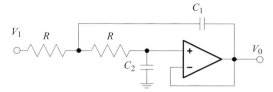

Laboratory Figure 6.2 Low-pass Butterworth two-pole filter.

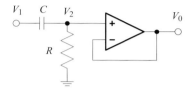

Laboratory Figure 6.3 High-pass one-pole filter.

2. Low-pass Butterworth two-pole filter

The low-pass Butterworth two-pole filter is shown in Laboratory Figure 6.2. See Chapter 2 for additional information.

The magnitude of the voltage gain is given by:

$$G = \frac{1}{\sqrt{1 + (f/f_c)^4}}$$

and the phase shift is given by:

$$\phi = \tan^{-1}(-2f/f_c)$$

See Chapter 2 for the relationship between f_c and R_1, R_2, and C.

3. High-pass one-pole filter

A high-pass one-pole filter is shown in Laboratory Figure 6.3. This filter consists of a passive high-pass filter followed by a buffer amplifier.

The magnitude of the voltage gain is given by:

$$|G| = \frac{\omega RC}{\sqrt{1 + (\omega RC)^2}} = \frac{f/f_c}{\sqrt{1 + (f/f_c)^2}}$$

and the phase shift is given by $\phi = \tan^{-1}(1/\omega RC) = \tan^{-1}(f_c/f)$.

Additional reading

Section 2.6 Analog filtering

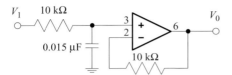

Laboratory Figure 6.4 Low-pass one-pole filter. Input requires a current path to ground.

Procedure

Warning 1: The inputs to op-amp filter circuits must be provided with a current path to ground, or input leakage currents will cause very large output offsets and possibly saturation.

Warning 2: Be sure that the oscilloscope is set for dc coupling. Ac coupling introduces a high-pass filter that distorts the slower waveforms.

1.   Low-pass one-pole filter

Set up the circuit shown in Laboratory Figure 6.4. The op-amp pinout is shown in Laboratory Figure 4.1. Connect pins 1, 4, 5, and 7 as shown in Laboratory Figure 4.2. Connect 0.1-μF capacitors between pin 4 (-12 V) and ground and between pin 7 ($+12$ V) and ground. The 20-kΩ trimpot used in Laboratory Exercises 4 and 5 to adjust the offset voltage will not be necessary.

1.1 Sine-wave response. Connect the output of the wave generator to V_1, the input of the filter. With a 1 V peak-to-peak (p–p) sine-wave input, observe both the filter input and output on your oscilloscope. Record: (i) input p–p amplitude; (ii) output p–p amplitude; and (iii) the phase shift between the input and output for frequencies of 10 Hz, 100 Hz, 1 kHz, 10 kHz, 100 kHz, 1 MHz, and some intermediate frequencies close to the corner frequency $f_c = 1/(2\pi RC)$. Note that the output amplitude of the wave generator may vary with frequency.

Measurement of phase shift. An accurate way to use an oscilloscope to measure the phase shift between two sine waves is the "zero crossing method," described in the following steps:

1. Display both waves in an alternating mode, always triggering on the positive slope of the reference wave (usually the filter input).
2. Choose a vertical amplification so that both waves span >50% of the screen.
3. Choose the vertical adjustment to center the waves approximately about the center horizontal line (the "zero" line).
4. As shown in Laboratory Figure 6.5, measure $B_1 - A_1$ (the difference between the upward crossings of the zero line), $B_2 - A_2$ (the difference between the downward crossings of the zero line), and $A_3 - A_1$ (the period).

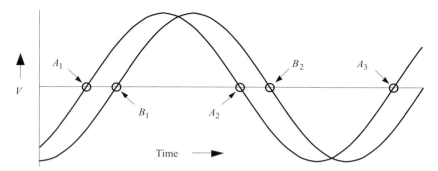

Laboratory Figure 6.5 "Zero crossing" method for using an oscilloscope to determine phase shift.

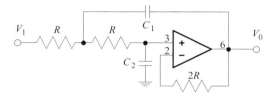

Laboratory Figure 6.6 Butterworth low-pass two-pole filter. Unity-gain Sallen–Key realization. Input requires a current path to ground.

5. Since $A_3 - A_1$ is one full period (2π), $(A_1 + A_2)/2$ is the position of the first peak of the reference wave, and $(B_1 + B_2)/2$ is the position of the first peak of the phase-shifted wave, the phase shift P is given by:

$$P = \left(\frac{2\pi}{A_3 - A_1}\right)\left(\frac{B_1 + B_2}{2} - \frac{A_1 + A_2}{2}\right) = \frac{\pi(B_1 - A_1 + B_2 - A_2)}{A_3 - A_1}$$

Note 1: This method uses measurements taken where dV/dt is large, and is more accurate than using the peaks themselves, where dV/dt is zero. Since the uncertainty in time is given by $\Delta t = \Delta V/(dV/dt)$, we want to make measurements where the slope dV/dt is as large as possible.

Note 2: The computed phase shift is relatively insensitive to the accuracy of the vertical adjustment in step 3.

1.2 Square-wave response. Observe the output waveform for an input square wave having a repeat frequency $f = 3f_c$, f_c, and $f_c/3$, where $f_c = 1/(2\pi RC)$, the corner frequency of the filter. Sketch the output waveforms.

2.   Butterworth low-pass two-pole filter

Set up the unity-gain Butterworth two-pole filter shown in Laboratory Figure 6.6. Connect pins 1, 4, 5, and 7 as shown in Figure 4.2. Connect 0.1-μF capacitors between pin 4 (−12 V) and ground and between pin 7 (+12 V) and ground. The 20-kΩ trimpot used in Laboratory Exercises 4 and 5 to adjust the offset voltage will not be necessary.

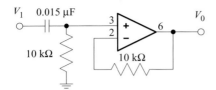

Laboratory Figure 6.7 High-pass one-pole filter.

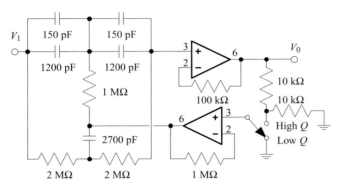

Laboratory Figure 6.8 Notch filter designed to suppress 60 Hz.

From Table 2.3, this filter has $k_1 = 1.414$ and $k_2 = 0.707$, where $k_1 = RC_1\omega_0$ and $k_2 = RC_2\omega_0$. Use $R = 10\,\text{k}\Omega$, $C_1 = 0.022\,\mu\text{F}$, and $C_2 = 0.011\,\mu\text{F}$ (0.01 μF in parallel with 1,000 pF). Set $V_1 = 1$ V p–p for 10 Hz, 100 Hz, 1 kHz, 10 kHz, 100 kHz, 1 MHz, and some frequencies close to f_c. Record sine-wave and square-wave data as in the previous procedure sections 1.1 and 1.2.

3.   High-pass one-pole filter

Set up the circuit shown in Laboratory Figure 6.7. Connect pins 1, 4, 5, and 7 as shown in Laboratory Figure 4.2. Connect 0.1-μF capacitors between pin 4 (-12 V) and ground and between pin 7 ($+12$ V) and ground. The 20-kΩ trimpot used in Laboratory Exercises 4 and 5 to adjust the offset voltage will not be necessary.

Set $V_1 = 1$ V p–p for 10 Hz, 100 Hz, 1 kHz, 10 kHz, 100 kHz, 1 MHz, and some frequencies close to $f_c = 1/(2\pi RC)$. Record sine-wave and square-wave data as in procedure sections 1.1 and 1.2. Sketch the output waveforms.

4.   Low-Q notch filter

Set up the circuit shown in Laboratory Figure 6.8 in low-Q mode, which is a twin-T notch filter followed by a unity-gain follower. Connect pins 1, 4, 5, and 7 as shown in Laboratory Figure 4.2. Connect 0.1-μF capacitors between pin 4 (-12 V) and ground and between pin 7 ($+12$ V) and ground. The 20-kΩ trimpot used in Laboratory Exercises 4 and 5 to adjust the offset voltage will not be necessary.

Set $V_1 = 1$ V p–p for 10 Hz, 100 Hz, 1 kHz, 10 kHz, 100 kHz, and 1 MHz. Find the notch frequency f_n and take sufficient data to plot the gain and phase shift from $f_n/4$ to $4f_n$. Record sine-wave and square-wave data as in procedure sections 1.1 and 1.2 (square-wave frequencies $f = 3f_n$, f_n, and $f_n/3$).

5.   High-Q notch filter

If time permits, switch the notch filter circuit to the high-Q mode and take data near the notch frequency. If the RC components are sufficiently matched, the notch should be narrower than it was in procedure section 4 above.

Laboratory report

1.   Setup

Sketch a block diagram of the major components used in this exercise.

2.   Data summary

2.1 Gain and phase shift. Tabulate data that you took using the four circuits. Include input p–p amplitude, output p–p amplitude, and phase shift at all frequencies measured.

2.2 Response to a square wave. For the high- and low-pass filters, sketch the output for a square-wave input whose fundamental frequency is $3f_c$ (low pass) or $f_c/3$ (high pass). For the notch filter, sketch the output for a square-wave input.

3.   Analysis

3.1 Bode amplitude plot. For each of the first three filters, tabulate the observed voltage gain as output/input for the frequencies used, and what you would predict from the formulas given in the background section. Plot gain versus frequency on log–log paper (Bode plot). Note that each factor of 10 in voltage gain corresponds to 20 dB.

3.2 Bode phase plot. For each of the first three filters, plot phase shift versus frequency on semilog paper (Δ phase versus frequency), and what you would predict from the formulas given in the background section.

3.3 Corner frequencies and slopes. Tabulate values of corner or notch frequencies, and roll-off slopes (decibel of gain per decade of frequency). Compare with expected values.

4.   Discussion and conclusions

4.1 Compare the characteristics of the filtering circuits and discuss situations where each would best be used.

4.2 Comment on any discrepancies between your observed values and the formulas given in the background section.

4.3 Comment on the shapes of the filter output for square-wave input.

5.   Questions

5.1 To double the corner frequency in the low-pass two-pole filter, how would you change the values of the resistors and capacitors?

5.2 How would you change the values of the resistors and capacitors in the low-Q notch filter for a notch frequency of 600 Hz?

6.   Laboratory data sheets

Include your handwritten data sheets (or a copy), which should consist of a log of the procedures that you used, any special circumstances, and the measurements you recorded manually.

3 Analog ⟷ digital conversion and sampling

3.1 Introduction

In this chapter, we discuss the components that convert data between the digital world of the microcomputer and its I/O ports and the analog world of continuously varying voltages. These are the digital-to-analog (D/A) converter, the analog-to-digital (A/D) converter, the sample-and-hold amplifier, and the comparator. We then go on to discuss some of the fundamental limits to sampling time-varying analog signals, including the role of the sample-and-hold amplifier, and the minimum sampling frequency.

Laboratory Exercise 7 is designed as an introduction to the characteristics of the D/A and A/D converters and uses an analog I/O circuit board. Laboratory Exercise 8 interfaces a D/A converter to the binary output port, measures its transfer characteristics, and uses it in waveform generation. Laboratory Exercise 9 interfaces an A/D converter to the binary input port and performs periodic sampling of sine waves. Laboratory Exercise 10 samples and recovers sine waves of various frequencies and explores the aliasing problems that arise when the input frequency is greater than one-half of the sampling frequency.

3.2 Digital-to-analog converter circuits

The digital-to-analog (D/A) converter changes an N-bit binary number to an analog output voltage that can have 2^N distinct values. It consists of resistors, switches, and an amplifier. Frequently, input latches are provided to retain the input number. Usually, the relationship between the input number and the output voltage is **linear**, but other relationships (e.g. logarithmic) are also used.

3.2.1 D/A converter characteristics

For an N-bit linear D/A converter, the **ideal** analog output voltage $V(n)$ is a linear function of digital input n between two reference voltages V_{ref}^- and V_{ref}^+:

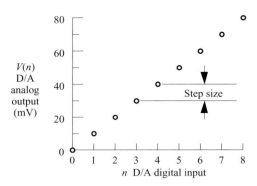

Figure 3.1 Ideal response of D/A converter. The analog output $= 0$ V for 0 input and increases in discrete 10-mV steps as the input is increased by successively larger integers.

$$V(n) = V_{\text{ref}}^{-} + n \left(\frac{V_{\text{ref}}^{+} - V_{\text{ref}}^{-}}{2^N} \right) = V_{\text{ref}}^{-} + n \Delta V \qquad (3.1)$$

Some D/A converters have fixed reference voltages established by internal circuits. On the other hand, the **multiplying D/A converter** permits V_{ref}^{-} and V_{ref}^{+} to be set by external circuits over a wide range of voltages and even vary with time. As we shall see, the output of this D/A converter is V_{ref}^{-} plus the product of $(V_{\text{ref}}^{+} - V_{\text{ref}}^{-})$ times the fractional equivalent of the digital input number.

At the minimum value $n = 0$, the output is V_{ref}^{-}, but the maximum value of n is $2^N - 1$ so the maximum output is $V_{\text{ref}}^{+} - \Delta V$. Figure 3.1 shows an example where $V_{\text{ref}}^{-} = 0$ V and $V_{\text{ref}}^{+}/2^N = 10$ mV.

The average **step size** is ΔV, which is also called the **least significant bit (LSB)** because it corresponds to a change in the least significant bit of the input number.

Some D/A converters have monopolar outputs (e.g. 0–5 V) and an input value $n = 0$ produces an output of 0 V. Other D/A converters have bipolar outputs (e.g. -5 to $+5$ V) and the input can either be offset binary (where $n = 0$ corresponds to the minimum output) or 2's complement.

As explored in Laboratory Exercises 7 and 8, the main characteristics of the D/A converter are:

1. The **absolute-accuracy error** is the difference between the measured output V_n and the ideal output (Eq. (3.1)), before adjustments for zero offset and gain. This error includes the zero-offset error, the gain error, and the linearity error. It is usually expressed in units of 1 LSB for comparison with the quantizing error.
2. The **relative-accuracy error** is the difference between the measured output V_n and the straight line passing through the measured end points $V_{\min} = V_0$ and $V_{\max} = V_{2^N-1}$. This straight line is given by:

$$V^{\text{rel}}(n) = V_{\min} + n \left(\frac{V_{\max} - V_{\min}}{2^N - 1} \right) \qquad (3.2)$$

Since this error includes the linearity error but not the zero-offset or gain errors, it is also called the **linearity error**. Note that calibration of the end points (by adjusting zero offset and gain) does not affect the relative-accuracy error. It is usually expressed in units of 1 LSB for comparison with the quantizing error.

3. The **resolution**, or **quantizing error**, of an N-bit D/A converter is the largest difference between any voltage within the full output range and the closest possible output voltage. In the case of the ideal D/A converter, this is one-half the step size, or 0.5 LSB. Because of the close association between resolution and the number of bits, the description "N-bit resolution" is frequently used.

4. The **zero-offset error** is the difference between the measured value of V_0 and V_{ref}^-.

5. The **gain error** is the difference between the measured slope $(V_{\text{max}} - V_{\text{min}})/(2^N - 1)$ and the ideal slope $(V_{\text{ref}}^+ - V_{\text{ref}}^-)/2^N$.

6. The **differential linearity error** is the difference between the output step sizes and the average step size. It is usually expressed in units of 1 LSB. If the differential linearity error is large enough, a nonmonotonic response can result (i.e. an increase in n can result in a decrease in V_n).

7. A **glitch** is a transient spike in the output of a D/A that occurs when more than one bit changes in the input code and the corresponding internal switches do not change simultaneously. For a short time, the switches contain an erroneous input number. The worst glitch usually occurs at the half-scale transition, when the bits change from $0111\ldots1111$ to $1000\ldots0000$ (Figure 3.2). The severity of the glitch is given by the product of the duration and magnitude, computed as the area under the time–amplitude curve. Note that a low-pass filter can reduce the magnitude of the glitch, but not the area. A better solution is a **deglitcher**, a sample-and-hold circuit that holds the output constant until the switches settle.

8. The **power-supply sensitivity** is the percentage change in output voltage per 1% change in supply voltage. It is also expressed as the change in the full-scale output (in LSB) for a standard (usually 3%) change in the power-supply voltage. This is an important specification for battery operation.

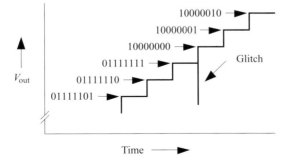

Figure 3.2 D/A output as a function of time for successively larger input values. A voltage spike (glitch) can occur when many bits change state, such as between 01111111 and 10000000.

9. The **settling time** is the time required for the output to settle to typically $1/4$ LSB after a large change in D/A output. It is expressed in units of nanoseconds or microseconds.

10. The **slewing rate** is the maximum rate of change in output voltage, usually imposed by the maximum driving current in the D/A output amplifier and the capacitive load. It is expressed in units of volts per microsecond.

11. The **rms linearity error** is the rms, or "root mean square" of the residuals $R_n = V(n) - V_n$, where V_n is the measured D/A output corresponding to input number n, and $V^{\text{rel}}(n)$ is the linear model given in Eq. (3.2):

$$V_{\text{rms}} = \sqrt{\frac{1}{2^N} \sum_{n=0}^{2^N - 1} R_n{}^2}, \qquad \text{where } R_n = V^{\text{rel}}(n) - V_n$$

12. The **temperature stability** is the insensitivity of the above characteristics (especially zero-offset error, gain error, and linearity error) to changes in temperature. It is expressed in units of percent per degrees celsius or LSB/°C. Good temperature stability is achieved by using voltage references, resistors, and amplifiers that have good temperature stability.

Figure 3.3 shows various combinations of absolute accuracy, relative accuracy, and differential linearity. Figure 3.3(a) is a case of excellent relative accuracy and differential linearity because the output values lie on a straight line and the step sizes are equal. But because the straight line does not lie on the ideal line, the absolute accuracy is poor. In Figure 3.3(b) the step size in the lower half is slightly larger than the step size in the upper half so the differential linearity is good, but the accumulated error results in a poor absolute and relative accuracies. In Figure 3.3(c) the differential linearity is poor, but since the large and small step sizes alternate, the output values lie close to the ideal line and the absolute and relative accuracies are good.

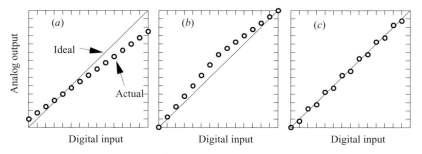

Figure 3.3 (a) Excellent relative accuracy and differential linearity but poor absolute accuracy. (b) Good differential linearity but poor absolute and relative accuracies. (c) Good absolute and relative accuracies but poor differential linearity.

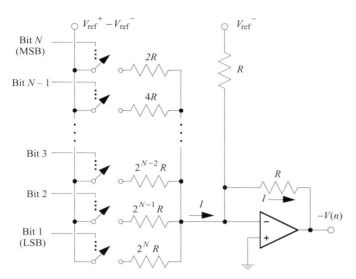

Figure 3.4 D/A converter circuit using a weighted adder. Currents through the binary sequence of resistors are summed at the input of the op amp.

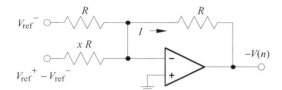

Figure 3.5 Op-amp circuit for converting the current through input resistor xR into an output voltage V_0.

3.2.2   Weighted-adder D/A converter

One of the simplest D/A designs uses a weighted adder, where current is individually switched through a set of parallel resistors and summed at the input of an op-amp circuit (Figure 3.4). The switches are usually field-effect transistors (FETs) that can be actuated in typically a few nanoseconds. Each closed switch represents a binary input one, and each open switch represents a binary input zero. The pattern of closed switches is equal to the pattern of ones in the number n to be converted:

$$n = \sum_{i=1}^{N} B_i 2^{i-1} \qquad B_i = 0 \text{ or } 1$$

We now review how the current is summed and converted into a voltage. Figure 3.5 shows a simplified schematic of Figure 3.4, where the resistance determined by the binary switches B_i has been combined into a single resistance xR. The op-amp circuit has two important properties, as discussed in Chapter 2: (1) no current enters either input terminal, so that the input current I passes through the feedback resistor R;

(2) negative feedback keeps the negative input terminal at the same potential as the positive input terminal, which in this case is zero:

$$I = \frac{V_{\text{ref}}^+ - V_{\text{ref}}^-}{xR} + \frac{V_{\text{ref}}^-}{R} = \frac{-V_0}{R} \qquad \frac{1}{xR} = \sum_{i=1}^{N} \frac{B_i}{2^{N+1-i}R}$$

As an example, consider the case where only the MSB is on and all other switches are off ($B_N = 1$, $B_i = 0$ for $i \neq N$). This gives $xR = 2R$ and the current I and the output V_0 are given by:

$$I = (V_{\text{ref}}^+ - V_{\text{ref}}^-)/(2R) + V_{\text{ref}}^-/R$$

$$-V_0 = IR = V_{\text{ref}}^- + (V_{\text{ref}}^+ - V_{\text{ref}}^-)/2$$

which is mid-range between the two reference voltages, as desired.

As another example, consider the case where only the LSB is on and all other switches are off ($B_1 = 1$, $B_i = 0$ for $i \neq 1$). This gives $xR = 2^N R$ and the current I and the output V_0 are then given by:

$$I = (V_{\text{ref}}^+ - V_{\text{ref}}^-)/(2^N R) + V_{\text{ref}}^-/R$$

$$-V_0 = IR = V_{\text{ref}}^- + (V_{\text{ref}}^+ - V_{\text{ref}}^-)/2^N$$

which is one D/A output step above V_{ref}^-, as desired.

In the most general case, we have an arbitrary pattern of open and closed switches and the output voltage is given by:

$$-V_0 = IR = V_{\text{ref}}^- + (V_{\text{ref}}^+ - V_{\text{ref}}^-)/x = V_{\text{ref}}^- + (V_{\text{ref}}^+ - V_{\text{ref}}^-) \sum_{i=1}^{N} \frac{B_i}{2^{N+1-i}}$$

$$V(n) = -V_0 = V_{\text{ref}}^- + \frac{(V_{\text{ref}}^+ - V_{\text{ref}}^-)}{2^N} \sum_{i=1}^{N} B_i 2^{i-1}$$

This gives the desired result.

Solving for V_0, we have:

$$\boxed{V(n) = V_{\text{ref}}^- + n\Delta V, \qquad \text{where } \Delta V = \frac{V_{\text{ref}}^+ - V_{\text{ref}}^-}{2^N}}$$

3.2.3   R-2R resistive-ladder D/A converter

One requirement of the weighted-adder D/A design shown before is a set of resistors that are both accurate and span a large range. While this does not present a problem when using discrete components, it is not possible to fabricate such a wide range of resistor

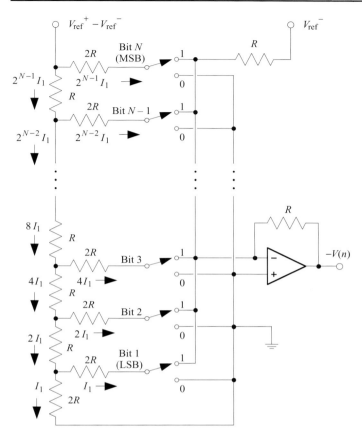

Figure 3.6 Digital-to-analog converter using the R–$2R$ resistor ladder. Note that at each node, the current is split into two equal parts to provide a binary series. The bit switches determine which currents are summed at the virtual ground of the amplifier.

values accurately in the same integrated-circuit chip. For this reason, the most common integrated-circuit D/A design uses the **R–$2R$ resistor ladder**, which establishes a binary sequence of currents that can be selectively summed to produce the analog output (see Figure 3.6).

Note that the same currents flow through the resistors to ground no matter how the bit switches are set. At each node along the left-hand side, the current is split into two equal parts so that the bit 1 switch controls the current I_1 and the bit i switch controls the current $2^{i-1} I_1$. Each switch sends its current either to ground or to the negative input terminal that is maintained at ground potential by the amplifier. The current passing through the feedback resistor is given by:

$$I = n\Delta I + V_{\text{ref}}^-/R \qquad \Delta I = \frac{V_{\text{ref}}^+ - V_{\text{ref}}^-}{2^N R} \qquad n = \sum_{i=1}^{N} B_i 2^{i-1}$$

The output voltage is therefore:

$$-V_0 = IR = V_{\text{ref}}^- + n\Delta IR = V_{\text{ref}}^- + n\Delta V$$

$$\Delta V = \frac{V_{\text{ref}}^+ - V_{\text{ref}}^-}{2^N}$$

The output voltage V_0 is therefore proportional to the digital value of the switch pattern.

For nonzero digital inputs, at least one bit switch is set to 1, and the negative feedback causes the potential of the negative op-amp terminal to be very close to V_{ref}^-. However, for zero digital input, all bit switches are set to zero, and the negative terminal of the op amp is disconnected. A large-value resistor connected between the negative op-amp terminal and V_{ref}^- defines the potential of this terminal in the special case of zero input without significantly affecting the accuracy for nonzero input.

3.2.4   Subranging D/A converter

In some applications, such as digital control of an analog process, it is only necessary that the D/A converter have many small steps and that the analog output be a monotonic function of the digital input. This can be achieved by using the **subranging D/A converter**, which uses a pair of accurate D/A converters to provide the reference voltages for a third. An example is shown in Figure 3.7. Two D/A converters are used for the N most significant bits and the third is used for the M least significant bits. The absolute accuracy error ε of the two N-bit converters limits the precision, absolute accuracy, and differential linearity of the combined converter.

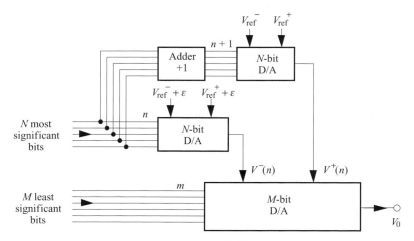

Figure 3.7 Subranging D/A converter with 2^{N+M} output steps. Output is monotonic, provided that the accuracy of the two N-bit D/A converters and the value of ε are such that $V^+(n) < V^-(n+1)$ for all n.

Table 3.1 *Some available monolithic D/A converters*

Model	AD7545	AD668K	CX20202A-1	CXA1236
Manufacturer	Analog Dev.	Analog Dev.	Sony	Sony
Number of bits	12	12	10	8
Relative accuracy	±0.5 LSB	±0.25 LSB	N/A	±0.5 LSB
Differential linearity	±1 LSB	±0.5 LSB	±0.5 LSB	±0.5 LSB
Maximum update rate	>500 kHz	15 MHz	160 MHz	500 MHz

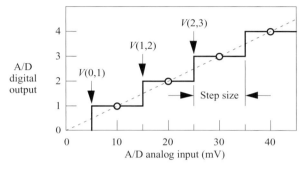

Figure 3.8 Ideal response of an A/D converter to a varying analog input for $V_{ref}^- = 0$ V. At the transition voltage $V(n, n + 1)$ the output toggles between n and $n + 1$.

Table 3.1 describes some commercially available monolithic (single integrated-circuit) D/A converters.

3.3 Analog-to-digital converter circuits

The analog-to-digital (A/D) converter changes a voltage level (the analog input) into a binary number (the digital output). Usually, the relationship between the input voltage and the output number is **linear** (Figure 3.8). The A/D converter divides the input voltage range into $2^N - 1$ bands, where N is the number of bits in the output word. The monopolar A/D converter (e.g. input range 0–5 V) converts an input voltage in the lowest band to an output of zero.

The bipolar A/D converter (e.g. −5 to +5 V) can produce an output which is offset binary or 2's complement. Some A/D converter circuits require that the input voltage remain steady during the conversion process, which can take many microseconds. To provide a steady level, the sample-and-hold amplifier (described in Chapter 2) is used. The A/D converter almost always requires a start conversion pulse, and often provides a busy status level that can be used to determine when conversion is complete and the data may be read. Internal latches and tri-state output are often provided for interfacing to a data bus.

3.3.1   A/D converter characteristics

For an **ideal** linear N-bit A/D converter, the digital output n is a linear function of the analog input voltage V between two reference voltages V_{ref}^- and V_{ref}^+. The range of analog inputs from V_{ref}^- to V_{ref}^+ is divided into $2^N - 1$ equal bands of width ΔV:

$$n = \left(\frac{V - V_{\text{ref}}^-}{\Delta V} + \frac{1}{2} \right)_{\text{INTEGER}}, \qquad \text{where } \Delta V = \frac{V_{\text{ref}}^+ - V_{\text{ref}}^-}{2^N - 1} \qquad (3.3)$$

The ideal response curve passes through the center of the steps (the dashed line in Figure 3.8). Because the same output is produced for a range of analog inputs, the response curve is best measured at the **transition voltages**, where the output changes by one bit. The center of any step can then be computed as the midpoint between the nearest transition voltages. In the ideal case, the first transition voltage $V(0, 1)$ occurs at $V_{\text{ref}}^- + 0.5\Delta V$, and the second $V(1, 2)$ occurs at $V_{\text{ref}}^- + 1.5\Delta V$. The nth transition voltage occurs at $V(n - 1, n) = V_{\text{ref}}^- + (n - 0.5)\Delta V$. The last transition voltage is $V(2^N - 2, 2^N - 1) = V_{\text{ref}}^- + (2^N - 1.5)\Delta V$. While a given value of n corresponds to a range of V that has width ΔV, the transition between $n - 1$ and n corresponds to a definite value of V. The ideal value of ΔV is given by:

$$\Delta V = \frac{V(2^N - 2, 2^N - 1) - V(0, 1)}{2^N - 2} = \frac{V_{\text{ref}}^+ - V_{\text{ref}}^-}{2^N - 1}$$

Linearity is measured by how well the transition voltages (or the midpoint of the steps) lie on a straight line. Differential linearity is measured by the equality of the step sizes. The examples of absolute accuracy, relative accuracy, and differential linearity shown in Figure 3.3 for the D/A converter apply also to the A/D converter.

As explored in Laboratory Exercises 7 and 9, the main characteristics of the A/D converter are:

1. The **resolution**, or **quantizing error**, of an N-bit A/D converter is the largest difference between any input voltage within the full range and the voltage corresponding to the output number (the midpoint of the step). In the case of the ideal A/D converter, this is one-half the step size, or 0.5 LSB. Because of the close association between resolution and the number of bits, the description "N-bit resolution" is frequently used.

2. The **zero-offset error** is the difference between the first measured transition voltage $V_{0,1}$ and $V_{\text{ref}}^- + 0.5\Delta V$.

3. The **gain error** is the difference between the measured slope:

$$(V_{2^N - 2, 2^N - 1} - V_{0,1})/(2^N - 2)$$

and the ideal slope:

$$(V_{\text{ref}}^{+} - V_{\text{ref}}^{-})/(2^{N} - 1)$$

4. The **absolute-accuracy error** is the difference between the measured input transition voltages $V_{n,n+1}$ and their ideal values $V(n, n+1)$, before adjustments for zero offset and gain. The ideal transition voltages are described by:

$$V(n, n+1) = V_{\text{ref}}^{-} + \left(n + \frac{1}{2}\right)\left(\frac{V_{\text{ref}}^{+} - V_{\text{ref}}^{-}}{2^{N} - 1}\right)$$

$$= V_{\text{ref}}^{-} + \left(n + \frac{1}{2}\right)\Delta V$$

This error includes zero-offset error, gain error, and linearity error.

5. The **relative-accuracy error** is the difference between the measured transition voltages $V_{n,n+1}$ and a straight line passing from the first to the last transition voltage. This line is described by:

$$V^{\text{lin}}(n, n+1) = V_{0,1} + n\left(\frac{V_{2^{N}-2, 2^{N}-1} - V_{0,1}}{2^{N} - 2}\right)$$

where $V_{0,1}$ and $V_{2^{N}-2, 2^{N}-1}$ are the measured end-point transition voltages.

Since this error includes the linearity error and not the zero-offset or gain errors, it is also called the **linearity error**. Note that calibration of the end points (by adjusting zero offset and gain) does not affect the relative-accuracy error. It is usually expressed in units of 1 LSB for comparison with the quantizing error.

6. The **differential-linearity error** is the difference between the spacing of neighboring transition voltages and their average spacing. It is usually expressed in units of 1 LSB. If the differential linearity error is large enough, **missed codes** (output numbers that cannot be produced by any input voltage) can result. Differential linearity error is usually the result of inaccurate resistor values.

7. The **conversion time** is the time required to produce the output number after the "start conversion" command has been given.

8. The **conversion rate (maximum)** is the largest rate that the A/D converter can perform conversions. For simple A/D converters, the maximum conversion rate is the inverse of the conversion time. More advanced converters can begin the next conversion before the previous one has completed, and their maximum conversion rate can be higher than the inverse of the conversion time.

9. The **power-supply sensitivity** is the percentage change in the transition voltages per 1% change in supply voltage. It is also expressed as the change in the last transition voltage (in LSB) for a standard (usually 3%) change in the power-supply voltage. This is an important specification for battery operation.

10. The **temperature stability** is the insensitivity of the above characteristics (especially zero-offset error, gain error, and linearity error) to changes in temperature.

It is expressed in units of percent per degrees celsius or least significant bit per degrees celsius. Good temperature stability is achieved by using voltage references, resistors, and amplifiers that have good temperature stability.

3.3.2   Relationship between A/D and D/A conversion

As explored in Laboratory Exercises 7, 8, 9, and 20, the ideal A/D converter produces an output n for input voltages in the range from $V(n - 1, n) = V_{\text{min}} + (n - 0.5)\Delta V$ to $V(n, n + 1) = V_{\text{min}} + (n + 0.5)\Delta V$. The midpoint of this range is $V_{\text{mid}} = V_{\text{min}} + n\Delta V$, which is the output voltage produced by the ideal D/A converter with input n. We see that the conversions in Eqs. (3.1) and (3.3) are defined so that if a set of analog signals are converted and stored digitally, the analog waveform can be recovered without systematic error. The analog output can differ from the corresponding analog input values by as much as $0.5\Delta V$, but the average error will be zero.

3.3.3   The comparator

The **comparator** is a high-gain differential amplifier whose output is limited between two logic levels (Figure 3.9). It is similar to the op amp without negative feedback, but it is not frequency compensated and has a much higher slewing rate. For inputs V_+ and V_-, the logic output L is high whenever $V_+ > V_-$ and L is low whenever $V_+ < V_-$. The output is clamped at logic voltages V_{low} and V_{high} by the diodes, and R_3 is a current-limiting resistor.

The output is poorly defined when V_+ and V_- are nearly equal, and the output may "chatter" between high and low states at the crossing point. A small amount of **positive feedback**, or **hysteresis** (provided by resistor R_2), is used to stabilize the comparator. The first output transition feeds back to reinforce the same logic state and can overcome moderate amounts of input noise that would otherwise reverse the state. Without positive feedback ($R_2 = \infty$), the op-amp output is low (V_{0L}) when $V_1 < V_2$ and high (V_{0H}) when $V_1 > V_2$. With positive feedback, the hysteresis current into R_1 is $(V_0 - V_+)/R_2$, which changes the effective value of V_1 by an amount $(R_1/R_2)(V_0 - V_2)$. To see this more clearly, consider two cases.

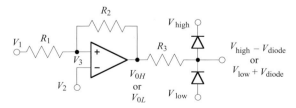

Figure 3.9 Voltage comparator with hysteresis, whose output is a logic V_{0H} when $V_1 > V_{1H}$ and V_{0L} when $V_1 < V_{1L}$. Between V_{1L} and V_{1H}, the output does not change.

Case I: $V_1 > V_2$ and $V_0 = V_{0H}$

$$\frac{V_3 - V_1}{R_1} = \frac{V_{0H} - V_3}{R_2} \qquad V_3 R_2 - V_1 R_2 = V_{0H} R_1 - V_3 R_1$$

$$V_3 = \frac{V_1 R_2 + V_{0H} R_1}{R_1 + R_2} = V_1 + \frac{(V_{0H} - V_1)R_1}{R_1 + R_2} > V_1$$

So $V_3 > V_1 > V_2$. If V_3 drops below V_2, then the output will flip to V_{0L}. How low does V_1 have to go for this to happen? To find the critical value of $V_1 = V_{1L}$, set $V_2 = V_3$ in the above equations and solve:

$$V_{1L} R_2 = V_2 R_2 - V_{0H} R_1 + V_2 R_1$$

$$V_{1L} = V_2 - (R_1/R_2)(V_{0H} - V_2) = V_2 - \Delta V_L$$

Case II: $V_1 < V_2$ and $V_0 = V_{0L}$

$$V_3 = \frac{V_1 R_2 + V_{0L} R_1}{R_1 + R_2} = V_1 - \frac{(V_1 - V_{0L})R_1}{R_1 + R_2} < V_1$$

The positive feedback makes V_3 higher than V_1. If V_1 drops so that $V_3 < V_2$, then the output will flip to V_{0L}. Setting $V_3 = V_2$ and solving for the threshold V_{1H}, we have:

$$V_{1H} = V_2 + (R_1/R_2)(V_2 - V_{0L}) = V_2 + \Delta V_H$$

For V_1 values between V_{1L} and V_{1H}, L does not change (Figures 3.10 and 3.11). As a result, the output state only changes when V_1 passes V_{1L} from above or V_{1H} from below, and does not change again until V_1 reverses by at least $|V_{1H} - V_{1L}|$. For the typical comparator, R_2 is much larger than R_1 and the hysteresis is small. The Schmitt trigger has a large amount of hysteresis, which is accomplished by making R_1 and R_2 approximately equal.

3.3.4 Tracking A/D converter

The **tracking A/D converter** repeatedly compares its input with the output of a D/A converter (Figure 3.12). If the analog input is larger, the D/A input is incremented. If the analog input is smaller, the D/A input is decremented.

Figure 3.10 Rule for determining comparator output L. When V_1 exceeds V_{1H}, L takes its high state value V_{high}. When V_1 is below V_{1L}, L takes its low state value V_{low}. When V_1 is between V_{1L} and V_{1H}, L retains its previous value.

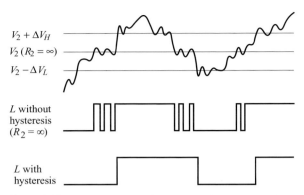

Figure 3.11 Effect of hysteresis on the logic output L when the input waveform is noisy. Without hysteresis, L toggles whenever the input crosses V_2. With hysteresis, the output L toggles only for larger changes in input.

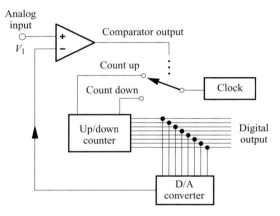

Figure 3.12 Tracking A/D converter. The voltage to be converted is compared with the output of a D/A converter that is connected to an up/down counter. If the voltage is larger, the counter is increased by one, otherwise the counter is decreased.

The conversion time is slow relative to the clock frequency. This may be seen by considering the Ferranti ZN433, a low cost 10-bit tracking A/D converter, with a clock rate of 1 MHz (1 count/μs). A minimum time of 1 ms is thus required to cover the full scale of 1,024 counts, which corresponds to the maximum slope of a sine wave with a period of 2π ms. Thus, the maximum frequency that can be reliably tracked is only 180 Hz.

Advantages:
- Simplicity.
- Low cost.

Disadvantages:
- Slow – requires 2^N cycles to slew full range.
- Accuracy, linearity, and differential linearity are limited by the accuracy of the D/A converter.

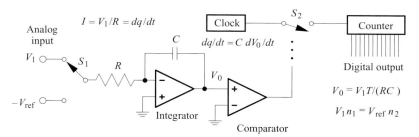

Figure 3.13 Simplified diagram of an integrating dual-slope A/D converter. Switch S_1 connects the input V_1 to the integrator for a set number of clock cycles, charging capacitor C to a voltage proportional to the time average of V_1. Then S_1 connects $-V_{\text{ref}}$ to the integrator to discharge C at a constant rate while S_2 closes to accumulate clock pulses in the counter. The number of clock pulses when the comparator input reaches zero is the digitized form of the integral of V_1.

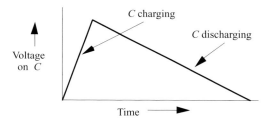

Figure 3.14 Voltage on capacitor C of the integrating dual-slope A/D converter during charging and discharging phases.

3.3.5   Integrating A/D converter

The **integrating, or dual-slope, A/D converter** accumulates the input on a capacitor for a fixed time and then measures the time it takes to discharge the same capacitor at a fixed discharge rate (Figures 3.13 and 3.14). A clock is used to measure the discharge time, and the number of clock pulses is the digital output. It is relatively slow, but has extreme accuracy and linearity. This method is most commonly used in pulse-height analyzers, which accumulate a distribution of pulse heights, such as in nuclear spectroscopy, because its high differential linearity avoids spectral distortion.

The steps are:

1. Connect switch S_1 to the analog input V_1 to integrate the input on the capacitor for n_1 clock ticks (fixed time T).
2. Connect switch S_1 to $-V_{\text{ref}}$ to discharge the capacitor at a fixed rate. Connect switch S_2 so that the counter determines the number of clock ticks n_2 needed to discharge C.
3. n_2/n_1 is the digital value of the V_1/V_{ref} integral.

The result is quite accurate, and the accuracy does not depend on knowing the exact clock rate or the exact value of the capacitor. In addition, high input frequencies are averaged to zero during the integration period, especially those that are an integral multiple of $1/T$.

Advantages:
- Provides a very accurate result that does not depend on the exact values of the capacitor or the clock rate.
- Integrates the input during step 1, which reduces noise and aliasing (complete cycles of higher frequencies average to zero).
- Gives excellent differential linearity, limited by op-amp linearity and uniformity of time between clock pulses.

Disadvantage:
- Slow – requires 2^N cycles to convert.

3.3.6 Successive-approximation A/D converter

The **successive-approximation A/D converter** uses a binary search to determine the bits of the output number sequentially.

The method is analogous to weighing an object using a balance and a binary sequence of known weights (e.g. 1, 2, 4, and 8 g) (Figure 3.15). In the first cycle, the analog input is compared with the 8-g weight. If the object is heavier, the weight is left and the 4-g weight is added. If the object is lighter, the 8-g weight is removed and the 4-g weight is added. The process continues, testing each weight in the descending sequence, leaving

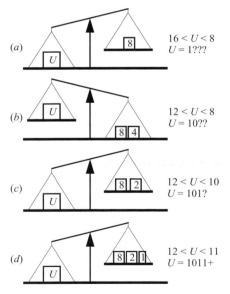

Figure 3.15 Example of 4-bit successive approximation using a binary series of weights and a balance. In (a), it is found that U is greater than 8, which determines that the fourth bit from the right is 1. In (b), it is found that U is less than 12, and the third bit is a 0. In (c), U is greater than 10, so the second bit is 1. In (d), U is greater than 11, so that bit 1 is 1.

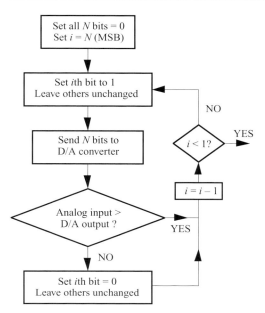

Figure 3.16 Flow chart for the method of successive approximation.

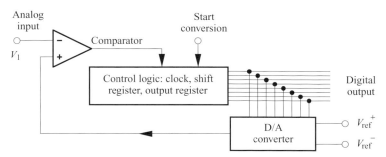

Figure 3.17 Block diagram for the method of successive approximation. The control logic is a hardware implementation of the flow diagram in Figure 3.16.

it in place if the object is heavier than the sum, and removing it if the object is lighter. The balance is analogous to the comparator, whose output is a logic 0 or 1, depending on the relative amplitudes of its two analog inputs (Figure 3.9). The binary set of weights is analogous to the internal D/A converter, whose analog output is proportional to the weighted sum of the binary input bits (Figure 3.15).

A flow chart for the method of successive approximation is shown in Figure 3.16 and a block diagram is shown in Figure 3.17.

Advantages:
- Low cost.
- Speed – conversion requires only 1 cycle per bit.
- Can handle a relatively large number of bits.

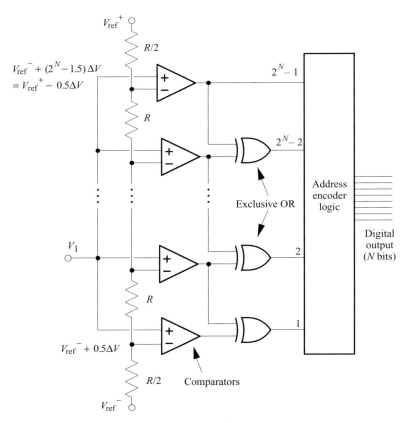

Figure 3.18 The N-bit flash converter uses $2^N - 1$ parallel comparators and exclusive-OR gates for high-speed conversion.

Disadvantages:
- Accuracy, linearity, and differential linearity are limited by the accuracy of the D/A converter.
- Requires a sample-and-hold amplifier to provide constant input during conversion.

3.3.7 Flash A/D converter

The **flash A/D converter** uses $2^N - 1$ comparators to determine all N bits of the digital output simultaneously (Figure 3.18). Since the comparators are constantly sampling the input, a sample-and-hold amplifier is generally not needed. A set of exclusive-OR gates (discussed in Chapter 1) determines the location of the comparator whose reference input most closely matches the analog input. Conversion is accomplished by latching the logic state of the exclusive-OR gates. The address of that comparator is the N-bit binary representation of the input voltage. The number of comparators needed is $2^N - 1$, which grows quite rapidly with the number of bits. Common units are 4 bit (15 comparators), 6 bit (63 comparators), and 8 bit (255 comparators).

Advantages:
- Very high speed: conversion is essentially continuous.
- Usually does not require a sample-and-hold amplifier.

Disadvantages:
- Expensive: $2^N - 1$ comparators required.
- Limited to 10 bits at present.
- Accuracy, linearity, and differential linearity are limited by the resistor chain accuracy.

3.3.8   Subranging flash A/D converter

The flash converter has the advantage of high speed, but it is costly to use this technique for a large number of bits. A practical solution to this problem is the **subranging flash A/D converter,** which is a hybrid between the successive approximation and the flash converters. One flash A/D converter determines the most significant bits, which are sent to a D/A converter. A differential amplifier computes the difference, which is used by a second flash converter to determine the least significant bits.

By way of example, let us examine the AD7820 8-bit "half-flash" converter, which is shown in Figure 3.19.

The sequence of operation follows (RD mode):
1. $\overline{\text{WR}}$/RDY is normally low and the comparators sample the input.
2. Conversion is initiated by taking $\overline{\text{RD}}$ low.
3. After 1.6–2.5 μs, $\overline{\text{WR}}$/RDY goes high and the four most significant bits are latched.

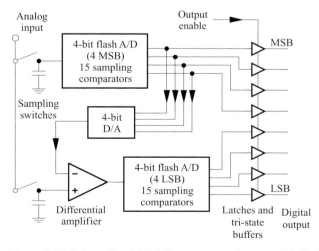

Figure 3.19 Subranging 8-bit A/D converter using two 4-bit flash A/D converters and a 4-bit D/A converter.

4. Some 20–50 ns later, the four least significant bits are latched and all eight output bits are valid.
5. After $\overline{\text{RD}}$ goes high, data are held for an additional 60–80 ns.
6. After 500–600 ns, conversion may be repeated by taking $\overline{\text{RD}}$ low.

Another example is the 10-bit 20-MHz Sony CX20220A-1 A/D converter. The A/D and D/A for the five most significant bits use the same resistor ladder to reduce error.

Advantages:
- High speed – conversion takes only two cycles.
- Reduced cost per 2^N steps – requires only two $N/2$-bit flash converters.

Disadvantages:
- Limited to 10–12 bits at present.
- Accuracy, linearity, and differential linearity are limited by the resistor chain accuracy.
- Requires a sample-and-hold amplifier to provide a constant input during the two-cycle conversion process.

3.3.9   1-bit oversampling sigma–delta A/D converter

This A/D converter consists of an analog filter and a comparator in a negative feedback loop (Figure 3.20). It operates by sampling the input waveform at a rate much higher than the highest frequency in the waveform (the oversampling ratio) and exchanges resolution in time for resolution in amplitude. The feedback loop serves to keep the difference between V_1 and V_5 small (i.e. a virtual short) at the lower frequencies of interest, and in the process increases the difference at higher frequencies. This is called noise shaping. The high-frequency digitization noise at V_4 (Figure 3.21) is eliminated from the output by the final low-pass digital filter. For a first-order converter such as that shown in Figure 3.20, each doubling of the sampling frequency can result in a gain in accuracy of 1.5 binary bits. So an oversampling ratio of 1,000 could provide

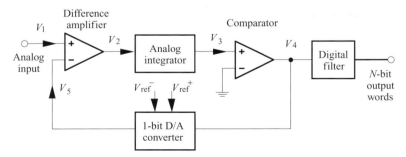

Figure 3.20 1-bit sigma–delta A/D converter schematic. Second-order converters use an additional difference amplifier and analog integrator at V_3.

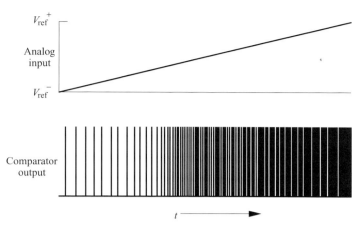

Figure 3.21 1-bit sigma–delta A/D converter operation.

15 bits of resolution. A more common second-order design uses two stages of difference amplification and analog integration to shape the noise further and can achieve 2.5 bits of accuracy for each factor of two of oversampling. The ADC16071 fifth-order converter provides 16 bits of resolution with an oversampling ratio of only 64. The output data rate is 192 kHz and the sampling rate is 12.3 MHz.

Advantage:
- Can achieve up to 24-bit accuracy without using precisely matched components.

Disadvantages:
- Requires sampling at typically 600 times the analog bandwidth (i.e. 12 MHz for 20 kHz).
- Requires many multistage digital circuits, each with many empirically determined coefficients.

3.3.10   Commercially available A/D converters

Table 3.2 lists some selected commercially available monolithic (single integrated-circuit) A/D converters.

3.4   The sample-and-hold amplifier

The **sample-and-hold amplifier** has an analog input, an analog output, and a digital control input. It consists basically of an amplifier, a switch, and a capacitor (Figure 3.22). The control input determines whether the device is operating in sample mode or hold mode. In the sample mode, the switch is closed and it operates as a typical amplifier or op amp. In the hold mode, the switch is open and the output is ideally constant,

Table 3.2 *Some available monolithic A/D converters*

Model	CX20220A-1	AD574	AD367	CXA1176
Manufacturer	Sony	Analog dev.	Analog dev.	Sony
Number of bits	10	12	16	8
Integral linearity	±1 LSB	±0.5 LSB	±2 LSB	±0.5 LSB
Differential linearity	±1 LSB	±1 LSB	±4 LSB	±0.5 LSB
Conversion method	Subranging flash	Successive approx.	Successive approx.	Flash
Input voltage range (V)	0 to −2	−10 to +10	0 to +20	0 to −2
Conversion time	50 ns	35 μs	17 μs	3.3 ns
Aperture jitter	*	*	*	3.6 ps
Max. conversion rate	20 MHz	29 kHz	59 kHz	300 MHz

* Requires an external sample-and-hold amplifier.

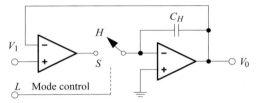

Figure 3.22 Unity-gain sample-and-hold amplifier. In sample mode S, $V_0 = V_1$. In hold mode H, V_0 is determined by the holding capacitor C_H.

independent of the input. This constant value is the output value present when the mode was last switched from sample to hold.

See Figure 3.23 for a schematic of the analog and digital delays that occur during the sample-to-hold transition, and Figure 3.24 for the response to an arbitrary waveform as the logic control is exercised. The sample-and-hold amplifier is used to provide a steady input during analog-to-digital conversion. We now discuss the properties of this device in the sample mode, the hold mode, and when making the transition between modes.

3.4.1   Sample mode

In the sample mode of operation, the sample-and-hold amplifier has properties similar to those or any amplifier or op amp:

1. The **output offset voltage** is the output voltage V_0 for zero input voltage V_1.
2. The **linearity error** is the deviation between the voltage gain $G = V_0/V_1$ and its average value over a specified range of V_1.
3. The **settling time** is the time required for the output to reach its final value (within a specified accuracy) after a full-scale analog input step.
4. The **bandwidth** is the highest frequency at which G has dropped less than 3 dB from its low-frequency average.

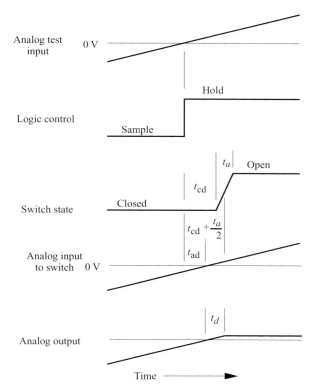

Figure 3.23 Analog and digital delays during the transition from sample mode to hold mode. The control delay t_{cd} is the time from the sample-to-hold logic control edge to when the switch begins to open. The aperture time t_a is the time required for the switch to open and the analog delay t_{ad} is the delay between the analog input and the output in sample mode. The aperture delay t_d is the time between the edge of the sample-to-hold edge and the time when the input was equal to the held value. In this example the analog test input is a ramp that crosses zero at the sample-to-hold edge. Only if the difference t_d between the analog delay t_{ad} and the digital delay $t_{cd} + t_a/2$ is zero will the held value at the output equal 0 V, the value of the input at the sample-to-hold logic control edge.

5. The **slewing rate** is the maximum rate of change dV_0/dt after an analog input step or after a hold-to-sample transition.

3.4.2   Hold mode

In hold mode, the output should remain constant and independent of the input:
1. **Droop** is the drift in output voltage V_0/dt due to charge leakage from the hold capacitor through the switch, the amplifier, or the capacitor itself. If the leakage current is I, $V_0/dt = I/C_H$.
2. **Feedthrough** is the fraction of input signal that appears at the output, caused primarily by the capacitance of the open switch.

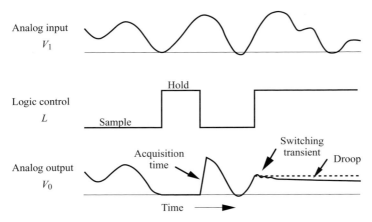

Figure 3.24 Output response of the sample-and-hold amplifier to an arbitrary waveform when the logic control is exercised.

3.4.3   Sample-to-hold transition

When sampling an analog waveform, the sample-to-hold transition is used to freeze the instantaneous value of the input waveform at the output so that the value can be converted into a digital form. Ideally, when the sample to hold edge is received, the output should hold at its current value without delay or amplitude shift. The types of deviations between this ideal behavior and the actual behavior are described below (and illustrated in Figure 3.23):

1. The **control delay** t_{cd} is the time between the edge of the sample-to-hold command and the time when the switch begins to open.
2. The **aperture time** t_a is the time required for the switch to open and characterizes only the response time of the switch. The value that is held at the output is determined by the average input value during the aperture time.
3. The **analog delay** t_{ad} is the delay between the analog input and the analog output in sample mode.
4. The **aperture delay** t_d is the time between the edge of the sample-to-hold command and the time when the input was equal to the held value. Because of the analog delay, this quantity can be negative:

$$t_d = t_{cd} + t_a/2 - t_{ad}$$

5. The **aperture jitter** Δt_d is the rms variation in aperture delay caused by noise in the control signal and switching circuit. It manifests itself as a random variation ΔV_0 in the held value and depends on the slope dV_0/dt:

$$\Delta V_0 = \Delta t_d (dV_0/dt)$$

6. The **charge transfer error** is the offset error in the held value caused by charge dumped onto the holding capacitor from the switching circuit. Ideally, this should be a fixed offset, independent of input voltage level. This error can be reduced by increasing the value of the holding capacitor C_H.

7. The **sample-to-hold offset** is the shift in output level during the transition from the sample mode to the hold mode after the charge transfer has been accounted for. It may depend on the characteristics of the input signal and is also called the offset nonlinearity.

8. The **switching transient time** is the time required for the output transient caused by charge transfer to settle to its final value within a specified error. In applications using high-accuracy analog-to-digital converters, conversion must be delayed during the switching transient.

3.4.4   Hold-to-sample transition

During the hold-to-sample transition, the output must abruptly change to agree with the input waveform. As a result, transients and offsets caused by charge transfer effects are relatively small and usually ignored.

The **acquisition time** is the time from the hold-to-sample edge to the time when output of the sample-and-hold amplifier has reached its final value (within a specified error). This includes the switch delay time, the slewing time, and the amplifier settling time. The acquisition time is reduced by reducing the value of the holding capacitor C_H.

EXAMPLE 3.1

Gain analysis of sample-and-hold amplifier in sample mode (Figure 3.25). The op-amp equations give:

$$V_2 = A(V_1 - V_0) \qquad V_0 = A(-V_2) = A^2(V_0 - V_1)$$
$$G = V_0/V_1 = A^2/(A^2 - 1) \approx 1$$

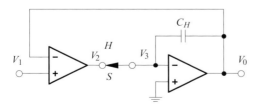

Figure 3.25 Unity-gain sample-and-hold amplifier in sample mode.

Table 3.3 *Some available monolithic sample-and-hold amplifiers*

Model	AD582	AD389	AD683	CXA1008P
Manufacturer	Analog Dev.	Analog Dev.	Analog Dev.	Sony
Linearity (%)	0.01	0.001	*	0.1
Unity gain (MHz) bandwidth	1.5	1.5	10	N/A
Acquisition time	6 µs (0.1%)	2.5 µs (0.003%)	500 ns (0.01%)	20 ns
Max slew rate (V/µs)	3	30	130	100
Settling time (µs)	0.5	1	*	*
Aperture delay (ns)	200	30	2.5	6
Aperture jitter	15 ns	0.4 ns	20 ps	*
Droop current or rate	< 0.1 nA	< 1 µV/µs	0.01 µV/µs	< 20 mV/µs

* Not provided in data sheets.

When this circuit is in hold mode, the first amplifier loses its negative feedback, the inputs become unequal, and the output goes into saturation. On switching back to sample mode, this op amp must get out of saturation and the capacitor C_H must be charged.

Table 3.3 lists some representative monolithic (single integrated-circuit) sample-and-hold amplifiers.

Note 1: Usually the combination of aperture delay and settling time is much shorter than the acquisition time, so in analog-to-digital conversion applications, the device is normally kept in the sample mode and only switched to the hold mode during conversion.

Note 2: A small value of C_H reduces the acquisition time, while a large value of C_H reduces charge-transfer error and droop. Fortunately, for most applications and sample-and-hold amplifiers, there is a range of good choices.

3.4.5   The role of the sample-and-hold amplifier

The successive approximation A/D converter requires a constant input during its conversion time T_{conv}. This is usually provided by the sample-and-hold amplifier. The sample-and-hold amplifier has a time uncertainty T_{jitt} between the logic input and the actual analog acquisition. It is important that the input amplitude not change significantly (say, less than 1/2 LSB) during T_{conv} if not using a sample-and-hold and during T_{jitt} if using a sample-and-hold.

The following analysis holds for either situation, and derives a relationship between an input sampling time T_{samp} (T_{conv} or T_{jitt}, whichever is relevant), the number of bits N, and the maximum input frequency f_{max} that insures that the input does not change by more than 1/2 LSB during T_{samp}.

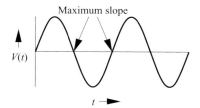

Figure 3.26 Sine wave, showing points of maximum slope.

Suppose we want to sample a sine wave with frequency f (cycles/second) and a peak–peak amplitude of $2V_0$ (Figure 3.26). The waveform is given by:

$$V(t) = V_0 \sin(2\pi f t)$$

Note that the 2π is needed to convert from cycles/second to radians/second. The first derivative gives the rate of change:

$$dV/dt = 2\pi f V_0 \cos(2\pi f t)$$

which has a maximum value of $2\pi f V_0$. If the A/D converter has N bits of resolution and an input sampling time T_{samp}, a conversion accuracy of $1/2$ LSB requires that the input not change by 1 part in $2^{N+1} = V_0 2^{-N}$ during the time T_{samp}:

$$2\pi f V_0 < \frac{V_0}{2^N T_{\mathrm{samp}}} \quad \text{or} \quad f_{\max} = \frac{1}{2^{(N+1)}\pi T_{\mathrm{samp}}}$$

If we do not use a sample-and-hold amplifier for the AD670 8-bit successive-approximation converter ($N = 8$, $T_{\mathrm{samp}} = 10\,\mu\mathrm{s}$), then $f_{\max} = 62$ Hz, which is very low. On the other hand, using the AD582 sample-and-hold amplifier reduces T_{samp} to 15 ns (the aperture jitter) and $f_{\max} = 41$ kHz, which is considerably higher. Note that this happens to be nearly equal to the Nyquist frequency limit of 50 kHz, corresponding to the AD670 maximum sampling rate of 100 kHz (see the following section).

The AD7820 8-bit half-flash converter, on the other hand, uses sampling comparators, and its input conversion time is less than 50 ns (the maximum delay between the latching of the four MSBs and the four LSBs). Even without the sample-and-hold amplifier, $f_{\max} = 12$ kHz. When used with a sample-and-hold having an aperture jitter of 1 ns, $f_{\max}$ is increased to 600 kHz. Since the conversion time is given as $1.4\,\mu\mathrm{s}$, the maximum sampling rate is 714 kHz and the Nyquist limit is 357 kHz.

The CXA1176 8-bit flash converter has a conversion time of 3.3 ns and an aperture jitter of 3.6 ps. The aperture jitter gives $f_{\max} = 170$ MHz, and the conversion time sets the Nyquist frequency limit at 150 MHz.

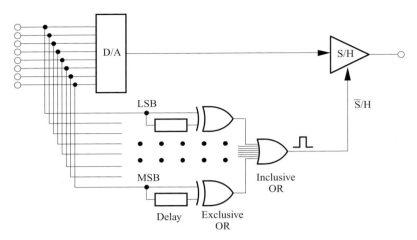

Figure 3.27 Circuit for deglitching the output of a D/A converter.

3.4.6   Using the sample-and-hold amplifier for D/A deglitching

A sample-and-hold amplifier can be used to eliminate glitches from the output of a D/A converter by switching to hold mode during the glitch (Figure 3.27). The delays and exclusive-OR circuits produce a fixed-width pulse whenever the corresponding bit changes. The OR of these pulses is used to put the sample-and-hold converter into hold mode until the switches have settled. The output takes the new analog value when the sample-and-hold is switched back into sample mode. The analog output is slightly delayed, but the glitch is eliminated.

3.5   Sampling analog waveforms

Waveform sampling requires the close coordination of a number of elements, especially when speed and accuracy are required. This section discusses several different sampling strategies and the situations that require a sample-and-hold amplifier.

3.5.1   Software-controlled sampling

Software-controlled sampling typically has the following steps:

1. The conversion command is initiated with a pulse produced by the program or by a digital timer. This switches the sample-and-hold amplifier (if used) from the sample to the hold mode and initiates A/D conversion. As discussed in Chapter 2, the sample-and-hold amplifier has a rapid response when switching from sample to hold, but it may be necessary to delay conversion while the charge transfer from the digital control pulse to the output settles.

2. When conversion is complete, an I/O status bit is set or an interrupt is produced that notifies the microcomputer that new data are available.
3. The microcomputer reads and stores the A/D output.

Software-controlled sampling is typically limited to a maximum rate of about 20 kHz with 12-bit accuracy.

3.5.2   Hardware-controlled DMA sampling

For higher speed operation, each A/D conversion is initiated by the digital timer and the data are transferred directly into memory under hardware control. This is called "direct memory access," or DMA sampling, and sampling speed is limited by A/D conversion speed and memory access, not by software. In addition, this mode permits the computer to perform other tasks simultaneously.

Hardware-controlled sampling is typically limited to a maximum rate of about 200 kHz with 12-bit accuracy.

3.5.3   Data-acquisition subsystem sampling

The highest possible sampling speeds are achieved by using additional hardware (a data-acquisition subsystem) that contains its own timer, high-speed A/D converters, and memory. Some may be plugged into the microcomputer, while others are external. The microcomputer: (i) serves as the human interface, (ii) sets up the subsystem for a particular number of samples and sampling rate, (iii) detects that all samples have been taken, and (iv) reads the data from subsystem memory for analysis and display. The disadvantage of this approach is cost of the subsystem, which may exceed the cost of the microcomputer system itself if high performance is required. At the low-performance end, sampling rates of 200 MHz are possible using integrated circuit A/D converters with 8-bit accuracy. At the high-performance end, sampling rates of many gigahertz are possible and the limitation is frequently the analog bandwidth of the input amplifiers.

Many digital oscilloscopes can be used as data-acquisition subsystems, provided that they permit computer control of the front panel settings and data transfer. Interfacing usually requires a special circuit board, which is plugged into the microcomputer. Maximum sampling rates are in the range of 10–500 MHz.

3.5.4   Pulse height analysis

In some cases information is not carried by a continuous waveform but by discrete pulses whose amplitudes convey information about the events that caused them. A pulse height spectrum is formed by dividing the range of pulse heights into a large number of narrow "bins" and plotting the number of pulses in each bin (Figure 3.28).

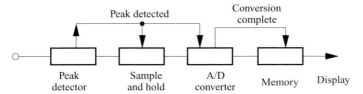

Figure 3.28 Pulse height analyzer circuit that detects a pulse peak, holds the peak amplitude for A/D conversion and adds one to the memory address given by the A/D output.

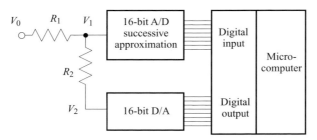

Figure 3.29 Sliding scale linearization used to average over the differential nonlinearity of the successive approximation A/D converter.

Historically, the dual-slope A/D converter was used in pulse height analyzers, because it has excellent differential linearity. As data-acquisition speeds increased over the years, the dual-slope conversion speed became a serious limiting factor. Using the faster successive approximation A/D converter had a serious problem because the poor differential linearity would introduce unwanted features in the pulse height spectrum. This problem is solved by a technique known as sliding scale linearization, where a small variable analog voltage is added to the input and the digital equivalent is then subtracted from the output of the A/D converter. The result is an average over the A/D converter response with very little loss in pulse height accuracy. See Figure 3.29 for a schematic block diagram. Note that $R_2 \gg R_1$ so that each A/D step corresponds to many D/A steps and that a calibration procedure is used to determine the ratio.

To see how the linearization procedure works, consider the following case where it is determined that a change in D/A input by 17,600 steps shifts the analog input so that the A/D output shifts by 200 steps, a ratio of 88. Periodically during data acquisition the D/A input is set to a multiple of 88. For a D/A input of $88n$, the number n is subtracted from the A/D output.

Most of the differential linearity in A/D converters that use internal D/A converters is caused by inaccuracies in the low-order bit resistor values. For example, changing the value of the resistor that controls the MSB shifts the upper half of the transition voltages but affects the size of only one voltage step. Changing the value of the resistor that controls the LSB affects all voltage steps. Therefore it is only necessary to slide the scale over a range of 32 A/D steps to remove most of the differential nonlinearity effects.

Historical note

An important clue to the cause of the extinction of the dinosaurs involved pulse height analysis in a crucial way. It was known that the mass extinction that included the dinosaurs occurred about 65 million years ago, at the boundary between the Cretaceous and the Tertiary periods (called the K–T boundary). Around most of the world, a clay layer exists at this boundary. It was speculated that this marked a period of destructive volcanic activity and the concentration of the element iridium could be used to determine how long the period of activity lasted. Because the Earth's iridium has concentrated in its iron core, this element is quite rare at the surface of the Earth but is slowly deposited by iron meteor dust. If the clay layer was due to volcanic activity lasting several thousands of years, the iridium concentration would be expected to be 0.1 ppb. If the layer were deposited more quickly, there would be less iridium. Walter Alvarez and others at UC Berkeley used neutron activation analysis, where neutrons are used to make any iridium radioactive, and pulse height analysis of the gamma rays produced to measure such small concentrations. To their surprise, they measured the unique gamma ray spectrum of iridium at 3 ppb. The discovery of such iridium concentrations at the K–T boundary all over the world led to the conclusion that the iridium was deposited by a large meteor, whose impact would have been violent enough to wipe out most of the life on Earth. Even at this level, the iridium gamma peaks are small compared to the peaks from the other elements in the sample and the measurement would have been impossible if the peaks were obscured by differential nonlinearity in the A/D converter.

3.6   Frequency aliasing

If a sine wave of frequency f is sampled at a frequency f_s such that its frequency is four cycles per 24 samples ($f/f_s = 1/6$), the samples are an accurate representation of the sampled wave (Figure 3.30). By this we mean that the lowest-frequency sine wave that can pass accurately through the sampled values has the same frequency, amplitude, and phase as the original wave.

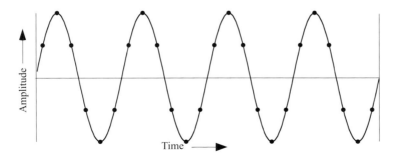

Figure 3.30 When the sine-wave frequency ratio f/f_s is four cycles per 24 samples (0.167), the wave is sampled six times per cycle, and the apparent frequency f_0 is equal to the true frequency f.

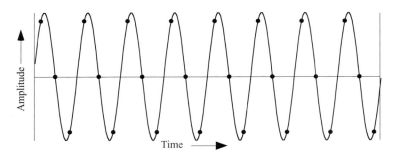

Figure 3.31 When the sine-wave frequency ratio f/f_s is eight cycles per 24 samples (0.333), the wave is sampled three times per cycle, and the apparent frequency f_0 is equal to the true frequency f.

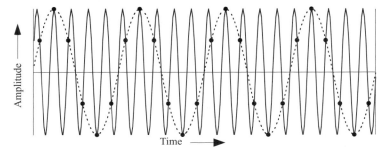

Figure 3.32 When the sine-wave frequency ratio f/f_s is 20 cycles per 24 samples (0.833), the wave is undersampled, and the apparent wave (dotted line) has a frequency ratio ($f_0/f_s = 0.167$) one-fifth the true frequency.

If we double the frequency f so that there are eight cycles per 24 samples ($f/f_s = 1/3$), the samples still represent the sampled wave (Figure 3.31). If we increase the frequency still further to 20 cycles per 24 samples ($f/f_s = 5/6$), the samples appear to represent a sine wave with an apparent frequency f_0 that is five times lower ($f_0/f_s = 1/6$) (Figure 3.32).

This misidentification of frequency is called "aliasing" and occurs whenever $f/f_s > 0.5$. As f/f_s approaches unity, the apparent sine wave has a frequency that approaches 0 Hz. Moreover, sampling a sine wave at any integer multiple of its frequency also yields a series of constant values. This is evident when one considers that such sampling picks off the wave at a fixed point of its phase. For all sine-wave frequencies, the apparent frequency always varies between 0 Hz and $0.5f_s$ (Figure 3.33). In general, f_0/f_s is the smaller of: (the fractional part of f/f_s) and (1 – the fractional part of f/f_s). Aliasing is demonstrated in Laboratory Exercise 10 in which the input sine-wave frequency ratio is varied from $f/f_s = 0.25$ to 2.0.

The primary conclusion here is that if any frequencies *above* $0.5 f_s$ exist in the sampled signal, they will appear in the sampled data as waves of *lower* frequency. This frequency aliasing is avoided only if the highest frequency in the waveform is sampled

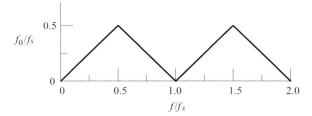

Figure 3.33 Apparent frequency f_0 as a function of true frequency f for a sine wave sampled at frequency f_s. The apparent frequency is equal to the true frequency only when $f/f_s < 0.5$.

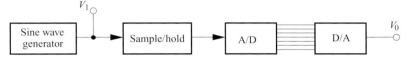

Figure 3.34 Data acquisition circuit for sampling and recovering an analog waveform.

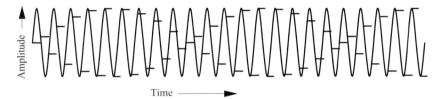

Figure 3.35 Output of the sampling circuit in Figure 3.22 when the sine-wave frequency is slightly less than one-half the sampling frequency. The curve is V_1, the input to the Sample/hold. The horizontal lines are V_0, the output of the D/A converter.

at least twice per cycle. This is the Nyquist sampling requirement and is discussed further in Chapter 5.

In the practical case, the sampling frequency f_s is limited and the input waveform may have components with frequencies greater than $0.5 \, f_s$. To avoid aliasing, these frequency components must be removed with a low-pass analog filter (such as those discussed in Chapter 2) *before* sampling.

EXAMPLE 3.2

What is the appearance of the samples for an input frequency just below the Nyquist frequency of half the sampling frequency? Figure 3.34 shows a simplified sampling circuit, where a waveform is sampled, digitized, and recovered in real time. Figure 3.35 shows the waveforms that result for an input frequency just below the Nyquist criteria (one-half the sampling frequency). Note that at this limit the D/A output resembles two separate slow sine waves rather a single sine wave of the input frequency.

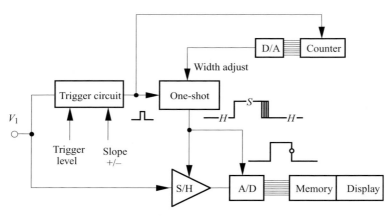

Figure 3.36 Sampling oscilloscope circuit. Each pulse increments the counter and causes the next pulse to be sampled at a slightly later time.

3.7   Available data-acquisition systems

For the highest performance in data acquisition, a number of systems are available. These are dedicated, self-contained systems that have an internal clock, sample-and-hold amplifiers, A/D converters, and high-speed memory. The microcomputer is used for interaction with the human operator, selecting the operating mode (e.g. sampling frequency, number of samples), starting, and reading the acquired data.

3.7.1   HP 54501A (Hewlett Packard)

The HP 54501A has 10-MHz real-time sampling, 1024 samples, 100-MHz effective sampling rate using 10 : 1 interleaving (Figure 3.36). Has self-contained keyboard and display – normally used as a stand-alone digital oscilloscope, but can be interfaced to a microcomputer using the IEEE 488.

3.7.2   Special-purpose external data-acquisition systems

These external systems contain timers, S/H amplifiers, A/D converters, digital memory, and control circuits. Typical units can take >1024 samples at sampling rates from 10 MHz to 50 GHz. The microcomputer initializes the data-acquisition system by loading numbers that specify the number of samples to be taken and the time interval between samples. Data acquisition is started either by an external event or by the computer program. After the samples have been taken, the computer program reads the data into memory and can then perform other functions, such as data analysis or display.

The advantages of this approach are: (i) the external data-acquisition circuit can take data much faster than the microcomputer; and (ii) as technology develops, higher sampling speeds can be obtained by keeping the same microcomputer and upgrading the external circuit.

3.7.3   Interface software

Points 1–4 below are text-based programming languages that allow creation of graphical user interfaces. With points 5–6, programming is graphical – instruments, actions, and data are objects on the screen
1. Visual Basic (Microsoft, Inc.) for IBM PC (DOS or Windows).
2. Labtech (Labtech Technologies) – data acquisition for IBM PCs.
3. LabWindows (National Instruments, Austin, TX) for IBM PC (DOS).
4. IBASIC (Hewlett Packard) for IBM PC (Windows).
5. LabView (National Instruments, Austin, TX) for both Macintosh and IBM PC (Windows) – allows programming of operations using IEEE 488 instruments and "virtual instruments" (disk files, statistical calculations, etc.).
6. HP-VEE. (Hewlett Packard)

3.8   Problems

3.1 You are given the following components:
- a sample-and-hold amplifier with 50-ns aperture jitter and a 0.5-μs acquisition time (see Section 3.4 for definitions),
- an 8-bit analog-to-digital successive-approximation converter with 5-μs conversion time,
- a 16-bit parallel input port with 10-μs read time.
  (a) If you did not use a sample-and-hold amplifier, what is the maximum input frequency such that the change is less than 0.5 LSB during the A/D conversion time?
  (b) Using the sample-and-hold amplifier, what is the maximum input frequency such that the aperture jitter corresponds to less than 0.5 LSB?
  (c) What is the maximum sampling frequency? What component would you change to improve it?
  (d) Considering both the S/H jitter and the Nyquist sampling requirement, what is the highest frequency in the input waveform that you can reliably sample and store?
  (e) For 10 V full scale, what is the resolution of the A/D?
  (f) How would your answers to (a)–(e) change if the A/D were 12 bit and had the same 5-μs conversion time?

3.2 You are using a 16-bit successive approximation A/D converter with an input range of 0–10 V and a conversion time of 100 μs. You want to use the AD582 sample-and-hold amplifier to provide an acquisition time $T_a < 20$ μs and a droop rate $R_d < 1$ LSB per 100 μs. Using the AD582 S/H data sheets, determine the maximum and minimum holding capacitance C_H that will satisfy the requirements.

3.3 Design a computer-based system for sampling *eight* analog input signals.
The design objectives are:
- The eight signals are to be sampled as simultaneously as possible (within 10 ns), then converted to digital form and stored under computer control. This constitutes one sampling step.
- The time interval between sampling steps and the number of sampling steps is entered by the user during initial setup.
- The minimum time interval between sampling steps is to be as small as possible.

The components provided are:
- A digital timing module with an internal 1-MHz clock, and an external output line. Periodic positive-going pulses can be produced on this line, where the pulse width, time interval between pulses, and number of pulses can be set by program control.
- Eight sample-and-hold amplifiers, with an aperture jitter of 1 ns (LO = sample, HI = hold).
- Eight 12-bit, successive-approximation A/D converters, which initiate conversion whenever they receive a LO-to-HI edge on a "start conversion" line, take 10 ± 1 μs to perform the conversion, assert the result on their output lines, and then set a "data ready" line HI. The "data ready" output is set LO by the next "start conversion" input. Data on the eight output data lines are valid until the next "start conversion."
- Eight tri-state 12-bit buffers. Each buffer asserts its input data onto its output lines whenever a single "select" input line is high. Unless selected, the outputs are in a high-impedance state and neither drive nor load any other circuit connected to them.
- *One* 16-bit parallel output port operates in transparent mode. The write operation takes 1 μs.
- *One* 16-bit parallel input port that you decide to operate in transparent mode by letting its "strobe" input float HI. The read operation takes 1 μs.
- An eight-input AND gate. Has eight inputs and one output. Output is HI only if all inputs are HI.

   Hint: Use this to determine when all A/Ds have finished converting.

Do the following:
(a) Draw a block diagram of all components and essential interconnections. Label all components, control lines, and data lines.

(b) Draw a timing diagram that shows the data-acquisition sequence for one set of eight analog inputs. For pulses that differ for the eight input channels, just show the first and eighth and omit channels 2–6.

(c) Draw a flow chart (or a list of what the steps do) for your microcomputer control program.

(d) In your design, what is the minimum time interval between sampling steps?

(e) What is the highest frequency input that you can reliably sample?

(f) What is the frequency limit in (e) if you do not use the eight sample-and-hold amplifiers?

3.4 You are designing a system that uses a sample-and-hold amplifier and a 13-bit successive-approximation A/D converter. The reference voltages have been trimmed to $V_{ref}^- = 0.0000$ V and $V_{ref}^+ = 8.1950$ V with an accuracy of 0.0001 V. The conversion time is $10\,\mu s$.

(a) Ideally, at what input voltage would you expect the first transition voltage $V(0, 1)$ to occur?

(b) Ideally, what is the input step size?

(c) What is the percent resolution for a 10-mV input?

(d) What is the percent resolution for a 5-V input?

(e) What is the maximum droop rate (volts per second) that the sample-and-hold amplifier can have for 1/2 LSB accuracy?

3.5 Design a **computer-controlled** system for the assembly line testing of *eight* units of a new type of 12-bit A/D converter.

You are provided with the following:

• eight sample A/D converters (to be tested eight at a time);

• a microcomputer equipped with 16-bit parallel I/O ports;

• eight 16-bit tri-state drivers;

• a microcomputer with:
    – a 16-bit D/A converter with 1/2 LSB absolute accuracy and 10-μs settling time,
    – two 16-bit parallel input ports,
    – two 16-bit parallel output ports.

You may assume the following:

• The 16-bit parallel output port is in "transparent" mode (no handshaking). New data can be written to the port every 2 μs.

• You have a timer function wait(N), that can delay program execution for N microseconds.

• The A/D converter requires a "start conversion" low-to-high edge signal and after conversion provides an "output data available" low-to-high edge. The A/D converter sets "output data available" low and resets all internal functions when "start conversion" goes low.

• For highest possible reliability, you must wait until the A/D has signaled that its data are ready before reading its output.

Do the following:

(a) Draw a block diagram of the major components, including two of the eight A/D converters being tested. Show and label all essential data and control lines.

   Hint: Think about Laboratory Exercise 9 (A/D converters) and how you would automate the measurement and data analysis procedures.

(b) List the steps your program must do to measure the nth transition voltage $V(n, n + 1)$ of the first A/D converter.

(c) How would you measure the maximum absolute accuracy error of the A/D? (Explain the procedure in steps or with a flow diagram.)

(d) How would you measure the maximum linearity error?

(e) How would you measure the maximum differential linearity error?

(f) With what accuracy could this system measure the quantities in (c)–(e) in units of 1 LSB of the A/D?

3.6 In your own words, describe the method of operation of the following A/D converters:

(a) integrating (or dual slope),

(b) flash,

(c) tracking,

(d) successive approximation,

(e) 1-bit sigma–delta.

3.7 The D/A has the problem that when a new input word is asserted, unwanted spikes (or "glitches") occur at the analog output. This is due to: (i) the inability to change all input bits at exactly the same time, and (ii) differences in response times of the switches inside the D/A. Even after the bits have stabilized, the effect of a glitch is present at the output during the settling time of the analog output amplifier. Design an 8-bit glitch-free D/A circuit with the following characteristics:

- output range of 0–10 V,
- accurate conversion at 0 Hz,
- minimum conversion time (time between new input and output accurate to 0.5 LSB),
- changing from any input number n_1 to any new input number n_2 results in a smooth output change from $V(n_1)$ to $V(n_2)$ – any output glitch outside this range has an amplitude less than 0.5 LSB.

You have available the following components:

1. A digital circuit that has eight inputs and one output (Figure 3.37 following). Whenever any of the inputs changes state, a pulse of width t_d is generated at the output. Conversely, if the output is low, then none of the input bits has changed during the last time interval t_d. (For a description of the exclusive and inclusive OR, see Figure 1.11.)

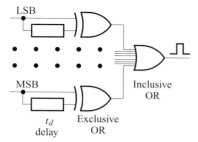

Figure 3.37 Circuit that produces an output pulse of width t_d whenever any of its input lines changes state.

Assume that:
  (i) the output is normally low,
  (ii) the leading edge of the output pulse is produced exactly 5 ns after the first input bit changes,
  (iii) the falling edge of the output pulse is produced exactly $t_d + 5$ ns after the last input bit changes,
  (iv) if only one bit changes, the output pulse has a width t_d, and
  (v) the pulse timing described above is the same for all bits.

2. A simple 8-bit R–$2R$ digital-to-analog converter circuit as shown in Figure 3.6 with an absolute accuracy of 0.25 LSB. The bit switches take a minimum of 20 ns and a maximum of 30 ns (as measured from the input bit edge to the start of the output change). The output op amp has a risetime of 10 ns and reaches 0.25% of its final output value in 40 ns.

3. A sample-and-hold amplifier as shown in Figures 3.22 and 3.23. The hold-to-sample acquisition time (time from the hold-to-sample edge to when the output is equal to the input to an accuracy of 0.25%) is 80 ns. The sample-to-hold switching time (control delay plus 1/2 aperture time) is 20 ns and the analog delay is 20 ns. Neglect charge transfer and settling time. The output droop is 100 mV/s.

4. Other basic circuit elements as needed, but keep it simple.

Do the following:

(a) Draw a block diagram of your circuit design, showing and labeling all circuit elements and interconnections between inputs and outputs.

(b) Draw the timing diagrams for the following signals in the system (show risetimes for analog signals, assume that digital signals transition in less than 1 ns):

    Figure 3.37 circuit input and output,
    R–$2R$ D/A input and output,
    S/H logic input,

S/H analog input and output,

input and output of any additional circuit elements used (if two or more of the above are identical, they need only be drawn once).

(c) Briefly describe the operation of your circuit (include how it will handle all possible bit transitions and anticipated variations in component timing).

3.8 Design a microcomputer-based data-acquisition system that does the following:

1. The microcomputer asks the user for the number of samples, the time interval between samples, and the name of the output data file.

2. The microcomputer initializes and triggers a special-purpose external circuit (to be designed by you).

3. The external circuit takes the specified number of samples and stores them in high-speed random-access memory.

4. The microcomputer reads the random-access memory and writes the data to an output file on disk.

Note: The computer is not used to take the individual samples – only to provide the user interface, to initialize and trigger the external circuit, and to store the data. The intent is for you to design the type of high-speed data-acquisition circuit that is used in digital oscilloscopes.

You are provided with the following components:

1. A microcomputer with disk drive and an I/O port having 16 input and 16 output lines.

2. A 9513 timing chip on an interface board.

3. An adder circuit that has two input lines and 20 output lines. A pulse on one input line causes the adder to be set to zero and a pulse on the other input line causes the number to be incremented (increased) by one. The number is always present on the 20 output lines.

4. A Sony CX20220A-1 subranging flash 10-bit A/D converter with 50-ns conversion time (Table 3.2). Conversion is initiated by a single input pulse and an output pulse is produced when conversion is complete.

5. A memory circuit with 20 address lines (input) and 16 data lines (input and output). A special input line (I/O control) is used to determine whether the address on the memory chip is being written to or read from.

Note: Items 1 and 2 are in the microcomputer box, and items 3–5 are used in the external circuit.

Do the following:

(a) Draw the block diagram for the microcomputer and external circuits and the lines that connect them. Label every essential item, control line, and data line.

(b) Describe in step-by-step sequence how your program and external circuit work. (There is no need to write detailed C code – just a flow chart in list form.)

(c) What are the overall characteristics of your system? (Minimum and maximum sampling rate, maximum number of samples, accuracy, etc.)

3.9 You are asked to sample the output of a device with an accuracy of 0.5% in amplitude over the frequency range from 0 to 10,000 Hz. Your sampling rate is 40,000 Hz.

(a) Determine the corner frequency f_c and number of poles n for a Butterworth anti-aliasing filter that attenuates all aliased portions of the signal by a factor of 100. Use the minimum number of poles. The frequency response of the Butterworth filter is given by:

$$|G| = \frac{1}{\sqrt{1 + (f/f_c)^{2n}}}$$

(b) In (a), the sampling frequency was four times the highest frequency of interest. If the sampling frequency were reduced to 30 kHz, approximately how many poles would the filter need?

3.10 Design a microcomputer-based system for the high-fidelity sampling and digital storage of music (like the compact audio disk or digital audio tape).

The design requirements are:

- The signal frequency range is from 10 to 20,000 Hz.
- Only consider music that is detected with microphones, such as classical or opera.
- The maximum microphone output signal is 10 mV peak-to-peak.
- The microphones and amplifiers produce white noise (constant noise power in each frequency band) from 0 Hz to 100 kHz.
- The A/D converter you are using has a -5 to $+5$ V input range.
- The digital resolution must be $<0.002\%(2 \times 10^{-5})$ of the full A/D range.
- Your design will include a device for storing large quantities of digital data. (In the recording industry, magnetic tape is used to store the digital master recordings.)
- The time interval between samples must be constant to one part in 10^6.
- You have rejected commercially available data-acquisition circuits as being too slow, too inaccurate, or too costly and have decided to design your own using parallel I/O ports and counters/timers circuits similar to those in the laboratory exercises.

(a) Draw a block diagram of your design, starting from one of the microphones and showing all necessary components and interconnections.

(b) According to the Nyquist theorem, what is the minimum sampling frequency necessary for subsequent recovery of the signal?

(c) Practically, what would be a good design value for the sampling frequency of the system? Explain your reasoning.

(d) Describe the A/D requirements in terms of number of bits and conversion time.

(e) Which type of A/D converter would be most appropriate? Explain. (Consider integrating, tracking, successive approximation, and flash.)

(f) What capacity (in megabytes) is required for the digital storing of 1 hour of music at the sampling rate from (c) above?

(g) What is the maximum aperture time jitter of the sample-and-hold amplifier that will guarantee 1/2 LSB accuracy at the maximum frequency of 20,000 Hz? (1 ns $= 10^{-9}$ s.)
Circle one choice:

121 ns   12.1 ns   1.21 ns   0.121 ns   0.0121 ns

3.11 Consider four types of analog-to-digital converters:
- integrating (or dual-slope, DS),
- successive approximation (SA),
- flash (FL),
- tracking (TR),

Write the best match or matches ($N =$ number of output bits):

	DS	SA	FL	TR
Requires 2^N clock cycles	—	—	—	—
Requires N clock cycles	—	—	—	—
Requires 1 clock cycle	—	—	—	—
Good accuracy and differential linearity	—	—	—	—
Low cost (per 2^N steps)	—	—	—	—
High speed	—	—	—	—
Slow	—	—	—	—

3.12 Design a microcomputer-based system for two-track (stereo) recording of music at a sampling rate of 50 kHz. Both tracks are to be sampled simultaneously, but the digital data must be read by the computer sequentially **because you have only one parallel input port**. You have the following components at your disposal:

- A microcomputer able to transfer 16 bits of data to and from its parallel I/O ports in 2 μs.
- A single 16-bit parallel input port. A BISTROBE line can be asserted by an external circuit and read by the program as a signal that new data are ready to be read. A low-to-high transition on an external $\overline{\text{BI HOLD}}$ line latches data onto the 16 internal registers. (Similar to the port used in Laboratory Exercises 3 and 9.)
- A 16-bit parallel output port, configured for transparent operation.
- A single timer that can be configured by the microcomputer to produce pulses

on an external line that are uniformly spaced in time. The pulse sense (high or low), width, and interval can be set by the computer program.

- Two sample-and-hold amplifiers with zero aperture delay and an acquisition time of 0.1 μs. High = hold, low = sample.
- Two 16-bit successive approximation A/D converters with a conversion speed of 15 μs. Conversion is initiated with a low-to-high transition on an input line. "End of conversion" is signaled by a low-to-high transition on an output line.
- Two 16-bit registers with tri-state output buffers. External data are latched onto internal registers with a low-to-high edge on a strobe line. When an additional "select" line is high, the 16 outputs agree with the 16 inputs. When the "select" line is low, the 16 outputs are in a high impedance state.
- One-shots and delays as needed.
- Additional components as needed, but keep it simple.

Do the following:

(a) Draw a block diagram of your recording system, showing and labeling all essential components and signal lines.

(b) Describe the operation of your system in flow chart or list of steps. Include both hardware and software processes.

(c) Show a timing diagram for one complete sampling cycle.

3.13 Design a microcomputer-based data-acquisition system that does the following:

1. The microcomputer asks the user for the number of samples (from 1 to 65,000), the time interval between samples (from 0.1 μs to 6 ms), and the name of the output data file.

2. The microcomputer interfaces with a special-purpose external circuit (to be designed by you).

3. The external circuit takes the specified number of samples and stores them in a high-speed random-access memory circuit.

4. The microcomputer reads the random-access memory and writes the data to an output file on disk.

You are provided with a computer with the following components:

1. Disk drive.

2. I/O port having 16 input and 16 output lines. Reading or writing takes 1 μs.

3. A special digital counters/timers interface circuit that has two 16-bit counters (A and B) that can be set to any number and then count down to zero.

   - Counter A is wired to count down at 10 MHz. When counter A reaches zero, it reloads its original number, generates an external pulse which is also sent to counter B, and resumes counting down.

   - Counter B is reduced by one count whenever it receives a pulse from counter A. All 16 bits of counter B are available on external lines, and the number is guaranteed to be valid only on the **leading** edge of the counter A external pulse.

You are also provided with the following components for your external circuit:

1. A 10-bit flash A/D converter with 50-ns conversion time. Conversion is initiated by a "start conversion" input pulse and the digital output data are guaranteed to be valid only on the **leading** edge of a "conversion complete" pulse.
2. A memory circuit with 16 address lines (input), 16 input data lines, 16 output data lines, and two logic lines, a "write" and "read." The leading edge of the "write" pulse latches data from the 16 input data lines into the memory location specified by the address lines. A pulse on the "read" line causes data to be read from the specified memory address and after a 10-ns delay, is present on the 16 output data lines for the duration of the "read" pulse.
3. A 16-bit D-type flip-flop, where all bits are latched on a common clock edge.
4. Any other components you may need, but keep it simple.

Do the following:

(a) Draw the block diagram for the microcomputer, counters/timers, external circuits, and the lines that connect them. Label all essential components, control lines, and data lines.
(b) Describe in step-by-step sequence how your program and external circuit work. (There is no need to write detailed C code – just a flow chart in list form.)
(c) Draw a timing diagram (logic level versus time) for all control and data lines.

3.14 After each of the six types of A/D converters listed below, write only the lettered characteristics that apply.

1. Successive approximation   _____
2. Tracking   _____
3. Dual slope (or integrating)   _____
4. Flash   _____
5. Half-flash   _____
6. Sigma–Delta (1-bit internal)   _____

(a) For N bits of resolution, conversion can take as many as 2^N steps.
(b) For N bits of resolution, conversion takes N steps.
(c) Conversion is done in two steps for any change in input voltage.
(d) Conversion is done in one step for any change in input voltage.
(e) Requires a sample-and-hold for rapid, accurate conversion.
(f) Limited to $N = 12$ bits at present.
(g) Conversion accuracy does not depend on resistor accuracy.
(h) Uses an internal D/A converter.
(i) Uses one or more internal comparators.
(j) Determines the output by counting pulses.

3.15 Describe how you would measure the frequency of the 60-Hz, 120-V power line to an accuracy of 0.001 Hz over a measurement period of 1 second. Assume that the frequency is in the 59–61 Hz range and that you have the following components:

- A microcomputer with a 9513 timer/counter circuit (like the one used in Laboratory Exercise 2) and an 8-bit parallel I/O port (like the one used in Laboratory Exercise 3).
- An analog comparator circuit.
- Additional components as needed, but keep it simple.

Do the following:

(a) Draw a block diagram of your measurement system, showing and labeling all essential components and signal lines.

(b) Describe the operation of your system (hardware and software) in a flow chart or list of steps.

(c) Draw a timing diagram for the important signals.

3.16 Some A/D converters average the input waveform over a time T_{ave}. Examples are the dual slope converter, and a S/H amplifier with finite switching time connected to a successive approximation converter. This introduces an error equal to $V_{ave} - V(t)$, where V_{ave} is $V(t)$ averaged from $t - T_{ave}/2$ to $t + T_{ave}/2$, and $V(t)$ is the instantaneous value at the center of the averaging interval.

(a) Derive an equation for determining the maximum frequency that can be converted with an error $<1/2$ LSB as a function of the number of bits N and T_{ave}. Assume that the waveform consists of a single harmonic with a peak-to-peak amplitude that covers the full input range of the A/D converter.

Hint: See the derivation in Section 3.4.5.

(b) For a S/H (aperture jitter $T_{samp} = 15$ ns, and switch averaging time $T_{ave} = 50$ ns) connected to an 8-bit successive approximation A/D converter, compute f_{max} due to T_{samp} and f_{max} due to T_{ave}.

3.17 Describe a step-by-step a method for measuring the analog delay and aperture delay of a sample-and-hold amplifier (see Section 3.4 for definitions). It is okay to use a human and an oscilloscope in the procedure.

Hint: Use a ramp waveform to determine time from a measurement of voltage.

3.18 You have been asked to design a peak-reading A/D converter system as part of a larger digital communication system. To overcome bandwidth limitations, discrete pulse heights are used to code digital information (Figure 3.38). Your system must: (1) hold the maximum input level, (2) determine when a peak has passed, (3) sample the held peak value, (4) store the digital value, and (5) reset the peak detector. For simplicity, assume that the peaks never overlap.

- A simple peak detector can be made using an op amp, a diode, and a capacitor as shown in Figure 3.39.
- Your system must also generate handshaking signals for the A/D conversion and provide a reset switch for the hold capacitor.
- You have decided to use the Precision Monolithics, Inc., PKD-01 monolithic peak detector with an external 1,000-pF holding capacitor.

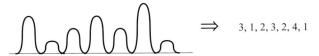

Figure 3.38 Series of pulses with discrete amplitudes used to encode digital information.

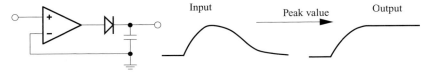

Figure 3.39 Circuit for determining and holding the peak value of a pulse. Additional circuit components are needed to reset the holding capacitor.

- The A/D conversion will be done with the analog input port plugged into an IBM PC, as used in the 145M laboratory. The input voltage range is −10 to +10 V, and the output is 12 bits. The circuit has a "ready for data" digital output (high = busy, low = ready) and a "data available" digital input (conversion starts on a low-to-high edge).
- The sampling time of the analog I/O port is 50 μs, including handshaking.

(a) Sketch a block diagram of your system. Clearly indicate any comparators, resistors, capacitors, etc. that you think are necessary for a working design.

(b) Describe the handshaking procedure that your system uses. Provide a timing diagram showing all important signal and control lines.

(c) Based on the conversion time of the analog I/O port board and the PKD-01, estimate the maximum allowable pulse rate.

(d) Estimate the error caused by capacitor droop for a 5-V input signal. Give your results both in units of volts and LSB.

3.19 Implement an 8-bit successive approximation A/D converter using a D/A converter, a comparator, and software control. Specifically, you are given only the following components:

1. A microcomputer with parallel I/O port (8 bits in, 8 bits out).
2. A D/A converter chip (8 bits) (*NOT* A/D).
3. A comparator chip. *Note*: A comparator has two analog inputs, $V+$ and $V-$, and a logic output L. If $V+ > V-$, then L is high. If $V+ < V-$, then L is low.

Show how you would use these to do "successive approximation" A/D conversion of analog input signals.

(a) Sketch a block diagram of all components and essential interconnections, and label same.

(b) Describe your control program, with a flow chart or in the language "C."

(c) If it requires 5 μs for 8-bit input or output, how long will your algorithm take to convert a dc signal?

3.20 Design a microcomputer-based system for generating two different analog waveforms using two D/A converters.

Assume:
- The microcomputer has **only one** 16-bit digital output port.
- The two D/A converters have 12 bits and you need all 12 bits for good accuracy.
- The D/As contain only resistors, bit switches, and an output amplifier.
- Do not worry about D/A output glitches (you will address that issue in the following two problems).
- You may use a number of 8-bit edge-triggered flip-flops or transparent latches, as your design requires.

Do the following:
(a) Draw a block diagram of your system, showing and labeling all essential components, connections, and signals. (Draw each D/A converter as a single box.)
(b) In proper time sequence, list the program and hardware steps necessary for your system to convert two different numbers (n_1 and n_2) rapidly into two different voltages at the output of the D/A converters.
(c) Draw a timing diagram for the signals described in (b).

3.21 After operating the system you designed in Problem 3.20, you discover that occasional output glitches occur because the D/A bit switches do not change state simultaneously. Design additional modifications to eliminate these glitches from the two analog outputs of your system.

Hint: Sample-and-hold amplifiers could be useful here.

Do the following:
(a) Draw the modifications to your Problem 3.20 design, showing and labeling all essential components, connections, and signals.
(b) In proper time sequence, list the program and hardware steps necessary for your system to convert n_1 and n_2 rapidly into two glitch-free analog voltages.
(c) Draw a timing diagram for the signals described in (b).

3.22 Design a microcomputer-based system for converting two lists of numbers val_right[] and val_left[] into corresponding analog waveforms.

Assume:
- The microcomputer has only one 16-bit digital output port.
- You will use a single 12-bit D/A converter and you need all 12 bits for good accuracy.
- The D/A contains only resistors, bit switches, and an output amplifier.
- Your computer has a 1-MHz clock that can be set to zero with the command reset_clock(); and provides the number of microseconds (as a 32-bit integer) with the command val_time = time().

- Both analog waveforms are to be glitch-free.
- Both analog outputs are to be updated as simultaneously as possible ($\ll 1$ μs) at a frequency as close to 40 MHz as possible.

Do the following:

(a) Draw a block diagram of your system, showing and labeling all essential components, connections, and signals.

(b) In proper time sequence, list the additional program and hardware steps necessary for your system to convert the arrays val_right[] and val_left[] into analog waveforms, where corresponding analog values are outputted as simultaneously as possible.

(c) Draw a timing diagram for one cycle of the signals described in (b).

(d) What difficulties would you encounter storing and playing 2 hours of stereo music using this system?

3.9   Additional reading

Walter Alvarez, *T. Rex and the Crater of Doom*, Vintage Books, New York, 1998.

James C. Candy and Gabor C. Temes, ed., *Oversampling Delta-Sigma Data Converters*, IEEE Press, Piscataway, NJ, 1992.

Daniel H. Sheingold, ed., *Analog-Digital Conversion Handbook*, Prentice Hall, Englewood Cliffs, NJ, 1986.

Laboratory Exercise 7

Introduction to A/D and D/A conversion

Purpose

To use an analog I/O board and to measure the characteristics of analog-to-digital and digital-to-analog converters.

Equipment

- IBM-compatible Pentium microcomputer with Windows NT operating system and Microsoft Visual C++ compiler
- Data Translation DT3010 PCI board (in microcomputer)
- Oscilloscope
- Digital multimeter
- ± 12-V power supplies
- Two 10-μF 25-V electrolytic capacitors (put between power and ground at circuit board binding posts)
- Superstrip circuit board
- One 10-kΩ 10-turn helipot (helical potentiometer)
- Two 1-MΩ resistors
- One 0.1-μF capacitor
- One LF356 op amp

Additional reading

Chapter 3 Analog $\leftrightarrow$ digital conversion and sampling
Appendix E Summary of Data Translation DT3010 PCI plug-in card

Procedure

1.   Programs

1. Firstly you must initialize the Data Translation I/O board, using InitAll() after all variables have been declared.
2. To read an analog value from the board, use:

   olDaGetSingleValue(hAd, &val, channel, gain)

3. To write an analog value to the board, use:

   olDaPutSingleValue(hDa, val, channel, gain)

4. The analog channels are:
   - Analog input channel 0 reads from board pins 1 and 2
   - Analog input channel 1 reads from board pins 3 and 4
   - Analog output channel 0 writes to board pins 41 and 42
   - Analog output channel 1 writes to board pins 43 and 44

The C programs ADC and DAC are listed in Appendix E. ADC repeatedly converts an external analog voltage into a number and displays the average value on the screen. DAC converts a number entered on the keyboard into an external voltage.

2.   Circuit construction

Construct a voltage divider as shown in Laboratory Figure 7.1, using a 10-turn 10-kΩ helipot connected between −11 and +11 V and an op-amp buffer. The 1-MΩ resistor and 0.1-μF capacitor act as a low-pass filter to reduce ac noise.

Connect the op-amp output to both your digital multimeter and the positive input of the data-acquisition circuit (Channel 0^+). Connect the negative input (Channel 0^-) and all digital and analog grounds of the IBM Data Acquisition and Control Adapter (ribbon cable lines 4, 5, 13, 15, and 60) to your external power-supply ground.

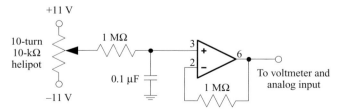

Laboratory Figure 7.1 Voltage source for A/D converter.

Laboratory Table 7.1 *Suggested neighboring A/D transition voltages to measure*

$V_{n,n+1}$	$V_{n+1,n+2}$
$V_{0,1}$	$V_{1,2}$
$V_{999,1000}$	$V_{1000,1001}$
$V_{1699,1700}$	$V_{1700,1701}$
$V_{2049,2050}$	$V_{2050,2051}$
$V_{2399,2400}$	$V_{2400,2401}$
$V_{2999,3000}$	$V_{3000,3001}$
$V_{3999,4000}$	$V_{4000,4001}$
$V_{4093,4094}$	$V_{4094,4095}$

3.   A/D conversion

The Data Translation I/O board uses a 12-bit analog-to-digital converter, connected for a ± 10-V input range. An input of -10 V produces a digital output of 0. An input of 0 V produces a digital output of about 2,047. An input of $+10$ V produces a digital output of about 4,095. One output count corresponds to an input change of about 5 mV.

You will use a computer program that controls the analog to digital conversion and prints the result on the screen in an endless loop. This program is started by typing "ADC" and then pressing the return key. The program takes a number of samples, prints them on the screen, and prints the average.

3.1 End-point voltages. Adjust the 10-kΩ trimpot and run "ADC" to find the *exact* voltage that gives an A/D average of 0.5. At this voltage the A/D output should be fluctuating evenly between 0 and 1 (the fluctuations are caused by noise). This is called the $V_{0,1}$ transition voltage. Record the multimeter value. Then find and record the exact voltage that gives an A/D average of 4,094.5, which is the highest transition voltage $V_{4094,4095}$. The transition voltages $V_{0,1}$ and $V_{4094,4095}$ span the full input range of the A/D and are called the end-point voltages.

3.2 Intermediate transition voltages. Observe and record the $V_{2047,2048}$ transition voltage. This will be indicated by an average of 2,047.5. Repeat for the $V_{2,048,2,049}$ transition. These values should occur near 0.0 V where the multimeter has an accuracy of about 0.2 mV. Repeat for several pairs of neighboring transition voltages spanning the full range, such as those shown in Laboratory Table 7.1.

4.   D/A conversion

The Data Translation I/O board uses a 12-bit digital-to-analog converter, connected for ± 10-V operation. In analogy with the A/D converter, digital inputs of 0, 2,047, and 4,095 produce corresponding analog outputs of about -10, 0, and $+10$ V.

Laboratory Table 7.2 *Suggested neighboring D/A input values*

n	$n + 1$
0	1
999	1,000
1,699	1,700
2,047	2,048
2,049	2,050
2,399	2,400
2,999	3,000
4,000	4,001

You will use a computer program that reads digital input (decimal) from the keyboard and converts the result to an analog output that appears on "D/A output 1." Start the digital-to-analog program by typing "DAC" and pressing return.

4.1 End point voltages. Record the output voltages corresponding to input numbers of 0 and 4,095.

4.2 Output voltages. Record the outputs for several pairs of neighboring input numbers spanning the full range, such as those listed in Laboratory Table 7.2.

Laboratory report

1.   Setup

Draw a simple block diagram of your experimental setup.

2.   Data summary

2.1 A/D transition voltages. Using your data from procedure section 2, tabulate transition voltages $V_{n,n+1}$ versus n.

2.2 A/D differential linearity. Using your data from procedure section 2.3, tabulate the differences $V_{n-1,n} - V_{n,n+1}$ between neighboring transition voltages. The uniformity of the intervals between transition voltages is a measure of differential linearity.

2.3 D/A response. Using your data from procedure section 3, tabulate output voltage $V(n)$ versus D/A input number n.

2.4 D/A differential linearity. Using your data from procedure section 3.2, tabulate the differences between neighboring output voltages $V_n - V_{n-1}$. The uniformity of output step size is a measure of differential linearity.

3.   Analysis

3.1 Plot of A/D measurements. Plot transition voltages versus digital output for the A/D converter and draw a best-fit line through the data (eyeball).

3.2 Table of A/D measurements and linear model. Tabulate your measured transition voltages $V_{n,n+1}$, the corresponding A/D linear model values, and the differences between them.

The linear model is given by:

$$V^{\text{lin}}(n, n+1) = V_{0,1} + n\Delta V, \qquad \text{where} \quad \Delta V = (V_{254,255} - V_{0,1})/254.$$

Note that this line passes through the measured end points, since $V^{\text{lin}}(0, 1) = V_{0,1}$ and $V^{\text{lin}}(254,255) = V_{254,255}$.

3.3 Plot of D/A measurements. Plot output voltages versus digital input for the D/A converter and draw a best-fit line through the data (eyeball).

3.4 Table of D/A measurements and linear model. Tabulate your measured values V_n, the corresponding D/A linear model values, and the differences between them.

The linear model output voltage $V^{\text{lin}}(n)$ is given by:

$$V^{\text{lin}}(n) = V_{\text{min}} + n\left(\frac{V_{\text{max}} - V_{\text{min}}}{255}\right)$$

where $V_{\text{min}} = V_0$ and $V_{\text{max}} = V_{255}$ are the measured end-point voltages.

4.   Discussion and conclusions

4.1 Discuss your measured values and compare these with the data sheets.

4.2 Discuss the application of these techniques to the digital recording and playback of music.

5.   Questions

5.1 How linear were the responses of the A/D and D/A converters?

5.2 Were the differential linearities of the A/D and D/A converters within specifications?

5.3 Were the slopes and zero intercepts what you expected?

5.4 For the A/D and D/A converters, what were the largest deviations between your data and the linear model values?

6.   Laboratory data sheets

Include your handwritten data sheets (or a copy), which should consist of a log of the procedures that you used, any special circumstances, and the measurements you recorded manually.

Laboratory Exercise 8

D/A conversion and waveform generation

Purpose

To interface a microcomputer to a D/A converter, and to write a C program that generates static voltage levels and time-varying waveforms. To determine the deviation between the output and a linear model.

Equipment

- IBM-compatible Pentium microcomputer with Windows NT operating system and Microsoft Visual C++ compiler
- Data Translation DT3010 plug-in circuit board
- Printer (shared with other laboratory stations)
- ± 12-V power supplies
- $+10$-V power supply
- Oscilloscope
- Digital multimeter
- Three 10-μF, 25-V electrolytic capacitors (put between power and ground at circuit board binding posts)
- Five 0.1-μF CK-05 capacitors (put between power and ground at integrated circuits)
- Parallel port terminal block for connecting binary I/O port to student circuit boards
- Superstrip circuit board
- One DAC0802 D/A converter integrated circuit chip
- One LF356 op amp
- One 0.01-μF capacitor
- Four 5.1-kΩ resistors

Background

The ideal D/A response is given by:

$$V(n) = V_{\text{ref}}^- + n\left(\frac{V_{\text{ref}}^+ - V_{\text{ref}}^-}{2^N}\right) = V_{\text{ref}}^- + n\Delta V$$

In this laboratory exercise, $V_{\text{ref}}^- = -10\,\text{V}\,(3.0\text{k}/5.1\text{k}) = -5.9\,\text{V}$,
$V_{\text{ref}}^+ = +10\,\text{V}\,(3.0\text{k}/5.1\text{k}) = +5.9\,\text{V}$, and $2^N = 256$.

The experimental D/A response to an input number n is the voltage V_n. In this laboratory exercise, V_0 is about $-5.9\,\text{V}$, V_{128} is about $0\,\text{V}$, and V_{255} is about $+5.9\,\text{V}$.

The linear D/A response is given by:

$$V^{\text{lin}}(n) = V_{\text{min}} + n\left(\frac{V_{\text{max}} - V_{\text{min}}}{255}\right) = V_{\text{min}} + n\Delta V$$

which is a straight line running between the minimum $V_{\text{min}} = V_0$ and the maximum $V_{\text{max}} = V_{255}$.

Additional reading

Section 3.2 Digital-to-analog converter circuits
DAC0802 D/A converter data sheets
Appendix E Summary of Data Translation DT3010 PCI plug-in card

Procedure

1.   Circuit construction

Connect the DAC0802 converter to the binary output bits 0–7 of the terminal block connected to the Data Translation I/O board. (See Laboratory Exercise 3 for pin assignments.) Connect the DAC0802 complementary current outputs to the op-amp circuit, which converts the differential currents to an output voltage. Refer to the DAC0802 and LF356 pinouts in Laboratory Figures 8.1 and 8.2.

- Connect the +12-V power supply to DAC0802 pin 13 and LF356 pin 7.
- Connect the −12-V power supply to DAC0802 pin 3 and LF356 pin 4.
- Connect the +10-V power supply to DAC0802 pin 14 (reference).
- Connect DT3010 pins 4 (analog ground), 13 (analog ground), 15 (digital ground), and 57 (digital ground) to the power-supply ground.
- Connect DAC0802 pin 5 (bit 1 MSB) to DT3010 pin 104 (binary output bit 7).

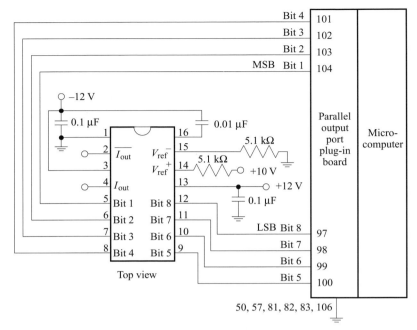

Laboratory Figure 8.1 Circuit diagram and output port connections for DAC0802 D/A converter.

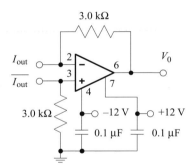

Laboratory Figure 8.2 Circuit diagram and pinouts for LF356 op amp. Output V_0 is about $0-6$ V for an input of 0 and about $+6$ V for an input of 255.

- Connect DAC0802 pin 6 (bit 2) to DT3010 pin 103 (binary output bit 6).
- Connect DAC0802 pin 7 (bit 3) to DT3010 pin 102 (binary output bit 5).
- Connect DAC0802 pin 8 (bit 4) to DT3010 pin 101 (binary output bit 4).
- Connect DAC0802 pin 9 (bit 5) to DT3010 pin 100 (binary output bit 3).
- Connect DAC0802 pin 10 (bit 6) to DT3010 pin 99 (binary output bit 2).
- Connect DAC0802 pin 11 (bit 7) to DT3010 pin 98 (binary output bit 1).
- Connect DAC0802 pin 12 (bit 8 LSB) to DT3010 pin 97 (binary output bit 0).

Using your digital multimeter, adjust all three power-supply outputs to their $+12$-, -12-, and $+10$-V values. Connect the digital grounds of the IBM Data Acquisition and Control Adapter (DT3010 cable pins 15 and 57) to your external power-supply ground.

2.   End-point voltages

Write a simple program that asks the user for a number between 0 and 255, writes that number into the D/A, and loops back for another number. See Laboratory Exercise 3 for the code that performs binary output without handshaking.

Write 0 to the D/A and record the minimum analog output voltage V_{min} (should be near -6 V).

Write 255 to the D/A and record the maximum analog output voltage V_{max} (should be near $+6$ V).

3.   Power-supply rejection ratio

Reduce the $+12$-V power supply to $+11.6$ V (a 3% decrease). Record the output voltages for D/A inputs of 0 and 255. Restore the power supply to $+12$ V.

4.   Glitches and settling time

Write a C program to output $0\times7F$ (127 decimal) and 0×80 (128 decimal) to the D/A alternately. Look at the output on the oscilloscope and sketch the appearance of the waveform. Observe the voltage steps, and record the height (in millivolts) and the width (in microseconds). Look at the step edges and record the settling time (time from the start of the step to when the output reaches a steady value). Record the amplitude, width, and shape of glitches (brief output spikes that occur when two or more D/A switches change at slightly different times).

5.   Static response

Write a C program that does the following:
1. Ask the user for a number n ranging from 0 to 255.
2. Write the number entered by the user to the parallel output port.
3. Ask the user to enter the voltage V_n measured by the digital multimeter to an accuracy of 0.001 V and then read and store the value as type "float."

Run the program for about 10 neighboring values of n spanning the full range. Example:

0	1	3	4
9	10	19	20
29	30	49	50
99	100	149	150
199	200	254	255

6.   Waveform generation

6.1 Ramp waveform. Change your program to send the series:

$$0, 1, 2, \ldots, 255, 0, 1, \ldots, 255, 0, 1, \ldots$$

to the D/A as rapidly as you can. Look at the output of the D/A with the oscilloscope to determine the time between peaks. Expand the horizontal (time) and vertical (amplitude) scales of the oscilloscope so that you can clearly see the individual "steps" of the ramp. Describe the waveform and record the step height (in millivolts), step width (in microseconds).

6.2 Other waveform. Devise some other waveform (sine, decaying sine, decaying exponential, etc.) and write a program to send it to the D/A converter rapidly. (Remember that you must output integers in the range from 0 to 255.) If your waveform is non-cyclic, you will have to compute each new value before you can send it to the D/A. If your waveform is cyclic, you can precompute the values and store them in an array. Then these values can be sent much more quickly to the D/A in an endless loop. Record the time interval between D/A output steps.

Laboratory report

1.   Setup

Draw a simple block diagram of your experimental setup, showing the components on your circuit board as well as the connections to the parallel output port.

2.   Data summary and analysis

2.1 Comparison between measurements and the linear model.
Summarize your results in a table with the following headings:

- n
- V_n measured D/A output voltages
- Linear model $V^{\text{lin}}(n)$
- Difference $V_n - V^{\text{lin}}(n)$ in millivolts
- Difference $[V_n - V^{\text{lin}}(n)]/\Delta V$ in LSB

2.2 Rms deviation. Compute the rms deviations V_{rms} between your data and the linear response given in the background section:

$$V_{\text{rms}} = \sqrt{\frac{1}{M} \sum_{i=1}^{M} \left[V^{\text{lin}}(n_i) - V_{n_i} \right]^2}$$

where the summation is carried out only over your M measured values of $V(n_i)$.

Note: Omit the end points in the summation, since they are used to define the linear response and contribute zero.

2.3 Differential linearity. Tabulate $V_n - V_{n-1}$ from your measured data for several values of n. Estimate differential nonlinearity in units of LSB.

2.4 Power-supply rejection ratio. For $n = 0$ and 255, compute the power-supply rejection ratio as $\Delta V_n / \Delta V_S$, where ΔV_n is the change in D/A output V_n for a power supply change ΔV_S.

2.5 Glitch description. Include a sketch of the $127 \leftrightarrow 128$ glitch you observed in procedure section 4. Label the voltage and time axes, and estimate the voltage and duration of the glitch. Also estimate the settling time.

2.6 Waveform generation. From your observations of the ramp waveform, compute the frequency that your program was able to send numbers to the D/A.

3.   Discussion and conclusions

3.1 Discuss the importance of good power-supply rejection for D/A converters used in battery-operated equipment.

3.2 Discuss the cause of the glitches observed in procedure section 4 in terms of the operation of the D/A converter.

3.3 Discuss the relative accuracy and the differential linearity measured in procedure section 5.

3.4 Describe how the properties you measured for your D/A differ from the data-sheet specifications in terms of relative accuracies, differential linearity, glitch amplitude, settling time, and power-supply sensitivity. Note that not all useful properties may be found in the data sheets.

4.   Questions

4.1 How would errors in the reference voltages affect the absolute accuracy and relative accuracies of the D/A?

4.2 What was the maximum millivolt deviation between your observed D/A output and the linear model?

4.3 How would you use a sample-and-hold amplifier to remove the glitches?

4.4 What is the clock speed of the microcomputer you used? What is the conversion time of the D/A converter? How fast were you able to convert digital numbers into analog outputs? Which of the following could account for the difference: your program, the I/O software, the output port, or the D/A converter?

4.5 What are the limitations of the method used in procedure section 6 to generate an arbitrary waveform (consider voltage accuracy, maximum frequency, system interrupts, glitches, amplitude quantization)?

5.   Program and laboratory data sheets

5.1 Include printouts of your program code, data, and output.

5.2 Include your handwritten data sheets (or a copy), which should consist of a log of the procedures you used, any special circumstances, and the measurements you recorded manually.

Laboratory Exercise 9

A/D conversion and periodic sampling

Purpose

To interface a microcomputer to an external A/D converter, to gain familiarity with the timing relationships between handshaking signals and converted data, to explore the properties of the A/D converter, and to sample a sine wave.

Equipment

- IBM-compatible Pentium microcomputer with Windows NT operating system and Microsoft Visual C++ compiler
- Data Translation DT3010 plug-in circuit board
- Printer (shared with other laboratory stations)
- Oscilloscope
- Two +5-V power supplies (one for circuit power and the other for a voltage reference)
- Parallel port terminal block
- Superstrip circuit board with ground plane and connections for ground, +5 V, ±12 V
- Digital multimeter
- Three 10-μF, 25-V electrolytic capacitors (put between power and ground at circuit board binding posts)
- Ten 0.1-μF CK-05 bypass capacitors (put between power and ground at all integrated circuits)
- One ADC0820 analog-to-digital converter chip
- One 1-kΩ potentiometer (to adjust analog input voltage)
- One 10-μF electrolytic capacitor (to suppress noise on analog input voltage)

Background

The A/D transition voltage $V_{n,n+1}$ is defined as the input voltage where the output changes between n and $n + 1$.

The ideal A/D transition voltages are given by:

$$V(n, n + 1) = V_{\text{ref}}^- + \left(n + \frac{1}{2}\right)\left(\frac{V_{\text{ref}}^+ - V_{\text{ref}}^-}{2^N - 1}\right)$$

In this laboratory exercise, use $V_{\text{ref}}^- = 0$ V, $V_{\text{ref}}^+ = 5.0$ V, and $2^N = 256$.

The linear A/D transition voltages are given by:

$$V^{\text{lin}}(n, n + 1) = V_{0,1} + n\left(\frac{V_{2^N-2,2^N-1} - V_{0,1}}{2^N - 2}\right) = V_{0,1} + n\Delta V$$

which is a straight line running between the first transition voltage $V_{0,1}$ and the highest transition voltage $V_{2^N-2,2^N-1}$.

Additional reading

Section 1.5.2 The parallel input port
Section 1.6 Digital data-acquisition procedures
Section 2.7 The sample-and-hold amplifier
Section 3.3 Analog-to-digital converter circuits
ACD0820 data sheets
Appendix E Summary of Data Translation DT3010 PCI plug-in card

Procedure

1.   Circuit connections

The ADC0820 A/D converter uses a subranging flash method, has eight output bits and converts in 1.5 μs. The connections in the list below, and in Laboratory Figure 9.1, establish an input range from 0 to 5 V, and RD mode, where conversion starts by pulling the input $\overline{\text{RD}}$ low, and the output $\overline{\text{INT}}$ going low indicates that conversion is completed and the data result is in the output latch. $\overline{\text{RD}}$ low also enables the tri-state outputs so that the result may be read by a parallel input port.

- Connect DT3010 pins 4 (analog ground), 13 (analog ground), 15 (digital ground), and 57 (digital ground) to the power-supply ground.
- Connect ADC0820 pin 7 (MODE) to ground for RD mode.

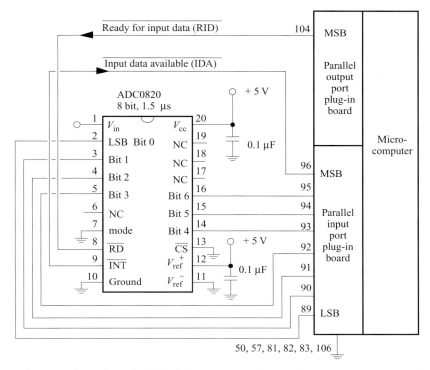

Laboratory Figure 9.1 ADC0820 A/D converter and external connections. Connect pin 1 to the output of the wave generator.

- Connect ADC0820 pin 8 ($\overline{\text{RD}}$) to pin 104 (MSB) of the DT3010 board digital output port. This will be used for the "ready for input data" handshaking line.
- Connect ADC0820 pin 9 ($\overline{\text{INT}}$) to pin 96 (MSB) of the DT3010 digital input port. This will be used for the "input data available" handshaking line.
- Connect ADC0820 pins 10 (ground), 11 ($V_{\text{ref}}{}^{-}$), and 13 ($\overline{\text{CS}}$) to the power-supply ground.
- Connect ADC0820 pin 12 ($V_{\text{ref}}{}^{+}$) to one $+5$-V power supply.
- Connect ADC0820 pin 20 (V_{cc}) to another $+5$-V power supply.
- Connect all DT3010 grounds to your power-supply ground.
- Connect ADC0820 pin 2 (bit 0 LSB) to DT3010 pin 89 (binary input bit 0).
- Connect ADC0820 pin 3 (bit 1) to DT3010 pin 90 (binary input bit 1).
- Connect ADC0820 pin 4 (bit 2) to DT3010 pin 91 (binary input bit 2).
- Connect ADC0820 pin 5 (bit 3) to DT3010 pin 92 (binary input bit 3).
- Connect ADC0820 pin 14 (bit 4) to DT3010 pin 93 (binary input bit 4).
- Connect ADC0820 pin 15 (bit 5) to DT3010 pin 94 (binary input bit 5).
- Connect ADC0820 pin 16 (bit 6) to DT3010 pin 95 (binary input bit 6).
- ADC pin 17 (bit 7 MSB) is not used.

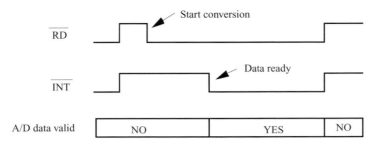

Laboratory Figure 9.2 Timing diagram for the ADC0820 data-acquisition circuit. Conversion begins when the program brings $\overline{RD}$ low and has ended when the ADC0820 has brought $\overline{INT}$ low. After reading data, the program brings $\overline{RD}$ high and the ADC0820 brings $\overline{INT}$ high.

2. ADC0820 circuit and program operation

The following steps describe the operation of the ADC0820 circuit and the data-acquisition program which you are to write as part of this laboratory exercise. See Laboratory Figure 9.2 for the timing diagram.

1. The program initializes the DT3010 board.
2. The program writes a 1 to the MSB of the digital output port, which brings $\overline{RD}$ high. (This resets the ADC0820, which brings $\overline{INT}$ high.)
3. To initiate a new conversion, the program writes a 0 to the MSB of the digital output port, which brings $\overline{RD}$ of the ADC0820 low.
4. The program reads the digital input port (all eight bits), waiting for the MSB to become zero.

   Note: this handshaking line has the opposite logic sense than that used in Laboratory Exercise 2.

5. When the ADC0820 has completed conversion, it brings $\overline{INT}$ low, which sets the MSB of the digital input port to zero.
6. The program detects that the MSB of the digital input port is zero, masks off the seven least significant bits, and stores them in memory.
7. Loop back to step 2 until the necessary samples have been taken.

   To sample analog waveforms rapidly, do steps 2–6 in a tight loop. Do not perform any printf operations within the data-acquisition loop or it will be slowed considerably.

3. Measurement of A/D characteristics

3.1 Program. Write a C program that loops endlessly over the following two steps:
1. Sample the analog input as described in procedure section 2.
2. Write the number to the terminal screen.

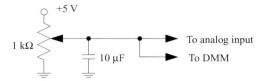

Laboratory Figure 9.3 Voltage input signal for the ADC0820 A/D converter.

3.2 Voltage divider. As shown in Laboratory Figure 9.3, construct a voltage divider using a 20-turn 1-kΩ trimpot connected between +5 V and ground. Connect the wiper to the input of the data-acquisition circuit and your digital multimeter. Bypass the sample-and-hold amplifier in this section and connect the wiper directly to the A/D converter if noise makes it difficult to measure transition voltages accurately.

Note: If you have a digital power supply whose output can be set to an accuracy of 1 mV or better, the above circuit is not needed.

3.3 Measurement of timing diagram. With your data-acquisition program running, display $\overline{RD}$ and $\overline{INT}$ on the oscilloscope, triggering on the falling edge of $\overline{RD}$. Record the waveforms on your data sheets and mark on the horizontal time axis the time taken to perform the A/D conversion ($\overline{INT}$ goes low), the time taken by your program to read the data and reset $\overline{RD}$ high, and the repetition period of the acquisition process.

3.4 Transition voltage measurements. Start with zero input voltage. Slowly increase until reaching the first transition voltage $V_{0,1}$ where the digital output toggles between 0 and 1 and has an average value of 0.5. Record the *exact* input voltage displayed on the digital multimeter. Record a table of neighboring pairs of transition voltages spanning the full range, such as:

$V_{0,1}$	$V_{1,2}$
$V_{2,3}$	$V_{3,4}$
$V_{24,25}$	$V_{25,26}$
$V_{49,50}$	$V_{50,51}$
$V_{74,75}$	$V_{75,76}$
$V_{99,100}$	$V_{100,101}$
$V_{125,126}$	$V_{126,127}$

3.5 Hysteresis. Slowly reduce the input voltage and record the transition voltages for $V_{2,1}$ and $V_{1,0}$. The reversed indices here mean that the transition voltage was approached from above.

3.6 Power-supply rejection ratio. Change the +5-V power supply (pin 20 of the ADC0820) to +4.85 V (a 3% decrease). Leave the +5-V reference supply (pin 12 of the ADC0820) unchanged. Record the $V_{0,1}$ and $V_{126,127}$ transition voltages. Restore the power supply to +5 V.

4.   Sampling a sine wave

Set the wave generator for a 1-kHz sine wave oscillating between about $+0.5$ and $+2.0$ V as measured on your oscilloscope. You will need to use the offset knob to set an average voltage of about 1.2 V. Connect the wave generator to pin 1 of the ADC0820 A/D converter. Connect the wave generator ground to your circuit ground. Change your program to sample 200 values in a tight data-acquisition loop and then store the values in a file for display and printing. The values must be separated by "\n." You should have enough data for at least two full cycles.

Laboratory report

1.   Setup

1.1 Draw a simple block diagram of your experimental setup.

1.2 Draw your timing diagram as you saw it on the oscilloscope. Include slightly more than one sampling cycle.

2.   Data summary and analysis

2.1 Comparison between measured transition voltages and the linear model. The linear model is given by:

$$V^{\text{lin}}(n, n+1) = V_{0,1} + n\Delta V, \qquad \text{where } \Delta V = [V_{126,127} - V_{0,1}]/126$$

Note that this line passes through the measured end points, since $V^{\text{lin}}(0,1) = V_{0,1}$ and $V^{\text{lin}}(126,127) = V_{126,127}$. Summarize your results in a table with the following headings:

- n
- $V_{n,n+1}$ measured transition voltages
- Linear model $V^{\text{lin}}(n, n+1) = V_{0,1} + n\Delta V$
- Difference $V_{n,n+1} - V^{\text{lin}}(n, n+1)$ in millivolts
- Difference $[V_{n,n+1} - V^{\text{lin}}(n, n+1)]/\Delta V$, where $\Delta V = 1$ LSB

2.2 Differential linearity. Tabulate $V_{n,n+1} - V_{n-1,n}$ from your data and compare with the average step size ΔV to determine the differential linearity error.

2.3 Rms deviation. Compute the rms deviation between $V_{n,n+1}$ and $V^{\text{lin}}(n,n+1)$:

$$V_{\text{rms}} = \sqrt{\frac{1}{M} \sum_{i=1}^{M} [V^{\text{lin}}(n_i, n_i+1) - V_{n_i,n_i+1}]^2}$$

where the summation is carried out only over your M measured values of V_{n_i,n_i+1}.

2.4 Sampling frequency. From the 100-Hz sine wave in procedure section 3, compute your sampling frequency in samples per second. (Frequency aliasing will not occur if you sample more than twice per sine wave.)

2.5 Power-supply rejection ratio. For the $V_{0,1}$ and $V_{126,127}$ transition voltages, compute the power-supply rejection ratio as:

$$\frac{\Delta V_{n,n+1}}{\Delta V_S}$$

where $\Delta V_{n,n+1}$ is the change in the transition voltage $V_{n,n+1}$ for a change in power-supply voltage ΔV_S.

3.   Discussion and conclusions

3.1 Discuss the timing diagram you measured in procedure section 3.3.

3.2 Discuss how well you were able to measure the transition voltages in procedure section 3.4.

3.3 Discuss any hysteresis you observed in procedure section 3.5 and possible causes.

3.4 Discuss the importance of good power-supply rejection for A/D converters used in battery-operated equipment.

3.5 Discuss procedure section 4. Consider how these techniques can be used to sample analog waveforms and their limitations.

3.6 Describe how your A/D differed from the specifications promised in the data sheets in terms of resolution, absolute (unadjusted) accuracy, linearity (relative accuracy), differential linearity, conversion speed, and power-supply rejection ratio.

4.   Questions

4.1 Was the $V_{0,1}$ transition voltage what you expected? Was the $V_{126,127}$ transition voltage what you expected?

4.2 What was the largest millivolt deviation between your transition data and the linear model?

4.3 What was your sampling frequency in procedure section 4? What is the maximum frequency sine wave that your system can sample without aliasing?

4.4 In characterizing an A/D converter, why is it better to measure transition voltages rather than recording digital output values for a number of random analog input voltages?

4.5 How would you use a high-resolution D/A converter and a computer with a digital I/O port to automate the measurement of A/D transition voltages?

5.   Program and laboratory data sheets

5.1 Include printouts of your program code, data, and output.

5.2 Include your handwritten data sheets (or a copy), which should consist of a log of the procedures you used, any special circumstances, and the measurements you recorded manually.

Laboratory Exercise 10

Frequency aliasing

Purpose

To use an analog I/O board to sample sine waves of various frequencies, to output the sampled data in analog form, and to observe frequency aliasing.

Equipment

- IBM-compatible Pentium microcomputer with Windows NT operating system and Microsoft Visual C++ compiler
- Data Translation DT3010 PCI plug-in board
- Printer (shared with other laboratory stations)
- Wave generator
- Oscilloscope

Additional reading

Section 3.5 Sampling analog waveforms
Section 3.6 Frequency aliasing
Appendix E DT3010 analog input and analog output

Procedure

1.   Circuit

Connect the output of the wave generator to the DT3010 analog input channel 0. The analog inputs are differential, so connect the signal to the + input (pin 1) and the wave

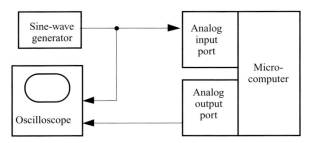

Laboratory Figure 10.1 Setup for digital sampling and recovery of waveforms.

generator ground to the − input (pin 2). Connect the wave generator to one input and the DT3010 analog output (pin 41) to the input of your oscilloscope. Connect DT3010 pins 2, 42, 50, 57, and 106 to the grounds of your wave generator and oscilloscope (Laboratory Figure 10.1).

2.   Programs

See Appendix E for a summary of the DT3010 capabilities and programming considerations.

Write a program that does the following:
1. Initializes the DT3010.
2. Converts analog input to digital data.
3. Stores the converted digital data as one element in a 200 array.
4. Converts the digital data to analog output.
5. Loops over steps 2 to 4 200 times to fill the array with 200 sequential samples.
6. Loops over steps 2 to 5 for a minute or two. Alternatively, loops over steps 2 to 5 until a key on the keyboard is depressed.
7. Writes the 200 sequential digital data values to a disk file for subsequent screen display, printing, and plotting.

For future reference, we will call steps 1 to 3 the "standard analog I/O loop." By using the same loop, analog input and output will occur at essentially the same data rate. This will be useful in future laboratory exercises, when a number of samples of an analog signal are to be acquired, stored, processed, and later output at the same rate that they were acquired.

The "standard analog I/O loop" has the following code:

```
#include <windows.h>
#include <stdio.h>
#include "DAboard.h"
int main
{
```

```
#define IMAX 200
unsigned int n, val[IMAX];
InitAll();
for (n=0; n<200; n++)
{
    olDaGetSingleValue(hAd, &val[n], 0, 1.0);
    olDaPutSingleValue(hDa, val[n], 0, 1.0);
}
```

Sampling rate for a 300 MHz Pentium is approximately 12 kHz.

3.   Sampling a sine wave with frequency $0.1 f_s$

Adjust the frequency of the sine wave until there are exactly 10 samples per cycle as seen on the oscilloscope. (Expand the time scale so that you see a single cycle.) Use the horizontal oscilloscope markers to measure the time it takes to acquire 10 samples. If the sine wave is significantly slower than 1 kHz, you may have unnecessary code in your acquisition loop. Note differences between the analog input and output waveforms. Pay particular attention to the detailed shape of the waveforms, phase shift between the waveforms, and relative p–p amplitudes. Print or plot the digital data values.

4.   Frequency aliasing

4.1 Increase the sine-wave frequency to $0.25 f_s$. Note how the sampled values are less able to recover the p–p amplitude of the sampled waveform. Observe and compare the raw and low-pass filtered D/A outputs. Print or plot the digital data values.

4.2 Increase the sine-wave generator frequency to $0.5 f_s$. Note that when sampling at exactly one-half the sine-wave frequency, it is possible to sample only at the zero crossings, or only at peaks and valleys, or anything in between, depending on the relative phase of the sine-wave generator and the sampling program. Since the frequency is not exactly $0.5 f_s$, the phase will drift at a fixed rate and the amplitude of the observed oscillations will slowly oscillate (see Figure 3.32). Observe and compare the raw and low-pass filtered D/A outputs. The rapid oscillations have the frequency of the sine-wave input while the slow oscillations have a "beat" frequency equal to the difference between the sine wave and $0.5 f_s$. Print or plot the digital data values.

4.3 Increase the sine-wave generator frequency to $0.75 f_s$. Measure the apparent frequency of the recovered analog output. Print or plot the digital data values.

4.4 Increase the sine-wave frequency to be close to but not exactly at f_s. The apparent frequency should be much slower than f_s. Measure the apparent frequency of the recovered analog output. A print or plot of digital data values is not required.

4.5 Increase the sine-wave frequency to 1.5, 1.75, and close to $2f_s$. Measure the apparent frequencies of the recovered analog output. A print or plot of digital data values is not required.

4.6 Repeat procedure steps 4.1, 4.2, 4.3, and 4.4 for a triangle wave (generator frequencies 0.25, 0.5, 0.75, and close to f_s). Print or plot the digital data values at 0.25 and $0.75 f_s$. Measure the observed frequencies and sketch the shapes at $0.5 f_s$ *and* close to f_s.

Laboratory report

1. Setup

Draw a simple block diagram of your experimental setup.

2. Data summary and analysis

2.1 From the $0.1 f_s$ sine wave in procedure section 3, determine your sampling frequency in hertz. Frequency aliasing will not occur with a slow sine wave.

2.2 For all the sine waves and triangle waves you sampled, prepare a table that compares the actual input frequency f with the apparent frequency f_0 of the sampled waveform. Include a column of the expected apparent frequency f_0 derived from Figure 3.32.

3. Discussion and conclusions

3.1 Describe and discuss your observations from procedure sections 3 and 4. Explain how you measured the sampling frequency in procedure section 3. Discuss the waveforms you observed in procedure sections 3 and 4. Pay particular attention to the detailed shape of the waveforms, phase shift between the waveforms, and relative p–p amplitudes.

3.2 Draw conclusions about what you have learned about sampling. For example, if you are only able to sample a few cycles of a pure sine wave, how frequently do you need to sample to get an accurate measurement of its p–p amplitude?

3.3 Discuss the ability and limitations of this technique to sample analog waveforms, store them digitally, and recover them at a later time.

Note: The audio compact disk is an important application of these techniques.

4. Questions

4.1 What was your sampling frequency in procedure section 3?

4.2 Were the apparent frequencies of the recovered analog waveforms approximately the same for sine-wave input frequencies of 0.25 and $0.75 f_s$? Explain.

4.3 What is the maximum frequency sine wave that your system can sample without aliasing?

4.4 Did aliasing at $f \approx f_s$ change the shape of the triangle wave? Explain your answer in terms of progressive shifts in the sampling time.

4.5 How would you sample a periodic 1-MHz waveform so that the sampled values had the same shape as the original waveform but occurred at a slower frequency of 1 kHz?

5. Program and laboratory data sheets

5.1 Include printouts of your program code, data, and output.

5.2 Include your handwritten data sheets (or a copy), which should consist of a log of the procedures you used, any special circumstances, and the measurements you recorded manually.

4 Sensors and actuators

4.1 Introduction

The **transducer** is a device that converts one form of energy to another. For interfacing to a microcomputer, we are primarily interested in the **electronic transducer**, which has an input or an output that is electrical in nature, such as a voltage, current, or resistance. The **sensor** is an electronic transducer that converts a physical quantity into an electrical signal. An **actuator** is an electronic transducer that converts electrical energy into a physical quantity, and is an essential element in control systems.

Sensors are used to detect displacement, temperature, strain, force, light, etc. Almost all sensors require additional circuits to produce the voltage and current needed for analog-to-digital conversion. As we shall see in this chapter, the thermistor changes its electrical resistance as a function of temperature, and a bridge is needed to produce a corresponding voltage, whereas the silicon photodiode produces a current, and a stage of amplification is needed to produce a voltage. Often, the term "sensor" includes both the transducer and the circuits needed to produce an output voltage.

Laboratory Exercise 11 uses a circular resistor and a computer to record angle and the oscillations of a damped pendulum. Laboratory Exercise 12 explores the measurement of temperature using the dial thermometer, a platinum resistance thermometer, the thermocouple, and the thermistor. Laboratory Exercise 13 measures force, using four metal foil strain gauges bonded to a plastic rod and wired in opposing pairs to form a bridge circuit. Laboratory Exercise 14 uses a silicon photodiode to measure light and the absorption of light by colored solutions. Laboratory Exercise 15 explores the thermoelectric heat pump and its ability to heat and cool a small system. Laboratory Exercise 16 measures the offset potential and frequency-dependent complex impedance for bare metal and Ag(AgCl) electrodes. Laboratory Exercise 17 measures the human electrocardiogram (ECG), phonocardiogram, and blood pressure. Laboratory Exercise 18 amplifies and processes the electromyogram (EMG) from the skin surface and relates it to the mechanical tension produced by the underlying muscles. Laboratory Exercise 19 measures the position of the eyes using the electrooculogram (EOG) to determine the maximum angular velocity of voluntary and involuntary eye motion.

Several important characteristics that are common to most sensors are now described.

1. The **transfer function** (or **response function**) of a sensor is its output as a function of the quantity being sensed. It is usually expressed in terms of a curve or a formula.

2. The **sensitivity** of a sensor is defined as the change in output for a unit change in the quantity being sensed. This is the first derivative of the response curve and generally depends on the value of the quantity being measured. For example, the iron–constantan thermocouple has a sensitivity of 50 μV/°C at 0 °C.

3. The **linearity error** of a sensor is the difference between the sensor response and either a best-fit straight line or a straight line passing through the end points. In either case, the value of the linearity will depend on the range of measurements used. Generally, sensors become more linear as the measurement range is restricted. For example, the thermocouple output depends almost linearly on temperature over a wide range. On the other hand, the thermistor resistance depends exponentially on temperature and the linear approximation is poor even over a small temperature range.

4. The **accuracy error** of a sensor is the difference between the measured quantity and the "true" value, as defined by accepted standards. The measured quantity is determined from the sensor output and the "ideal" relationship between output and input.

5. The **precision** of a sensor is the ability to detect small changes in the measured quantity reliably, and the ability to measure the same value under repeated identical conditions.

6. The **stability** of a sensor is the ability to maintain the same response and noise level, despite the effects of time and usage.

7. The **noise** of a sensor is any component of the output that would be interpreted as a signal but does not depend on the quantity being sensed. This includes thermal noise in resistors, shot noise in amplifier elements, external electrical interference, etc.

8. The **response curve** of a sensor is the output versus time curve after an abrupt change in the input. Usually, the sensor output changes quickly at first, and then more slowly as it asymptotically reaches a final value. For a simple system, this response curve is often described by:

$$V(t) = V_2 + (V_1 - V_2)e^{-t/\tau}$$

where V_1 is the initial output value, V_2 is the final (asymptotic) value, and τ is the exponential response time. To estimate τ, plot $V(t)$ and draw a horizontal line 63.2% of the way from V_1 to V_2. The curve will pass through this line at $t = \tau$. Several time constants may be required before the output agrees with its final value within the desired accuracy. For example, if an accuracy of 0.5% of $V_2 - V_1$ is desired (and the previous formula is appropriate), then it is necessary to wait 5.3τ.

Table 4.1 *Examples of sensors and actuators*

Real-world quantity	Sensor	Actuator
Motion	Digital encoder	Stepping motor
Temperature	Thermocouple	Resistor
Strain	Resistive wire	Piezoelectric
Force	Load cell	Motor
Light	Photocell	Light bulb
Image	CCD camera	Ink jet printer
Pressure	Strain gauge membrane	Pump
Radiation (π, α, β, γ, etc.)	Geiger counter	Cyclotron
Radio waves	Radio receiver	Radio transmitter

9. The **response time** of a sensor is the time required for its output to change from 10 to 90% of its final value change ($V_2 - V_1$) in response to an abrupt change in the quantity being sensed. In terms of the example in (8), the 10% point is reached in 0.1τ and the 90% point is reached in 2.3τ. The response time is then 2.2τ.

10. The **temperature coefficient** of a sensor is the change in a quantity per unit temperature change. This is an important characteristic of sensors used to measure quantities other than temperature but are nonetheless sensitive to temperature. For example, the temperature coefficient applies to the leakage current and offset voltage of an amplifier, the dark current of a silicon photodiode, the change in resistance of a strain gauge, etc.

11. The **hysteresis** of a sensor is the dependence of its output on previous history. It is very common in magnetic and mechanical systems (backlash).

Table 4.1 lists some examples of physical quantities and the associated sensors and actuators.

4.2   Position and angle sensors

Computer-readable position sensors are used in a variety of applications, including:
- determining the relative position and angle of machine tools and parts during machining,
- determining the altitude and azimuth angles of a solar collector,
- providing a robot with the ability to sense the angle and position of its "limbs."

4.2.1   The potentiometer

The potentiometer consists of a resistive element distributed along a line or around an arc and a sliding contact connected to the point whose position or angle is to be

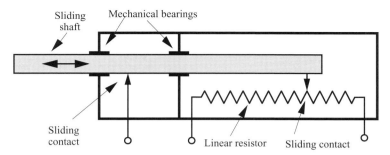

Figure 4.1 Linear potentiometer, which relates a position to a contact point along a linear resistor.

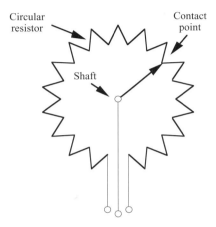

Figure 4.2 Rotary potentiometer, which relates a shaft angle to a contact point along a circular resistor.

sensed. Electrical connections are provided to each end of the resistor and the sliding contact (Figures 4.1 and 4.2). By providing a fixed voltage across the entire resistor, the potential of the sliding contact depends on the position or angle.

The common three-quarter-turn potentiometer is the lowest-cost angle sensor, but has limited range and accuracy. It is used in Laboratory Exercise 11 to measure angle and the dynamic response of the damped pendulum.

For improved accuracy, 5- to 20-turn rotary potentiometers are made with accuracies and linearities of 0.1%. In many cases, this is sufficient for the application and much simpler and less costly than the digital encoders discussed in the next section.

For sensing linear motion, an alternative to the linear potentiometer (Figure 4.1) is an accurate multi-turn rotary potentiometer connected to the point to be sensed with tensioned steel cables and pulleys. A significant advantage of this approach is that it provides a linear range that greatly exceeds the length of available linear potentiometers.

Figure 4.3 shows the loading effect when the voltage sensor has an input impedance comparable to the linear resistor. It is helpful to add another resistor (Figure 4.4) so

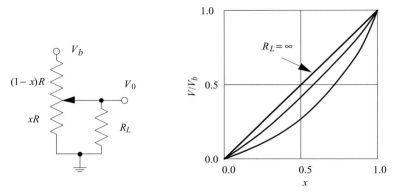

Figure 4.3 Linear resistive position sensor without linearization.

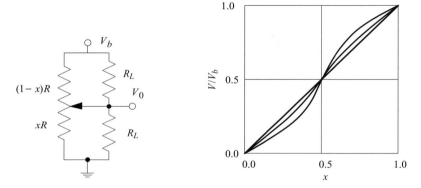

Figure 4.4 Linear resistive position sensor with linearization.

that the output is constrained to read correctly at mid-range.

$$V_0 = V_b \frac{xR \| R_L}{(1 - x)R + xR \| R_L} = V_b \frac{xR_L}{R_L + xR(1 - x)}$$

When $x = 0.5$, then $V_0 = V_b \frac{2R_L}{4R_L + R}$

$R_L = R$ and $x = 0.5$, then $V_0 = 0.4V_b$

4.2.2   The digital encoder

The digital encoder provides a very precise conversion from a shaft angle (or a linear position) to a digital number. There are two basic types. The **relative digital encoder** consists of a disk (or strip) with a pattern of uniformly spaced marks and a sensor that detects the marks and produces pulses as the strip is rotated (or translated). The pulses are counted to give a number whose value is proportional to relative angle (or position). A battery-powered circuit can be used to keep track of absolute position during power shutdowns.

Decimal	Binary		Gray code	
15		1111		1000
14		1110		1001
13		1101		1011
12		1100		1010
11		1011		1110
10		1010		1111
9		1001		1101
8		1000		1100
7		0111		0100
6		0110		0101
5		0101		0111
4		0100		0110
3		0011		0010
2		0010		0011
1		0001		0001
0		0000		0000

Figure 4.5 Digital encoder patterns for linear position sensing. The dark portion of the patterns represents a "zero" (blocks light) and the white portion of the patterns represents a "one" (lets light through). The pattern on the left is binary code, where the rightmost strip is the LSB and the leftmost strip is the MSB. The pattern on the right is Gray code, where only one bit switches at a time to avoid ambiguities at transition points.

The **absolute digital encoder** has a series of patterns of marks that can be uniquely related to the absolute angle (or position). The absolute-position encoder keeps track of absolute position during power shutdowns without the need for additional circuitry.

Early designs of absolute-position encoders used circular patterns of insulating and conducting regions. The conducting regions were detected by electrical contacts made by "brushes" (similar to those in small electric motors), and these encoders are called "brush encoders." The primary problem with the brush encoder is bit errors due to dust, oil, oxide layers, and the occasional loss of electrical contact with the rotating surface. Modern "optical encoders" use patterns of opaque and transparent regions, where the transparent regions are detected optically (using light-emitting diodes and silicon photodiodes).

Digital linear encoder

If the pattern is arranged in a binary code, as shown on the left-hand side of Figure 4.5, and an array of light-emitting diodes and photodiodes are placed on opposite sides of the pattern as shown in Figure 4.6, the angular position is transformed to a series of "zeros" and "ones," depending on whether light can pass through the segments of the pattern. However, many bits can switch at a single transition (e.g. between 01111 and 10000), and since there is no way of ensuring that all bits will switch at exactly the same instant, it is possible to read an incorrect code at the transition. Adding an

Light-emitting diodes

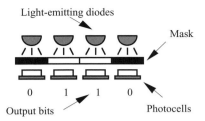

Mask

0   1   1   0

Output bits   Photocells

Figure 4.6 Arrangement of light-emitting diodes and photocells for reading the digital encoder pattern of Figure 4.5.

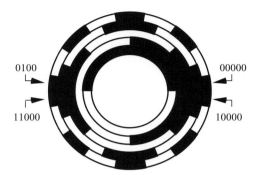

0100   00000

11000   10000

Figure 4.7 Circular optical-encoder pattern. Gray code is shown here, which has the property that only one bit switches at a time. The code starts at 0000 and advances counterclockwise to 1000.

additional timing bit to the pattern will provide a pulse that can be used to latch the data into a buffer only when all bits are stable, but it is more common to overcome this problem by arranging the pattern in the Gray code, where only one bit switches at a time (Figure 4.5, Table 1.3). With Gray code, all bits are always valid and may be read at any time. However, the circuit that converts from Gray to binary code (Figure 1.3) does not have reliable data during the conversion and should be read only after conversion is complete.

In both cases, 0000 is at the bottom and the code advances upward. Whereas five bits are shown here as an example, optical encoders range from 10 to 20 or more bits. They are more expensive than the simple linear potentiometer described in the previous section, but are used extensively whenever high accuracy is required. Applications include hand-held precision digital calipers, which include batteries, Gray code to BCD conversion circuits, and digital display in English or metric units. The light emitters and sensors are shown in Figure 4.6.

Digital rotary encoders

The rotary version of the digital encoder is shown in Figure 4.7. The code 00000 is to the right and advances counterclockwise. Applications include numerically controlled milling machines and optical and radio telescopes.

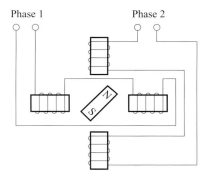

Figure 4.8 Electromagnets and coils used in the stepping motor. By sequencing the direction of current, the central permanent magnet can be rotated in precise, discrete angular increments.

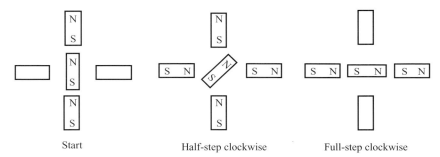

Start   Half-step clockwise   Full-step clockwise

Figure 4.9 Sequence of magnetic polarizations that rotate the permanent magnet clockwise in half-step increments.

One technical problem that arises is the difficulty of making the photoemitters and photodetectors small enough to fit in a line across the optical mask. Usually, these elements and their encoder disk circles are rotated to different angles to increase the distance between them.

4.2.3   The stepper motor

The stepper motor rotates in discrete angular increments, and is highly accurate and reproducible. It is the most commonly used mechanical actuator in digital control applications.

It consists of a central, magnetized rotor that moves from one magnetic well to the next magnetic well, controlled by switching current through electromagnets. See Figure 4.8 for a simplified four-step, two-phase stepping motor, and Figure 4.9 for the rotor positions involved in the first half-step and the first full step. Figure 4.10 shows the phase polarities for a complete sequence of eight half-steps.

Figure 4.11 shows a possible toothed-pole construction for a 16-step, four-phase stepper motor. Practical stepping motors have typically 200 steps per turn, 400 with

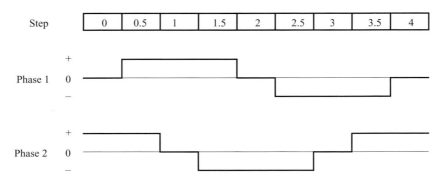

Figure 4.10 Sequence of currents that rotate the permanent magnet four steps (one complete rotation) in half-steps.

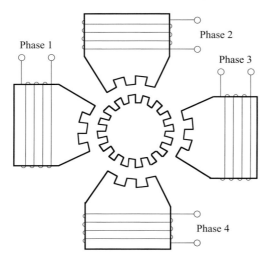

Figure 4.11 Possible toothed-pole construction for a 16-step, four-phase stepper motor.

half-stepping. A solid-state controller is usually used to convert clockwise and counterclockwise control pulses into the necessary electromagnet currents. At slow speeds (<50 steps/s), it is possible to control the stepping motor using individual steps. At higher speeds, momentum causes the rotor to advance, even after the pulses have been stopped. Precise, high-speed operation is only possible by varying the pulse rate so that the speed gradually increases to the maximum rate (typically 200 steps/s) and then gradually decreases as the final desired angular position is approached.

4.3   Temperature transducers

One of the most common temperature sensors that you have seen is the mercury or alcohol thermometer, which consists of a bulb connected to a sealed capillary tube. As

Table 4.2 *Temperature standards*

Phenomena	kelvin	°C
Minimum possible thermal energy	0	−273.15
Triple point of hydrogen	13.81	−259.34
Triple point of oxygen	54.36	−218.79
Boiling point of oxygen at 760 mm Hg	90.19	−182.96
Triple point of water (4.58 mm Hg pressure)	273.16	0.01
Freezing point of water at 760 mm Hg pressure	273.15	0
Boiling point of water at 760 mm Hg pressure	373.15	100
Freezing point of zinc	692.73	419.58
Freezing point of gold	1,337.58	1,064.43

the temperature of the bulb changes, the liquid expands or contracts and the change in volume is transduced to a change in length in the capillary tube.

Note that there are two transduction processes: the change in temperature to a change in volume (which assumes that the temperature coefficient of the liquid is nonzero and does not change its sign over the temperature of interest) and a change in volume to a change in length (which assumes that the volume of the thermometer changes less than the volume of the liquid). Another nonelectric temperature transducer is the liquid crystal thermometer, which can be purchased in drugstores and placed on the forehead to indicate fever. The most important electric temperature transducers (the platinum resistance thermometer, the thermocouple, and the thermistor) are explored in Laboratory Exercise 12.

4.3.1   Temperature standards

The primary temperature standards are fixed temperatures produced by physical phenomena (Table 4.2).

Examples are the triple point of hydrogen at 13.81 K, the triple point of water at 273.16 K, and the freezing point of gold at 1,337.58 K. The triple point is a specific point in the temperature versus pressure plane where all three phases (solid, liquid, and gas) coexist in equilibrium (Figure 4.12). These are expressed in the **Kelvin** absolute thermodynamic temperature scale, where 0 K corresponds to the minimum possible thermal energy. It was developed in the early 1800s by Lord Kelvin and based on the coefficient of expansion of an ideal gas.

In science and engineering, the **Celsius** temperature scale is also used and is defined with 0 °C at the freezing point and 100 °C at the boiling point of water at standard pressure. The triple point of water occurs at a lower pressure (6.11 mbar or 4.58 mm Hg) and is slightly warmer (0.01 °C). The freezing point of water is at 273.15 K, and to convert from Celsius to Kelvin scales, add 273.15. This scale was developed around 1742 by Anders Celsius.

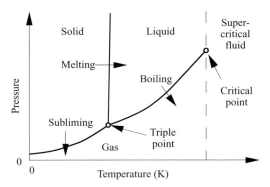

Figure 4.12 Phase diagram for water, showing the regions of pressure and temperature where water exists in the solid, liquid, gaseous, and super-critical states.

The **Fahrenheit** temperature scale is defined by 32 °F at the ice point and 212 °F at the boiling point of water at standard pressure. Its use is discouraged for scientific writing. To convert from Fahrenheit to Celsius, subtract 32 and multiply by 5/9. This scale was developed in the early 1700s by Gabriel Fahrenheit, a Dutch instrument maker, who based it on a mixture of ice, water, and ammonium chloride (the lowest temperature that he could reliably produce) at 0 °F and the temperature of the human body at about 99 °F.

The **Rankine** temperature scale is the Fahrenheit equivalent of the Kelvin temperature scale. To convert from Rankine to Fahrenheit, add 459.67.

Absolute zero is the temperature corresponding to the minimum possible thermal energy and is defined as 0 K, −273.15 °C, −459.67 °F, and 0 °R.

The **calorie** is the quantity of thermal energy required to raise 1 g of water 1 °C at 15 °C. It is equal to 4.186,8 watt seconds. The Calorie used in nutrition (note the capital C) is 1,000 times larger.

Between these fixed temperature standards, interpolation standards are used. The most practical of these is the platinum resistance thermometer.

4.3.2   Platinum resistance thermometer

Platinum is a noble metal that can withstand high temperatures and harsh chemical environments with good stability. The standard **platinum resistance temperature detector** (PRTD), or **platinum resistance thermometer**, is a fine platinum wire carefully trimmed to a resistance of 100.0 Ω at 0 °C. It is used as an interpolation standard from −183 to +631 °C and will withstand temperatures as high as 800 °C. The resistance is given as a function of temperature by the equation:

$$R = R_0[1 + \alpha T - \alpha \delta (T/100 - 1)(T/100)]$$

where R is the resistance in ohms, and T is the temperature in degrees celsius,

Table 4.3 *Platinum resistance R versus temperature T*

$T(°C)$	$R(\Omega)$	$T(°C)$	$R(\Omega)$	$T(°C)$	$R(\Omega)$
−200	18.49	−100	60.25	0	100.00
−190	22.80	−90	64.30	10	103.90
−180	27.08	−80	68.33	20	107.79
−170	31.32	−70	72.33	30	111.67
−160	35.53	−60	76.33	40	115.54
−150	39.71	−50	80.31	50	119.40
−140	43.87	−40	84.27	60	123.24
−130	48.00	−30	88.22	70	127.07
−120	52.11	−20	92.16	80	130.89
−110	56.19	−10	96.09	90	134.70
−100	60.25	0	100.00	100	138.50

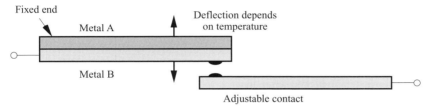

Figure 4.13 Bimetallic switch, whose deflection depends on the temperature. Contact is made or broken at a temperature value set by the adjustable contact.

$\alpha = 0.003,92/°C$, and $\delta = 1.49$. This expression simplifies to:

$$R = R_0(1 + 0.003,98T - 5.8 \times 10^{-7}T^2)$$

See Table 4.3 for R versus T over a large temperature range. Usually, the platinum is used in a resistance bridge, but a digital multimeter may be used as well. For faster response and improved ruggedness, the platinum resistance thermometer is also available in the form of a thick film of platinum metal on an alumina (Al_2O_3) substrate.

4.3.3 The bimetallic switch and the dial thermometer

One of the most common temperature sensors is the bimetallic switch, which opens and closes an electrical contact, depending on whether the temperature is above or below a set value (Figure 4.13). It consists of two layers of materials (usually metal alloys) with different thermal-expansion coefficients (Table 4.4), so that mechanical bending occurs as the temperature changes.

Applications include temperature control for rooms and ovens (the switch actuates an electric heater or flame), temperature control in a mixing faucet (the deflection controls

Table 4.4 *Linear thermal expansion coefficients α of materials at* 20 °C. $\alpha = dL/(LdT)(10^{-6}/°C)$

Material	α	Material	α
$Al_2O_3(\parallel c$ axis)	5.6	Platinum	8.9
$Al_2O_3(\perp c$ axis)	5.0	Silicon	2.5
Aluminum	23.0	$SiO_2(\parallel$ to axis)	7.4
Carbon	7.8	$SiO_2(\perp$ to axis)	13.1
Copper	16.7	SiO_2 (vitreous)	0.35
Gold	14.2	Silver	19.0
Iron	11.8	Stainless steel 304	15.9
Invar	1.2	Tungsten	4.5
Lead	28.7	Yellow brass	19.0
Nickel	12.8		

the ratio of hot to cold water), and flashers for decorative lights and automotive turn signals. If the bimetallic strip is wound into a helix, with one end fastened and the other end connected to a needle indicator, the angle of the needle will depend on the temperature. Equipped with a temperature scale, this device is the "dial thermometer" and is commonly found in kitchens for measuring the temperature of meat or candy. It is used to determine approximate temperature values in Laboratory Exercises 4, 5, 12, 13, 15, 25, 26, and 27. If the rotating end is connected to an angle sensor (described in the previous section), the result is a rugged temperature sensor with good sensitivity and linear response. The main disadvantages are the length and thermal mass of the sensing element.

4.3.4   The thermocouple

Thompson emf

When a bar of metal or semiconductor is heated at one end, some of the conduction electrons diffuse from the warmer end to the colder end. Although the electrons are free to move throughout the bar, they spend more time at the cooler end where their agitation velocity is lower. The positive ions, on the other hand, are immobile. The **Thompson emf** is the electrostatic potential difference that results after this thermal rearrangement. At equilibrium, the diffusion force (caused by the difference in temperature) and the electrostatic force (caused by the charge separation) are balanced (Figure 4.14). Note that the actual number of electrons separated is very small compared to the number of conduction electrons in the metal. The Thompson emf depends both on the material and the temperature difference between its ends.

If you connect the ends of the bar to a voltmeter, the voltage will read zero (electric fields are zero in conductors), but if you connect the metal bar to a battery, the resistance will be lower in the direction of electron flow that is facilitated by the thermal diffusion.

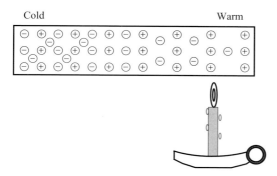

Figure 4.14 Distribution of charge in a conductor that is warmer at one end. The positive ions are locked in place but the electrons are free to move and spend most of their time at the colder end.

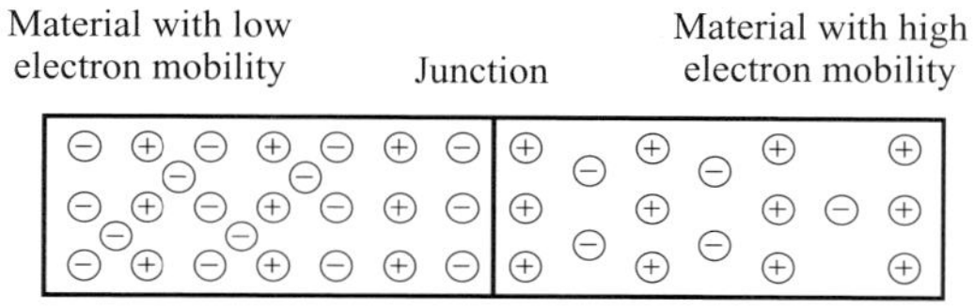

Material with low   Material with high
electron mobility   Junction   electron mobility

Figure 4.15 Distribution of charge when two materials of dissimilar electron mobility are brought into contact. The positive ions are locked in place but the electrons are free to move and spend most of their time in the material of lower electron mobility.

This current will transport warm electrons to the cooler end of the bar and *convert heat to electrical energy.* On the other hand, if the current opposes the thermal diffusion, the resistance will be higher and *electrical energy will be converted into heat.*

Historical note

The Thompson emf was predicted by William Thompson (Lord Kelvin) in 1854.

Peltier emf

When two unlike conductors are brought into electrical contact, there will be a diffusion of electrons from the high-mobility metal to the low-mobility metal (Figure 4.15). The **Peltier emf** is the resulting diffusion potential. Again, zero voltage will be measured across the opposite ends of the two conductors, but the electrical resistivity will be different in the two directions of current flow.

If electrons are driven from the low-mobility conductor to the high-mobility conductor, the effect is like an expanding gas, and the junction will be cooled. If the electrons are driven from the high-mobility conductor to the low-mobility conductor, the effect is like compressing a gas, and the junction will be heated. The Peltier emf depends on the two materials and the temperature of the junction.

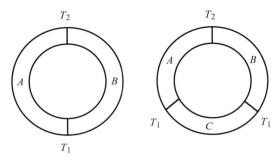

Figure 4.16 Circuits formed when dissimilar conductors A and B are joined and the junctions are at different temperatures T_1 and T_2. In the thermocouple temperature sensor, the conductor is broken at C and the voltage developed is approximately proportional to the temperature difference, $T_1 - T_2$.

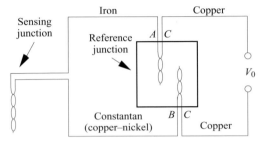

Figure 4.17 Iron–constantan thermocouple for sensing the temperature difference between the reference junction and the sensing junction. V_0 is approximately 50 μV for each degree celsius difference between the sensing and reference junctions.

Historical note

The Peltier effect was discovered by Jean Peltier in 1834 when he thought that he had discovered a violation of Ohm's law.

Seebeck emf

If a closed circuit is formed by two different metals (Figure 4.16) and if the junctions are at different temperatures, the net result is the Seebeck emf (the sum of two Thompson emfs and two Peltier emfs), and a current will flow around the circuit.

The thermocouple as a temperature sensor

If the circuit in Figure 4.16 is broken, a potential difference will develop (the thermocouple potential) that depends on the temperature difference between the two junctions. In practice, the two dissimilar thermocouple wires are welded together at one end to form the sensing junction and welded at the other two ends to copper wires for connection to a voltmeter (Figure 4.17). Note that it is easier to measure the thermocouple

Table 4.5 *Physical properties of thermocouples*

+ Element	− Element	Type	Temp. range (°C)	Sensitivity (μV/°C)	Environment
Copper	Constantan	T	−200 to 370	40.5	Reducing, inert, or vacuum
Chromel	Constantan	E	−200 to 900	67.9	Oxidizing or inert
Iron	Constantan	J	0 to 760	52.6	Reducing, inert, or vacuum
Chromel	Alumel	K	−200 to 1,250	38.8	Oxidizing or inert
Pt(13%Rh)	Platinum	R	0 to 1,450	12.0	Oxidizing or inert
Pt(10%Rh)	Platinum	S	0 to 1,450	10.6	Oxidizing or inert
Pt(30%Rh)	Pt(6%Rh)	B	0 to 1,700	7.6	Oxidizing or inert
W(5%Rh)	W(26%Rh)		0 to 2,320	16.6	Vacuum, inert, or hydrogen
W	W(26%Rh)		0 to 2,320	16.0	Vacuum, inert, or hydrogen
W(3%Rh)	W(25%Rh)		0 to 2,320	17.0	Vacuum, inert, or hydrogen

Source: *Omega Temperature Measurement Handbook and Encyclopedia*. Reproduced with the permission of Omega Engineering, Inc., Stamford, CT.

potential after the circuit is broken than it is to measure the Seebeck current that exists before the circuit is broken.

Use of a high-impedance meter keeps the current very low and resistive heating (Joule) and heat transfer between the junctions (Peltier effect) is then negligible. During use, the connections to the copper wire (which constitute a single junction) are kept at a known reference temperature. To justify the inclusion of the copper wire, note that the Thompson emf in the copper wires cancels (both have the same temperature difference across them) and the Peltier emf across an iron–constantan reference junction is the same as the sum of the Peltier emfs across the iron–copper and constantin–copper junctions (both junctions are at the same temperature).

Table 4.5 lists the commonly available thermocouples, which cover a wide variety of temperature ranges. Table 4.6 lists the Seebeck emf as a function of temperature of the more commonly used thermocouples. Alumel is an aluminum–nickel alloy, chromel is a chromium–nickel alloy, and constantan is a copper–nickel alloy.

If the data-acquisition system is capable of accurately recording the thermocouple voltage V, the National Bureau of Standards (NBS) polynomials (Table 4.7) can be used to estimate the temperature T (°C):

$$T = a_0 + a_1 V + a_2 V^2 + \cdots + a_n V^n \tag{4.1}$$

For iron–constantan, T is given to an accuracy of 0.1 °C over the 0–760 °C range by the coefficients in Table 4.7. As seen from Table 4.8, the fit is excellent within the

Table 4.6 *Thermoelectric output (mV) of common thermocouples*

$T(°C)$	Copper–constantan	Chromel–constantan	Iron–constantan	Chromel–alumel	Platinum–Pt (10%Rh)
−260	−6.232	−9.797			
−240	−6.105	−9.604			
−220	−5.889	−9.274			
−200	−5.603	−8.824	−7.890	−5.891	
−180	−5.261	−8.273	−7.402	−5.550	
−160	−4.865	−7.631	−6.821	−5.141	
−140	−4.419	−6.907	−6.159	−4.669	
−120	−3.923	−6.107	−5.426	−4.138	
−100	−3.378	−5.237	−4.632	−3.553	
−80	−2.788	−4.301	−3.785	−2.920	
−60	−2.152	−3.306	−2.892	−2.243	
−40	−1.475	−2.254	−1.960	−1.527	
−20	−0.757	−1.151	−0.995	−0.777	
0	0.000	0.000	0.000	0.000	0.000
20	0.789	1.192	1.019	0.798	0.111
40	1.611	2.419	2.058	1.611	0.232
60	2.467	3.683	3.115	2.436	0.363
80	3.357	4.983	4.186	3.266	0.501
100	4.277	6.317	5.268	4.095	0.647
120	5.227	7.683	6.359	4.919	0.800
140	6.204	9.078	7.457	5.733	0.959
160	7.207	10.501	8.560	6.539	1.124
180	8.235	11.949	9.667	7.338	1.294
200	9.286	13.419	10.777	8.137	1.468
250	12.011	17.178	13.553	10.151	1.923
300	14.860	21.033	16.325	12.207	2.400
350	17.816	24.961	19.089	14.292	2.896
400	20.869	28.943	21.846	16.395	3.407
450		32.960	24.607	18.513	3.933
500		36.999	27.388	20.640	4.471
600		45.085	33.096	24.902	5.582
700		53.110	39.130	28.128	6.741
800		61.022		33.277	7.949
900				37.325	9.203
1,000				41.269	10.503
1,100				45.108	11.846
1,200				48.282	13.224
1,300				52.398	14.624
1,400					16.035
1,500					17.445
1,600					18.842
1,700					20.215

Source: *Omega Temperature Measurement Handbook and Encyclopedia*. Reproduced with the permission of Omega Engineering, Inc., Stamford, CT.

Table 4.7 *Polynomial coefficients for estimating temperature* (°C) *as a function of thermocouple output* (mV). *(See Eq. (4.1))*

	Copper–constantan	Chromel–constantan	Iron–constantan	Chromel–alumel	Platinum–Pt (10%Rh)	Platinum–Pt (13%Rh)
Range	−160 to 400 °C	−100 to 1,000 °C	0 to 760 °C	0 to 1,370 °C	0 to 1,750 °C	0 to 1,000 °C
accuracy	±0.5 °C	±0.5 °C	±0.7 °C	±0.7 °C	±1 °C	±0.5 °C
a_0	0.100,860,910	0.104,967,248	−0.048,868,252	0.226,584,602	0.927,763,167	0.263,632,917
a_1	25,727.943,69	17,189.452,82	19,873.145,03	24,152.109,00	169,526.515,0	179,075.491
a_2	−767,345.829,5	−282,639.085,0	−218,614.535,3	67,233.424,88	−31,568.363,94	−48,840.341,37
a_3	78,025,595.81	12,695,339.5	11,569,199.78	2,210,340.682	8,990,730.663	$1.900,02 \times 10^{10}$
a_4	−9,247,486,589	−448,703,084.6	−264,917,531.4	−860,963,914.9	$−1.635,65 \times 10^{12}$	$−4.827,04 \times 10^{12}$
a_5	$6.976,88 \times 10^{11}$	$1.108,66 \times 10^{10}$	2,018,441,314	$4.835,06 \times 10^{10}$	$1.880,27 \times 10^{14}$	$7.620,91 \times 10^{14}$
a_6	$−2.661,92 \times 10^{13}$	$−1.768,07 \times 10^{11}$		$−1.184,52 \times 10^{12}$	$−1.372,41 \times 10^{16}$	$−7.200,26 \times 10^{16}$
a_7	$3.940,78 \times 10^{14}$	$1.718,42 \times 10^{12}$		$1.386,90 \times 10^{13}$	$6.175,01 \times 10^{17}$	$3.714,96 \times 10^{18}$
a_8		$−9.192,78 \times 10^{13}$		$−6.337,08 \times 10^{13}$	$−1.561,05 \times 10^{19}$	$−8.031,04 \times 10^{19}$
a_9		$2.061,32 \times 10^{13}$			$1.695,35 \times 10^{20}$	

Source: Omega Temperature Measurement Handbook and Encyclopedia. Reproduced with the permission of Omega Engineering, Inc., Stamford, CT.

Table 4.8 *Thermoelectric voltage V of iron–constantan as a function of temperature T and the NBS polynomial (Table 4.7, Eq. (4.1))*

Temp. (°C)	Output (mV)	Polynomial temp. (°C)	Temp. (°C)	Output (mV)	Polynomial temp. (°C)
−200	−7.890	−177.23	160	8.560	159.97
−180	−7.402	−164.66	180	9.667	179.94
−160	−6.821	−150.05	200	10.777	199.93
−140	−6.159	−133.84	220	11.887	219.91
−120	−5.426	−116.40	240	12.998	239.92
−100	−4.632	−98.07	260	14.108	259.93
−80	−3.785	−79.08	280	15.217	279.95
−60	−2.892	−59.65	300	16.325	299.98
−40	−1.960	−39.93	350	19.089	350.06
−20	−0.995	−20.05	400	21.846	400.09
0	0.000	−0.05	450	24.607	450.06
20	1.019	19.99	500	27.388	499.98
40	2.058	40.02	550	30.210	549.91
60	3.115	60.06	600	33.096	599.92
80	4.186	80.08	650	36.066	650.02
100	5.268	100.07	700	39.130	700.10
120	6.359	120.05	750	42.283	749.99
140	7.457	140.01	760	42.922	759.93

0–670 °C range and rather poor below 0 °C, where the polynomial was not fit to the data.

Peltier thermoelectric device

By running a thermocouple backwards, it is possible to convert electrical energy into a temperature difference. Usually, these are made using *p*- and *n*-doped alloys of semiconductors such as Bi_2Te_3 and Sb_2Te_3 which have: (i) carriers with good mobility and heat capacity, and (ii) low thermal conductivity. For this application, metals are poor because they have high thermal conductivity and insulators are poor because they have low electron mobility.

Thinking of electrons as a refrigerant "gas," the electrical energy is used to compress electrons at the "hot" junction from *n*-type into *p*-type material and the electrons expand at the "cold" junction as they move from *p*-type to *n*-type material (Figure 4.18). If one side is kept at a fixed temperature by means of a heat reservoir or flowing water or air, the other side can be used for heating or cooling, depending on the direction of the electric current. The maximum cooling that can be achieved is limited by the rate at which the hot junction can be cooled, the heat load from the current passing through the device, and the heat from the surrounding medium. Single-stage units are capable

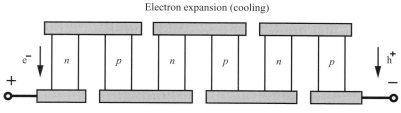

Electron expansion (cooling)

Electron compression (heating)

Figure 4.18 Peltier thermoelectric heat pump. Usually, the materials are p- and n-doped alloys of semiconductors such as Bi_2Te_3 and Sb_2Te_3 which have: (i) carriers with good mobility and heat capacity, and (ii) poor thermal conductivity. Reversing the flow of carriers (by reversing the power supply leads) reverses the flow of heat.

of boiling or freezing water (depending on the direction of heat flow) and cascaded triple-stage units can achieve temperatures as low as $-80\,^{\circ}C$.

The rate Q at which an object gains heat energy is given by:

$$Q = \pi I + I^2 R/2 + K_p(T_s - T_o) + K_a(T_a - T_o)$$

where π is the Peltier coefficient, I is the current through the Peltier device, R is the electrical resistance of the device, K_p is the thermal conduction coefficient of the device, K_a is the thermal convection coefficient to the air (or surrounding medium), T_o is the temperature of the object, T_s is the temperature of the heat sink, and T_a is the temperature of the air (or surrounding medium). The first term is the Peltier effect, the second is Joule heating, the third is conductive transfer through the device, and the fourth is conductive and convective transfer to the surrounding medium.

When used to cool an object, $I < 0$, $T_o < T_s$, and $T_o < T_a$. In this case only the Peltier term acts to cool the object – all other terms heat the object. At low currents, the first term can be larger than the second, resulting in a net cooling ($Q < 0$). However, as the current is increased, the I^2 term can dominate, resulting in net heating ($Q > 0$).

When used to heat an object, $I > 0$, $T_o > T_s$, and $T_o > T_a$. In this case both the Peltier and Joule terms heat the object, while the transfer to the heat sink and the surrounding medium cool the object.

When a constant current I is established, the object will approach an equilibrium temperature T_{equ} where heat transfer with the heat pump and the surroundings are equal. At this point $Q = 0$ and T_{equ} is given by:

$$T_{\text{equ}} = \frac{\pi I + I^2 R/2 + K_p T_s + K_a T_a}{K_p + K_a}$$

The relationship is parabolic and the minimum equilibrium temperature is achieved at a specific current $I_{\text{min}} < 0$ (Figure 4.19).

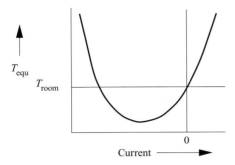

Figure 4.19 Equilibrium temperature versus thermoelectric device current.

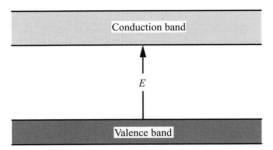

Figure 4.20 Diagram of energy levels of the valence and conduction bands and the bandgap E. At higher temperatures, more electrons are in the conduction band, and the electrical resistance decreases.

Design tip

The Peltier heat pump cannot produce cold, it can only pump heat from one object to another. It is most effective in cooling an object if:
- the hot side of the heat pump has good thermal contact with a heat sink;
- the object has good thermal contact with the cold side and good thermal insulation to reduce heat flow from the outside world;
- the current is optimally chosen – generally the equilibrium temperature obeys the equation $T_{equ} = AI + BI^2$, where A and B depend on the system.

4.3.5 The thermistor

Practical thermistors for temperature measurement consist of a piece of sintered metal oxide that exhibits a large decrease in electrical resistance with increasing temperature. In semiconductors, electrical conductivity is due to the electrons in the conduction band. If the temperature is increased, some electrons are promoted from the valence band into the conduction band (Figure 4.20), and the conductivity also increases. The electrical

Table 4.9 *Typical thermistor resistance R versus temperature T ($\beta = 3,000$ K, $R = 10$ kΩ at 0 °C)*

T (°C)	R (kΩ)	T (°C)	R (kΩ)	T (°C)	R (kΩ)
−50	117.2	−10	15.2	30	3.4
−40	65.8	0	10.0	40	2.5
−30	38.8	10	6.8	50	1.8
−20	23.8	20	4.7	60	1.4

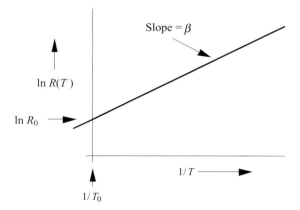

Figure 4.21 Relationship between $1/T$ and ln R for the thermistor, where the $(\ln R)^3$ term in Eq. (4.2) has been neglected.

conductivity is described by the Boltzmann relation, which states that the concentration of electrons in the conduction band depends on temperature as $\exp(-E/kT)$, where E is the bandgap, typically 0.3 eV; and k is Boltzmann's constant, equal to $8.617,09 \times 10^{-5}$ eV/K. Since the resistance is the inverse of the conductivity, resistance is proportional to $\exp(+E/kT) = \exp(3,500/T)$.

To the first order in $1/T$, the relationship between resistance R and temperature T is given by:

$$R(T) = R(T_0) \exp[\beta(1/T - 1/T_0)]$$

where T is in degrees kelvin, T_0 is the reference temperature, and β is the temperature coefficient of the material (Figure 4.21, Table 4.9). The rapid exponential drop in resistance with increasing temperature is due to the increase in the concentration of electrons in the conduction band with temperature. A more accurate description is given by:

$$1/T = A + B(\ln R) + C(\ln R)^3 \tag{4.2}$$

where A, B, and C are empirical constants determined by a best fit to measured data.

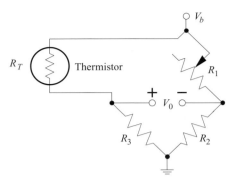

Figure 4.22 Thermistor in bridge circuit. Variable resistor R_1 may be adjusted to produce a zero output voltage at a selected temperature.

Since Eq. (4.2) is linear in A, B, and C the least-squares fitting techniques of Chapter 5 can be used. As an initial approximation, $A = 1/T_0$, $B = 1/\beta$, and $C = 0$.

The typical **time constant** for a thermistor in an insulating liquid (such as oil) is several seconds, but for use in aqueous media (such as in water or in the mouth), a Teflon tube is used for electrical insulation and to maintain sterility, which increases the response time by about a factor of 10. The **dissipation constant** is the power required to raise the temperature 1 °C above the surrounding media. For a thermistor suspended by its leads in a "well-stirred" oil bath, the dissipation constant is about 10 mW/°C and a factor of 10 less in still air. It is therefore important that the current through the thermistor be kept sufficiently small so that Joule heating does not affect the temperature measurement.

Conversely, self-heating can be used to determine the level of a liquid. Through the self-heating effect, the equilibrium temperature of a thermistor (and its resistance) will depend on whether the thermistor is in the liquid or in the vapor above.

To sense temperature, thermistors are used in a bridge circuit such as that shown in Figure 4.22. The bridge equation is given by:

$$V_0 = V_+ - V_- = V_b \left(\frac{R_3}{R_T + R_3} - \frac{R_2}{R_1 + R_2} \right)$$

Solving for R_T:

$$R_T = R_3 \left[\frac{V_b R_1 - V_0 (R_1 + R_2)}{V_b R_2 + V_0 (R_1 + R_2)} \right]$$

The bridge output V_0 is only linear in the temperature range where R_T is close to R_3 in value. For low temperatures, $R_T \gg R_3$ and $V_0 = -V_b R_2/(R_1 + R_2)$. For high temperatures, $R_T \ll R_3$ and $V_0 = +V_b R_1/(R_1 + R_2)$. At these extremes, the bridge output does not depend on temperature.

Optimization of thermistor-bridge sensitivity

The sensitivity Q in this case is the change in bridge output per unit change in temperature, and we now ask what choice of bridge resistors maximizes Q:

$$Q = \frac{dV_0}{dT} = R_T \frac{dV_0}{dR_T} \frac{1}{R_T} \frac{dR_T}{dT}$$

From the thermistor relationship, we have:

$$\frac{1}{R_T} \frac{dR_T}{dT} = -\frac{\beta}{T^2}$$

From the bridge relationship, we have:

$$V_0 = V_b \left(\frac{R_3}{R_T + R_3} - V_- \right)$$

$$R_T \frac{dV_0}{dR_T} = -\frac{V_b R_3 R_T}{(R_T + R_3)^2} = -\frac{V_b \alpha}{(1 + \alpha)^2}$$

where $\alpha = R_3/R_T$.

Combining, we have:

$$Q = \frac{V_b \alpha \beta}{(1 + \alpha)^2 T^2}$$

To maximize Q, we set $dQ/d\alpha = 0$ and find that the maximum occurs at $\alpha = 1$:

$$\frac{dQ}{d\alpha} = \frac{V_b \beta (1 - \alpha)}{T^2 (1 + \alpha)^3} = 0$$

The conclusion is that to maximize the sensitivity of the thermistor bridge in Figure 4.22 at temperature T (which maximizes the change in bridge output voltage for a given temperature change around the value T), we want to select $R_3 = R_T$. This is a surprisingly simple result, considering the nature of the equation for V_0 just given. Figure 4.23 shows a plot of bridge voltage versus temperature for a typical example.

Thermal dynamics

When a thermal mass that is in thermal equilibrium with a medium at temperature T_1 is suddenly moved to a medium at temperature T_2, its temperature changes with time t according to:

$$T = T_2 + (T_1 - T_2)e^{-t/\tau}$$

where τ is the exponential time constant.

4.3.6   The solid-state temperature sensor

The current I_D through a silicon diode is given by:

$$I_D = I_S[\exp(qV_D/kT) - 1]$$

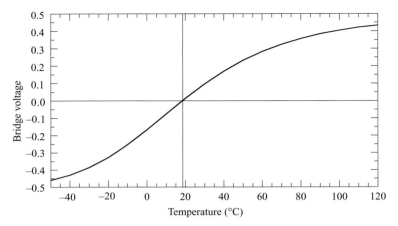

Figure 4.23 Plot of bridge voltage as a function of temperature for $\beta = 3,000$ K and $R_1 = R_2 = R_3 = 5\,\text{k}\Omega$.

where V_D is the voltage across the diode, I_S is the saturation current when the diode is strongly reverse biased, q is the electron charge, k is Boltzmann's constant, T is the temperature in kelvins, and $k/q = 86.170{,}9\ \mu\text{V/K}$. Solving this equation for V_D as a function of T, we obtain:

$$V_D = (86.170{,}9\ \mu\text{V/K})T\ln(1 + I_D/I_S)$$

For typical values of $I_D = 1$ mA and $I_S = 1$ nA:

$$V_D = (1.2\ \text{mV/K})T$$

Unfortunately, V_D cannot be used to determine the absolute temperature because I_S depends on temperature.

This problem is solved by using a matched pair of transistors with a well-known current ratio passing through them. The difference in base-emitter voltages is then a function of the absolute temperature and the logarithm of the ratios of the currents:

$$V_T = V_{\text{BE2}} - V_{\text{BE1}} = \frac{kT}{q}\ln\left(\frac{I_1}{I_2}\right)$$

where I_1 and I_2 are the collector currents, T is the absolute temperature, and $k/q = 86.170{,}9\ \mu\text{V/K}$. By accurately dividing a current so that the ratio I_1/I_2 is known, the voltage V_T becomes proportional to the absolute temperature.

The Analog Devices AD590 solid-state temperature sensor consists of a pair of transistors that divides a current I_T into two equal parts (Figure 4.24). One-half of I_T passes through a transistor and the other half of I_T passes through eight transistors connected in parallel. All nine transistors are identical. The difference in base-emitter voltages $V_T = (86.1\ \mu\text{V/K})\ (\ln 8)\ T = (179\ \mu\text{V/K})\ T$. This voltage appears across a 358-Ω resistor through which one-half of I_T flows. The total device current is

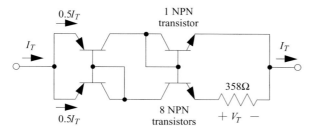

Figure 4.24 Analog Devices AD590 solid-state temperature sensor. The device is biased at 4–30 V and the resulting current is proportional to the absolute temperature. Typical sensitivity is 1 μA/K.

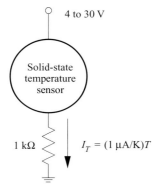

Figure 4.25 Equivalent circuit for Analog Devices AD590 solid-state temperature sensor. The device acts as a high-impedance, constant current regulator passing 1 μA/K.

$I_T = (1 \; \mu\text{A/K}) \, T$, and is proportional to the absolute temperature (Figure 4.25). This sensor is operated by applying a bias in the range from 3 to 40 V and measuring the current. It can be used with an external 1-kΩ resistor to give a sensitivity of 1 mV/K, which is about 20 times larger than the thermocouple. For measurements of temperature relative to 0 °C, a difference amplifier is used to subtract 273 mV. Over the temperature range from -55 to $+150$ °C, this device has an absolute error of ± 2 °C after room-temperature calibration, a linearity error of ± 0.8 °C, and a repeatability of ± 0.1 °C. The power-supply rejection ratio over the 5–15-V bias range is 0.2 μA/V or 0.2 °C/V.

4.3.7 Automatic thermocouple reference junction compensation (electronic "ice point")

Rather than maintaining the thermocouple reference junction at a specific temperature (e.g. by using ice), it is often more convenient to measure the temperature of the reference junction by using a solid-state temperature sensor or thermistor. To do this, two steps are necessary:

1. Convert the reference junction sensor (solid-state temperature sensor or thermistor) to a signal whose sensitivity in millivolts per degrees celsius matches that of the thermocouple.
2. Add the converted signal to the thermocouple signal.

4.3.8   Infrared temperature sensors

One very useful (but not electronic) temperature sensor consists of a telescope with a temperature calibrated filament at its internal focus. It is useful for measuring at a distance the surface temperature of objects hot enough to glow visibly. If the filament is cooler than the glowing surface, the filament will appear dark. If the filament is hotter than the glowing surface, the filament will appear bright. When the filament temperature is adjusted to be equal to the glowing surface to within 5 or 10 °C, the filament will disappear. This depends on the fact that black-body radiators at the same temperature have the same color spectrum. The lower temperature limit is 800 °C (the onset of visible incandescence) and the upper limit is 3,000 °C (the maximum temperature that a filament will survive).

A much more sensitive (and electronic) method for measuring temperature by infrared emission is an array of solid-state (CdS or Si) diodes at the focus of a telescope. Infrared radiation below 20 μm heats the diode elements and the change in conductivity is measured with a sensitive self-scanned circuit or a scanned electron beam to produce a video signal. This type of thermal imaging device is used to detect the increase in body temperature caused by infection or tumors and the leakage of heat from buildings due to insufficient thermal insulation. The accuracy is ±5 to 10 °C at room temperature.

One of the most sophisticated thermal imaging systems ever devised is the silicon avalanche photodiode array, which is operated in Earth orbit at a temperature of 10 K. It is capable of detecting individual photons at wavelengths from 0.3 to 17 μm and (by using engine heat) can image the location and motion of individual vehicles on the surface of the Earth at night.

4.3.9   Summary of temperature sensors

Table 4.10 lists the temperature sensors described in this section, their useful temperature range and typical accuracy.

Table 4.10 *Summary of temperature sensors*

Temperature sensor	Range (°C)	Accuracy (°C)
Platinum resistance	−220 to +850	1
Iron–constantan thermocouple	0 to +760	0.2
Chromel–alumel thermocouple	−200 to +1,250	1
W–W(26% Rh)	0 to +2,320	3
Thermistor	−80 to +100	0.05 (calibrated)
Solid state	−55 to +150	2 (calibrated)

4.4   Strain-sensing elements

When a mechanical structure is subject to a force, it undergoes deformation. **Stress** describes the intensity of the force on the structure as the force per unit area (F/A) and **strain** describes the deformation as the fractional change in length ($\Delta L/L$). **Young's modulus** describes the stiffness of the structure as the ratio of stress to strain. If the structure is very stiff, a large stress is required to produce a given strain and the Young's modulus is large.

One of the simplest resistive strain gauges is a rubber tube filled with mercury. As the rubber tube is stretched, its length L increases and its cross-sectional area A decreases. The electrical resistance R is given by $R = \rho L/A$, where ρ is the resistivity in ohm centimeter, and for constant volume $V = AL$, $R = \rho L^2/V$. For small changes in length ΔL, and a constant electrical resistivity ρ, $\Delta R = 2\rho L \Delta L/V$ and $\Delta R/R = 2\Delta L/L$ is the measure of the strain. The mercury tube strain gauge has been used for respiration monitoring.

4.4.1   The bonded resistance strain gauge

The metal-foil strain gauge consists of a pattern of metal on a Mylar backing (Figure 4.26) that changes its resistance as it is placed under tension or compression (Figure 4.27). This sensor has a very low cost and can be cemented to structural members of buildings, bridges, boilers, hulls of submarines and ships, etc. It can be used to detect strain in real-time and warn against excessive strain that could cause material failure. A typical maximum safe strain is $\Delta L/L = 0.5\%$.

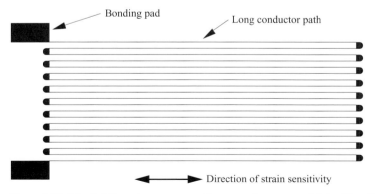

Figure 4.26 Metal-foil strain gauge, usually a metal pattern on Mylar film. For most metals, the fractional change in electrical resistance ($\Delta R/R$) is approximately twice the fractional change in length ($\Delta L/L$).

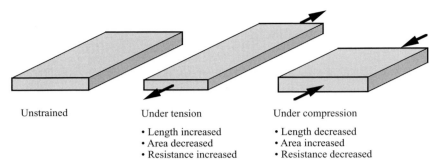

Unstrained	Under tension	Under compression
	• Length increased	• Length decreased
	• Area decreased	• Area increased
	• Resistance increased	• Resistance decreased

Figure 4.27 The metal-foil element in the strain gauge changes its shape during tension and expansion, resulting in a change in electrical resistance.

When metals are placed under tension, their length L will increase but the girth D will not necessarily decrease to keep the volume a constant, as it was for the mercury tube strain gauge. Defining the longitudinal strain $\varepsilon_l = \Delta L/L$ and the transverse strain $\varepsilon_t = \Delta D/D$, **Poisson's ratio** υ is defined as $-\varepsilon_t/\varepsilon_l$. Because the cross-sectional area A varies as the square of the girth D, $\Delta A/A = 2\Delta D/D = -2\upsilon(\Delta L/L)$.

The electrical resistance of a rectangular bar of metal of volume resistivity ρ is given by:

$$R = \frac{L\rho}{A}$$

Taking partial derivatives:

$$dR = \frac{\rho}{A}\partial L - \frac{\rho L}{A^2}\partial A + \frac{L}{A}\partial \rho = R\frac{\partial L}{L} - R\frac{\partial A}{A} + R\frac{\partial \rho}{\rho}$$

In differential form:

$$\frac{\Delta R}{R} = \frac{\Delta L}{L} - \frac{\Delta A}{A} + \frac{\Delta \rho}{\rho} = (1 + 2\upsilon)\frac{\Delta L}{L} + \frac{\Delta \rho}{\rho}$$

The term $\Delta L/L$ is dimensional and the term $\Delta \rho/\rho$ is piezoresistive. For the constant-volume mercury strain gauge, $\sigma = 0.50$ and the **gauge factor** G_S in the expression $\Delta R/R = G_S \Delta L/L$ has the value $G_S = 2$. For most metals, $\upsilon = 0.30$ (increase in volume with strain), but the piezoresistive term increases G_S so that it is also about 2. The gauge factor ranges from 2 to 4.5 for metals and can be as high as 150 for semiconductors (Table 4.11). The high gauge factor for semiconductive strain elements is because the resistivity ρ is a strong function of the strain.

Table 4.11 *Strain gauge materials and gauge factors*

Material	$G_S = (\Delta R/R)/(\Delta L/L)$	$(\Delta R/R)/\Delta T$ $(10^{-5}/°C)$	$(\Delta L/L)/\Delta T$ $(10^{-5}/°C)$
$Ni_{45}Cu_{55}$ alloy	2.1	± 2	1.6
Silicon (n-type)*	100 to 170	70 to 700	0.23
Silicon (p-type)†	-100 to -140	70 to 700	0.23

*For n-type silicon ($\rho = 3 \times 10^{-4}$ Ω-cm), $\Delta R/R = -110\Delta L/L + 10{,}000(\Delta L/L)^2$.
†For p-type silicon ($\rho = 2 \times 10^{-2}$ Ω-cm), $\Delta R/R = 120\Delta L/L + 4{,}000(\Delta L/L)^2$.

4.5   Force and pressure transducers

4.5.1   Force transducers

A common method of measuring the force of gravity F on a mass m uses a spring and position sensor. $F = mg = kx$, where g is the gravitational acceleration, k is the spring constant, and x is the difference between the loaded and unloaded lengths of the spring. Since the gravitational force does not depend on displacement, we can use a spring that permits a significant displacement, which helps the accuracy of the measurement. In many cases, however, a displacement in the direction of the force (i.e. "yielding" to the force) significantly reduces the magnitude of the force, so it is important to measure the force with a small displacement. One of the best methods for accomplishing this uses the piezoelectric crystal, which produces a voltage that is proportional to the force. Another method uses one or more strain gauges cemented to a flexible rod (Figure 4.28). One end of the rod is fixed and the other is attached to the force. The stiffness of the rod is chosen to provide an accurate measurement over the loads of interest without excessive displacement.

The resistance of the strain gauge is measured by placing it in one arm of an initially balanced Wheatstone bridge. If a single strain element is used (left-hand side of Figure 4.29), the output will be sensitive to thermal expansion of the elastic element. A better design uses four strain elements in opposing pairs (right-hand side of Figure 4.29). This results in a four-fold improvement in sensitivity and a relative insensitivity to temperature changes.

For the single-element bridge (left-hand side of Figure 4.29) with an excitation voltage V_b, the output is given by:

$$V_+ - V_- = V_b \left(\frac{R}{2R} - \frac{R}{2R + \Delta R} \right) = V_b \left(\frac{\Delta R}{4R + 2\Delta R} \right)$$

$$\approx V_b \left(\frac{\Delta R}{4R} \right) = V_b \left(\frac{G_S}{4} \frac{\Delta L}{L} \right)$$

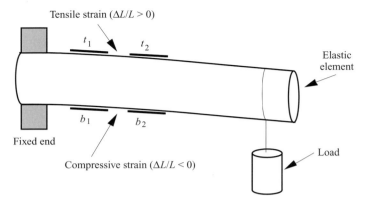

Figure 4.28 Force transducer consisting of four strain gauges mounted in opposing pairs to an elastic element.

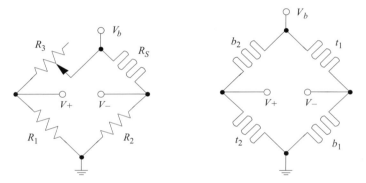

Figure 4.29 Force-transducer bridge circuits using a single strain element R_S, or four strain elements, t_1, t_2, b_1, and b_2, in opposing pairs. The four-element bridge has four times the sensitivity of the single-element bridge.

For the four-element bridge (right-hand side of Figure 4.29), assuming that all four strain elements have the same unstrained resistance R, we have:

$$V_+ - V_- = V_b \left(\frac{R + \Delta R}{R + \Delta R + R - \Delta R} - \frac{R - \Delta R}{R + \Delta R + R - \Delta R} \right)$$

$$= V_b \left(\frac{2\Delta R}{2R} \right) = V_b \left(G_S \frac{\Delta L}{L} \right)$$

Note that the relationship for the four-element bridge is naturally linear, whereas that of the single-strain-element bridge is not. In addition, the four-element force transducer has four times the sensitivity. A typical maximum strain $\Delta L/L$ would be 0.5%, so that for a gauge factor $G_S = 2$, $\Delta R/R$ would be only 1% and V_0 would be 1% of V_b.

It should also be noted that a uniform change in temperature will cause a change in the length of the bar and put an equal strain on all the gauges. Therefore, the single-element force transducer is sensitive to temperature changes, while the four-element force transducer is not.

4.5.2   Pressure transducers

The pressure of a gas or a fluid is the force per unit area exerted perpendicularly on the surface of the surrounding container. One of the most sensitive pressure transducers is the piezoresistive diaphragm, which consists of four nearly identical semiconductor piezoresistors buried in the surface of a thin circular silicon diaphragm. Pressure causes the diaphragm to bend, inducing a stress on the diaphragm and the buried piezoresistors. Two of the resistors increase in value and two decrease, depending on their orientation with respect to the crystalline direction of the silicon material. Gold pads attached to the silicon diaphragm provide connection from the piezoresistors to a full bridge similar to the force transducer described in the previous section. Silicon is elastic throughout its operating range and fails by rupturing. Units are available for measuring pressures from 1 to 15,000 psi and have typical accuracies from 0.1 to 1% of full scale. For measuring absolute pressure, one side of the diaphragm is evacuated and sealed. For measuring differential pressure, both sides are used.

4.5.3   Piezoelectric transducers

Piezoelectric crystals have the special property of generating a voltage when pressure is applied, and undergoing a small deformation when a voltage is applied. Although the deformations are small, they can be used at high frequency as oscillators or as ultrasonic emitters and sensors. Common applications include timekeeping, underwater sonar ranging, and medical imaging. The most commonly used piezo-electric materials are barium titanate and single crystal quartz. One of the oldest examples is the sodium potassium tartrate crystal used in phonograph styluses to convert the small undulations in the groves of phonograph records into a usable electronic signal.

Figure 4.30 shows a piezoelectric crystal and its equivalent circuit. The charge separation produced in the crystal is given by:

$$q = S_q F = S_q P A$$

where S_q is the charge sensitivity.

The voltage V produced by this charge is given by:

$$V = q/C, \quad \text{where } C = k A/d$$

and A is the area, d is the thickness, and k is the dielectric constant.

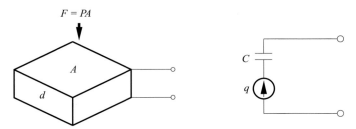

Figure 4.30 Piezoelectric crystal and equivalent circuit.

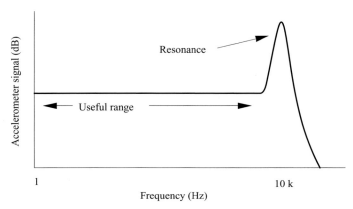

Figure 4.31 Piezoelectric crystal accelerometer.

Combining these, we have

$$V = \frac{qd}{kA} = \frac{S_q P d}{k} = S_v P d$$

and S_v is the voltage sensitivity.

For sensing acceleration, an inertial mass is used to convert acceleration into a force. Typical sensitivities are 10 pC/g or 5 mV/g. Larger units have a frequency response from 1 Hz to 10 kHz (Figure 4.31). Smaller units can respond to 1 MHz.

EXAMPLE 4.1

For a typical barium titanate piezoelectric crystal of thickness 1 mm and area 1 cm², what is the output voltage for a pressure of 1 atmosphere?

Assume:

- charge sensitivity $S_q = 1.5$ pC/N;
- dielectric constant $k = 1.25 \times 10^{-8}$ F/m;
- voltage sensitivity $S_v = 1.2 \times 10^{-4}$ Vm/N;
- $A = 1$ cm², $d = 1$ mm, $V = S_v P d$;
- pressure $P = 1$ atmosphere $= 10.1 \times 10^{-4}$ Nm^{-2}.

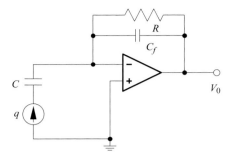

Figure 4.32 Piezoelectric crystal integrating amplifier readout.

For $A = 1$ cm^2, $d = 1$ mm, we have $V = S_v P d$, which evaluates to:

$$V = (1.2 \times 10^{-4} \text{ Vm/N})(10^{-3}\text{m})(10.1 \times 10^4 \text{ Nm}^{-2}) = 12.1 \text{ mV}$$

Figure 4.32 shows a charge amplifier commonly used to produce a step output V_0 proportional to an input charge q. $V_0 = q(C/C_f)$, where C_f is the value of the feedback capacitor. The resistor R has a large value and is used to restore the output to zero over long time periods.

Inertial navigation

The purpose of the inertial navigation system is to measure accelerations accurately in three directions and integrate them to determine velocity and position. Drift and noise produces errors that increase with time. It is necessary to recalibrate the system at a known location to remove these errors. This approach has been largely replaced by the global positioning system (GPS), which uses the transit time of radio signals from artificial Earth satellites in known orbits to determine position.

4.5.4   Vacuum sensors and pumps

The vacuum has many practical applications, including:
- cathode ray tube – TV and computer display;
- power amplifiers – radio, TV, and radar transmitters;
- ion implantation – integrated circuit production;
- electron, proton, and ion accelerators – medical therapy beams;
- proton and deuteron cyclotrons – medical isotope production.

The following are techniques for measuring vacua (see Table 4.12 for summary):

Capacitance

The parallel plate capacitor is used in a resonant circuit. The resonance frequency can be measured very accurately, and depends on the dielectric constant (proportional to gas pressure) between plates.

Table 4.12 *Pressure sensors and their pressure ranges*

Pressure sensor	Pressure range
Silicon disk with strain gauges	1,000 atm to 1 torr*
Capacitance	1,000 to 1 torr
Thermocouple	1 to 0.01 torr
Ion gauge[†]	10^{-2} to 10^{-4} torr
Vac-ion pump	10^{-4} to 10^{-12} torr

*1 atm = pressure of a 760 mm column of Hg = 760 torr.
[†]Bayart–Alpert gauge can record 10^{-6} torr.

Thermocouple

The heater element is placed below the sensing junction of a thermocouple. Convective heat transfer to the thermocouple is proportional to gas pressure.

Ion gauge

A heated filament generates electrons that are accelerated to a (+) collection plate. These can collide with gas atoms in the way to produce (+) ions that are collected on a (−) collection plate. That current is proportional to gas pressure.

Vac-ion pump

Electrons are generated and spiral in a combined electric and magnetic field. These can collide with gas atoms to make (−) ions that are driven into a (+) plate. That current is proportional to gas pressure.

The following are means for producing vacua:

Aspirator

Uses the entrapment of air by a rapidly flowing fluid, such as water. Limited to 10 mm Hg (10 torr) and is used by chemists in filtering solutions. The solution is placed above the filter paper, in contact with atmospheric pressure, and the aspirator provides a vacuum in the beaker below the filter.

Bellows and valve

Limited to 1 mm Hg (1 torr).

Roughing pump

Mechanical pump (pistons and valves) – can achieve 0.01 torr. Lubrication oil a potential contamination – an oil trap is usually placed at the pump inlet.

Turbomolecular pump

High-speed turbine blades collide with gas molecules and drive them into a roughing pump – this operates below 0.1 torr and can achieve 10^{-5} torr.

Vac-ion pump

When the gas atoms are driven into the (+) plate, they stick to provide a pumping action. This continues until the plate is saturated. When this occurs the trapped atoms can be driven from the plate in a "degassing" process that consists of heating the plate while the roughing pump is operating. The pump operates below 10^{-4} torr and can achieve 10^{-12} torr.

Diffusion pump

(This pump is not used in modern systems.) Boiling oil or Hg is sprayed to entrain gas molecules and drive them to the roughing pump. The pump operates below 0.1 torr and can achieve 10^{-6} torr. Hot oil can contaminate the system and limits achievable pressure.

Cryopump

Conventional pumping techniques (such as the vac-ion pump) are limited by the vapor pressure of residual gases. These residual gases can be frozen onto a cryogenically cooled surface to produce pressures below 10^{-12} torr.

Note: The 1994 *Guinness Book of Records* states that the best vacuum ever achieved was 10^{-14} torr (using 4K cryopumping and an outer guard vacuum) at the IBM Watson Research Center in 1976.

4.6   Measuring light

4.6.1   The silicon photodiode

The best commercially available photodiode for the measurement of low light levels is the *pin* photodiode, which is manufactured by diffusing donor (*n*-type) impurities (usually phosphorous or arsenic) and acceptor (*p*-type) impurities (usually boron) into opposite sides of a high-purity silicon crystal (Figure 4.33). The crystal is typically 50–500 μm thick and the *p* layer transmits light in the wavelength range from 400 to 800 nm. The relationship between photon wavelength λ and energy E is given by:

$$E = hc/\lambda, \quad \text{where } hc = 1,240 \text{ eV nm}$$

Photons of energy 2 eV have λ = 620 nm and appear red in colour. Photons of energy 3 eV have λ = 413 nm and appear blue in colour.

The *n*-type material is doped by replacing some Si atoms with valence 5 atoms, such as phosphorous, to produce bound P^+ plus mobile electrons. The *p*-type material is doped by replacing some Si atoms with valence 3 atoms, such as boron, to produce bound B^- plus mobile holes.

Some of the mobile holes and electrons recombine, leaving the *n*-type material positively charged and the *p*-type material negatively charged. This produces an internal

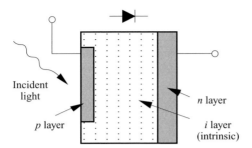

Figure 4.33 Electronic structure of the *pin* photodiode. Photons pass through the *p* layer and produce electron–hole pairs in the *i* layer.

electric field that causes the *n*-type material to attract negative carriers and the *p*-type material to attract positive carriers.

The *i*-type material is high purity, high resistivity, and is relatively free from carriers. It is essential that either the *n*-type or the *p*-type doping is transparent, so that photons can enter the *i*-type material

Photovoltaic mode

The doping of the *pin* diode produces free negative carriers (and fixed positive ions) in the *n* layer and free positive carriers (and fixed negative ions) in the *p* layer. These free carriers (in the absence of any other forces) will diffuse to fill the space available. This driving force is the **diffusion emf**. However, since these carriers leave their oppositely charged fixed ions behind, the diffusion emf is opposed by a **coulomb emf**. At equilibrium, the two emfs are balanced and do not produce an external voltage or current, but the charge separation produces an electric field in the *i* layer. In this field, excess free negative carriers will drift toward the *n* layer and excess free positive carriers will drift toward the *p* layer.

Photons shining into the intrinsic layer (through the thin transparent *p* layer) produce electron–hole pairs that are separated by the internal electric field to produce an external emf (called the **photovoltaic potential**). In an open circuit, the potential developed is in the range of 200–600 mV over a wide range of light intensities (shown as the intercept with the $+V$-axis in Figure 4.34). On the other hand, the short-circuit current is linearly proportional to the light intensity over many decades (shown as the intercept with the $-I$-axis in Figure 4.34). If a load resistor is used, then the voltage–current relationship is given by the load line (Figure 4.35).

Considering the energy flow of the process, the energy of the interacting photon breaks the bond between a valence electron and a silicon atom, creating an electron and a hole. The internal field of the diode separates the carriers and they drift to the *n*- and *p*-type electrodes. If the circuit is completed, the electron can do electrical work on its way to recombining with the positive carrier. In this way light is converted into useful electrical energy.

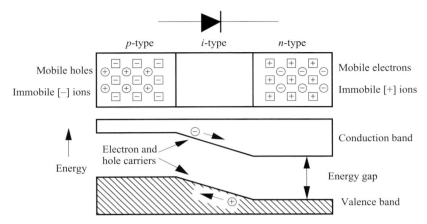

Figure 4.34 Electronic structure of the *pin* photodiode, showing how donor (*n*-type) atoms and acceptor (*p*-type) atoms produce an internal potential.

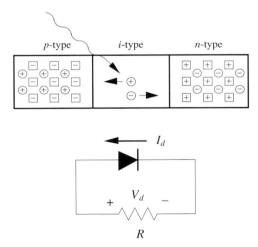

Figure 4.35 Photovoltaic mode of the *pin* photodiode, showing how electrons and holes produced in the intrinsic layer by light can do work in an external circuit.

This also makes it clear why series-connected diodes add their voltage and not their current – the electrons from one diode recombine with the holes from the next diode. The current in the external circuit is the holes from the diode on one end and the electrons from the diode on the other end.

Photoconductive mode

When a reverse bias is applied, and if the material is sufficiently pure, the *i* layer is almost totally depleted and devoid of charge carriers. This results in a lowering of capacitance (by typically a factor of 4) and an increase in speed, which is especially important in communication applications. In darkness, the V–I curve for the photodiode is similar

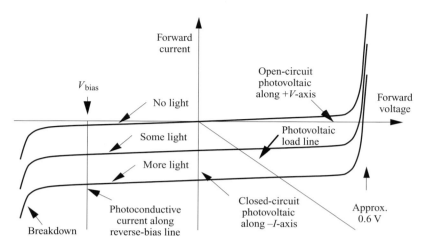

Figure 4.36 V–I characteristics of the *pin* photodiode for various incident-light levels.

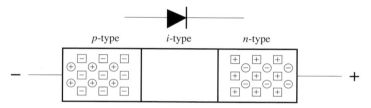

Figure 4.37 *pin* photodiode in darkness with reverse bias. Internal and external fields add to sweep the intrinsic region clear of free charges.

to a conventional diode, except that the reverse-bias current is much lower, typically a few nanoamps for the S1723. Illumination mainly shifts the curve *downward* in the direction of reverse current. As seen in Figure 4.36, and explored in Laboratory Exercise 14, this current is the best indicator of light intensity.

In darkness, an external reverse bias pulls the negative and positive carriers farther away from each other, the *i*-type material is swept clear of free charges, and only a small current is produced (Figure 4.37).

When photons interact in the *i*-type material, an external reverse bias quickly sweeps the holes and electrons into the corresponding materials to produce an increased reverse current (Figure 4.38). In darkness, an external forward voltage must exceed the internal potential of about 0.6 V before a significant current is produced (Figure 4.39).

The I–V characteristic of an ideal photodiode obeys the equation:

$$I_D = I_S\left(e^{qV_D/kT} - 1\right) + I_P$$

where V_D is the voltage across the diode, I_D is the current though the diode, $kT/q = 0.026$ V is the thermal voltage at room temperature, I_S is the saturation current, which depends on the area of the diode junction, and I_P is the current generated by the photons.

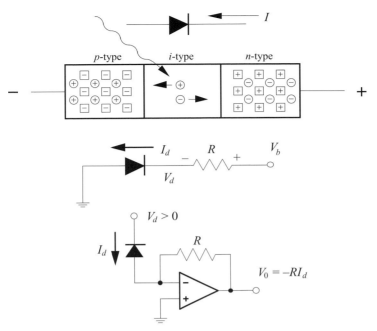

Figure 4.38 *pin* photodiode in light with reverse bias. Electrons and holes produced in the intrinsic region by light produce a reverse current that is proportional to the intensity of the light (number of photons/s).

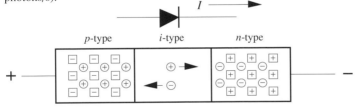

Figure 4.39 *pin* photodiode with foward bias. If the forward external potential exceeds the internal potential (about 0.6 V for silicon), a large current flows.

Amplification

To convert current to voltage, a resistive load and a noninverting amplifier can be used (Figure 4.40).

$$V_0 = I_{photo} R_1 (R_2 + R_3)/R_3$$

The voltage-to-current converter shown in Figure 2.6 is not recommended because op-amp leakage currents can develop a large voltage drop across the photodiode (typical reverse-bias impedance is 10^{12} Ω) and saturate the amplifier:

4.6.2 Lambert–Beer law

The photodiode is often used to measure transmission through a colored solution of unknown concentration. The optical transmission is defined as the intensity $I(L)$ measured

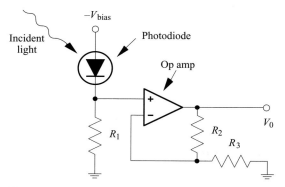

Figure 4.40 Circuit for biasing a photodiode and converting the photoconduction current into a voltage. $V_0 = I R_1 (R_2 + R_3)/R_3$, where I is the photodiode current. For $R_1 = 1$ MΩ, $R_2 = 900$ kΩ, and $R_3 = 100$ kΩ, the op-amp output V_0 is 10 mV/nA of photodiode current.

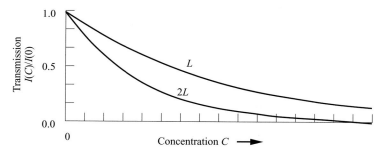

Figure 4.41 Transmitted intensity as a function of concentration C of a colored solution for path lengths L and $2L$.

through path length L divided by the intensity $I(0)$ at zero path length:

$$I(L) = I(0)e^{-kCL} \tag{4.3}$$

where C is the concentration of the solution, and L is the path length through the solution (Figure 4.41). The constant of proportionality k is usually determined by using known standards.

4.6.3   Summary of solid-state photodetectors

The spectral response is 190–1,100 nm for silicon and 190–760 nm for GaAs. The radiant sensitivity is defined as the photocurrent produced per luminous watts received. The quantum efficiency is the number of electron–hole pairs generated per incident photon. The energy per photon is given by $E = hc/\lambda$, where $hc = 1,240$ eV nm and λ is the wavelength. A 2-eV photon has a wavelength of 620 nm and appears red. A 3-eV photon has a wavelength of 413 nm and appears blue. The relationship between photocurrent I_{photo}, power received P_{photo}, quantum efficiency QE, and wavelength λ

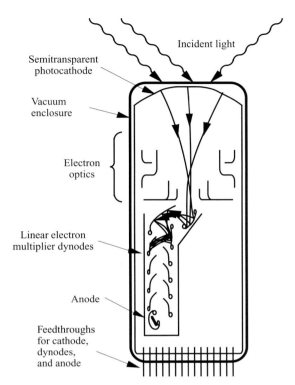

Figure 4.42 Photomultiplier tube. For clarity, only the electron amplification at the first three dynodes is shown.

is given by:

$$I_{\text{photo}} = \text{QE}\,\lambda\,P_{\text{photo}}/(1{,}240 \text{ eV nm})$$

Design tip

When using a photodiode to convert light intensity into a voltage, remember that the photocurrent is proportional to the rate at which photons make electron–hole pairs. A current-to-voltage converter circuit is needed.

4.6.4   The vacuum photomultiplier tube

The vacuum photomultiplier is one of the most sensitive photodetectors and combines a special photosensitive layer called the photocathode with a high-gain electron multiplier. The photocathode converts incident photons in the wavelength range from 200 to 500 nm into photoelectrons released into the vacuum with typically 15–25% efficiency. The electron multiplier consists of a series of 8–14 plates (called dynodes) coated with high secondary emission material so that the electrons multiply as they cascade from plate to plate (Figure 4.42). The overall electron gain in this process is typically 10^5–10^7. When operating at low levels of illumination, photons can be detected individually as a pulse 10^5–10^7 electrons in size and 3 ns in duration. An important feature is the absence of Johnson noise, which makes it impossible for room-temperature electronics to detect individual photons.

4.7   Producing visible light

There are two primary mechanisms for producing visible light: (1) **incandescence**, caused by the thermal agitation of electrons; and (2) **luminescence**, caused by electrons dropping from one atomic energy level to a lower level.

4.7.1   Incandescence

Due to the random motion of the electrons, two characteristics of incandescent radiation are: (1) an intensity that varies as the fourth power of the temperature, and (2) a broad wavelength spectrum that peaks at $\lambda_{\max} = (2.897{,}8 \times 10^6 \text{ nm K})/T$. Examples of incandescence include:

1. The electric light bulb, in which an electric current heats a coil of fine tungsten wire to $3{,}000\,°\text{C}$ in an inert gas.
2. The radiant space heater, where an electric current heats a nichrome alloy wire or quartz filament to typically $1{,}000\,°\text{C}$.
3. A candle flame, where a chemical reaction heats gases to about $2{,}000\,°\text{C}$.
4. The Sun, where internal thermonuclear reactions produce a surface temperature of $6{,}000\,°\text{C}$.

The radiant power is proportional to the surface area A and varies as the fourth power of the temperature in kelvin:

$$P_R = \sigma A T^4, \quad \sigma = 5.669{,}6 \times 10^8 \text{ W m}^{-2} \text{ K}^4$$

6,000 K (surface of the Sun): $P = 7{,}300$ W/cm
3,000 K (incandescent filament): $P = 460$ W/cm
300 K (objects at room temperature): $P = 46$ mW/cm
(actually heat transfer depends on temperature difference with surroundings: $\Delta T = 20\,°\text{C}$ gives $P = 1\ \mu\text{W/cm}$).

The Planck function gives the amount of radiant power per unit wavelength and per unit area as a function of temperature:

$$W(\lambda, T) = \frac{c_1}{\lambda^5 \left(e^{c_2/\lambda T} - 1\right)}$$

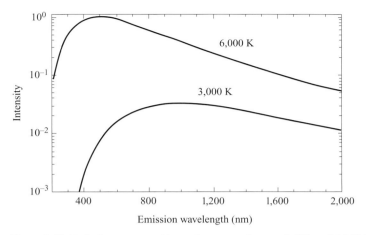

Figure 4.43 Emission spectra of incandescent surfaces at 3,000 and 6,000 K.

$c_1 = 2\pi h c^2 = 3.742 \times 10^{-16} \text{ Wm}^2$
$c_2 = hc/k = 1.439 \times 10^{-2} \text{ mK}$

See Figure 4.43 for a plot of emission intensity (W/m^2/nm) for surfaces at 3,000 and 6,000 K.

The wavelength spectrum is broad and peaks at:

$$\lambda_{\text{max}} = (2.897{,}8 \times 10^6 \text{ nm K})/T$$

6,000 K (surface of the Sun): $\lambda_{\text{max}} = 480$ nm
3,000 K (incandescent filament): $\lambda_{\text{max}} = 960$ nm
300 K (objects at room temperature): $\lambda_{\text{max}} = 9.6$ μm

4.7.2   Luminescence

Because the light is produced when electrons drop from one energy level to another, two characteristics of luminous radiation are: (1) a relatively low temperature, and (2) a well-defined wavelength. It is essential that a "forbidden" energy band lies between the two levels. Although there are several mechanisms that broaden the wavelength spectrum (such as in high pressure plasmas and in the solid state), the spectrum is narrower than that of incandescence. Examples of luminescence include:

1. The fluorescent lamp, where ultraviolet light from a heated mercury vapor excites the electrons in a phosphor coated into the inside of the lamp. When the electrons in the phosphor return to their ground state, visible light is produced. This is called **spontaneous emission** because it occurs without any other influence.

2. The neon advertising sign, where high voltage (typically 15 kV ac) is used to produce an excited plasma. Atoms in the plasma produce spontaneous emission when they return to their ground state.

3. The light emitting diode (LED), where a forward external voltage applied to a GaAs diode promotes electrons from the valence band to the conduction band and these electrons emit spontaneous light when they return to the valence band.

4. The cathode ray tube (used as the display device in most television sets and desktop computers), where a narrow beam of electrons is steered by electromagnets and accelerated by high voltage (typically 20 kV) to strike a phosphor screen in a desired pattern. The electrons in the beam transfer energy to the electrons in the phosphor, which raises them to a number of different excited states. The electrons in the phosphor then emit spontaneous visible light when they return to the ground state.

5. The **laser**, where electrons are excited to a higher energy level by a plasma discharge or the absorption of photons, and then return to a lower energy state by **stimulated emission**. Stimulated emission occurs when a photon interacts with an excited electron to produce two photons with the same energy, direction, and phase. The best wavelength for stimulated emission is the same as that of spontaneous emission. The characteristics of laser emission are: (1) a relatively low temperature, (2) a well-defined wavelength, (3) an intense pulse of light, and (4) a narrow angle of emission.

The **light emitting diode** or **LED** is made from a high bandgap semiconductor such as GaAs (red), GaAsP (green), or GaN (blue). A forward bias promotes some electrons to the conduction band and they produce spontaneous emission when then return to the ground state. The minimum pulse width is typically 100 ns (1 ns for GaN). The **laser diode**, on the other hand, produces light by stimulated emission and is capable of producing pulses as short as 30 ps.

Design tip

To convert a voltage into a light intensity using a light emitting diode (LED), remember that light intensity is proportional to current. A voltage-to-current driver is needed.

4.7.3   Luminous efficiency

Most light sources are designed for the human eye, such as computer screens, room lights, indicator lights. The efficiency of such light sources is rated as lumens/watt, where the lumen is weighted by the response curve of the human eye (Figure 4.44). Table 4.13 lists commonly available light sources and their luminous efficiency.

Theoretically, the highest possible efficiency in a "white" lamp would be achieved by using three colors with an average luminous efficiency of about 300 lumens/watt. Available lamps fall well below this level.

Table 4.13 *Common lamps and their luminous efficiencies*

Lamp	Efficiency (lumens/W)
60 W tungsten (1000 hr)	14
5,000 W tungsten (75 hr)	33
Fluorescent	54–76
Mercury vapor*	65
Metal halide	100
High pressure sodium*	100
Low pressure sodium*	150

*Poor color rendition.

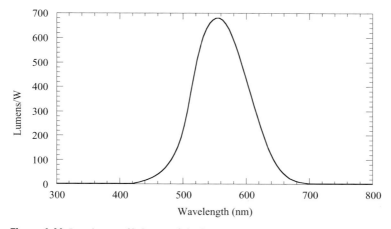

Figure 4.44 Luminous efficiency of the human eye.

4.8   Ionic potentials

4.8.1   Origin of the ionic potential

Two important examples of ionic potentials are those that appear on the surface of the skin as a result of the electrochemical activity of neurons and muscles, and those that are measured in test tubes to determine ionic concentration.

In the first case, ionic potentials are measured at the surface of the skin to study the electrical activity of the heart (the electrocardiogram, or ECG), the brain (the electroencephalogram, or EEG), the skeletal muscles (the electromyogram, or EMG), and the position of the eyes (the electrooculogram, or EOG).

It is important to note that these ionic potentials are not the result of a concentration of electrons, but a concentration of ions that are not free to travel down wires to our

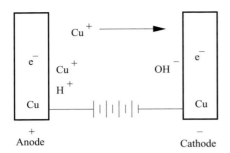

Figure 4.45 Conduction using bare copper electrodes in pure water by the reactions Cu (anode) → e^- (anode) + Cu^+ (solution) and Cu^+ (solution) + e^- (cathode) → Cu (cathode). This process is hindered by the low solubility of copper in water.

amplifiers. As a result, it is necessary to transduce a current of ions in solution to a current of electrons in a conductor.

4.8.2   Bare-metal electrodes

Bare-metal electrodes are poor for the detection of ionic potentials because:
1. The skin provides a barrier to charge transport.
2. The ionic potentials are small (1 mV for the ECG and much less for the EEG) and can only transform a small amount of ion charge into electrons in the metal before the electrode polarizes.
3. Reactions such as:

   Cu (anode) → e^- (in anode) + Cu^+ (in solution) and
   Cu^+ (in solution) + e^- (in cathode) → Cu (solid)

   are limited at low voltages (mV) by the low solubility of copper and the binding of Cu^+ ions by OH^- ions (Figure 4.45). At higher voltages (0.1 V) the current is limited by the rate of diffusion. Even when no current flows, there is a Cu^+ space charge around the wire which can produce a transient potential on the electrode whenever the solution is disturbed (microphonic effect).
4. Reactions such as the following require potentials of several volts (Figure 4.46):

   $2H_2O → O_2 + 4H^+ + 4e^-$ (into anode)

   $2H^+ + 2e^-$ (from cathode) → H_2

   $2Cl^-$ (in solution) → $2e^-$ (into anode) + Cl_2 (gas)

4.8.3   Ag(AgCl) electrodes

The Ag(AgCl) gel electrode has a silver core surrounded by a sintered AgCl layer (Figure 4.47). The reaction:

Ag (solid) + Cl^- (aqueous) ↔ AgCl (solid) + e^- (in AgCl)

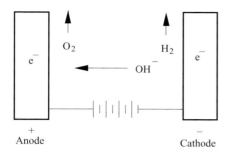

Figure 4.46 Electrolysis of water – $2H_2O + 2e^- \rightarrow 2OH^- + H_2$ (gas); $4OH^- \rightarrow 2H_2O + 4e^- + O_2$ (gas).

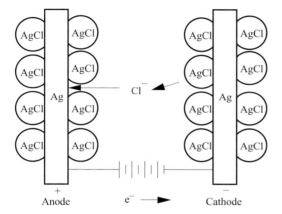

Figure 4.47 Ag(AgCl) electrodes. A current of electrons is transduced into a current of Cl^- ions in solution, using the reaction $AgCl + e^- \leftrightarrow Ag + Cl^-$.

allows an efficient transfer of e^- charge in the Ag wire to and from the Cl^- ion charge in solution for the following reasons:

1. AgCl is insoluble and stays with the Ag core.
2. AgCl is porous, so that the Cl^- ions in solution can reach the Ag in the core.
3. AgCl is slightly conductive, and can receive e^- from the Ag core.
4. NaCl in the gel reduces the skin resistance to the passage of ions.

Note that while Ag and AgCl are both present in the electrode, it can act either as a positive terminal (Ag and Cl^- are converted to AgCl and e^-) or as a negative terminal (AgCl and e^- are converted to Ag and Cl^-).

Due to the cost of the Ag(AgCl) electrode, copper electrodes smeared with a special ECG paste are also used, but this is much more subject to offset potentials (i.e. battery potentials) and polarization.

The dc electrical resistance of the skin varies greatly from person to person and depends on how "dry" the skin is. A pair of bare dry metal electrodes applied to the skin about 15 cm apart will generally show a resistance in the $1–10$-$M\Omega$ range. By using

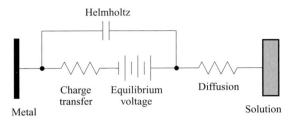

Figure 4.48 Equivalent circuit for a metal electrode in an ionic solution.

water, the resistance will drop to the 100–1-MΩ range. Ag(AgCl) floating electrodes typically show a resistance in the 20–50-kΩ range.

A commonly used model for electrodes is shown in Figure 4.48. It has several components:

1. a resistance due to charge transfer from electrode into solution,
2. a resistance due to diffusion velocity of ions away from electrode,
3. an equilibrium voltage emf,
4. a Helmholtz capacitance due to the ion charge layers.

4.9   The detection and measurement of ionizing radiation

4.9.1   The electromagnetic spectrum

The electromagnetic spectrum ranges from energetic gamma rays found in the cosmic radiation to low-frequency radio waves (Table 4.14). The visible spectrum covers only a very narrow band.

4.9.2   Applications of radiation

Non-medical applications include:
- metal gauging in roller mills (betas);
- monitoring liner thickness in metal production furnaces (betas);
- tribology (measuring wear)(put tracer in steel of engine cylinder – measure radio-activity in oil);
- smoke detectors (alphas are stopped by smoke particles);
- imaging boilers and suspected bombs (gammas);
- drug detection (by density – gammas);
- chemical analysis by neutron activation (neutron in, gammas out) – used in oil exploration, geology;
- dating materials in archaeology and anthropology (C-14);
- "cold pasturization" (sterilization) of food (Co-60 gammas).

Table 4.14 *Sixteen decades of the electromagnetic spectrum*

Wavelength	Energy	Radiation
0.003,1–0.000,031 nm	0.1–10 MeV	Nuclear gamma rays
0.006,2–0.031 nm	40–200 keV	Diagnostic X-rays
0.124–1.24 nm	1–10 keV	Low-energy X-rays
200–400 nm	3.1–6.2 eV	Ultraviolet light
400–700 nm	1.8–3.1 eV	Visible light
0.7–10 μm	0.12–1.8 eV	Infra red
2.7–3.5 m		87–108-MHz fm band
187–560 m		530–1610-kHz am band

Medical applications include:

- medical radiology and CT scanning (X-rays),
- nuclear medicine (gamma ray tracer compounds),
- measuring proteins in the blood (radioimmunoassay – betas),
- determining DNA code in plants and animals,
- biochemical pathway research in animals (e.g. determining fat and carbohydrate metabolism) and plants (e.g. photosynthesis).

4.9.3   X-Rays

Sources: medical X-ray tubes, energetic atomic transitions

Typical energies: 1–100 keV

Interactions in matter: discrete interactions by total absorption (photoelectric effect) or by Compton scattering

Number surviving after distance x: $N(x) = N(0)e^{-x\mu}$, where $\mu^{-1} = 7$ cm water at 100 keV

4.9.4   Gamma rays

Sources: decay of radioactive nuclei (radioactive materials), cosmic radiation, nuclear reactors

Typical energies: 0.1–10 MeV

Interactions in matter: discrete interactions by total absorption (photoelectric effect), by Compton scattering, or (above 1 MeV) by electron-positron pair production $\mu^{-1} = 15$ cm water at 1 MeV

4.9.5   Neutrons

Sources: cosmic radiation, nuclear reactors

Typical energies: 0.025 eV to 10 MeV

Interactions in matter: discrete interactions by interactions with nuclei or scattering from nuclei

The number surviving after distance x is given by: $N(x) = N(0)e^{-x\mu}$

$\mu^{-1} = 1$ cm for slow neutrons (<0.5 eV) in water

$\mu^{-1} = 10$ cm for fast neutrons (>1 MeV) in water

4.9.6   Betas

Beta particle = electron or positron

Sources: decay of radioactive nuclei (radioactive materials), cosmic radiation, nuclear reactors

Typical energies: 0.02 keV to 10 MeV

Interactions in matter: gradual energy loss by scattering on electrons, abrupt changes in direction by scattering on nuclei – result is a distribution of ranges

Range of 1 MeV beta: 1 m in air, 1 mm in water, 0.1 mm in steel

Positron: after stopping in matter, annihilates with an electron, produces two 511-keV photons emitted in opposite directions

4.9.7   Alphas

Alpha particle = helium nucleus (two protons + four neutrons)

Sources: decay of radioactive nuclei of high atomic number (natural U, Th in rocks or fission products)

Typical energies: 3–6 MeV

Interactions in matter: gradual energy loss by scattering on nuclei – result is a well-defined range

Range of 5 MeV alpha: 3.5 cm in air, 20 μm in silicon

4.9.8   Radioactive isotopes

Table 4.15 lists commonly used radioactive isotopes, their half-lives, and their emissions. See the previous sections for properties and useful applications.

4.9.9   Radiation detectors

Cloud chamber (droplets form on ionizing track)

Bubble chamber (bubbles form on ionizing track)

Gas-filled (multiplication of ionization electrons in an electric field) – (Geiger, proportional, multi-wire)

Scintillation (ionization produces light)
- Organic (plastic, liquid)
- Inorganic (crystals of NaI, CaF_2, etc.)

Table 4.15 *Radioactive isotopes, their half-lives, and emissions*

Isotope	Half-life (yr)	Emissions
H-3	12.3	18.6 keV beta
C-14	5730	156 keV beta
Co-60	5.26	0.3 MeV beta + 1.17, 1.33 MeV gammas
Sr-90	28.1	0.5, 2.3 MeV betas
I-131	8.07[†]	0.6 MeV beta + 0.36 MeV gamma
Cs-137	30.0	0.5 MeV beta + 0.66 MeV gamma
Ra-226	1,600	4.7 MeV alpha*
Pu-239	2.44×10^4	5.1 MeV alpha* (used in nuclear weapons)
Pu-238	87.7	5.5 MeV alpha* (used in nuclear batteries)
Th-232	1.41×10^{10}	4.0 MeV alpha*
U-238	4.51×10^9	4.2 MeV alpha*

*Plus emissions from daughter isotopes.
[†]Half-life measured in days.

Solid-state (collection of ionization charge)
- (Si, Ge, CdTe, PbI_2, HgI_2)

4.10   Measuring time

4.10.1   Traditional time measurement

The need to measure the passage of time arose early in the history of civilization, and both the sundial and the water clock were developed in Egypt around 1600 BC. While a properly constructed sundial has an absolute accuracy of about 1 minute, it is only useful when the Sun is bright enough to cast a shadow, and a correction factor (called "the equation of time") is needed to account for the fact that the Earth's orbit around the Sun is not a perfect circle. The water clock is not as accurate, but is useful when clouds hide the Sun and stars, and can be read directly. The water clock was used in Greek law courts around 300 BC.

The first mechanical clocks (developed in China in AD 725 and in Europe around 1350) were weight driven and were regulated by an oscillating horizontal bar. In the 1600s, the pendulum clock was developed. By using a pendulum as the oscillator and by improving the escapement, accuracy was greatly improved. An escapement rotates a gear one tooth per swing of the pendulum and also provides impulses to keep the pendulum swinging. For three centuries, the pendulum clock was the most common timepiece used. Accuracies of several seconds per year (1 part in 10^7) are possible.

The development of the spring drive and oscillating wheel escapement made the mechanical pocket watch possible, as well as the far more accurate maritime chronometer developed by John Harrison in 1764. The chronometer (accurate to 0.1 s/day or 1 part in 10^6) and a measurement of the position of the Sun or a star allows ships at sea to determine their position on the Earth.

4.10.2 Modern time measurement

The two most commonly used timepieces are the electric wall clock, which depends on the frequency of its ac power source, and the quartz clock. The quartz clock uses the piezoelectric effect to establish an electric/mechanical resonance in a quartz crystal. It has an accuracy of typically 0.5 s/day or 1 part in 2×10^5. The clocks used in microcomputers to control the sequential execution of instructions, and in digital counter/timer chips (such as the AM9513) also use a quartz crystal. Recently, quartz wrist watches and wall clocks have almost completely replaced mechanical clocks.

The cesium atomic clock is based on the resonance absorption of cesium atoms at 9,192,631,770 Hz. This clock has an accuracy of 1 s/300,000 years (1 part in 10^{12}). An international agreement in 1967 based the definition of time on the resonance frequency of the cesium atom. The cesium clock is used as a laboratory standard and in navigation systems. Another atomic clock, the rubidium clock, operates at a resonance frequency of 6,834,692,608 Hz and is used as a laboratory standard and as a frequency standard for radio transmitters.

The most accurate timepiece is the hydrogen maser, developed in 1964 at the US Naval Research Laboratory in Washington, DC. It is accurate to <1 s/3 million years (<1 part in 10^{14}).

4.11 Problems

4.1 Consider a position sensor using a 10-kΩ linear resistor with 0.1% absolute accuracy and a full range of 10 cm. One end of the resistor is connected to +10 V, the other end is connected to ground, and the wiper is connected to a computer-based data-acquisition system with an input impedance of 10 kΩ (assume purely resistive), and 12-bit analog-to-digital conversion (0 = 0 cm, 4,095 = 10 cm). See Figure 4.49.

   (a) Derive an expression for the output voltage V_0 as a function of displacement x. (Assume that $V_0 = 0$ V at $x = 0$ cm and $V_0 = 10$ V at $x = 10$ cm.)
   (b) What is the displacement accuracy of the system as limited by the accuracy of the linear resistor?
   (c) What is the displacement accuracy of the system as limited by the accuracy of the analog-to-digital conversion?
   (d) Do parts (a)–(c) above, with the addition of a 10-kΩ resistor between V_0 and +10 V.

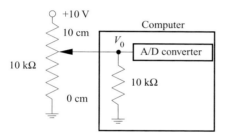

Figure 4.49 Position sensor using a linear potentiometer connected to an A/D converter.

4.2 Consider another position sensor using a linear digital encoder such as that shown in Figure 4.5, but with 16 bits rather than 4, and 2^{16} binary numbers rather than 2^4. The length of the Gray-code pattern (distance from first to last number) is 10 cm. Each number in the pattern is produced with an absolute position error less than one-half the distance between neighboring numbers.
  (a) What displacement produces a change from one number to the next?
  (b) Which position sensor is more accurate, the analog system described in Problem 4.1 or the digital system described in this problem?

4.3 You wish to measure air temperatures over the 0–50 °C temperature range using the thermistor bridge shown in Figure 4.22 with $R_1 = R_2 = R_3 = 1$ kΩ. The bridge output is amplified by an instrumentation amplifier with a gain of 5. In this problem we investigate the thermal dissipation constant of the thermistor.

   The thermistor manufacturer gives the following relationship between temperature and resistance:

T (°C)	R_T (kΩ)	T (°C)	R_T (kΩ)
0	2	10	1.33
20	1	30	0.667
40	0.5	50	0.333

   With the thermistor in air and a bias voltage $V_b = 1$ V, you adjust R_1 to make $V_0 = 0.00$ V.

   Increasing the bias voltage to $V_b = 10$ V, and after waiting for the thermistor temperature to stabilize, you measure $V_0 = 5.00$ V.
  (a) What are the thermistor resistances at these two bias voltages?
  (b) What are the thermistor temperatures at these two bias voltages?
  (c) What power is dissipated by the thermistor at $V_b = 1$ V?
  (d) What power is dissipated by the thermistor at $V_b = 10$ V?
  (e) What is the thermal dissipation constant for the thermistor in air?
  (f) Comment on the best bias voltage for this application.

4.4 You have just joined a team to design incubators (temperature-controlled enclosures) for hatching chicken eggs and raising baby chickens. Your part of the project is to develop a temperature sensor that can be read by a data-acquisition board connected to a microcomputer. The data-acquisition board has a 10-bit A/D converter with a full-scale input range from 0 to 5 V and an input impedance of 1 kΩ. The temperature range of interest is 20–60 °C, with maximum sensitivity at 40 °C, near the optimum temperature of the incubators. The thermistor has a calibration scale shown, in part, in the following table:

T (°C)	R (Ω)	T (°C)	R (Ω)	$\Delta R/\Delta T$ (Ω/°C)
20	10,000	21	9,700	300
40	5,000	41	4,800	200
60	2,500	61	2,425	75

You also use the standard bridge circuit shown in Figure 4.22, with $V_b = 1$ V.
(a) For maximum bridge output sensitivity dV_0/dT at 40 °C, and zero output at 20 °C, what should be the values of resistors R_1 and R_2?
(b) What is the sensitivity of your bridge circuit (in millivolts per degree celsius) at 20, 40, and 60 °C?

   Hint: If $F = A/(A + x) + B$, then $dF = [-A/(A + x)^2] dx$.

(c) What is the bridge output voltage V_0 at 20, 40, and 60 °C?
(d) What type of amplifier would you use to provide the input signal to the A/D converter? What gain would you choose? What would be reasonable values for the specifications on input impedance, output impedance, and common-mode rejection?
(e) What limit does the resolution of the A/D place on the precision of the temperature measurements near 40 °C? (Give your answer in degrees celsius.)
4.5 You are given the following components and asked to design the hardware for a temperature-measurement system:
   1. A solid-state temperature sensor and amplifier whose output is proportional to the absolute temperature from 200 to 400 K (-73 to $+127$ °C). The output voltage is 2.000 V at 200 K, 4.000 V at 400 K, and perfectly linear in between.
   2. A successive-approximation 12-bit A/D converter with an input range from 0.000 to 4.095 V. Conversion time is 10 μs. Conversion is initiated 100 ns after a low-to-high edge on the "conversion start" input line. The "conversion status" output line is high only when a conversion is in progress and the digital output data are valid 100 ns before the "conversion status" goes from high to low.
   3. A sample-and-hold amplifier that is in the hold mode when its input control line is high and is in the sample mode when its control line is low. Assume that the output is stable 100 ns after the transition from sample to hold.

4. A 12-bit parallel input port that simultaneously: (i) latches data onto its input registers; and (ii) sets bit 0 of a status register to 1, whenever a high-to-low edge appears on its "input strobe" line. Your computer program can read the 8-bit status register with the statement "inp(3);", the low eight bits on its data registers with the statement "inp(1);" and the high four bits with the statement "inp(2);". If all 12 bits are ones, then the "inp(1);" will read FF and the "inp(2);" will read 0F. The read process causes the port to reset the status register.

5. A 1-bit output port that produces a 1-μs positive pulse using the statement "outp(1,1);".

6. A microcomputer with keyboard, CRT screen, with components 4 and 5 as plug-in boards.

(a) Draw a block diagram of all essential components and interconnections, and label same.

(b) What voltage transformation is needed to match the range of the temperature sensor to the A/D converter?

(c) Draw timing diagrams for the following lines and registers:
- A/D conversion start input (show logic level),
- A/D conversion status output (show logic level),
- sample-and-hold control input (show logic level),
- parallel port input strobe (show logic level),
- parallel port data registers (show new versus old data),
- bit 0 of the parallel port status word (show zero or one).

4.6 Design a computer code for the hardware system of Problem 4.2 to display the temperature (in degrees celsius) on the screen whenever the user asks for it.

(a) Show the computer code (or a list of what the program steps do) for the following:
- receive a prompt from the user,
- initiate data conversion,
- read the temperature data,
- convert to degrees celsius,
- display the answer on the user's screen (*do not omit any essential steps*).

(b) If the A/D has an absolute accuracy of 1 LSB, how accurate is the temperature value that is displayed?

(c) What is the smallest temperature change that the system can reliably detect?

4.7 You are given a force transducer consisting of a single-element strain gauge cemented to the top of an aluminum bar (Figure 4.50). The gauge factor of the strain gauge is $G = 2$. The flexibility of the rod is such that the strain at the top of the bar (where the strain gauge is located) is 1 microstrain per gram of mass. Aluminum has a thermal-expansion coefficient of 23 ppm/°C.

(a) Draw the bridge circuit you would use (excitation = 1 V).

> *Hint*: The equations are easier if all resistors have the same value as the unloaded strain gauge and one end of the strain gauge is connected to ground.

(b) Derive the bridge equation for this transducer (i.e. relate output voltage to strain $\Delta L/L$).

(c) Give the sensitivity in microvolts per gram for a very small mass on the bar.

(d) What is the bridge output for a 10-kg mass on the bar? What would you have expected from your answer to (c), assuming perfect linearity?

(e) With a fixed mass on the bar, how will the bridge output change if the temperature changes by 10 °C? To what load does this correspond?

(f) How could you use a second identical strain gauge to compensate for such temperature-caused errors?

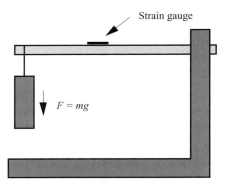

Figure 4.50 Force transducer using a flexible bar and a single strain gauge.

4.8 You are given four identical resistive strain gauges with the following resistance/strain relation: $\Delta R/R = G(\Delta L/L)$, where $G = 2$. These are cemented to opposite sides of a flexible bar: two on one side and two on the other as shown in Figure 4.51. Using the assembly as a force transducer, we have $\Delta L/L = KF = Kmg$, where K is a constant, m is the mass, g is the acceleration of gravity, and $\Delta L/L$ is the strain experienced at the top of the bar.

(a) Draw the bridge circuit you would use.

(b) Derive the bridge equation for 1 V excitation and solve for the output voltage V_0 as a function of m.

(c) If $V_0 = 1$ mV for 1 kg, what is the value of K?

(d) Is this force transducer linear?

(e) What limits the maximum m that can be measured?

(f) How sensitive is this force transducer to temperature?

(g) What would be the effect of a second power term? $\Delta R/R = G(\Delta L/L) + \alpha(\Delta L/L)^2$.

Hint: This new term does not depend on the sign of ΔL.

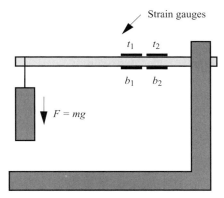

Figure 4.51 Force transducer consisting of a flexible bar and two opposing pairs of strain gauges.

4.9 A railroad track support column has four solid-state strain gauges cemented to it, as shown in Figure 4.52. The force on the column is straight down (no bending, only compression). Your goal is to be able to measure the strain on the column as trains pass over it. Two of the strain gauges are p-type, with the strain relationship: $\Delta R/R = 100\Delta L/L + 10,000(\Delta L/L)^2$, and the other two are n-type,

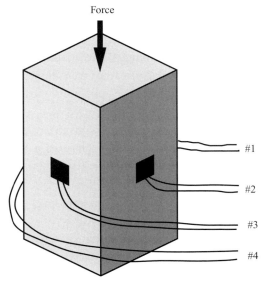

Figure 4.52 Four strain gauges mounted on a column.

with the strain relationship: $\Delta R / R = -100 \Delta L / L + 10,000(\Delta L / L)^2$. Assume that all gauges have the same unstrained resistance R.

(a) Draw a bridge circuit to measure the compressive strain on the beam. Indicate the positions of the p- and n-type strain gauges.

(b) Derive an expression relating the bridge output to the compressive strain on the column.

(c) What is the sensitivity of the bridge in terms of millivolts per microstrain at very small strain? (Assume a 1-V bridge excitation.)

(d) For a strain of $\Delta L / L = 0.1\%$ (1,000 microstrains), how far does the actual output deviate from a straight line passing through zero with the slope given in (c)? (Assume that the gauge factor remains constant.)

4.10 Design a thermocouple-based system for measuring the Space Shuttle main engine exhaust temperature with the following requirements:

1. The response time is 1 s or less.
2. The temperature range is 20–2,300 °C.
3. The electronics must be located in a forward compartment, 20 m from the sensing junction.
4. There is considerable electromagnetic radiation in the 1-kHz to 100-MHz range from other circuits in the spacecraft.
5. The electrical output of your circuit will be sampled at 10 Hz with a data-acquisition circuit similar to the one you built in the laboratory.

(a) Draw a block diagram, labeling all essential components and wires.

> *Hint* 1: Your circuit must reject interference that appears on both thermocouple wires.
> *Hint* 2: You will need a low-pass filter so that the data-acquisition circuit does not see unwanted frequencies.

(b) Label the drawing with the typical voltages that would be present at various points in your circuit when the sensing junction is at a temperature of 2,000 °C.

(c) What sensor would you use to measure the temperature inside an astronaut's space suit?

(d) Estimate the accuracy and precision of the systems in (a) and (c).

4.11 You are developing a spectrophotometer system that uses a *pin* photodiode to measure the concentration of a colored solution in a test tube. The op amp you are using has an input impedance of 10^{14} Ω. With a reverse bias of 10 V, the photodiode dark current is 10 nA and the maximum light level you need to detect produces a photodiode current of 1 μA.

You first try the circuit shown in Figure 4.53, with $R_1 = 9 \, k\Omega$ and $R_2 = 1 \, k\Omega$:

(a) With this circuit, the output saturates. What is the problem with the circuit?

(b) Draw a modification of the circuit that would make it work so that the output voltage is a linear function of the intensity of the light falling on the photodiode (i.e. $V_0 = a + bI$).

(c) Describe briefly how you would bias the circuit for the photovoltaic mode and for the photoconductive mode.

(d) Draw the block diagram for a complete microcomputer-based system for measuring the concentration of a colored solution. Label all essential components.

(e) How would the output voltage depend on the concentration of the solution? (Give a formula with an unknown constant multiplier.)

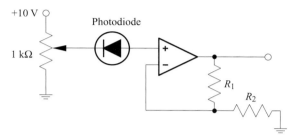

Figure 4.53 Faulty photodiode amplifier circuit.

4.12 You are developing a spectrophotometer system that uses a *pin* photodiode, and want to design a circuit that produces an output that is linearly proportional to the *concentration* of a colored solution.

(a) First, derive the closed-loop gain V_0/V_1 for the op-amp circuit shown in Figure 4.54 that uses a standard diode (not a photodiode) as a feedback element.

Assume an ideal op amp, $V_1 > 0$, and a diode current $I_D = I_S \exp(V_D/V_T)$, where I_S and V_T are constants, and V_D is the voltage across the diode ($V_D > 0$ for forward bias) (Figure 4.55).

(b) Draw a modification of the circuit of Figure 4.49 that could be used for measuring the concentration of solutions such as you did in Laboratory Exercise 14.

Hint: The photodiode acts as a current source, not a voltage source.

(c) Assuming that the photodiode current is directly proportional to incident light, and neglecting dark current, how would this circuit aid data processing for the measurement of concentration?

Hint: Derive an expression for V_0 as a function of concentration.

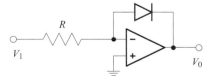

Figure 4.54 Op-amp circuit with a diode as a feedback element.

Figure 4.55 Forward-biased diode.

4.13 Give two significant sources of error for the following sensors or transducers:

> *Hint*: Call upon your experiences in the laboratory.

(a) thermistor in a bridge circuit,
(b) thermocouple and differential amplifier,
(c) a force transducer consisting of four strain gauges mounted in opposing pairs and used in a bridge circuit,
(d) a potentiometer used to measure angle,
(e) *pin* photodiode with current-to-voltage amplifier.

4.14 A test kit is available for measuring the levels of lead in eating utensils (cups, bowls, plates, etc.). The utensil is first soaked in hot acetic acid (vinegar) and the acid is mixed with a reagent. If no lead is present, the mixture is clear. If a small amount of lead is present, the mixture is yellow. If a dangerous amount of lead is present, the mixture is dark orange.

Design a system for determining the concentration of lead in ppm, using a green LED, a photodiode, and a microcomputer with A/D converter (input range -10 to $+10$ V). (It is not necessary to include analog filtering.)
Assume the following:

- The light intensity A passing through the solution is given by $A = A_0 e^{-kLC}$, where C is the lead concentration in parts per million, L is the thickness of the solution in centimeters, and the extinction coefficient for green light is $k = 1/\text{ppm/cm}$.
- The thickness of the solution $L = 1$ cm.
- The LED shining through a clear solution produces a photodiode current of $100\ \mu A$, and your design should convert this into a signal of 5 V at the A/D converter of the microcomputer.
- The entire system has a measured noise level of $10\ \text{mV}$ rms at the A/D converter.
- The amplifier offset output voltage is V_B.
- You operate the photodiode in photovoltaic mode.

(a) Sketch a block diagram including and labeling all essential components. (You can show the A/D and microcomputer as a single block.)
(b) Derive an expression for the A/D input voltage as a function of lead concentration C, and amplifier output offset V_B.
(c) Describe how a user would calibrate the system.

    (d) Derive an expression for the uncertainty ΔC as a function of concentration C.

       Hint: $\Delta C = \Delta I/(dI/dC)$.

    (e) What is the uncertainty ΔC at $C = 0.1$, 1, and 3 ppm?

    (f) What are the lowest and highest concentrations of lead that you can reliably measure?

       Hint: Find the two values of concentration C at which $\Delta C = C$.

4.15 More than one-half of our electrical energy is produced by the burning of fossil fuels. This results in the depletion of nonrenewable resources and in the production of greenhouse gases. The future of society depends on the development of nonpolluting, sustainable sources of energy. One approach is a solar converter consisting of a parabolic mirror and a high-temperature GaAs photocell at the focus of the mirror (Figure 4.56). (A silicon photocell such as the one you used in Laboratory Exercise 14 would be destroyed by the intense heat.) On a clear day the Earth receives about 1,000 W/m^2 of solar power.

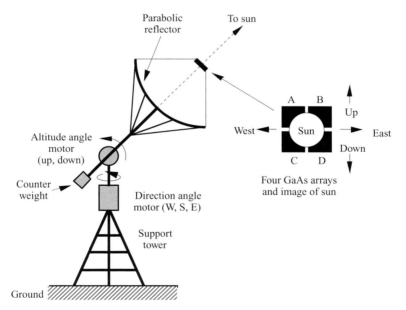

Figure 4.56 Solar concentrator and energy converter using a mirror and GaAs high-temperature solar cell. The altitude and direction angle motors allow the mirror to track the Sun during the day.

    Your job is to design an ***analog control system*** that tracks the Sun and keeps its image focused on the GaAs photocell. Once the system has been aligned with the Sun, your system should automatically track the Sun throughout the day without any previous knowledge of the path of the Sun in the sky.

The photoconverter consists of four GaAs arrays A, B, C, and D. Each array has 200 elements connected in series. When the mirror is aimed directly at the Sun, one-quarter of the solar image falls on each array, producing a voltage of 100 V and a current of 100 A. When the mirror is not aimed directly at the Sun, its image shifts and some arrays produces less current and voltage while others produce more current and voltage.

Each of the four GaAs arrays is connected to a high-power dc-to-ac converter circuit. The ac voltages are combined and increased by a step-up transformer so the electric power can be sent over high-voltage transmission lines to an energy-hungry world (Figure 4.57).

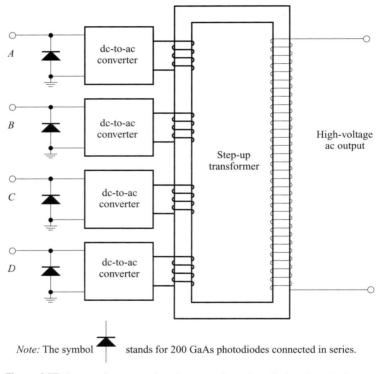

Note: The symbol ◤◥ stands for 200 GaAs photodiodes connected in series.

Figure 4.57 System for converting dc power from four GaAs photodiode arrays into high-voltage ac power that can be transmitted long distances.

- The altitude angle and direction angle motors use direct current and can rotate in either direction, depending on the polarity of the input voltage. These motors require a minimum of 5 V and 5 A before they can begin to move the mirror.

Do the following:

(a) Design a system that senses the voltage produced by the four GaAs arrays and produces a dc voltage to control the direction and altitude motors and keep the mirror accurately pointed at the Sun during the course of the day.

Sketch and label all essential components and connecting wires.

(b) Describe how your system responds to the motion of the Sun in the sky.

4.16 You are planning to detect light using a photodiode and an op amp in the circuit shown in Figure 4.58.

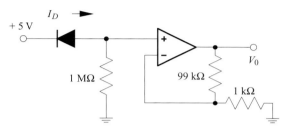

Figure 4.58 Photodiode circuit.

The specifications for the op amp are:
- input leakage current difference $= 0.1$ nA;
- input offset voltage $= 1$ mV;
- rms input noise $= 10$ nV$/\sqrt{Hz}$ at 90 kHz, 20 nV$/\sqrt{Hz}$ at 9 MHz;
- rms output noise $= 100$ nV$/\sqrt{Hz}$ at 90 kHz, 200 nV$/\sqrt{Hz}$ at 9 MHz;
- $R_{\text{in}} = 10^9$ Ω;
- unity-gain bandwidth $= 9$ MHz.

The specifications for the photodiode are:
- dark current at 5 V reverse bias $= 0.9$ nA.

(a) Explain how the circuit converts the current I_D into the output voltage V_0 and give a formula for V_0 as a function of I_D.

(b) What happens if R_1 is taken out of the circuit? Justify your answer.

(c) What is the amplifier output V_0 (compute both the average value and rms noise) when no light reaches the photodiode? Show work.

4.17 The circuit shown in Figure 4.59 was designed to amplify the differential signal from a thermocouple temperature sensor. However, when built and tested, a serious design error was discovered.

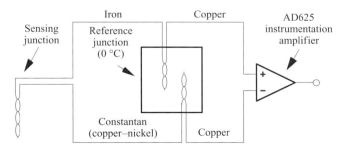

Figure 4.59 Thermocouple connected directly to an instrumentation amplifier.

(a) What symptom would have developed during the test? What caused this unwanted behavior?

(b) Design and draw a new circuit that works as originally intended.

4.18 A thermistor is set up to measure the temperature in still air using the bridge circuit shown in Figure 4.22. Under these conditions, self-heating causes the temperature of the thermocouple to rise 1 °C for every 10 mW of power. The thermistor resistance is 10 kΩ at 0 °C, 5 kΩ at 20 °C, and 1 kΩ at 60 °C. The resistor values are $R_1 = 10$ kΩ, and $R_2 = R_3 = 5$ kΩ.

(a) At 20 °C, and 1 V bridge bias V_b, how much power is dissipated by the thermistor? What is the temperature rise?

(b) At 60 °C and 1 V bridge bias V_b, how much power is dissipated by the thermistor? What is the temperature rise?

(c) At 20 °C, when the bridge bias is increased to 10 V, how much power is dissipated by the thermistor? What is the temperature rise?

4.19 There are almost always tradeoffs associated with any given design consideration. Before deciding which tradeoffs to make, the specific application must be considered:

(a) For a photodiode, when would you use a 5-V reverse bias and when would you use zero volt bias?

(b) For a force transducer using an elastic rod, when would you use four strain elements and when would you use one strain element?

(c) For a bandpass filter to extract a signal in a noisy environment, when would you use a high-Q (narrow) filter and when would you use a low-Q (wide) filter?

(d) For sensing temperature, when would you use a thermistor and when would you use a thermocouple?

4.20 Answer the following:

(a) What thermocouple gives the highest sensitivity at 300 °C?

(b) What thermocouple gives the highest sensitivity at 1,000 °C?

(c) What thermocouple gives the highest sensitivity at 1,700 °C?

(d) You want to design a thermocouple-based system with a response time of 0.1 s and where the electronics must be located a considerable distance from the sensing junction. Describe three techniques that you can use to reduce the effect of induced signals from nearby ac currents.

4.21 You have just been hired to help test a new type of large windmill. Large windmills generate considerable power at high wind speed, but their blades can be destroyed if the wind speed becomes too great. Your job is to instrument a test windmill so that you can measure the backward bending of the moving blades and determine whether the wind speed is within safe limits (Figure 4.60). At dangerous wind speeds, the windmill can be "shut down" by rotating the generator housing 90° on a vertical axis so that the wind strikes the blades from the side.

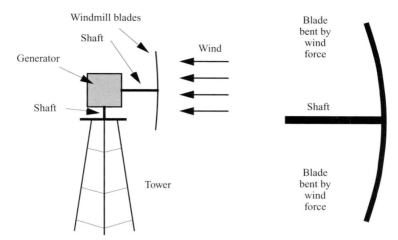

Figure 4.60 Windmill to be instrumented with strain gauges to determine whether the wind speed is within safe limits.

Assume the following:
- You have decided to use metal-foil strain gauges with gauge factor $G_S = 2$ (similar to those used in Laboratory Exercise 13).
- You have decided to mount the strain gauges, necessary electronics, and a small, rugged radio transmitter on the moving blade and to transmit the information to a radio receiver on the ground.
- The radio transmitter can accept signals in the -10 to $+10$ V range and the radio receiver reproduces the transmitted signal with the same amplitude.
- The radio system has a frequency response of 0 Hz to 10 kHz, and adds 10 mV of white noise in this frequency band.
- The windmill normally swings to face the wind.

Design a system that senses the backward bending of the wind on the moving blade, transmits the signal to a radio receiver on the ground, and interfaces the signal to a microcomputer.
- Your design should reject noise outside of the 0–100-Hz frequency band of interest
- Your design should compensate for temperature variations

Do the following:
(a) Sketch where you would attach the strain gauges and other components to the blade.
(b) Draw a block diagram of your system, showing and labeling all essential components and connecting wires both on the blade and on the ground.
(c) Show on your block diagram above, the voltages that would occur for a strain $|\Delta L/L| = 0.1\%$.

4.22 You are designing a thermocouple-based system for measuring the temperature of a furnace over the temperature range from 25 to 500 $°C$ with an absolute accuracy of 2 °C and do not want to provide ice to stabilize the temperature of the reference junction at 0 °C. Instead, you decide to leave the reference junction in the air of the room and measure the temperature of the room (maximum range 10–45 °C) with a thermistor, which would provide sufficient accuracy. The correction of the thermocouple output for room temperature will be done by a microcomputer program:

(a) Design a circuit that converts the thermocouple output into a suitable voltage V_{tc} (−5 to +5 V) for input to a microcomputer. Draw a block diagram and label all necessary analog circuit elements and signal lines. Include the thermocouple wires. (It is not necessary to include analog filtering.)

(b) Design a circuit that converts the thermistor resistance into a suitable voltage V_{tm} (−5 to +5 V) for input to a microcomputer. Draw a block diagram and label all necessary analog circuit elements and signal lines. Show where the thermistor is placed in the diagram of (a) above. (It is not necessary to include analog filtering.)

(c) Sketch the thermocouple voltage V_{tc} as a function of the temperature difference $\Delta T = T_{sens} - T_{ref}$. Label the axes with numbers and units.

(d) Sketch the thermistor voltage V_{tm} as a function of the thermistor temperature T_{tm}. Label the axes with numbers and units.

(e) Describe in a flow chart, a list of steps, or in a sentence or two, what the microcomputer program would have to do to convert V_{tc} and V_{tm} into the temperature T_{sens} of the sensing junction in the furnace.

4.23 You work as a design engineer for Thermal Manufacturing, Inc. Frequently, prospective customers ask your professional advice whether to use a thermistor or a thermocouple in various situations. For each application below, mark an X to indicate which device is the better choice.

	Thermistor	Thermocouple
Measuring the temperature in a glass furnace		
Measuring small variations in body temperature over 24 hours		
Measuring the temperature difference between sunlit and shaded surfaces on the Moon over periods of several years		
A case where linear sensor output (volts versus temperature) is required over a large temperature range		
Accurate temperature measurement where the thermal conductivity of the surrounding medium is frequently changing		
Sensor with the maximum sensitivity at 1,700 °C		
Sensor for a temperature control system (0–50 °C) in an electrically noisy environment		

4.24 You have just been hired by an engineering firm that provides instrumentation to large industrial corporations. Your first problem is designing a system for measuring the liquid level in a large tank.

- The liquid level is to be sensed electronically and the electrical signal is to be connected to the analog input of a microcomputer for display and storage.
- The tank is 10 m in diameter and 10 m high.
- The liquid absorbs green light with an absorption of 10% per meter:

$$I(L) = I(0)e^{-kCL}, \quad kC = 0.1/\text{m}$$

- The liquid is slightly conductive. (A column of liquid with area A and length L has resistance $R = \rho A/L$.)
- The liquid is non-flammable.
- The liquid level is to be measured to an accuracy of 0.1 m.

(a) Describe in about 50 words and/or a simple sketch how you would measure the liquid level using light sensors.

(b) Describe in about 50 words and/or a simple sketch how you would measure the liquid level using strain gauges.

(c) Describe in about 50 words and/or a simple sketch how you would measure the liquid level using a digital angle sensor.

(d) Describe in about 50 words and/or a simple sketch how you would measure the liquid level using ideal electrodes.

> *Hint*: Ideal electrodes transform ionic conductivity in a solution into simple electrical conductivity.

(e) Describe in about 50 words and/or a simple sketch how you would measure the liquid level using thermistors.

(f) Describe in about 50 words and/or a simple sketch how you would measure the liquid level using sound (speaker and microphone).

4.25 You have just been hired to design a sensor for a new class of icebreaker ship. These ships are used to smash their way through sheets of floating ice, either to get to a destination, or to clear a passage for other ships (Figure 4.61). The hull (outer structural surface) is very thick and strong, but the ship's engines are powerful and it is possible to damage or destroy the hull by driving it against large floating slabs of thick ice. The plan is to provide the captain of the ship with a continuous monitor of the **shear** on the hull at the water line. The **shear** is the difference between the strain a few feet below the water line (where the ice is pushing against the moving ship) and the strain a few feet above the water line (where there is no ice). When the ice becomes thicker or the slabs become larger, the engine speed can be reduced to avoid damage to the hull.

Design a system that:

- senses the difference in strain $S_+ - S_-$ on the hull between one meter above the water line S_+ and one meter below the water line S_-;

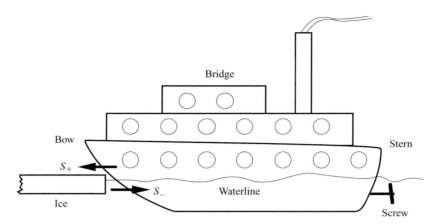

Figure 4.61 Icebreaker in the process of breaking ice with its prow.

- provides an analog signal proportional to $S_+ - S_-$ such that an output of 5 V corresponds to $S_+ - S_- = 0.1\%$;
- provides filtering suitable for the signal and noise frequencies that you would anticipate;
- displays the value of the shear on the ship's bridge (control room), and sounds an alarm when $S_+ - S_- \geq 0.1\%$.

Assume that the temperature of ice-laden sea water is constant ($-3\,°C$) so you do not have to compensate for changes in temperature.

(a) Sketch and label all essential components and connecting wires.

(b) Show on your block diagram the typical voltages that would occur for a compressive strain $S_- = \Delta L/L = 0.1\%$ below the water line and $S_+ = 0$ above the water line.

(c) What is the maximum change in amplifier output offset voltage that your system could tolerate?

4.26 You are designing a system for measuring temperature over the range from 20 to 60 °C using a thermistor in a voltage divider (Figure 4.62). The voltage is to be read by the analog interface of a computer. At 0 °C, the thermistor resistance is 10 kΩ, and $\beta = 3{,}500$ K. The A/D input range is -10.24 to $+10.24$ V and the output range is 0–2,047 (11 bits).

(a) Sketch a block diagram of your design, showing the resistor divider, the microcomputer with analog input port, and anything else needed. (The best value of the resistor R will be explored below.)

(b) Write an expression for the output V_0 as a function of thermistor resistance R_T.

(c) Determine the resistor value R that provides the maximum sensitivity dV_0/dT at a temperature of 40 °C.

Hint 1: Maximize $dV_0/dT = (dV_0/dR_T)(dR_T/dT)$.
Hint 2: The maximum value of $x/(1+x)^2$ occurs at $x = 1$.

(d) What is V_0 at 20, 40, and 60 °C?
(e) What is the change in A/D input voltage ΔV that corresponds to a change of one unit in the A/D output?
(f) At 40 °C, what is the temperature change ΔT that corresponds to a change of one unit in the A/D output?

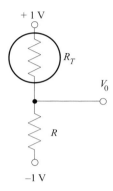

Figure 4.62 Thermistor resistor divider circuit.

4.27 For the following physical quantities, describe: (1) typically what causes them, (2) their nature (the qualities that define them, typical magnitude, etc.), (3) a means for sensing them and producing a useful ≈5-V electrical signal:
(a) the electrocardiogram (ECG),
(b) visible light,
(c) strain.

4.28 Describe two sensors and two actuators for each of the following physical quantities:
(a) temperature,
(b) visible light,
(c) position,
(d) force.

4.29 Describe in one or two sentences how you would measure:
(a) temperature over the range 0–600 °C with a precision of 0.2 °C and an accuracy of 0.5 °C,
(b) temperature over the range 0–100 °C with a precision of 0.05 °C and an accuracy of 0.1 °C,
(c) voltage in the range 0–1 mV across a 1-MΩ resistor, only in the frequency band from 100 Hz to 10 kHz, and where neither end of the resistor is grounded.

4.30 You have a light sensor at the end of a long metal pole that produces a signal in the 100-Hz to 100-kHz frequency range (Figure 4.63). The sensor also receives 60-Hz interference from nearby power lines and 1-MHz interference from a nearby radio station. In addition, the sensor output also has an additive component that depends on temperature.

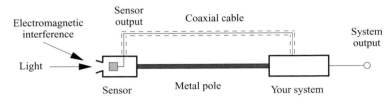

Figure 4.63 Light sensor connected to a circuit.

Design two systems that each meet the following requirements:
- Amplifies a 1-mV sensor output signal in the 100-Hz to 100-kHz frequency range to produce a 10-V system output with an accuracy of 1%.
- All unwanted signals (0 Hz to 1 MHz) must contribute less than 0.1 V to the system output.

Assume:
- The unwanted 60-Hz background produces a sensor output of ± 1 mV.
- The unwanted 1-MHz background produces a sensor output of ± 10 mV.
- The maximum temperature variation produces an unwanted sensor output from -10 to $+10$ mV (assume a maximum frequency of 0.1 Hz).
- The sensor output is connected to the input of your circuit with a coaxial cable that effectively shields the internal signal wire from external interference.

Do the following:
(a) Sketch the design of a system (system No. 1) that uses analog filtering to accomplish the design objectives. Specify general characteristics such as number of stages and corner frequencies, but you do not need to show individual resistors and capacitors. Show sufficient detail that a skilled technician can build it and understand how it meets the design objectives.
(b) Sketch the voltage gain of your system No. 1 from 0.001 Hz to 10 MHz.
(c) Sketch the design of a system (system No. 2) **that uses two identical sensors and differential amplification by an instrumentation amplifier**. Show sufficient detail that a skilled technician can build it and understand how it meets the design objectives.
(d) What are the common-mode-rejection requirements of the instrumentation amplifier in system No. 2 at 0 Hz, 60 Hz, and 1 MHz? Which would be the most difficult and why?

4.31 Design a system for converting sunlight into electrical energy stored in a battery.

Assume:

- You have a 100 large-area photodiodes, to be connected in series.
- You have a clever battery charger circuit (which you do not have to design) whose effective input resistance automatically adjusts to different light levels to extract the largest electrical power from the photodiodes. The batteries are part of this circuit.
- The I–V characteristic of each solar cell is shown in Figure 4.64.

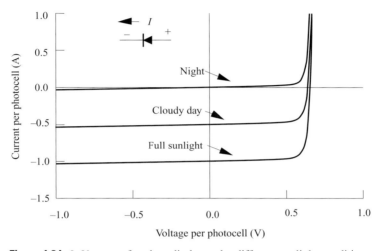

Figure 4.64 I–V curves for photodiodes under different sunlight conditions.

Do the following:

(a) Sketch a block diagram of your design. Include and label all essential components and signals.

(b) Describe briefly how the photodiode converts the energy of a photon into electrical energy.

(c) On a cloudy day, what is the approximate input to the battery charger circuit in terms of voltage, current, and power? (Show your work.)

> *Hint*: $I = P/V$ is a curve of constant power.

(d) Under full sunlight, what is the approximate input to the battery charger circuit in terms of voltage, current, and power? (Show your work.)

4.32 Design a system for controlling the thickness of sheet metal in a rolling mill. To produce sheet metal, thick sheets are pressed between two rollers (see Figure 4.65). The distance between the rollers is adjusted to control the thickness of the final product. You plan to use a beta source and a solid-state detector to sense the thickness of the sheet at the output of the rollers and control the distance between the rollers *so that the thickness is the same as that of a reference sheet*. Since the sheet absorbs some of the beta energy, a thicker sheet will produce less current in the solid-state detector than a thinner sheet.

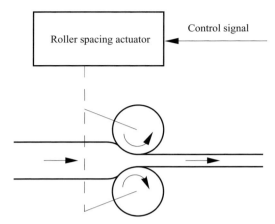

Figure 4.65 Roller mill for reducing the thickness of metal sheets to a desired value.

You have:

- two identical beta sources,
- two identical solid-state detectors that produce a current that is proportional to the beta energy deposited in them,
- an interface to the roller mill so that a positive control signal decreases the spacing between the rollers and a negative control signal increases the spacing between the rollers,
- the control signal only needs to provide a small current (a few milliamps),
- electronic components necessary to generate the control signal.

Do the following:

(a) Sketch your design. Provide enough detail so that a skilled technician would be able to build it and understand how it works.

(b) Describe how the control system responds when the reference sheet is replaced by a thinner sheet.

4.12   Additional reading

George C. Barney, *Intelligent Instrumentation*, Prentice Hall, Englewood Cliffs, NJ, 1985.

Richard S. C. Cobbold, *Transducers for Biomedical Measurements*, Wiley, New York, 1974.

Leslie Cromwell, Fred J. Weibell, and Erich A. Pfeiffer, *Biomedical Instrumentation and Measurements*, Prentice Hall, Englewood Cliffs, NJ, 1980.

Glenn F. Knoll, *Radiation Detection and Measurement*, John Wiley & Sons, 2000.

Omega Engineering, *Omega Temperature Measurement Handbook and Encyclopedia*, Stamford, CT, 2001.

Robert L. Powell, William J. Hall, Clyde H. Hyink, et al., *Thermocouple Reference Tables Based on the IPTS-68*, National Bureau of Standards Monograph 125, National Institute of Standards and Technology, Gaithersburg, MD.

Daniel H. Sheingold, ed., *Transducer Interfacing Handbook*, Analog Devices, Norwood, MA, 1981.

Peter Strong, *Biophysical Measurements*, Tektronix, Beaverton, OR, 1970.

Willis J. Thompkins and John G. Webster, *Interfacing Sensors to the IBM PC*, Prentice Hall, Englewood Cliffs, NJ, 1988.

Laboratory Exercise 11

Measuring angular position

Purpose

To use a potentiometer to measure angle and to determine the linearity, accuracy, and backlash. To use the microcomputer to measure the angle of the decaying pendulum as a function of time. To analyze the system as an underdamped oscillator.

Equipment

- IBM-compatible Pentium microcomputer with HP VEE
- Digital oscilloscope
- Printer (shared with other laboratory stations)
- Pendulum and 1-kΩ potentiometer mounted on a wooden frame
- ± 5-V power supply
- Digital multimeter

Background

1.   Damped-harmonic oscillator

For small displacements, the pendulum used in this laboratory exercise is a damped-harmonic oscillator consisting of a mass that is subject to two forces: (1) the force of gravity, which is proportional to the displacement x and oppositely directed; and (2) friction, which exerts a force proportional to the velocity v and oppositely directed. The force equation is:

$$F = ma = -kx - cv$$

which results in the equation of motion:

$$m(d^2x/dt^2) + c(dx/dt) + kx = 0$$

where k is the restoring-force constant, and c is the friction-force constant.

Note that the restoring force is not actually proportional to the displacement x, but to the vertical component $\sin(\theta)$.

This differential equation has the characteristic form:

$$mr^2 + cr + k = 0, \qquad \text{with the solution } r = \frac{-c - \sqrt{c^2 - 4km}}{2m}$$

There are three cases:

Case 1: the underdamped oscillator
When $c^2 < 4km$, the solution is:

$$x = e^{-\alpha t}[A\cos(\omega t) + B\sin(\omega t)] = Re^{-\alpha t}\cos(\omega t + \delta)$$

where $R = \sqrt{A^2 + B^2}$, $\quad \tan(\delta) = -B/A$, $\quad \alpha = c/2m$, $\quad$ and $\omega = \left(\sqrt{4km - c^2}\right)/2m$.

The undamped natural frequency is given by $\omega_0\sqrt{k/m}$ and the damping factor α reduces the frequency of oscillation:

$$\omega^2 = \omega_0{}^2 - \alpha^2$$

The constants A and B are determined from the initial position x_0 and velocity v_0:

$$x_0 = A, \qquad v_0 = \omega B$$

Case 2: the critically damped oscillator
When $c^2 = 4km$, the solution is:

$$x = (A + Bt)e^{-\alpha t}, \qquad \text{where } \alpha = \frac{c}{2m} = \frac{2k}{c} = \sqrt{k/m}$$

The constants A and B are determined from the initial position x_0 and velocity v_0:

$$x_0 = A, \qquad v_0 = B - \alpha A$$

Note: Case 2 is of particular importance in electrical and mechanical engineering because when a harmonic oscillator (such as a circuit or building) is critically damped, it recovers from an impulse disturbance more quickly than with any other damping.

Case 3: the overdamped oscillator
When $c^2 > 4km$, the solution is:

$$x = Ae^{-\alpha t} + Be^{-\beta t}$$

where $\alpha = \dfrac{c + \sqrt{c^2 - 4km}}{2m}$ and $\beta = \dfrac{c - \sqrt{c^2 - 4km}}{2m}$

The constants A and B are determined from the initial position x_0 and velocity v_0:

$$x_0 = A + B, \qquad v_0 = -\alpha A - \beta B$$

    Note: When the damping term c is large, β is small, and after a disturbance the system returns slowly to $x = 0$.

Additional reading

Section 4.2 Position sensors
Appendix F Using the digital oscilloscope to record waveforms

Procedure

1.   Setup

Attach $+5$ and -5 V to opposite contacts of the potentiometer mounted on the pendulum board (Laboratory Figure 11.1). Connect the wiper (center contact) to your digital multimeter and to the red lead of a coaxial cable micrograbber adapter. Connect the coaxial cable to channel 1 of the digital oscilloscope. Connect the black lead of the micrograbber adapter (outer shield of the coaxial cable and oscilloscope ground) to the power-supply ground. Set the oscilloscope input for high impedance (1 MΩ) and dc coupling.

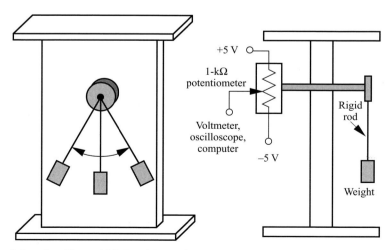

Laboratory Figure 11.1 Pendulum with a potentiometer for the measurement of angle.

2.   Static measurement of angle

Place the wooden frame flat and measure the output for five different angles using the polar graph paper attached. Check reproducibility and see if the result depends on whether you approach the angle from the right or from the left (backlash). Record the output from the potentiometer for each angle measured using the HP oscilloscope and the HP VEE panel driver window on the computer (see Appendix F Using the digital oscilloscope to record waveforms).

3.   Dynamic measurement of angle

3.1 Oscilloscope output. Set the wooden frame upright and set the oscilloscope for manual triggering and a sweep speed of 1 s per division. Release the pendulum at the same instant that the oscilloscope is manually triggered.

3.2 Computer output. Run the HP VEE panel driver (see Appendix F Using the digital oscilloscope to record waveforms) to sample and print the amplitude of the decaying pendulum as a function of time.

3.3 Baseline calibration. Repeat procedure sections 3.1 and 3.2 but with the pendulum at its resting position. This gives the zero baseline for your data.

Laboratory report

1.   Setup

Draw a simple block diagram of your experimental setup.

2.   Data summary and analysis

2.1 Tabulate output voltage and computer output as a function of angle (data from procedure section 2).

2.2 From the plot of output voltage versus time ($V(t)$) produced in procedure section 3.2, do the following:

- Draw a horizontal line at the output voltage for the resting pendulum (the baseline from procedure section 3.3).
- Estimate the apparent frequency $f = \omega/2\pi$.

  Hint: Average the time between baseline crossings to get the period $P = 1/f$.

- Estimate the decay time $\tau = 1/\alpha$.

  Hint: The distance from the peaks and valleys of the output to the baseline should fall as $e^{-\alpha t}$.

- Estimate the natural frequency $f_0 = \omega_0/2\pi$.

    Hint: For the underdamped oscillator, $\omega^2 = \omega_0^2 - \alpha^2$.

3.   Discussion and conclusions

3.1 Discuss the principles covered in procedure section 2 (static measurement of angle). Consider sensitivity, hysteresis (backlash), linearity error, precision.

3.2 Discuss the principles covered in procedure section 3 (dynamic measurement of angle). Consider the apparent frequency of the pendulum, the decay time, and the undamped natural frequency.

3.3 Discuss applications such as measuring position, velocity, voltage waveforms.

4.   Questions

4.1 If the potentiometer had a resistance of 10 kΩ and the digital oscilloscope input impedance was set to 50 Ω, would you expect the digitized values to be a linear function of angle? Justify your answer.

4.2 If the potentiometer had a resistance of 10 kΩ and if the digital oscilloscope had an input impedance of 50 Ω that could not be changed, how would you change the laboratory exercise so that the digitized values would be a linear function of angle?

4.3 If you used an oscilloscope whose input coupling was set to ac rather than dc, what effect would this have on your data?

4.4 If the pendulum were frictionless (did not decay), had a period of 0.5 s, and you took three samples per second, what would your data look like? Using only those data, what would the apparent frequency be?

5.   Laboratory data sheets

Include your handwritten data sheets (or a copy), which should consist of a log of the procedures you used, any special circumstances, and the measurements you recorded manually.

Laboratory Exercise 12

Measuring temperature

Purpose

To use three important temperature transducers: the thermocouple, the thermistor, and the platinum resistance thermometer (or dial thermometer). To measure and compare their response time, sensitivity, linearity, precision, and accuracy.

Equipment

- IBM-compatible Pentium microcomputer with HP VEE
- Digital oscilloscope
- Platinum resistance thermometer (or dial thermometer)
- +5-V, ±12-V power supplies
- Three 10-μF, 25-V electrolytic capacitors (put between power and ground on the superstrip breadboard binding posts)
- Two 0.1-μF, CK-05 capacitors (put between power and ground on all chips)
- Iron–constantan thermocouple
- Superstrip circuit board
- Digital multimeter with 0.1-mV sensitivity and 0.1-Ω accuracy on 200-Ω full scale
- Two 500-ml Pyrex beakers and stirring rod (wood or glass)
- Crushed ice
- Hot plate
- Precision thermistor (Omega type YSI 44004 (1207), 2,252 at 25 °C) coated with insulator for water immersion
- Two 1-MΩ resistors
- Two 2.4-kΩ resistors (fixed bridge resistors)
- One 20-kΩ trimpot (bridge adjust resistor)
- AD625 or LH0036 instrumentation amplifier

For AD625 instrumentation amplifier, use resistors below:

- One 39-Ω resistor (R_G for gain of 1,000)
- One 390-Ω resistor (R_G for gain of 100)
- One 2-kΩ resistor (R_G for gain of 20)
- Two 20-kΩ resistors (R_F)
- One 25-kΩ trimpot (offset adjust)

For LH0036 instrumentation amplifier, use resistors below:

- One 51-Ω resistor (R_G for gain of 1,000)
- One 510-Ω resistor (R_G for gain of 100)
- One 2.4-kΩ resistor (R_G for gain of 20)
- One 100-kΩ trimpot (offset adjust)
- One 3.3-kΩ resistor (offset adjust)
- One 33-kΩ resistor (offset adjust)

Additional reading

Section 2.4.1 Instrumentation amplifiers
Section 4.3 Temperature transducers
Appendix F Using the digital oscilloscope to record waveforms

Procedure

1. Setup

1.1 Circuit construction. Set up your instrumentation amplifier with a gain of about 100. For the AD625, $R_F = 20$ kΩ and $R_G = 390\,\Omega$. For the LH0036, $R_G = 510\,\Omega$. You will use it to amplify the thermocouple signal, which is quite small. Set up the thermistor bridge, as shown in Laboratory Figure 12.1 with $V_b = 1$ V, $R_1 = 25$ kΩ

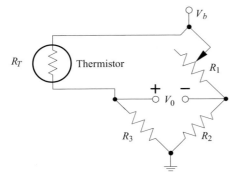

Laboratory Figure 12.1 Thermistor bridge circuit. The thermistor resistance is 2,252 Ω at 25 °C, the trimpot R_1 is 25 kΩ, and the two resistors R_2 and R_3 are 2.4 kΩ.

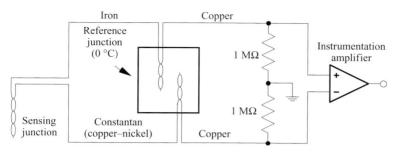

Laboratory Figure 12.2 Thermocouple junctions and instrumentation amplifier.

(trimpot), and $R_2 = R_3 = 2.4$ kΩ. Measure the amplifier voltage gain, which will be needed for your thermocouple data analysis. The bridge output may be accurately read with the digital multimeter without amplification.

1.2 Reference bath. Fill a small beaker with water and crushed ice and place the small beaker in a larger beaker filled with ice. This will get closer to 0 °C than using only water and ice in a single beaker. You will also use the platinum resistance thermometer (or dial thermometer) to provide a standard temperature measurement. See Table 4.1 for platinum resistance versus temperature.

2.   Thermocouple

2.1 Connections and reference junction. Connect the copper leads of the thermocouple to the differential inputs of the instrumentation amplifier. The external connections for the AD625 and LH0036 are shown in Laboratory Exercise 5. Connect the amplifier output to the digital multimeter and immerse the copper reference junctions in the 0 °C ice bath, as shown in Laboratory Figure 12.2.

2.2 Calibration of sensing junction. Place the sensing junction and a small amount of crushed ice in a second beaker and stir. When the standard thermometer reads 0 °C, measure the output of the thermocouple amplifier with your digital voltmeter (DVM) and record the value. Now place the second beaker on the hot plate and warm slowly while stirring. At 5- or 10-°C intervals, record the standard thermometer reading and the thermocouple DVM reading. Continue until 100 °C is reached.

3.   Thermistor

3.1 Thermistor bridge. You will be using a thermistor with a resistance of 2,252 Ω at 25 °C (Laboratory Figure 12.1). We select $R_2 = R_3 = 2.4$ kΩ, since the bridge output is in its most linear range when R_2 and R_3 are both equal to the thermistor resistance. R_1 is a 25-kΩ variable resistor, sufficiently high to balance the thermistor at 0 °C.

The bridge equation is:

$$V_0 = V_+ - V_- = V_b \left(\frac{R_3}{R_T + R_3} - \frac{R_2}{R_1 + R_2} \right)$$

Solving for R_T we have:

$$R_T = R_3 \frac{V_b R_1 - V_0(R_1 + R_2)}{V_b R_2 + V_0(R_1 + R_2)}$$

$$V_T = V_b \left(\frac{R_T}{R_3 + R_T} \right), \qquad I_T = V_T/R_T, \qquad P_T = V_T I_T$$

3.2 Thermistor calibration. Place the thermistor in the 0 °C bath and adjust R_1 so that V_0 (measured by your DVM) is zero volts on the most sensitive scale. As before, record the temperature and the DVM reading as you heat a beaker of water from 0 to 100 °C. Leave the water boiling gently for the next step.

> *Note*: You will need to measure the values of R_1, R_2, and R_3 accurately for your analysis. *Suggestion*: To save time, calibrate the thermocouple and the thermistor simultaneously.

3.3 Self-heating. With the thermistor in still air, let the bridge output come to equilibrium and record the output voltage and the air temperature.

4. Dynamic measurement of temperature

4.1 Thermistor. Prepare two beakers of water, one at room temperature, the other at about 50 °C. (*Caution*: If the temperature difference is too great, the thermistor may crack when it is rapidly transferred between the two beakers.) Set up the thermistor bridge and the instrumentation amplifier circuit with a gain of about 20. For the AD625, $R_F = 20\,\mathrm{k\Omega}$ and $R_G = 2\,\mathrm{k\Omega}$. For the LH0036, $R_G = 2.4\,\mathrm{k\Omega}$. Adjust the gain and offset as needed to provide an output swing of several volts from the cooler beaker to the warmer beaker. Connect the amplifier output to analog input 0^+ and ground analog input 0^-. Use the microcomputer to record the transient that occurs when the thermistor is rapidly transferred from one beaker to the other. After equilibrium is established, rapidly transfer to the first beaker. Print the results.

4.2 Thermocouple. Repeat procedure section 4.1 for the thermocouple. To provide a 2-V output change for the data-acquisition board, you will need a voltage gain of 1,000. For the AD625, $R_F = 20\,\mathrm{k\Omega}$ and $R_G = 39\,\Omega$. For the LH0036, $R_G = 51\,\Omega$.

5. Reproducibility

To check reproducibility, remeasure the 50 °C bath with the standard thermometer, thermistor, and thermocouple.

Laboratory report

1.   Setup

Draw a simple block diagram of your experimental setup.

2.   Data summary and analysis

2.1 Thermocouple data. From the data of procedure section 2 (thermocouple), tabulate and plot the amplifier output and thermocouple output (computed by using the measured gain) as a function of temperature. Compute and tabulate the thermocouple sensitivity before and after amplification, in millivolts per degrees celsius.

2.2 Thermistor sensitivity. From the data of procedure section 3 (thermistor), plot the bridge output as a function of temperature. Compute and tabulate the thermistor bridge sensitivity in millivolts per degrees celsius as a function of temperature.

2.3 Thermistor resistance and power dissipation. Use the bridge equation to compute the thermistor resistance as a function of temperature. Tabulate the following quantities:

Temperature, T
Bridge output voltage, V_0
Thermistor resistance, R_T
Voltage across the thermistor, V_T
Current through the thermistor, I_T
Power dissipated by the thermistor, P_T

2.4 Thermistor beta. Tabulate and plot ln (log base e) resistance versus inverse temperature (in degrees kelvin!) and estimate (by eyeball) the best-fit value and uncertainty in β for your thermistor.

2.5 Thermistor self-heating. From the self-heating data of procedure section 3.3, compute the thermistor resistance and power dissipation in still air. Use your resistance versus temperature data from procedure section 3.2 (when the thermistor was in water and self-heating can be neglected) to determine the temperature of the thermistor in air. Estimate the dissipation constant as $P/\Delta T$, where P is the power dissipated by the thermistor in air and ΔT is the temperature difference between the air and thermistor temperatures.

2.6 Dynamic response. Plot temperature as a function of time for the dynamic temperature data of procedure section 4. Estimate the thermal time constant of the thermistor.

2.7 Reproducibility. Tabulate the thermistor and thermocouple outputs measured at the same temperature but at different times (procedure section 5).

3.   Discussion and conclusions

3.1 Discuss the relative merits of the thermocouple (before amplification) and the thermistor (in a bridge) for measuring temperature. Consider temperature range, linearity, sensitivity (in millivolts per degrees celsius), response time, ruggedness, accuracy (degrees celsius), and precision (degrees celsius).

Note: Accuracy means adherence to a standard, and precision means reproducibility and the ability to reliably detect small changes.

3.2 Discuss briefly how you would use the thermocouple and a metal–ceramic resistor to control the temperature of a small oven.

4.   Questions

4.1 What was the range of power dissipated by your thermistor during your measurements? Considering the dissipation constant in air estimated in analysis section 2.4 and assuming that it is ten times larger in water, do you think that self-heating in water affected your measurements in procedure sections 2 and 3?

4.2 Assuming that the dissipation constant is ten times larger in water than in air, what would the actual thermistor temperature be in 20 °C water if the bridge supply voltage were increased from 1 to 10 V?

4.3 What value of $R_2 = R_3$ gives the largest bridge sensitivity to temperature?

Hint: Read the section in Chapter 4 that discusses the resistor values that maximize the quantity dV_0/dT.

4.4 What are the two most important thermal characteristics that determine the response time of a temperature transducer?

5.   Laboratory data sheets

Include your handwritten data sheets (or a copy), which should consist of a log of the procedures you used, any special circumstances, and the measurements you recorded manually.

Laboratory Exercise 13

Measuring strain and force

Purpose

To investigate the sensitivity, linearity, hysteresis, and temperature dependence of two force transducers. The first uses a single-strain element cemented to a lucite rod. The second uses four strain elements mounted in opposing pairs.

Equipment

- Two 120-Ω 1% resistors (or two 200-Ω trimpots)
- One 200-Ω trimpot
- +1-V, ±12-V power supplies
- Superstrip circuit board
- Three 10-μF, 25-V electrolytic capacitors (put between power terminals and ground on superstrip circuit board binding posts)
- Two 0.1-μF CK-05 bypass capacitors (put between power and ground on instrumentation amplifier)
- Digital multimeter
- Strain gauge unit (lucite rod with four strain elements)
- Set calibration weights (50–1,000 g, shared with other laboratory groups)
- Heat gun (shared with other laboratory groups)
- Dial thermometer
- Four strain gauges:
  BLH Electronics, 42 Forth Ave, Waltham MA 02254
  Type FAE-25-35-SO (SR-4) resistance 120.0 Ω, gauge factor 2.04
- Aluminum plate, vertical support rod, and clamps to hold the strain gauge unit as shown in Laboratory Figure 13.1
- AD625 or LH0036 instrumentation amplifier

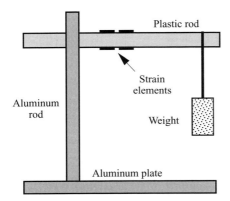

Laboratory Figure 13.1 Setup for force measurement.

For the AD625 instrumentation amplifier, use resistors below:
- One 390-Ω resistor (R_G for gain of 100)
- Two 20-kΩ resistors (R_F)
- One 25-kΩ trimpot (offset adjust)

For the LH0036 instrumentation amplifier, use resistors below:
- One 510-Ω resistor (R_G for gain of 100)
- One 100-kΩ trimpot (offset adjust)
- One 3.3-kΩ resistor (offset adjust)
- One 33-kΩ resistor (offset adjust)

Background

1. Single-strain-element force transducer

In Section 4.5.1 (Force transducers) the single-strain-element bridge equation was derived as:

$$V_+ - V_- = V_b \left(\frac{G_S}{4} \frac{\Delta L}{L} \right)$$

2. Four-strain-element force transducer

In Section 4.5.1 (Force transducers) the four-strain-element bridge equation was derived as:

$$V_+ - V_- = V_b \left(G_S \frac{\Delta L}{L} \right)$$

3.   Instrumentation amplifiers

As demonstrated in Laboratory Exercise 5, the gain of the instrumentation amplifier is given by:

$$G = \frac{V_0}{V_+ - V_-} = 1 + \frac{2R_F}{R_G}$$

Additional reading

Section 2.4.1 Instrumentation amplifiers
Section 4.4 Strain-sensing elements
Section 4.5.1 Force transducers

Procedure

1.   Single-strain-element force transducer

Set up the single-strain-element bridge circuit shown in Laboratory Figure 13.2. Measure the resistance of all four strain elements. If any show an infinite resistance, a wire is broken and you will need to choose another strain gauge bar.

Use two 120-Ω 1% resistors or two 200-Ω trimpots adjusted to read 120 Ω on the multimeter. Set up the instrumentation amplifier for a gain of about 100. For the AD625, use $R_F = 20\,\text{k}\Omega$ and $R_G = 390\,\Omega$. For the LH0036, use $R_G = 510\,\Omega$. See Laboratory Exercise 5 for connections. With no load on the rod, adjust the 200-Ω trimpot for $V_0 = 0$.

1.1 Amplifier gain. Use a voltage divider to produce a 10-mV dc signal and apply it to the input of your instrumentation amplifier circuit. Accurately measure the input and the output with your multimeter and determine the amplifier gain.

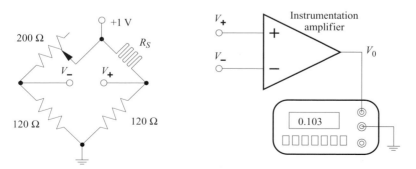

Laboratory Figure 13.2 Single-strain-element bridge and AD625 instrumentation amplifier.

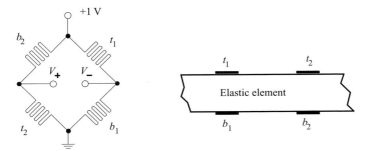

Laboratory Figure 13.3 Force transducer bridge using four strain elements.

1.2 Force calibration. Hang 0, 50, 100, 200, 300, 500, 700, and 1,000 g weights and for each; record V_0.

1.3 Drift. Remove the 1,000 g weight, and immediately record V_0. Record every minute for 5 min.

1.4 Reproducibility. Replace the 1,000 g weight, and record V_0. Remove the weight, and immediately record V_0.

1.5 Effect of heating. Gently heat the *top* of the rod (no load). Record any changes in V_0 while heating. Estimate the temperature rise using a dial thermometer. Do not heat more than 10 °C. Repeat, heating the *bottom* of the rod. Repeat, heating the top and bottom at the same time.

2. Four-strain-element force transducer

Set up the four-element bridge circuit shown in Laboratory Figure 13.3. With no load, adjust the instrumentation amplifier offset for $V_0 = 0$. Repeat procedure section 1 for this force transducer.

Laboratory report

1. Setup

Draw a simple block diagram of your experimental setup.

2. Data summary and analysis

2.1 Measured output versus load. For the single-element and four-element transducers, tabulate load, amplifier output V_0, sensitivity (in millivolts per gram), and $\Delta L / L$. (Use the bridge equation in the background section and the known values of amplifier gain, bridge bias voltage, and G_S to compute $\Delta L / L$.)

2.2 Derivation of a straight-line model from the measured data. For the single-element and four-element transducers, plot output voltage versus load. Draw straight lines through each of the data sets. Adjust the slope and intercept of the lines to minimize the deviations from the data. Derive algebraic expressions for these straight lines.

2.3 Comparison between measurements and the linear model. For the single-element and four-element transducers, tabulate load, measured output voltage, straight-line model, and the difference between them.

3.   Discussion and conclusions

3.1 Compare the relative merits of the single-element and the four-element force transducers. Include sensitivity (millivolts per gram), effect of uniform temperature changes, linearity, reproducibility, accuracy, and precision. Express errors both in terms of amplifier output V_0 and load in grams. (A user would be more interested in errors in load than errors in V_0.)

3.2 Compare the measured ratio of sensitivities of the four-element and the single-element bridges with the expected ratio.

3.3 Discuss the advantages of a single-element bridge over a simple voltage divider.

3.4 Discuss the effect of uniform and nonuniform heating that you observed on the two force transducers. Express in terms of amplifier output V_0 and load in grams.

4.   Questions

4.1 Why was it important to limit the temperature rise in procedure sections 1.5 and 2.5?

4.2 For both transducers, what strain $\Delta L/L$ occurred for a 1,000 g load?

4.3 How would you change the setup to measure 0–1,000 kg loads and 0–1 g loads?

4.4 What was the largest deviation between your data and your straight-line fit? What does this mean in terms of load in grams?

Laboratory data sheets

Include your handwritten data sheets (or a copy), which should consist of a log of the procedures you used, any special circumstances, and the measurements you recorded manually.

Laboratory Exercise 14

Measuring light with a photodiode

Purpose

To measure the $I-V$ characteristics, dark current, noise, and linearity of a high-sensitivity *pin* silicon photodiode and to use it in a simple photometer to measure the concentration of solutions.

Equipment

- +5-V, ±12-V power supplies
- Three 10-µF, 25-V electrolytic capacitors (put between power terminals and ground at circuit board binding posts)
- Four 0.1-µF, CK-05 bypass capacitors (put between power and ground at all integrated circuits)
- Superstrip circuit board
- DMM (digital multimeter)
- Wooden block for *pin* photodiode, LED source, and test tube
- Hollow black cylinder, taped on one end, to cover top of test tube
- Solid black cylinder to block light when no test tube is present
- Red, light emitting diode (LED) (20 mA maximum current)
- *pin* silicon photodiode
- Two 10-kΩ trimpots
- One 1,300-pF capacitor
- One 20-kΩ trimpot
- One 330-Ω resistor
- Two 10-kΩ resistors
- Two 1-MΩ resistors
- One 1,200-pF capacitor

- Two LF 356 op amps
- One sealed glass tube of 1.0-molar red food coloring
- One sealed glass tube of 0.5-molar red food coloring
- One sealed glass tube of 0.2-molar red food coloring
- One sealed glass tube of 0.1-molar red food coloring
- One sealed glass tube of 0.05-molar red food coloring
- One sealed glass tube of red food coloring of "unknown" concentration
- One sealed glass tube of water

Additional reading

Section 2.2 Operational amplifier circuits
Section 4.6 Measuring light

Procedure

1.   Setup

The silicon *pin* photodiode and the LED are mounted in a wooden block as shown in Laboratory Figure 14.1.

Note: To reduce variations in light transmission, the glass tubes should have black tape around the top for a snug fit in the hole, and the hole should be deep enough so that the LED beam shines through the tube in its cylindrical region (above the curved bottom).

1.1 Photodiode biasing and amplifier circuits. Construct the photodiode biasing and current amplifier circuit shown in Laboratory Figure 14.2. A photodiode current

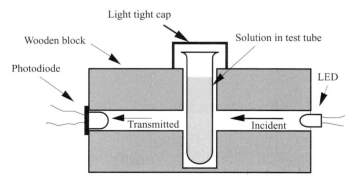

Laboratory Figure 14.1 Setup for measuring light transmission through a solution.

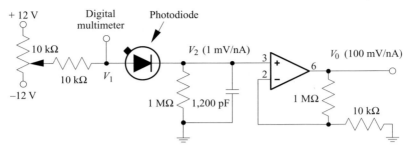

Laboratory Figure 14.2 Circuit for biasing a photodiode and converting the photoconduction current into a voltage. Metal tab on the case is toward the anode lead.

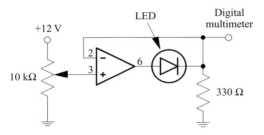

Laboratory Figure 14.3 Op-amp circuit for providing a controlled current through a light emitting diode. Output is 3 mA/V.

of 1 nA should produce an op-amp output voltage of 100 mV. The 1,200-pF capacitor shunts high frequencies to ground. The LF 356 op-amp pinout and connections to the external components are shown in Laboratory Figures 4.1 and 4.2. Connect 0.1-μF capacitors between pin 4 (−12 V) and ground and between pin 7 (+12 V) and ground. Adjust the 20-kΩ trimpot for zero output offset voltage.

1.2 LED biasing circuit. Construct the voltage-controlled current driver circuit shown in Laboratory Figure 14.3. The op-amp offset adjustment will not be necessary. In this circuit the op amp produces the current necessary to make the voltage across the 330-Ω load resistor equal to the input voltage. The current passing through the LED is thus 3 mA per input volt, and the circuit compensates for the nonohmic nature of the LED.

2.   Photodiode *I*–*V* characteristics and noise with dark conditions

To exclude room light, tape the test tube hole with black tape.

2.1 Reverse bias. Using your multimeter, measure both V_1 and op-amp output voltage as you vary the 10-kΩ trimpot. Suggested values for V_1 are 0.0, −0.1, −0.3, −1, −2, and −5 V.

2.2 Reverse bias noise. Remove the 1,200-pF capacitor and use your oscilloscope to estimate the op-amp output rms noise at V_1 values of 0.0 and -5 V (use aluminum foil for shielding if necessary). Replace the 1,200-pF capacitor and repeat.

2.3 Forward bias. Determine the values of V_1 required to produce forward currents with approximate values of 1, 2, 5, 10, 20, and 50 nA.

Note: Due to the steepness of the curve, it is difficult to adjust the voltage to achieve a specific current – just make the measurements needed to define the curve.

3.   Photodiode photovoltaic and photoconductive currents as a function of LED current

3.1 Photovoltaic mode. With a photodiode bias of 0 V, measure photodiode current versus LED current for seven values of LED current: 0, 1, 2, 5, 10, 15, and 20 mA (the maximum output of the op amp).

3.2 Photoconductive mode. Repeat procedure section 3.1 with a reverse bias of -5 V.

4.   Photodiode *I–V* characteristics and noise with illumination

Repeat procedure section 2 with an LED current of 20 mA.

5.   Measuring the concentration of a solution

Set the LED current to 20 mA. Set the photodiode reverse bias to 0 V (photovoltaic mode). Record the photodiode output for the four known solutions and for the unknown. Due to irregularities in the glass surface, the output may depend on tube rotation – average the output over several orientations.

The photodiode current is given by $I = I_0 e^{-kLC}$, where L is the path length, C is the concentration, and k is the extinction coefficient. The photodiode voltage is given by $V_1(C) = I(1 \text{ M}\Omega)$ and the op-amp output is given by $V_0(C) = I(100 \text{ M}\Omega) + V_b$, where V_b is the amplifier output offset voltage. Use the water-filled test tube ($C = 0$) to measure $V_0(0) = I_0(100 \text{ M}\Omega) + V_b$, and turn off the LED ($C = \infty$) to measure $V_0(\infty) = V_b$. To use Beer's law, plot $\log((V_0(C) - V_0(\infty))/(V_0(0) - V_0(\infty)))$ versus C.

Beer's law:
$$\frac{\text{transmitted light}}{\text{incident lignt}} = e^{-kLC}$$

Laboratory report

1.   Setup

Draw a simple block diagram of your experimental setup.

2.   Data summary and analysis

2.1 Photodiode $I–V$ curves. Tabulate and plot photodiode current as a function of voltage for LED currents of 0 (procedure section 2) and 20 mA (procedure section 4). Be sure to include both forward and reverse biased data, and to consider any voltage drop across the 1-MΩ input resistor (between op-amp pin 3 and ground):

$I_{\text{diode}} = V_2/1\,\text{M}\Omega$
$V_0 = 100V_2$
$I_{\text{diode}} = V_0/100\,\text{M}\Omega$
$V_{\text{diode}} = V_1 - V_2 = V_1 - V_0/100$

2.2 Noise versus dark current. Tabulate rms noise as a function of photodiode current (procedure sections 2 and 4).

2.3 Noise model. Devise an empirical expression that describes the noise and incorporates Johnson and shot noise as separate terms. (Remember that shot noise varies as $\sqrt{I}$, whereas Johnson noise is independent of I.)

2.4 Photodiode versus LED currents. Tabulate and plot photodiode current versus LED current (procedure section 3).

2.5 Beer's law plot. Using your photovoltaic measurements from procedure section 5, tabulate and plot the natural logarithm of photodiode current versus solution concentration. Use Beer's law to estimate the concentration of your unknown solution. Estimate your uncertainty.

3.   Discussion and conclusions

3.1 Discuss the variety of physical quantities that can be sensed by using a photodiode with other components.

3.2 Compare the advantages and disadvantages of the photovoltaic and photoconductive modes. Consider dark current, speed, circuit complexity.

3.3 Discuss the underlying mechanism that makes the LED light intensity proportional to LED current and not LED voltage.

3.4 Discuss the underlying mechanism that makes the photodiode current and not the photodiode voltage proportional to light level.

4.   Questions

4.1 Was the photodiode current linearly proportional to the LED current? If not, what could be the reasons?

4.2 If the maximum photodiode power dissipation is 0.1 W, what is the maximum safe current in forward biased mode and in reverse biased mode?

4.3 Does the plot from analysis section 2.1 agree with the graph in Figure 4.39?

4.4 Given the bandwidth of your circuit in Laboratory Figure 14.2 without the 1,200-pF capacitor and the relationship between photodiode current and the voltage across the 1-MΩ resistor, what photodiode current is equal to the rms Johnson voltage noise in the 1-MΩ resistor? How would you expect the capacitor to change this number?

> *Hint* 1: Look up the gain–bandwidth product for your op amp in the data sheets.
> *Hint* 2: Derive the equation that relates photodiode current to voltage across the RC parallel combination as a function of frequency.

4.5 Assuming a quantum efficiency of 80%, what are the minimum and maximum photon signal currents (photons/s) that the photodiode circuit could accurately detect (with and without the 1,200-pF capacitor)?

5.   Laboratory data sheets

Include your handwritten data sheets (or a copy), which should consist of a log of the procedures you used, any special circumstances, and the measurements you recorded manually.

Laboratory Exercise 15

The thermoelectric heat pump

Purpose

To investigate the application of a thermoelectric device as a heat pump to heat and cool a small object. To measure the relationship between electrical energy absorbed and heat energy pumped for heating and cooling.

Equipment

- Digital multimeter
- Cambion No. 801-3959-01 thermoelectric device (maximum current 10 A at 5 V)
- High-current power supply: 5 V at 10 A
- Thick aluminum plate (heat sink)
- Small beaker, approx. 10 ml
- Heat sink compound: zinc oxide in silicone paste
- Small Mylar sheet
- Platinum or dial thermometer

Background

1.   Peltier and Seebeck effects

While the Peltier and Seebeck effects have been known for over 100 years, practical applications for the production of electrical power and as a heat pump have been exploited relatively recently. Examples include power generators in space vehicles, cooling devices for low-temperature instrumentation, precision temperature references, temperature stabilization of precision instrumentation, etc.

You will be using a Bi_2Te_3 semiconductor Peltier heat pump with p and n doping as shown in Figure 4.18. The heat pump will be mounted on an aluminum heat sink to maintain the temperature of one of its surfaces and a small beaker of water will be placed on its other surface. The rate Q at which the beaker gains heat energy is given by:

$$Q = \pi I + I^2 R/2 + K_p(T_s - T_o) + K_a(T_a - T_o)$$

where π is the Peltier coefficient, I is the current through the Peltier device, R is the electrical resistance of the device, K_p is the thermal conduction coefficient of the device, K_a is the thermal convection coefficient to the air, T_o is the temperature of the beaker, T_s is the temperature of the heat sink, and T_a is the temperature of the air. The first term is the Peltier effect, the second is Joule heating, the third is conductive transfer through the device, and the fourth is conductive and convective transfer to the air.

When $I < 0$, only the Peltier term acts to cool the beaker – all other terms heat the beaker. At low currents, the first term can be larger than the second, resulting in a net cooling ($Q < 0$). However, as the current is increased, the I^2 term can dominate, resulting in net heating ($Q > 0$).

When $I > 0$, both the Peltier and Joule terms heat the beaker, while the transfer to the heat sink and the surrounding medium cool the beaker.

2.   Thermoelectric efficiency

When the beaker is near room temperature, and a current I is abruptly applied to the heat pump, the third and fourth terms are small and the initial rate of temperature rise of the beaker will be given by:

$$\frac{dT}{dt} = (0.238 \text{ g } {}^\circ\text{C/J})\frac{Q}{m}$$

where Q is the rate of thermal energy transfer to the beaker in joules per second (i.e. watts) and m is the mass of the beaker and water in grams. The electrical power consumed by the heat pump is given by:

$$P = IV = I^2 R$$

The thermoelectric efficiency is given by $|Q/P|$, the absolute value of the ratio of the rate of thermal heat energy pumped to the electrical power consumed:

$$\left|\frac{Q}{P}\right| = \frac{\pi I + I^2 R/2}{I^2 R}$$

In the heating mode ($I > 0$), the additive combination of Joule heating and heat pumping can be sufficiently large that the thermoelectric efficiency exceeds unity.

However, in the cooling mode ($I < 0$), the Joule heating reduces the thermoelectric efficiency to a value below unity.

3.   Equilibrium temperature T_{equ}

When a constant current I is established, the beaker will approach an equilibrium temperature T_{equ} where heat transfer with the heat pump and the surroundings are equal. At this point $Q = 0$ and T_{equ} is given by:

$$T_{equ} = \frac{\pi I + I^2 R/2 + K_p T_s + K_a T_a}{K_p + K_a}$$

The relationship is parabolic in I and the minimum equilibrium temperature is achieved at a specific current $I_{min} < 0$.

4.   Estimation of T_{equ} from T versus time measurements

For a simple (one compartment) thermal system, the time rate of change in temperature dT/dt is proportional to the difference between the current temperature T and some equilibrium temperature T_{equ}:

$$\frac{dT}{dt} = k(T_{equ} - T)$$

The solution is:

$$T = T_{equ} - (T_{equ} - T_0)e^{-t/\tau}$$

where $\tau = 1/k$, the initial temperature is T_0 at $t_0 = 0$, and the temperature asymptotically approaches T_{equ} as t becomes very large (Laboratory Figure 15.1).

In a realistic thermal system, after a change in the heat input or the system being heated, the initial change in temperature does not necessarily obey the above equation, but after a few time constants τ, the equation is usually accurate enough to predict the equilibrium value T_{equ} without having to wait forever!

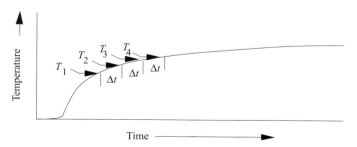

Laboratory Figure 15.1 Measurement of temperature at regular time intervals for the estimation of the equilibrium temperature.

We now describe a method that has two objectives: (i) determining when the above equation describes the measured asymptotic behavior of the thermal system, and (ii) estimating T_{equ} from the data.

If the equation is valid, we expect that the ratio of successive differences will be constant:

$$\eta = \frac{T_{n+2} - T_{n+1}}{T_{n+1} - T_n} = \frac{(T_{equ} - T_0)(e^{-t_{n+2}/\tau} - e^{-t_{n+1}/\tau})}{(T_{equ} - T_0)(e^{-t_{n+1}/\tau} - e^{-t_n/\tau})} = e^{-\Delta t/\tau}$$

where the t_n are equally spaced in time and $\Delta t = t_{n+1} - t_n$.

So the procedure is to compute η for successive time intervals and when it is reasonably constant, use the following analysis to estimate T_{equ}:

$$T_{equ} = T_{n+1} + (T_{n+2} - T_{n+1}) + (T_{n+3} - T_{n+2}) + (T_{n+4} - T_{n+3}) + \cdots$$

$$= T_{n+1} + (T_{n+2} - T_{n+1}) + (T_{n+2} - T_{n+1})\frac{(T_{n+3} - T_{n+2})}{(T_{n+2} - T_{n+1})}$$

$$+ (T_{n+2} - T_{n+1})\frac{(T_{n+3} - T_{n+2})}{(T_{n+2} - T_{n+1})}\frac{(T_{n+4} - T_{n+3})}{(T_{n+3} - T_{n+2})} + \cdots$$

$$= T_{n+1} + (T_{n+2} - T_{n+1})(1 + \eta + \eta^2 + \cdots)$$

the working result is:

$$\boxed{T_{equ} = T_{n+1} + \frac{T_{n+2} - T_{n+1}}{1 - \eta}} \qquad \boxed{\eta = \frac{T_{n+2} - T_{n+1}}{T_{n+1} - T_n}}$$

Additional reading

Section 4.3.4 The thermocouple

Procedure

1.   Setup

As shown in Laboratory Figure 15.2, paste a sheet of Mylar or paper to the aluminum heat sink plate with ZnO cream, paste the thermoelectric device (TED) to the top of the Mylar or paper, and then paste a 10-ml beaker of water on top of the TED. Fill the beaker about half-full of water and place a dial thermometer into the beaker. Record the amount of water in the beaker for later calculations.

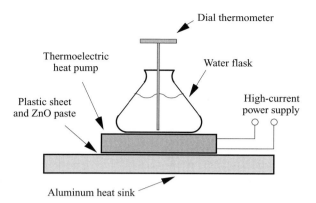

Laboratory Figure 15.2 Setup for using a thermoelectric heat pump to heat and cool a beaker of water.

2. Thermoelectric efficiency

In this section, the beaker is close to room temperature, and the initial rise in temperature is used to measure the thermal energy pumped by the TED.

Abruptly apply 5 A of current to the TED and record the voltage and temperature every 30 s for 5 min. Observe whether the water temperature is increasing or decreasing. These data will be used in the analysis section to compare the power consumed by the TED with the caloric power received (or extracted from) the water in the beaker.

Turn off the power supply and wait 10 min for the system to approach thermal equilibrium with the room. Then reverse the polarity of the power leads and apply 5 A. Record the voltage and temperature every 30 s for 5 min.

3. Thermoelectric equilibrium temperature

In this section, the beaker is allowed to approach the equilibrium temperature, where the heat pumped by the TED is equal to that lost by convection and conduction.

Set the voltage for +4 A and record the temperature of the water in the beaker every 2 min for 10 min. You may need to adjust the voltage during this time to keep the current constant. Alternatively, use a power supply that has a constant current mode. In your analysis, use the preceding equations to determine T_{equ}. Repeat for +2, −2, and −4 A.

Laboratory report

1. Setup

Draw a simple block diagram of your experimental setup.

2.   Data summary and analysis

2.1 Initial slope. Tabulate and plot the temperature-versus-time data from procedure section 2. Estimate the initial slope (in degrees celsius per minute) as the steepest linear portion of your curve after the initial lag, but before the curve begins to flatten out to its asymptotic limit.

2.2 Thermoelectric efficiency. From your initial slope data of procedure section 2 and the known weight of the water, compute the caloric thermal power received by (or extracted from) the beaker and the electrical power consumed by the thermoelectric device. The thermal power needed to change the temperature of 1 g of water at the rate of 1 °C/s is 4.19 W. Use these to compute the thermal power that must have been transferred to the heat sink. (Assume that the beaker temperature was sufficiently close to room temperature, so that you can ignore the heat transfer with the surroundings.) Compute the thermoelectric efficiency as a ratio of the power received by (or extracted from) the beaker to the electrical power consumed by the thermoelectric device.

2.3 Thermoelectric equilibrium temperature. Tabulate and plot the equilibrium temperature versus current (procedure section 3). Include a column for power input. Do an "eyeball" fit of the model:

$$T_{\text{equ}} = T_{\text{room}} - AI + BI^2$$

to your data.

3.   Discussion and conclusions

3.1 Compare the advantages and disadvantages of a thermoelectric heat pump versus an electric resistor for heating.

3.2 Compare the advantages and disadvantages of a thermoelectric heat pump versus a mechanical refrigerator for cooling.

3.3 Discuss how the thermoelectric heat pump can heat with an efficiency that exceeds unity. Where does the extra thermal energy come from?

3.4 Discuss how you would use a thermistor, a thermoelectric heat pump, a difference amplifier, a power amplifier, and thermal insulation to control the temperature of an object at a desired temperature. Draw a simple block diagram. Discuss the minimum and maximum temperatures that you could achieve and the reasons for these limits.

4.   Questions

4.1 What power-supply wattage was required to achieve a beaker equilibrium temperature 10 °C above ambient and 10 °C below ambient?

4.2 When you cooled the beaker you may have noticed a point for which further increases in current (and power to the TED) did not result in further cooling. Why did this occur? What are the major factors (in the real world) that determine the lowest temperature that you can achieve?

4.3 How would thermal insulation around the beaker affect the equilibrium temperature-versus-current curve and the thermal time constant of the system.

4.4 Was the thermoelectric efficiency the same for cooling as for heating? Why?

5.   Laboratory data sheets

Include your handwritten data sheets (or a copy), which should consist of a log of the procedures you used, any special circumstances, and the measurements you recorded manually.

Laboratory Exercise 16

Electrodes and ionic media

Purpose

To measure and compare several important properties of Ag(AgCl) electrodes and bare-metal electrodes: offset potential, stability, microphonics, and complex impedance as a function of frequency.

Equipment

- Superstrip circuit board
- Digital multimeter
- Sine-wave signal generator
- Oscilloscope
- Two 100-Ω resistors
- One 1-μF capacitor
- Two Ag(AgCl) electrodes (pressed Ag and AgCl powders)
- Two stainless steel or copper plates
- One 500-ml Pyrex beaker
- 1% NaCl solution
- Tap water

Background

1. Electrodes used as sensors and actuators

Electrodes are used as sensors to record potentials from the heart (electrocardiogram, or ECG), from the muscles (electromyogram, or EMG), from the eyes (electrooculogram or EOG), from the brain (electroencephalogram, or EEG), and even from individual cells. Electrodes are essential components in electromagnetic flowmeters, pH meters, and ion meters. They are also used as actuators to stimulate nerve conduction and muscle contraction. The skin electrodes studied in this laboratory exercise will also be used for the ECG in Laboratory Exercise 17, the EMG in Laboratory Exercise 18, and the EOG in Laboratory Exercise 19.

An understanding of the electrical nature of the recording or stimulating electrodes and the properties of the ionic media (the tissue fluids or salt solution) is necessary for a better understanding of the resultant effects or measurements.

2. Complex impedance analysis

To understand the nature of your electrode measurements better, we review some basic material relating phase shifts with reactive components. The generalized complex impedance and the graphical relationship between resistive and reactive impedance are shown in Laboratory Figure 16.1.

Exciting potential: $V(t) = V_0 \sin(\omega t)$
Resulting current: $I(t) = I_0 \sin(\omega t + \phi)$

$$\tan(\phi)\frac{X}{R}, \qquad I_0 = \frac{V_0}{\sqrt{R^2 + X^2}}$$

Resistor: $Z = R$
Inductor: $Z = j\omega L = jX_L$
Capacitor: $Z = -j/\omega C = -jX_C$
Pure R: $\phi = 0°$, $V(t)$ and $I(t)$ in phase
Pure L: $\phi = 90°$, $I(t)$ lags $V(t)$ by $90°$
Pure C: $\phi = -90°$, $I(t)$ leads $V(t)$ by $90°$
R, L, and C in series: $Z = R + jX$, $X = \omega L - 1/\omega C$
Add Zs in series: $Z_{12} = Z_1 + Z_2$
Add Zs in parallel: $1/Z_{12} = 1/Z_1 + 1/Z_2$

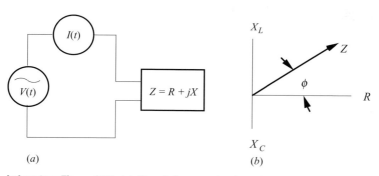

(a) (b)

Laboratory Figure 16.1 (a) Circuit for complex impedance measurement. (b) Graphical relationship between phase angle and amplitudes of resistive and reactive impedance.

Additional reading

Section 4.7 Ionic potentials
Walter A. Getzel and John G. Webster, "Minimizing silver–silver chloride electrode impedance," *IEEE Trans. Biomed. Eng*. BME-23, No. 1, (1976): pp 87–88.
Peter Strong, *Biophysical Measurements*, Tektronix, Beaverton, OR, 1970, Chapter 16.

Procedure

1.   Electrode offset potential, stability, and microphonics

Metals in contact with ionic media often undergo chemical reactions that produce a potential difference in the absence of a current, very much like a weak battery. This potential may depend on layers of charge near the surface that can change with time and can be displaced by a mechanical disturbance.

1.1 Preparation of ionic medium. Fill a 500-ml beaker with 1% NaCl solution.

1.2 Two Ag(AgCl) electrodes. Immerse two Ag(AgCl) electrodes (but not their connecting wires) in the solution (Laboratory Figure 16.2) and use your digital multimeter to record the offset potential every 30 s for 5 min. Then determine how microphonic the electrodes are by observing the oscilloscope deflection (slow sweep) when the table is given a "standard" tap.

1.3 Ag(AgCl) versus bare-metal electrodes. Repeat procedure section 1.2 using one Ag(AgCl) electrode and a bare-metal electrode (a copper alligator clip).

1.4 Two bare-metal electrodes. Repeat the measurement using two bare-metal electrodes.

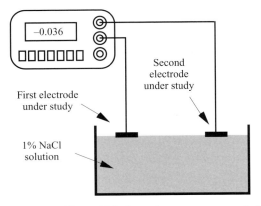

Laboratory Figure 16.2 Setup for measurement of offset potential.

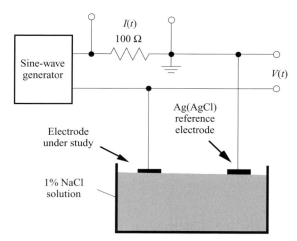

Laboratory Figure 16.3 Setup for measuring the complex impedance of electrodes as a function of frequency.

2.   Ac electrode impedance

In tissue stimulation, knowledge of the electrical impedance of the stimulation electrode is important, because it will determine the actual level of current passed into the tissue for a given voltage. In another type of study, measurement of the actual impedance across or through some types of tissue or organ is desired. In this case, the impedance of the electrode–tissue interface may be significantly greater than the impedance of the tissue itself. For both of these reasons, we need to measure electrode impedance. In addition, under normal conditions, we want to minimize this impedance.

Set up the electrode-impedance measuring apparatus as shown in Laboratory Figure 16.3. Test the circuit by measuring the impedance of a 100-Ω resistor and of a 100-Ω resistor in series with a 1-μF capacitor.

Choose one of your Ag(AgCl) electrodes as a reference, and at $V(t) = 0.1$ V measure $I(t)$ and the $V(t) - I(t)$ phase difference for the other Ag(AgCl) and bare-metal electrodes at 1, 10, and 100 Hz, and at 1, 10, and 100 kHz. For each electrode and at each frequency, you will need to adjust the oscillator output for $V(t) = 0.1$ V amplitude. This is necessary because different electrode impedances load the oscillator differently.

Laboratory report

1.   Setup

Draw a simple block diagram of your experimental setup.

2.   Data summary and analysis

2.1 Offset potential versus time. Plot the offset potential as a function of time for the three electrode pairs.

2.2 Microphonics. Tabulate the effect of a standard tap for the three electrode pairs.

2.3 Complex impedance versus frequency. For the three electrode pairs, tabulate the amplitude of $V(t)$ and $I(t)$ and the phase difference between them as a function of frequency. Compute the real (resistive) and imaginary (reactive) parts of the impedance and enter them into the table.

2.4 Equivalent circuits. From the results of analysis part 2.3, deduce the equivalent circuits that describe the individual electrodes.

Hint: See Figure 4.25. Use your 1-Hz data to determine the sum of the charge transfer and diffusion resistance, and use your 100-kHz data to estimate the diffusion resistance.

3.   Discussion and conclusions

3.1 Discuss procedure sections 1 and 2.

3.2 Compare the characteristics of the electrodes used.

4.   Questions

4.1 From your measurements of the complex impedance of the Ag(AgCl) and bare-metal electrodes, what can you conclude about the relative capacitive, inductive, and resistive components as a function of frequency?

4.2 Were any of your electrodes microphonic? Were any not microphonic? Explain your answers in terms of ionic events.

4.3 Which of your electrodes had the best properties (low resistive impedance, low offset potential, low noise, good stability)? Which was worst?

5.   Laboratory data sheets

Include your handwritten data sheets (or a copy), which should consist of a log of the procedures you used, any special circumstances, and the measurements you recorded manually.

Laboratory Exercise 17

The human heart

Purpose

To gain familiarity with basic cardiovascular physiology and the origin of the most commonly used noninvasive indicators of heart activity. To measure the electrocardiogram (ECG), the phonocardiogram, and blood pressure; and to observe the effect of light exercise on these indicators.

Equipment

- IBM-compatible Pentium microcomputer with HP VEE
- Cambridge ECG unit with strip chart recorder
- Ag(AgCl) ECG skin electrodes
- Digital oscilloscope
- Headphone to record heart sounds
- Stethoscope
- Pressure cuff and sphygmomanometer
- Instrumentation amplifier or differential op-amp circuit (see Laboratory Exercises 4 and 5 for components)

Background

1.   Ag(AgCl) electrodes

We will be using "floating" skin electrodes, consisting of a sintered plug of Ag and AgCl in contact with a piece of foam containing electrolyte gel. An adhesive pad keeps the

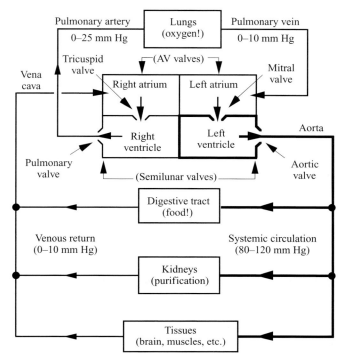

Laboratory Figure 17.1 The human circulatory system (simplified). Heavy lines show the high-pressure portion of the system.

foam pad in contact with the skin and allows for small amounts of movement without breaking electrical contact.

2.   The circulatory system

Laboratory Figure 17.1 shows a simplified schematic of the human circulatory system, including the heart, lungs, and organs. Note that oxygenated blood is so important that the lungs are in "series" with all the other organs, and have their own pumping chambers.

3.   The cardiac cycle

Laboratory Figure 17.2 and Laboratory Table 17.1 give the times, signals, and arterial pressures associated with the major events in the cardiac cycle at 75 beats/min.

4.   Sequence of cardiac depolarization

1. SA node fires.
2. Depolarization proceeds right to left over both atria.
3. Delay at AV node during atrial contraction.

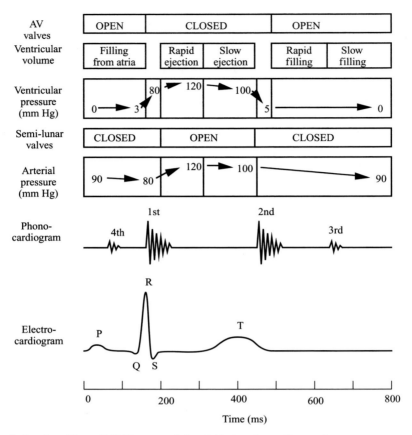

Laboratory Figure 17.2 Events and signals during the cardiac cycle.

4. Bundle of His (pronounced "hiss") at top of ventricular septum.
5. Bundle branches along inner surfaces of ventricles down the septum, around the apex, and back up toward the base.
6. Purkinje network from inner to outer surfaces of ventricle.

Note: The sequence of ventricular contraction is controlled by the bundle branches.

5.   Glossary

apex: bottom of the heart.

atrial systole: period of active atrial contraction between the P wave and the closing of the AV valves.

AV (atrioventricular) node: a cluster of cells leading from the lower portion of the right atrium to the ventricular septum that conducts the depolarization wave very slowly (70 ms) to allow the atrial systole to reach completion before the ventricular systole begins.

AV (atrioventricular) valves: the tricuspid valve between the right atrium and the right ventricle; and the mitral valve between the left atrium and the left ventricle.

base: top of heart, where valves and great vessels are located.

Laboratory Table 17.1 *Events in the cardiac cycle at 75 beats/min*

Time (ms)	Event	Signal	Arterial pressure (mm Hg)	Ventric. pressure (mm Hg)
0	Depolarization SA node		90	0
0–50	Depolarization atria	P-wave	85	0
50–120	AV node delay; atrial contraction	P–R-segment; 4th heart sound	80	3
130–200	Depolarization ventricles; repolarization atria	QRS complex	80	2–80
170–200	AV valves close; isovolumetric ventricular contraction; end ventricular diastole; begin ventricular systole	Start 1st heart sound	80	2–80
200	Semilunar valves open		80	80
200–300	Ventricular contraction; rapid ejection	S–T-segment	80–120	80–120
320–480	Ventricular repolarization	T-wave	120–100	120–100
440	Semilunar valves close; end ventricular systole; begin ventricular diastole	Dichrotic notch; start 2nd heart sound	100	100
480	AV valves open		100	5
480–600	Rapid ventricular filling	3rd heart sound at end of rapid filling	100–95	0
600–800	Slow ventricular filling		95–90	0

bundle branches: Conduct the depolarization wave from the bundle of His around the inner (endo-cardial) surface of the ventricles.

diastole: when used alone, means ventricular diastole, the period of ventricular relaxation between the closing of the semilunar valves and the closing of the AV valves.

dichrotic notch: brief drop in arterial pressure due to backflow associated with the closing of the semilunar valves.

heart murmur: abnormal heart sound, produced by blood passing through deformed cardiac valves.

heart sounds: first (lub) is due to closure of AV valves, second (dub) is due to closure of semilunar valves.

P-wave: electrical signal produced by the depolarization of the atria.

Purkinje network: conducts the depolarization wave through the ventricular wall from the inner (endocardial) to the outer (epicardial) surfaces.

QRS-complex: electrical signals produced by the depolarization of the ventricles and the repolariza-tion of the atria.

SA (sinoatrial) node: cluster of highly conductive cells in the back wall of the right atrium where the cardiac depolarization wave is initiated.

semilunar valves: the aortic valve between the left ventricle and the aorta; and the pulmonary valve between the right ventricle and the pulmonary artery.

septum: muscular wall between the left and right ventricles. Contains the bundle of His where ventricular depolarization originates.

systole: when used alone, means ventricular systole, the period of active ventricular contraction between the closing of the AV valves and the closing of the semilunar valves. See also atrial systole.

T-wave: electrical signal produced by the repolarization of the ventricles.

Additional reading

Leslie Cromwell, Fred J. Weibell, and Erich A. Pfeiffer, *Biomedical Instrumentation and Measurements*, Prentice Hall, Englewood Cliffs, NJ, Chapter 5, Chapter 6: Sections 6.1 and 6.2.

Michael Rudd, *Basic Concepts of Cardiovascular Physiology*, Hewlett-Packard, Waltham, MA, 1973, Chapters 8 and 9.

R. F. Rushmer, *Cardiovascular Dynamics*, W. B. Saunders Co., Philadelphia, 1970.

Peter Strong, *Biophysical Measurements*, Tektronix, Beaverton, OR, 1970, Chapters 2, 5, and Section 8.2.

Procedure

1.   Electrocardiogram (ECG) (demonstrated by a physician)

Attach four electrodes to both wrists and ankles, as shown in Laboratory Figure 17.3. Record the ECG, leads I, II, and III, with the subject at rest and during *light* exercise (*walking* up and down two flights of stairs).

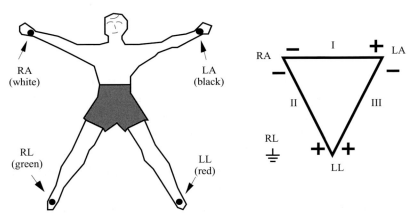

Laboratory Figure 17.3 Limb lead placement for the electrocardiogram and the Einthoven triangle. RA = right arm, LA = left arm, RL = right leg, LL = left leg. Lead II lies along the heart axis and is the most commonly displayed ECG signal.

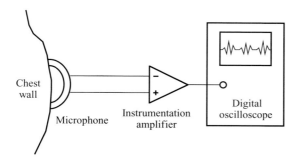

Laboratory Figure 17.4 Procedure for recording a phonocardiogram, using a large foam-covered dynamic headphone as a microphone.

2.   Phonocardiogram

Clean the ear pieces of the stethoscope with cotton and alcohol and listen to your heart sounds (pick a quiet place). Describe what you hear in terms of pitch, quality, duration, and repetition of sounds. Note that the heart sounds are loudest after an exhalation, when the heart is pressed against the chest wall.

Connect a dynamic microphone (one side of the headphones) to the digital oscilloscope, as shown in Laboratory Figure 17.4. Since the microphone output is on the order of 5 mV, you will need to use an instrumentation amplifier or differential op-amp circuit to increase the signal level. Hold the microphone over your chest and record the phonocardiogram on the oscilloscope.

For best results, sit or stand perfectly still while a laboratory partner triggers the digital oscilloscope. (Set the trigger source to "line.") The first and second heart sounds should be visible. Do at rest and after light exercise. Use the HP VEE panel driver to capture and plot the data from the oscilloscope (see Appendix F Using the digital oscilloscope to record waveforms).

3.   Blood pressure

Have the subject sit down next to a table so that the left arm can rest level with the heart. Make sure that the garment sleeve is not too tight. Circle the arm with the pressure cuff midway between the shoulder and elbow and fasten it by touching the Velcro surfaces together. The cuff should be snug but not uncomfortable.

Locate the brachial artery. It is above and slightly to the right of the bend in the elbow. Feel for its pulse with the first two fingers of the right hand. Measure and record the pulse rate. Note that the normal pulse variations in this artery produce audio frequencies that are in the 1-Hz range and very little in the 60-Hz–10-kHz range that can be heard by the human ear.

Tighten the screw valve and inflate the cuff to 150 mm Hg. Place the stethoscope head firmly over the brachial artery and listen with the stethoscope. You should hear

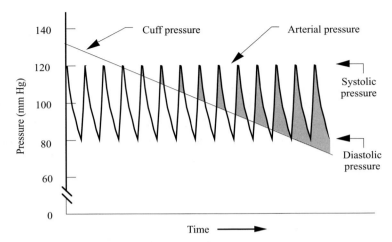

Laboratory Figure 17.5 Pulsatile flow occurs when the cuff pressure is between the systolic and the diastolic pressures.

nothing. Open the screw valve a bit and deflate the cuff at a rate of 2–3 mm/s. Laboratory Figure 17.5 shows the pulsing arterial pressure and the decreasing cuff pressure.

As the pressure falls, sounds (called Korotkoff sounds) become audible and pass through the following five phases:

Phase I. Faint, clear tapping sounds that gradually increase in intensity, caused by the abrupt distention of the arterial wall as a jet of blood surges under the cuff. These short duration pulses have higher harmonics that can be heard by the human ear.

Phase II. A swishing quality is heard, due to turbulence of the jet of blood in the artery.

Phase III. The sounds become crisper and increase in intensity, as the volume of blood in the jet increases.

Phase IV. A distinct, abrupt muffling of sound so that a soft blowing quality is heard. At this point, the blood flow in the artery is not interrupted, but is restricted enough to produce an audible turbulent flow.

Phase V. The point at which sound disappears completely because the cuff pressure is too low to restrict the flow of arterial blood.

Systolic blood pressure

The point in phase I where the initial tapping sound is heard for two consecutive beats.

Diastolic blood pressure

The point in phase IV where there is a distinct muffling of sound.

Potential sources of error

If the limb is thick in relation to the width of the cuff or if the cuff is loosely applied so that the rubber bag must be partially inflated before it exerts pressure on the tissues, which reduces the area of contact, the pressure in the cuff will significantly exceed the pressure in the tissues surrounding the artery. The result is an overestimate of the systolic and diastolic pressures.

In the general population, the systolic pressure can be as high as 180 mm Hg, a starting pressure of 200 mm Hg is generally used.

In some individuals, the sounds emitted by the artery disappear over a fairly large range of pressure between the systolic and diastolic pressures. This is called an "auscultatory gap" and the cause is not known. If the cuff pressure is initially set to a pressure within this gap, the pressure at the lower end of this silent range may be mistaken for a normal systolic pressure when, in fact, the actual systolic pressure is considerably higher.

Measurement of blood pressure

Measure the systolic and diastolic blood pressure for the following conditions:
1. subject is seated and has the arm at heart level,
2. subject is seated and has the arm above the head,
3. repeat condition 1 after light exercise.

Caution

When fully inflated, the cuff stops blood flow completely. Do not leave in this condition for more than 30 seconds.

Laboratory report

1.   Setup

Draw a simple block diagram of your experimental setup.

2.   Data summary and analysis

2.1 Electrocardiogram. The ECG record is traditionally calibrated at 0.1 mV per vertical mm and 40 ms per horizontal mm. On your ECG recordings, label the P-wave, QRS complex, and T-wave. For both resting and exercise conditions measure and tabulate the lead II amplitude of the P-, R-, and T-waves. Mark the time of the peaks of the P-, R-, and T-waves and tabulate the R–R, P–R, R–T, and T–P intervals. (The R–R interval is the cardiac period.) Compute the percentage change in these four intervals due to light exercise.

2.2 Phonocardiogram. From your phonocardiogram, measure and tabulate the duration of the first and second heart sounds and the interval between them. Compare the resting with the exercise values.

2.3 Blood pressure. Tabulate all systolic and diastolic blood pressures that you measured as well as the pulse pressure (systolic minus diastolic pressure). Estimate the pressure difference due to arm elevation and compare with the observed difference.

Hint: 1 atmosphere pressure = 760 mm Hg = 10.3 meters of water.

3.   Discussion and conclusions

3.1 Discuss how you would use sensors, amplifiers, and a microcomputer to measure the ECG and display useful results.

3.2 Discuss how you would use sensors, amplifiers, and a microcomputer to measure the phonocardiogram and display useful results.

3.3 Discuss how you would use sensors, amplifiers, and a microcomputer to measure the blood pressure and display useful results.

4.   Questions

4.1 Will a heart murmur show up on an ECG? Explain.

4.2 What events produce the P-, R-, and T-waves?

4.3 When are the highest and lowest pressures reached in the arteries?

4.4 What events produce the first and second heart sounds?

4.5 When you measured the blood pressure using the cuff, why was it best to have the arm at the same level as the heart?

4.6 Which interval was affected most by light exercise: P–R, R–T, or T–P?

4.7 How much work (in joules) is done by the heart per beat? What is the power level (watts)?

Hints for Question 4.7:
- work = mean ventricular pulse pressure × ejection volume + 1/2 × mass × velocity^2
- 1 joule (work) = 1 newton meter = 1 kg m^2/s^2 = 10^7 g cm^2/s^2
- 1 newton (force) = 1 kg × 9.8 m/s^2
- 760 mm Hg (pressure) = 76 cm × 13 g/cm^3 × 980 cm/s^2
    = 96,800 kg/(m s^2) = 968,000 g/(cm s^2)
- mean ventricular pulse pressure = 1/2 (systolic + diastolic)
- ejection volume = 80 cm^3, aortic velocity = 10 cm/s

5.   Laboratory data sheets

Include your handwritten data sheets (or a copy), which should consist of a log of the procedures you used, any special circumstances, and the measurements you recorded manually.

Laboratory Exercise 18

The electromyogram (EMG)

Purpose

To investigate the electrical potentials produced by skeletal muscles; to build a circuit that amplifies, rectifies, and filters these signals; to relate the processed signal to the force of muscular contraction; and to determine the possibilities and limitations of using the EMG for control purposes.

Equipment

- IBM PC with data-acquisition and control adapter
- Set of weights with finger loop
- Three floating Ag(AgCl) skin electrodes
- Isolation amplifier circuit with two triax cables for connection to skin electrodes, AD625 instrumentation amplifier, Burr–Brown 3656 isolation amplifier, powered by a wall-plug sealed power supply
- ± 12-V power supply
- Two 10-μF, 25-V electrolytic capacitors (put between power and ground at circuit board)
- Six 0.1-μF bypass capacitors (put between power and ground on all chips)
- Superstrip circuit board
- Digital oscilloscope
- Two coaxial cables
- Headphone
- Two 1N914 diodes
- Three LF356 op amps
- Three 1-kΩ resistors
- Three 5.1-kΩ resistors
- Three 10-kΩ resistors

- Two 100-kΩ resistors
- Two 20-kΩ trimpots
- Capacitors (one each): 1, 3.3, and 10 μF

Background

1.   The motor unit and its action potential signal

The motor unit is the smallest unit of the skeletal muscle that can be controlled by the nervous system (Laboratory Figure 18.1). The motor neuron causes a single twitch of its associated muscle fibers by sending an action potential down one of its axons (Laboratory Figure 18.2). When that action potential reaches a motor end plate, a small amount of acetylcholine is released, which produces an action potential in the muscle cells. Individual muscle cells conduct action potentials similar to those conducted by axons.

In an actively contracting muscle, the many parallel fibers (cells) making up the muscle will be conducting such action potentials at various repetition rates. With suitable electrodes, this barrage of electrical activity can be detected and recorded as the

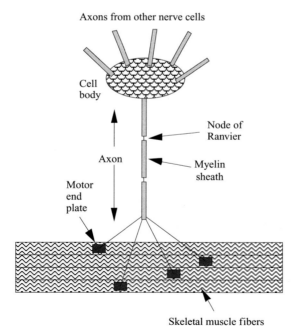

Laboratory Figure 18.1 Motor unit consisting of a nerve cell, axon, motor end plates, and skeletal muscle fibers. The motor unit is the smallest element of the skeletal muscle that can be controlled by the nervous system.

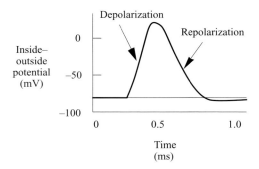

Laboratory Figure 18.2 Action potential as measured in an axon or muscle cell.

electromyogram (EMG). In some cases, needle electrodes are inserted through the skin directly into the muscle under study. If the needles are fine enough, little damage results. For most practical purposes, however, the activity is detected by electrodes placed on the skin surface over the muscle. After pickup by the electrodes, the signal can be processed and used for control purposes. EMG activity has commonly been used in the control of orthotic–prosthetic devices such as artificial hands and arms but it need not be limited to control aids for the physically handicapped.

The EMG is the result of many motor units producing action potentials without close synchronization and its frequency content is in the 100 Hz to 10 kHz region. The ECG (observed in Laboratory Exercise 17 The human heart) is the result of large masses of cardiac muscle depolarizing synchronously for efficient pumping action.

2.   Summary of steps in skeletal muscle contraction and recovery

The following steps present a simplified sequence of events in skeletal muscle contraction:
1. Higher brain function decides which muscles to contract and how hard.
2. A number of motor units (spatial summation) are pulsed at a chosen frequency (temporal summation) to produce the desired force.
3. To pulse a motor neuron, several axons from other nerve cells liberate acetylcholine at synaptic junctions on the cell body, which triggers an action potential that travels down the motor axon (1–100 m/s) to the motor end plate.
4. An action potential is due to the sudden inrush of Na^+ ions, which are normally excluded from the interior. The increase in membrane potential (from -85 to $+20$ mV internal) momentarily opens up Na^+ channels and so propagates the action potential along the length of the axon.
5. When the action potential reaches the motor end plate, acetylcholine is again liberated, which triggers an action potential along the length of the muscle fiber itself.
6. In the muscle fiber, the action potential triggers the release of Ca^{++} ions, which cause actin and myosin filaments to slide forcefully and produce mechanical contraction.

This process uses the high-energy molecule ATP (adenosine triphosphate) to provide the energy for the contraction.

The following steps present a simplified sequence of events in skeletal muscle recovery:

1. Cholinesterase destroys the acetylcholine ($\approx$2 ms).
2. The Na^+ channels close, resting potential returns to -85 mV ($\approx$1 ms, the duration of the action potential).
3. The Na^+ is pumped back out of the cell ($\approx$100 ms).
4. The Ca^{++} ions are removed ($\approx$20 ms, the duration of the contraction).
5. Food is burned to replace the ATP (minutes).

3.   Applications of the processed EMG signal

In almost all practical control uses to date, the EMG activity has been detected by electrodes placed on the skin surface over the contracting muscle and then differentially amplified to reject the much stronger power-line interference present in normal environments. Because of the electrode placement and the relatively large recording areas, the EMG waveform is a spatial–temporal summation of the many muscle-fiber potentials reaching the electrodes. As a result, the waveform observed at the amplifier output is primarily marked by a peak-to-peak amplitude that increases with increasing contraction. In addition, the number of peaks per second similarly increases. For even a constant contraction level, the peak-to-peak amplitude can only be defined in a statistical sense.

The dc component of the raw EMG signal is dominated to an electrode offset that drifts randomly and is not a useful measure of the strength of contraction. It should be blocked by high-pass filtering.

Because of the nature of the raw EMG waveform, some initial analog-signal processing is done to extract a feature that is monotonically related to contraction intensity. This means that the processed EMG signal increases with increasing contraction intensity, but the relationship is not necessarily linear. In this laboratory exercise, you will differentially record this signal, amplify it, rectify it, and then smooth it with a low-pass filter. You should consider how this processed signal depends on contraction force and what problems you would encounter in using this signal to control a powered prosthetic device.

Additional reading

Leslie Cromwell, Fred J. Weibell, and Erich A. Pfeiffer, *Biomedical Instrumentation and Measurements*, Prentice Hall, Englewood Cliffs, NJ, Chapter 16.

J. Duchene, and J.-Y. Hogrel, A model of EMG generation. *IEEE Trans Bio-Medical Engineering*, Vol. 47 (2) (2000): pp 192–201.

Peter Strong, *Biophysical Measurements*, Tektronix, Beaverton, OR, 1970, Chapter 3 and Chapter 12, Sections 12.5 and 12.6.

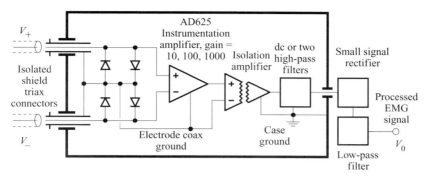

Laboratory Figure 18.3 Block diagram of EMG amplification and processing circuit.

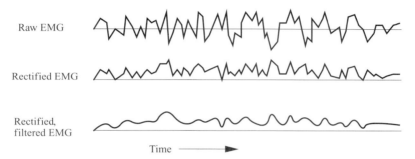

Laboratory Figure 18.4 Schematic of raw, rectified, and filtered EMG signals.

Procedure

1.   EMG signal-processing circuit – construction

The overall circuit you will use is shown in Laboratory Figure 18.3. The first stage consists of an AD625 instrumentation amplifier followed by a Burr–Brown 3656 isolation amplifier in a metal box. Triax cables are used to keep the electrode ground isolated from the power-supply ground. Power is supplied by a sealed dc power supply in a plastic case (the type commonly used to recharge hand calculators). Connect the three electrodes (V_+, V_-, and the intermediate ground) to the input of the isolation amplifier. Switch the isolation amplifier to ac-coupled mode.

Use this circuit in its ac-coupled mode. Use a gain setting that provides an output of at least 1 V peak-to-peak. Laboratory Figure 18.4 shows a schematic of raw, rectified, and filtered EMG signals.

Build the full-wave rectifier (Laboratory Figure 18.5) and a single-pole low-pass filter (Laboratory Figure 18.6) on your circuit bread-board. The high-pass filter is used to block electrode drift and other noise below the frequencies in the EMG signal (10 Hz

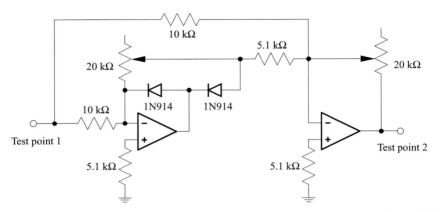

Laboratory Figure 18.5 Full-wave rectifier circuit used after the isolation amplifier. The 20-kΩ trimpots are nominally set at 10 kΩ.

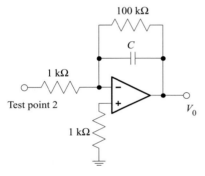

Laboratory Figure 18.6 Low-pass filter for the output of the full-wave rectifier. Use $C = 1$, 3.3, and 10 μF to explore the tradeoff between the noise and the response time of the processed EMG signal.

to 10 kHz). See Laboratory Figure 4.2 for the LF356 op-amp external connections. Connect 0.1-μF capacitors between pin 4 (−12 V) and ground and between pin 7 (+12 V) and ground.

The purpose of the high-pass filter is to reject electrode drift. The purpose of the full-wave rectifier is to make positive and negative excursions of the raw EMG signal count the same, rather than cancel. The purpose of the low-pass filter is to take a moving average of the positive excursions plus the sign-flipped negative excursions.

2.   EMG signal-processing circuit – testing

Test the rectifier circuit with a 1-V p–p sine wave at test point 1. If alternate lobes of the output waveform (test point 2) are not equal, adjust the first 20-kΩ trimpot. If the lobes do not have 0.5 V amplitude, adjust the second 20-kΩ trimpot.

To test the entire circuit, use a 100-kΩ/1-kΩ voltage divider to provide a 20-mV sine wave into the isolation amplifier. The output of the low-pass filter should be a slowly

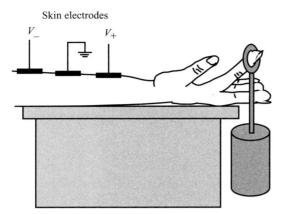

Skin electrodes

V_- V_+

Laboratory Figure 18.7 Experimental setup relating force and the EMG signal.

varying waveform with a level that depends on input wave amplitude and offset level, but does not depend on frequency.

3.   Skin electrodes

Use the bare-metal probes of your digital multimeter to record the dry and wet electrical resistances between two points about 10 cm apart on the forearm of one of your laboratory partners. Obtain three Ag(AgCl) skin electrodes and apply them as demonstrated on the forearm. They should be placed to record the EMG signal from the group of muscles that control the flexion of the long finger. Record the electrical resistance between the two electrodes.

4.   Experimental setup

Position the forearm with the palm up and the long finger in the lifting hole of one of the weights provided, as shown in Laboratory Figure 18.7. Practice raising the finger to pick up one of the weights provided.

5.   The unprocessed EMG

Attach the skin electrodes to the isolation amplifier with the pair of triaxial cables provided. The two green buttons are connected to the V_- and V_+ skin electrodes. The black button is connected to the center skin electrode. Observe the output on your oscilloscope at test point 1 and describe the amplitude, average value, and quality of the "raw" EMG signal as a function of finger tension. For a maximum signal, remove the finger from the ring and clench the fist as tight as possible.

Listen to the amplified raw EMG signal with the headphones during high and low muscle tension. Describe what you hear.

Check that the rectified signal is present at test point 2.

6.   EMG versus mechanical load

Using the largest low-pass filter time constant, measure the average EMG signal (output of the low-pass filter) for zero load and when lifting each weight with the finger. Use either the oscilloscope or the digital multimeter. Some EMG signal will occur at zero load due to muscle tone.

7.   Response time and noise

Connect the processed EMG signal to the digital oscilloscope. For three low-pass filter time constants, pick up the weight abruptly as your laboratory partner triggers the oscilloscope. Choose a horizontal sweep speed that allows you to measure both the response time and the noise level after the signal has leveled off. Use the HP VEE panel driver to print the waveform, which will be used in your analysis of the tradeoff between response time and noise (see Appendix F Using the digital oscilloscope to record waveforms).

8.   Voluntary control of EMG

Under no-load conditions, flex your forearm muscles without moving your fingers and see how well you can produce EMG-processing circuit outputs that correspond to actually lifting weights. Record the largest processed EMG signal you can produce this way.

Laboratory report

1.   Setup

Draw a simple block diagram of your experimental setup.

2.   Data summary and analysis

2.1 Summarize your data from procedure sections 2, and 4–7.
2.2 Plot the processed EMG signal as a function of load.
2.3 Plot the response time as a function of the filter time constant for the three capacitors.

3.   Discussion and conclusions

3.1 Discuss the processes that occur after you decide to contract a skeletal muscle and how they result in the raw EMG signal that you observed.

3.2 Discuss the tradeoffs between response time and noise in the final low-pass filter.

3.3 Discuss how an artificial hand would be controlled using skin electrodes on the forearm.

3.4 Signals from nerves going to muscles have been rejected by prosthetics designers as a possible source of control signals. Discuss any possible reasons for this.

4.   Questions

4.1 Why is the amplified EMG signal rectified before filtering?

Hint: How would the output change if the rectification were omitted?

4.2 What are the high and low corner frequencies at test point 1 (the output of the isolation amplifier)?

Hint: Consult the data sheets for the high-frequency responses of the Analog Devices AD625 instrumentation amplifier and the Burr–Brown 3656 isolation amplifier.

Using the 10-μF capacitor, what are the high and low corner frequencies for the low-pass filter circuit?

4.3 Is the processed EMG noise level greater for greater finger forces?

Hint: Explain in terms of the motor units involved.

4.4 What are the tradeoffs of using lower and higher corner frequencies in the low-pass filter that follows the rectifier?

5.   Laboratory data sheets

Include your handwritten data sheets (or a copy), which should consist of a log of the procedures you used, any special circumstances, and the measurements you recorded manually.

Laboratory Exercise 19

The electrooculogram (EOG)

Purpose

To record the human EOG during various sequences of horizontal eye movements; to identify smooth pursuit and saccadic eye movements.

Equipment

- IBM-compatible Pentium microcomputer with HP VEE
- Digital oscilloscope
- Isolation amplifier circuit
- Digital multimeter
- Gear motor with controller for the range 10–300 rpm. Mount with vertical axis and attach lucite rod with battery above center of rotation and small light at outer edge
- Cardboard with 2.5-cm wide black strips on 15-cm centers
- Board with three LEDs on 45-cm centers

Background

1.   Origin of the EOG

There exists between the cornea and retina of the eye a potential difference thought to be due to the difference in potential between the interior ends of the photoreceptors (rods and cones) and the pigment epithelium in which they are imbedded. This potential is relatively steady, although changes do occur in response to different light levels. Under constant background illumination, this potential may be considered an electrical dipole that rotates about the center of orbit as the eye rotates. The potential at any point on the head may be estimated by volume conductor theory. In practice, the best placement of

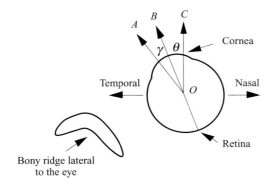

Laboratory Figure 19.1 The direction AO of the ocular dipole and the line of vision BO relative to the direction CO of the head. The angle BOC is θ, the direction of vision. The angle $AOB = \gamma$ is about 14° in a human.

the sensing electrodes are on the skin just over the bony ridge just lateral to the eye. A center electrode can be used as the ground for the coaxial cables that carry the differential signal to an instrumentation amplifier. A horizontal plane through the eye and recording electrodes is shown in Laboratory Figure 19.1, where O is the center of rotation, OA is the direction of the dipole, OB is the direction of the eyeball, OC defines the straight-ahead position, and angle AOB is about 14° in a human. By using electrodes on both temples, the observed differential potential is given by:

$$V_{EOG} = K \cos 14° \sin \theta$$

2.   Smooth pursuit versus saccadic motion

When the head is held steady and the eye follows a slowly moving object, the eye muscles can be controlled so that the object appears stationary in the field of view. This is called smooth pursuit. When the object is moving too rapidly for this type of muscle control, a different strategy is used where the eye muscles apply a brief, strong force to realign the object in the field of view. This motion is called a saccade, and has two measurable properties in this laboratory exercise, $\Delta\theta$ (the angular displacement of the eye) and Δt (the time interval). For small angular displacement, the moving object often appears fixed in the field of view and the subject is not aware that a saccade has occurred.

Additional reading

Theodore C. Ruch and Harry D. Patton, *Physiology and Biophysics*, Saunders, Philadelphia, PA, 1982, Chapter 3, pp 85–91.

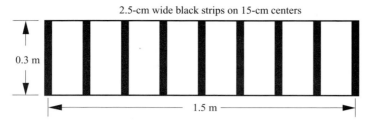

Laboratory Figure 19.2 Bar pattern for determining the relationship between EOG amplitude and eye angle.

Procedure

1.   EOG signal versus eye position

Place the subject 1.5 m away from the bar pattern shown in Laboratory Figure 19.2. Measure the distance between bars and calculate the angular separation as seen by the subject.

Place two skin electrodes on either side of the forehead and connect to the signal leads of two triaxial cables. Place a third skin electrode between the two and connect to the ground shield of both coaxial cables. Connect both triaxial cables to the isolation amplifier. Switch the isolation amplifier output to dc-coupled mode. Connect the output of the isolation amplifier to the oscilloscope. Use the HP VEE panel driver for printing (see Appendix F Using the digital oscilloscope to record waveforms).

1. Fixate on the center bar and record the EOG signal for a few minutes to estimate drift. Fixate on each bar in turn and measure the EOG signal versus angle. Periodically return to the center bar to recheck the zero angle.
2. Shift from one bar to another bar at the edge of the pattern *as rapidly as possible.* Use the oscilloscope to record the EOG signal, V_{EOG}, as a function of time. Choose a sweep speed that allows you to measure the maximum voluntary angular velocity. Print the waveform for later analysis. Repeat for a smaller angular shift.

2.   Response time

Place the subject 1.5 m away from the LED board (Laboratory Figure 19.3). Perform as many of the following experiments as time permits:

1. The subject fixates on the center LED (on) with earphones on to prevent hearing switch noise. A partner switches to the right or left LED. Trigger the scope on the switch pulse and record the EOG. Measure the latency time.
2. The subject knows which way the light will "move" but not when and cannot hear switch noise.

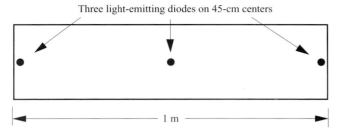

Laboratory Figure 19.3 Three LEDs mounted in a meter stick for measuring maximum voluntary angular speed.

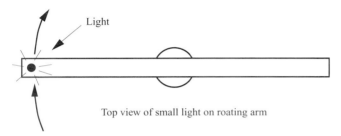

Laboratory Figure 19.4 Rotating light for determining maximum smooth pursuit and involuntary saccade velocities.

3. As step 1 but the subject can hear the switch closure by electrical connection to the earphones through a capacitor.
4. Change the lit diode back and forth at random. Measure the latency as the subject tries to fixate on the lit LED. See if there is a minimum interval between fixation shifts. See if you can devise a pattern that produces an exceptionally long latency.

3.   Smooth pursuit and saccadic motion

Place the subject 1.5 m away from the center of the revolving light (Laboratory Figures 19.4 and 19.5).
1. Record the EOG at slow speed when it is a smooth curve without any sharp dislocations (Laboratory Figure 19.6).
2. Slowly increase the speed and record the EOG at the maximum velocity at which the EOG is a smooth curve. Determine the maximum angular velocity for smooth pursuit (Laboratory Figure 19.6).
3. Increase the speed still further and record the EOG when you can see some saccadic motion (brief steep sections in an otherwise smooth curve) (Laboratory Figure 19.7).
4. Increase the speed still further and record the EOG when the curve is dominated by saccades (Laboratory Figure 19.8).

For the case of limited angular acceleration (Laboratory Figure 19.9), the angular velocity increases linearly with time and a doubling of the time interval $\Delta t_2 = 2\Delta t_1$

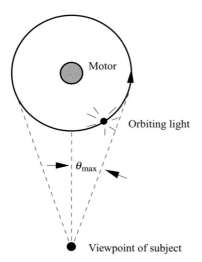

Laboratory Figure 19.5 Angle θ as see from the viewpoint of the subject.

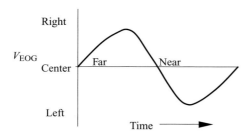

Laboratory Figure 19.6 Smooth pursuit.

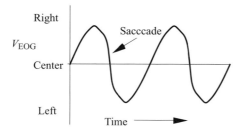

Laboratory Figure 19.7 Onset of saccades.

produces a four-fold increase in angular displacement $\Delta\theta_2 = 4\Delta\theta_1$. The result is that the angular velocity is proportional to the square root of the angular displacement.

For the case of limited angular velocity (Laboratory Figure 19.10), the angular velocity does not depend on the angular displacement.

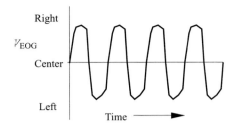

Laboratory Figure 19.8 Case of extreme saccades.

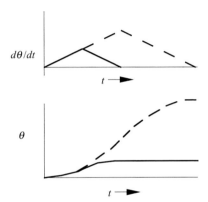

Laboratory Figure 19.9 Angular velocity $d\theta/dt$ and displacement θ during a saccade, where the acceleration is limited.

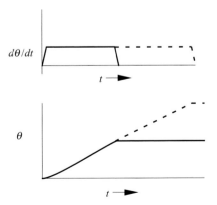

Laboratory Figure 19.10 Angular velocity $d\theta/dt$ and displacement θ during a saccade, where the velocity is limited.

Laboratory report

1.   Setup

Draw a simple block diagram of your experimental setup.

2.   Data summary and analysis

2.1 From your data of procedure section 1, print a plot of EOG versus angle (V_{EOG} versus θ).

2.2 From your data of procedure section 1, tabulate the $\Delta\theta$ and Δt values for the most rapid voluntary shifts.

2.3 Plot EOG versus time (V_{EOG} versus t) from your data of procedure section 3. Using the relationship between V_{EOG} and θ from procedure section 1, add a θ scale alongside the V_{EOG} scale.

2.4 For the data of procedure section 3, determine and tabulate the maximum smooth pursuit velocity $d\theta/dt$, and the $\Delta\theta$ and Δt values for all saccades.

3.   Discussion and conclusions

3.1 Discuss your measurement of EOG signal versus eye angle. Discuss the relative contributions of linearity, amplifier noise, 60-Hz interference, electrode drift, etc. on the overall error.

3.2 Discuss the relative values of the EOG signal and the common-mode 60-Hz pickup and the advantage of differential amplification in this application.

3.3 Discuss procedure section 2 (response time).

3.4 Discuss procedure section 3 (smooth pursuit and saccadic motion). Consider whether the saccades are velocity limited, acceleration limited, or some combination (for example, acceleration limited for small $\Delta\theta$, but velocity limited for large $\Delta\theta$).

4.   Questions

4.1 Since it is known from other better methods of measuring eye position that humans can maintain fixation for long periods within 15 min of arc, the EOG signal drift must come from your recording setup. What source do you hypothesize? Is the drift random?

4.2 What is the maximum smooth-pursuit angular velocity observed (procedure section 3)?

4.3   What are the maximum angular velocities of the involuntary (procedure section 3) and voluntary (procedure section 2) saccades?

4.4   High-pass filtering is often used to counteract electrode drift. Why could this not be used in this laboratory exercise?

5.   Laboratory data sheets

Include your handwritten data sheets (or a copy), which should consist of a log of the procedures you used, any special circumstances, and the measurements you recorded manually.

5 Data analysis and control

5.1 Introduction

One of the major advantages of the microcomputer is its use to analyze data quickly even while new data are being taken. The result of the analysis can be a meaningful display or control of the system being measured. This chapter discusses data analysis, including statistical analysis, least-squares fitting, fast Fourier transforms, and control.

5.2 The Gaussian-error distribution

5.2.1 Repeated measurements of the same quantity

Ideally, if repeated measurements are made of the same quantity under identical conditions, the results should be the same. In the real world, however, all measurements are influenced to some extent by random factors, and repeated measurements exhibit unavoidable and unpredictable fluctuations.

The measured values will cluster about some special value – most of them will be close to the average value, but some will lie farther away. An example of such a distribution is shown in Figure 5.1. An alternative method for showing such data is to divide the horizontal axis into "bins," and use the vertical axis to show the number of events in each bin, as shown in Figure 5.2.

If a very large number of measurements are made, and if the bins are made very narrow, the famous "bell-shaped" curve results, also called the Gaussian curve of error (Figure 5.3). This distribution is given analytically by:

$$G(x) = \frac{1}{\sqrt{2\pi\sigma^2}} \exp\left[-\frac{(x-\mu)^2}{2\sigma^2}\right]$$

where $G(x)$ is dN/dx, dN is the number of measurements in the interval (or very small bin) between x and $x + dx$; and σ is a constant called the "root mean square" or "rms" (the average of the squares of the deviations from the mean μ). $G(x)$ is the relative

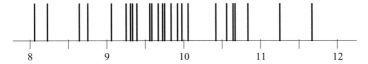

Figure 5.1 Distribution of 25 measurements. Each vertical line represents a single measurement placed along the horizontal axis according to the measured value. The average of all the measurements is very close to 10.

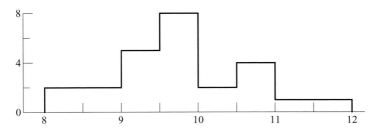

Figure 5.2 Data of Figure 5.1 plotted in bins 0.5-units wide. The vertical axis is the number of measured values in each bin.

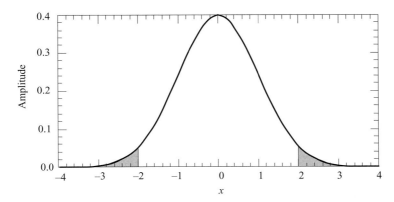

Figure 5.3 The Gaussian curve of error, shown with a mean of 0, a standard deviation of 1, and unit area. The shaded region $|x| > 2$ is the 4.55% probability of exceeding two standard deviations.

probability of obtaining a particular measured value x, given that the average is μ and the rms is σ.

While there is no upper bound to the magnitude of this random error, it is comforting to know that very large errors are extremely unlikely (Table 5.1). While the probability of exceeding one standard deviation is 32%, a random process that occurs 10^{16} times per second would be expected to exceed 12 standard deviations only once in the entire history of the universe (10^{10} years, or 3.2×10^{17} seconds).

The Gaussian distribution arises quite naturally when random processes are combined. For example, the set of random numbers uniformly distributed between -0.5 and $+0.5$ has mean $\mu = 0$ and rms $\sigma = \sqrt{1/12} = 0.289$, and looks quite different

Table 5.1 *Probability of exceeding ±x standard deviations*

x	Probability	x	Probability	x	Probability
0	1.00	5	5.73×10^{-7}	15	7.34×10^{-51}
1	3.17×10^{-1}	6	1.97×10^{-9}	20	5.51×10^{-89}
2	4.55×10^{-2}	8	1.24×10^{-15}	30	9.81×10^{-198}
3	2.70×10^{-3}	10	1.52×10^{-23}	50	2.16×10^{-545}
4	6.33×10^{-5}	12	3.55×10^{-33}	100	$2.69 \times 10^{-2,174}$

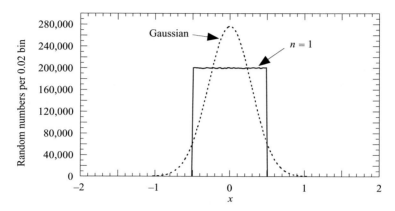

Figure 5.4 Distribution of 10 million random numbers uniformly distributed between −0.5 and +0.5 (standard deviation = 0.289) compared with a Gaussian distribution having the same mean, area, and standard deviation.

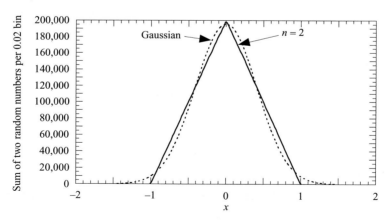

Figure 5.5 Distribution of 10 million sums of two random numbers (standard deviation = 0.408) compared with a Gaussian distribution having the same mean, area, and standard deviation.

than a Gaussian distribution with the same mean and rms (Figure 5.4). The sum of two such random numbers has a triangular distribution with $\mu = 0$ and $\sigma = \sqrt{1/6} = 0.408$ and only vaguely approximates a Gaussian distribution with the same μ and σ (Figure 5.5).

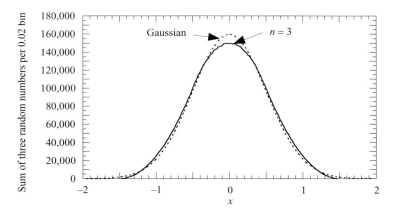

Figure 5.6 Distribution of 10 million sums of three random numbers (standard deviation = 0.500) compared with a Gaussian distribution having the same mean, area, and standard deviation. Note that the sums are distributed only between -1.5 and 1.5, and outside of this range the Gaussian has 0.27% of its area.

However, the sum of three such random numbers is a bell-shaped curve with $\mu = 0$ and $\sigma = 0.500$ that begins to resemble a Gaussian distribution with the same μ and σ (Figure 5.6). In general, if n such random numbers are added, the sum has $\mu = 0$, $\sigma = \sqrt{n/12}$, and resembles the Gaussian distribution quite closely as n becomes large.

We now present a useful random variable that approximates the Gaussian distribution with $\mu = 0$ and $\sigma = 1$, and takes under consideration the fact that almost all random number generators provide random numbers R_i uniformly distributed between 0 and 1:

$$\sqrt{12/n} \sum_{i=1}^{n} (R_i - 0.5)$$

In common practice, n is usually taken to be 12. The expression above then simplifies to the rule:

Computation tip

To generate a random Gaussian distribution with mean $= 0$ and rms $= 1$, add 12 random numbers and subtract 6. To transform to any other mean μ and rms σ, firstly multiply by σ and then add μ.

While this rule can only produce random numbers distributed between -6σ and $+6\sigma$, the approximation is quite accurate, as the true Gaussian distribution has only 2.0×10^{-9} of its area outside that range. A far more rigorous approximation is provided by the choice $n = 48$, which gives a distribution between -12σ and $+12\sigma$. The true Gaussian has only 3.6×10^{-33} of its area outside that range.

5.2.2   Estimating the sample mean and standard deviation

Suppose we make repeated experimental measurements $a_i (i = 1 \rightarrow m)$ of some quantity whose "true" value is μ. As mentioned before, these repeated measurements will not all be exactly equal, but will have some average value that is presumably close to the true value μ. Two important questions we can ask are: "How can we best estimate the quantity μ from our measurements?" and "How reliable is that estimation?". The estimator of μ is simply the average of the a_i (called the **sample mean**):

$$\mu \approx \bar{a} = \frac{1}{m} \sum_{i=1}^{m} a_i$$

To answer the question about the reliability of the average, we must firstly examine an important index of how much variability exists in our measurements a_i. To do this, we define a quantity called the **residual** R_i, which is the difference between the measurements a_i and their average:

$$R_i = a_i - \bar{a}$$

Whereas the sum of the R_i is zero, the average of the sum of the squares of R_i (the square of the **standard deviation**) is an important statistical quantity called the **variance**:

$$\sigma_a^2 = \frac{1}{m-1} \sum_{i=1}^{m} R_i^2 = \frac{1}{m-1} \sum_{i=1}^{m} (a_i - \bar{a})^2$$

$$\sigma_a^2 = \sum_{i=1}^{m} \frac{(a_i - \bar{a})^2}{m-1} = \frac{\sum a_i^2 - 2\bar{a} \sum a_i + m\bar{a}^2}{m-1} = \frac{\sum a_i^2 - m\bar{a}^2}{m-1} \tag{5.1}$$

The latter form is preferred for programming because it can be formed by summing the data and the square of the data in a single loop.

Note: It is tempting to use m for the denominator in Eq. (5.1), like a simple average, but this assumes that the average value of a is a constant, not a random variable. Because we have estimated the variance of a from the same data that was used to compute the average value of a, the residuals are "biased" toward lower values. To correct for this, the denominator $(m - 1)$ is used.

Another important question is how variances combine when the quantity of interest is a function of independent random variables. Consider the function $f(a_1, a_2, \ldots, a_n)$. If the variables $a_1, a_2, \ldots, a_n$ are independently random, that is if the random fluctuations in any variable have no influence on the random fluctuations in any other variable, then variances may be combined using the formula:

$$\sigma_f^2 = \left(\frac{\partial f}{\partial a_1}\right)^2 \sigma_{a1}^2 + \left(\frac{\partial f}{\partial a_2}\right)^2 \sigma_{a2}^2 + \cdots + \left(\frac{\partial f}{\partial a_n}\right)^2 \sigma_{an}^2 \tag{5.2}$$

This formula will be used in the following sections to estimate the standard error of the mean and the standard error of the standard deviation.

EXAMPLE 5.1

Combine variances for the simple linear function $f = a_1 - 4a_2 + 3a_3$. Since we have $\partial f/\partial a_1 = 1$, $\partial f/\partial a_2 = -4$, $\partial f/\partial a_3 = 3$:

$$\sigma_f{}^2 = \sigma_{a1}{}^2 + 16\sigma_{a2}{}^2 + 9\sigma_{a3}{}^2$$

Note that whether random quantities are added or subtracted, their variances add. This has considerable significance when subtracting nearly equal, noisy measured values.

EXAMPLE 5.2

Combine variances for the function $f = a_1 a_2/a_3$:

$$\partial f/\partial a_1 = a_2/a_3, \qquad \partial f/\partial a_2 = a_1/a_3, \qquad \partial f/\partial a_3 = -a_1 a_2/a_3{}^2$$

$$\sigma_f{}^2 = (a_2/a_3)^2 \sigma_{a1}{}^2 + (a_1/a_3)^2 \sigma_{a2}{}^2 + \left(a_1 a_2/a_3{}^2\right)^2 \sigma_{a3}{}^2$$

dividing both sides by f^2, we have

$$\sigma_f{}^2/f^2 = \sigma_{a1}{}^2/a_1{}^2 + \sigma_{a2}{}^2/a_2{}^2 + \sigma_{a3}{}^2/a_3{}^2$$

Note that when random quantities are multiplied or divided, their fractional variances add.

5.2.3   Estimating the standard error of the mean

Let us apply this to the computation of the variance in the average of the measurements a_i, noting that each $\partial f/\partial a_i$ is equal to $1/m$, and that $\sigma_{ai}{}^2$ represents the variance in each of the measurements a_i:

$$\bar{a} = \frac{1}{m}\sum_{i=1}^{m} a_i, \quad \partial \bar{a}/\partial a_i = 1/m$$

$$\boxed{\sigma_{\bar{a}}{}^2 = \frac{1}{M^2}\sum_{i=1}^{m}\sigma_{ai}{}^2 = \frac{\sigma_a{}^2}{m}} \tag{5.3}$$

The variance in the mean is equal to the sample variance divided by the number of samples.

$\sigma_{\bar{a}}$ is also called the **standard error of the mean** (sem):

$$\boxed{\sigma_{\bar{a}} = \frac{\sigma_a}{\sqrt{m}}}$$

For binned data, where f_i is the number of measurements falling into the ith bin centered at a_i, the weighted average is given by:

$$\bar{a} = \frac{1}{M} \sum_{i=1}^{m} a_i f_i$$

The weighted total M is given by:

$$M = \sum_{i=1}^{m} f_i$$

The variance of the weighted data is given by

$$\sigma_a{}^2 = \frac{1}{M-1} \sum_{i=1}^{m} f_i (a_i - \bar{a})^2 = \frac{1}{M-1} \left(\sum_{i=1}^{m} f_i a_i{}^2 - M \bar{a}^2 \right) \tag{5.4}$$

5.2.4 Estimating the standard error of the standard deviation

Since the standard deviation σ derived from a particular set of experimental measurements is a random variable, it has its own standard deviation, called the **standard error of the standard deviation**. If the underlying random distribution is Gaussian, the standard error of the standard deviation is given by:

$$\delta\sigma = \frac{\sigma}{\sqrt{2(m-1)}}$$

where m is the number of measurements.

If σ_0 is the true standard deviation and σ is the standard deviation that varies from experiment to experiment, the quantity $(m-1)\sigma^2/\sigma_0^2$ has a chi-squared distribution with $m-1$ degrees of freedom. Example 5.7 describes how to use the chi-squared table to determine confidence intervals in σ.

5.3 Student's *t* test

5.3.1 Unpaired data

Suppose we make m_a repeated measurements a_i under some experimental condition A and m_b repeated measurements b_i under condition B. In the unpaired case, there is no particular correspondence between a_i and b_i. (In the next section, we discuss the paired-data case, where a_i and b_i are measurements of the same m individuals under conditions A and B.) The means (averages) are given by:

$$\bar{a} = \frac{1}{m_a} \sum_{i=1}^{m_a} a_i \quad \text{and} \quad \bar{b} = \frac{1}{m_b} \sum_{i=1}^{m_b} b_i$$

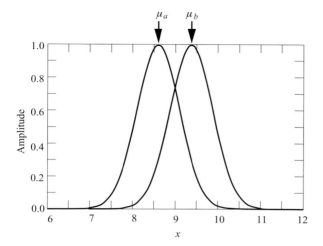

Figure 5.7 Case where two distributions have different means μ_a and μ_b but overlap considerably. The Student's t test is used to determine if the difference in the means of measurements made during two different experimental conditions could have occurred purely by chance.

Clearly, if the distributions of the a_i and the b_i are so far apart that there is very little overlap, then we may safely conclude that there is a clear difference between condition A and condition B, as far as the measurements are concerned. However, we must frequently deal with the case where the a_i and b_i distributions have a significant overlap, as shown in Figure 5.7.

The question we ask in this case is whether the difference $\Delta = \bar{a} - \bar{b}$ is statistically significant. In other words, what is the probability that the difference arose by chance? To answer this question, we must know the variance in the difference Δ itself. From Eq. (5.2) for combining variances:

$$\sigma_\Delta{}^2 = \sigma_{\bar{a}}{}^2 + \sigma_{\bar{b}}{}^2$$

Student's t is defined as the difference in the averages divided by the standard error of that difference, and is the number of standard deviations that Δ differs from zero:

$$t = \frac{\Delta}{\sigma_\Delta} = \frac{\bar{a} - \bar{b}}{\sqrt{\sigma_{\bar{a}}{}^2 + \sigma_{\bar{b}}{}^2}} = \frac{\bar{a} - \bar{b}}{\sqrt{\sigma_a{}^2/m_a + \sigma_b{}^2/m_b}}$$

where σ_a and σ_b can be computed from Eqs. (5.1) or (5.4), depending on whether the data are individual measurements or binned.

If we can assume that the variances of the a and b distributions are the same, we estimate the variance in Δ by combining both data sets:

$$\sigma_a{}^2 = \sigma_b{}^2 = \frac{1}{m_a + m_b - 2}\left(\sum_{i=1}^{m_a} a_i{}^2 + \sum_{i=1}^{m_b} b_i{}^2 - m_a\bar{a}^2 - m_b\bar{b}^2\right)$$

Combining, we have:

$$t = \frac{\bar{a} - \bar{b}}{\sqrt{\left(\dfrac{1}{m_a + m_b - 2}\right)\left(\dfrac{1}{m_a} + \dfrac{1}{m_b}\right)\left(\sum_{i=1}^{m_a} a_i^2 + \sum_{i=1}^{m_b} b_i^2 - m_a\bar{a}^2 - m_b\bar{b}^2\right)}}$$

Since the two sample means were determined from the data, the number of degrees of freedom is the number of independent data values minus 2, $n_f = m_a + m_b - 2$.

5.3.2 Paired data

In the case where different conditions are applied to an assortment of individuals, Student's t is computed using the differences between the individual pairs of measurements, rather than the sample means. For $i = 1$ to m, measurements a_i and b_i are made under conditions A and B, respectively. There may be relatively large differences between the set a_i and the set b_i due to differences among the individuals; however, the paired differences $d_i = a_i - b_i$ may have much less variation because the individuals are being used as their own controls. For paired data, the number of degrees of freedom $n_f = m - 1$, and Student's t is given by

$$t = \frac{\bar{d}}{\sigma_{\bar{d}}} = \frac{\bar{d}}{\sqrt{\left(\dfrac{1}{m}\right)\left(\dfrac{1}{m-1}\right)\left(\sum d_i^2 - m\bar{d}^2\right)}}$$

5.3.3 Using Student's t test

Now t is a random variable that will be different every time the experiment is performed, and can take any value even if the experimental conditions A and B did not cause a difference between the data sets a_i and b_i (Figure 5.7). Initially, we must assume that differences in the measured values are due to chance alone (the "**null hypothesis**"). Only if $|t|$ is so large that its probability is sufficiently small (say, less than 0.1%), can we reject the null hypothesis and conclude that something about the experimental conditions A and B *caused* the measured values to differ. Probabilities above 1% are not considered statistically significant.

Because of the inability to reject or accept the null hypothesis with perfect certainty, two types of errors are possible. A **Type I error** is committed if the null hypothesis is rejected when it is true. A **Type II error** is committed if the null hypothesis is accepted when it is false. The history of science is littered with the tarnished reputations of researchers who claimed an effect that was not borne out by subsequent experiments (the Type I error). On the other hand, the Type II error is more acceptable and often the mark of a careful researcher who would rather wait to see the results of a more sensitive experiment than run the risk of making a false claim.

Table 5.2(a) *Student's t test of significance (see Table 5.2(b) for $n_f > 15$)*

$P(>|t|)$	$n_f = 1$	2	3	5	7	10	15
1.000,00	0.000,0	0.000,0	0.000,0	0.000,0	0.000,0	0.000,0	0.000,0
0.900,00	0.158,4	0.142,1	0.136,6	0.132,2	0.130,3	0.128,9	0.127,8
0.800,00	0.324,9	0.288,7	0.276,7	0.267,2	0.263,2	0.260,2	0.257,9
0.700,00	0.509,5	0.444,7	0.424,2	0.408,2	0.401,5	0.396,6	0.392,8
0.600,00	0.726,5	0.617,2	0.584,4	0.559,4	0.549,1	0.541,5	0.535,7
0.500,00	1.000,0	0.816,5	0.764,9	0.726,7	0.711,1	0.699,8	0.691,2
0.400,00	1.376,4	1.060,7	0.978,5	0.919,5	0.896,0	0.879,1	0.866,2
0.300,00	1.962,6	1.386,2	1.249,8	1.155,8	1.119,2	1.093,1	1.073,5
0.200,00	3.077,7	1.885,6	1.637,7	1.475,9	1.414,9	1.372,2	1.340,6
0.100,00	6.313,8	2.920,0	2.353,4	2.015,0	1.894,6	1.812,5	1.753,1
0.050,00	12.706,2	4.302,7	3.182,4	2.570,6	2.364,6	2.228,1	2.131,4
0.020,00	31.820,5	6.964,6	4.540,7	3.364,9	2.998,0	2.763,8	2.602,5
0.010,00	63.656,7	9.924,8	5.840,9	4.032,1	3.499,5	3.169,3	2.946,7
0.005,00	127.321	14.089,0	7.453,3	4.773,3	4.029,3	3.581,4	3.286,0
0.002,00	318.308	22.327,1	10.214,5	5.893,4	4.785,3	4.143,7	3.732,8
0.001,00	636.619	31.599,1	12.924,0	6.868,8	5.407,9	4.586,9	4.072,8
0.000,50	1,273.23	44.704,6	16.326,3	7.975,7	6.081,8	5.049,0	4.416,6
0.000,20	3,183.09	70.700,1	22.203,7	9.677,6	7.063,4	5.693,8	4.880,0
0.000,10	6,366.19	99.992,5	28.000,1	11.177,7	7.884,6	6.211,1	5.239,1
0.000,05	12,732.3	141.416	35.297,9	12.892,8	8.782,5	6.756,8	5.607,0
0.000,02	31,830.9	223.603	47.927,7	15.546,9	10.102	7.527,0	6.108,9
0.000,01	63,661.9	316.225	60.396,8	17.896,9	11.214	8.150,3	6.501,7

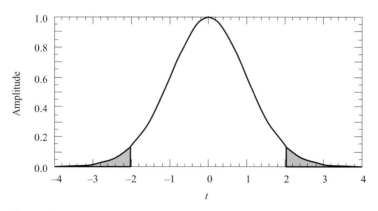

Figure 5.8 Distribution of Student's *t* for 30 degrees of freedom. Shaded areas show the probability that the absolute value of *t* will exceed 5% purely by chance.

Figure 5.8 shows the distribution of Student's t, and the tails of the distribution are shaded to show the values of t that correspond to a probability below 5%. Tables 5.2(a) and 5.2(b) list the values of t corresponding to the probability factors in the first column. The number of degrees of freedom ($n_f = m_a + m_b - 2$ for unpaired data or $n_f = m - 1$ for paired data) determines the column to be used.

Table 5.2(b) *Student's t test of significance (see Table 5 2(a) for $n_f < 20$)*

$P(>\lvert t\rvert)$	$n_f = 20$	30	50	100	200	500	∞
1.000,00	0.000,0	0.000,0	0.000,0	0.000,0	0.000,0	0.000,0	0.000,0
0.900,00	0.127,3	0.126,7	0.126,3	0.126,0	0.125,8	0.125,7	0.125,7
0.800,00	0.256,7	0.255,6	0.254,7	0.254,0	0.253,7	0.253,5	0.253,3
0.700,00	0.390,9	0.389,0	0.387,5	0.386,4	0.385,9	0.385,5	0.385,3
0.600,00	0.532,9	0.530,0	0.527,8	0.526,1	0.525,2	0.524,7	0.524,4
0.500,00	0.687,0	0.682,8	0.679,4	0.677,0	0.675,7	0.675,0	0.674,5
0.400,00	0.860,0	0.853,8	0.848,9	0.845,2	0.843,4	0.842,3	0.841,6
0.300,00	1.064,0	1.054,7	1.047,3	1.041,8	1.039,1	1.037,5	1.036,4
0.200,00	1.325,3	1.310,4	1.298,7	1.290,1	1.285,8	1.283,2	1.281,6
0.100,00	1.724,7	1.697,3	1.675,9	1.660,2	1.652,5	1.647,9	1.644,9
0.050,00	2.086,0	2.042,3	2.008,6	1.984,0	1.971,9	1.964,7	1.960,0
0.020,00	2.528,0	2.457,3	2.403,3	2.364,2	2.345,1	2.333,8	2.326,3
0.010,00	2.845,3	2.750,0	2.677,8	2.625,9	2.600,6	2.585,7	2.575,8
0.005,00	3.153,4	3.029,8	2.937,0	2.870,7	2.838,5	2.819,5	2.807,0
0.002,00	3.551,8	3.385,2	3.261,4	3.173,7	3.131,5	3.106,6	3.090,2
0.001,00	3.849,5	3.646,0	3.496,0	3.390,5	3.339,8	3.310,1	3.290,5
0.000,50	4.146,0	3.901,6	3.723,1	3.598,3	3.538,7	3.503,7	3.480,8
0.000,20	4.538,5	4.234,0	4.014,0	3.861,6	3.789,1	3.746,8	3.719,0
0.000,10	4.837,3	4.482,4	4.228,3	4.053,3	3.970,5	3.922,2	3.890,6
0.000,05	5.138,8	4.729,2	4.438,5	4.239,6	4.145,8	4.091,3	4.055,6
0.000,02	5.542,8	5.054,0	4.711,0	4.478,4	4.369,4	4.306,2	4.264,9
0.000,01	5.853,7	5.299,5	4.913,9	4.654,2	4.533,0	4.462,9	4.417,2

5.3.4   Computing the probability of exceeding $\lvert t\rvert$

After acquiring data, it is natural to compute not only the value of Student's t, but also the probability that that value (or larger) could have arisen by chance. The probability of exceeding $\lvert t\rvert$ is given by:

$$P(>\lvert t\rvert) = \frac{2\Gamma\left(\dfrac{n+1}{2}\right)}{\sqrt{n\pi}\,\Gamma(n/2)} \int_{t}^{\infty} \left(1 + \frac{x^2}{n}\right)^{-(n+1)/2} dx$$

For a very large number of degrees of freedom, it is difficult to compute the gamma functions, and the approximation:

$$\Gamma(x) \approx x^x e^{-x}\sqrt{2\pi/x} \qquad (x\ \text{large})$$

yields the more convenient form:

$$P(>\lvert t\rvert) \approx \sqrt{\frac{2}{e\pi}}\left(\frac{n+1}{n}\right)^{n/2} \int_{t}^{\infty} \left(1 + \frac{x^2}{n}\right)^{-(n+1)/2} dx$$

which has an error $\leq 1.5 \times 10^{-8}$ for $n \geq 3000$.

In the limit $n \to \infty$, $P(>|t|)$ simplifies to:

$$P(>|t|) = \sqrt{\frac{2}{\pi}} \int_{t}^{\infty} e^{-t^2/2} \, dt$$

See Appendix D for the numerical methods (adaptive quadrature integration and Newton's method) used in computing Table 5.2(*a*, *b*).

EXAMPLE 5.3

Suppose that we have taken 20 measurements under condition *A* and 32 measurements under condition *B*. We analyze the two sets of measurements and find the following:

$$a_i, i = 1, 20 \quad \bar{a} = 573 \quad \sigma_a = 12 \quad \sigma_{\bar{a}} = 2.7$$
$$b_i, i = 1, 32 \quad \bar{b} = 578 \quad \sigma_b = 14 \quad \sigma_{\bar{b}} = 2.5$$
$$t = -1.36$$

How likely is it that this *t* value could have occurred by chance? Is it significant?

The number of degrees of freedom is $20 + 32 - 2 = 50$. From Table 5.2 we find that $P(>|t|) \approx 0.15$, which means that *t* values outside the range from -1.36 to $+1.36$ occur 15% of the time. Therefore we cannot rule out the null hypothesis and must accept the possibility that the difference in the means occurred by chance.

EXAMPLE 5.4

We notice that for a given difference in the means, the *t* value increases as the square root of the number of measurements. We therefore expand the experiment of Example 5.3 by taking 200 measurements under condition *A* and 320 measurements under condition *B*. We average the two sets of measurements, find that they are similar to the previous result, but that the denominator is $\sqrt{10}$ smaller, and that $t = -3.5$. How likely is it that this new *t* value could have occurred by chance? Is it significant?

$$a_i, i = 1, 200 \quad \bar{a} = 573 \quad \sigma_a = 12 \quad \sigma_{\bar{a}} = 0.85$$
$$b_i, i = 1, 320 \quad \bar{b} = 577 \quad \sigma_b = 14 \quad \sigma_{\bar{b}} = 0.78$$
$$t = -3.48$$

The number of degrees of freedom is $200 + 320 - 2 = 518$. From Table 5.2 we find that $P(>|t|) = 0.000,5$, which means that *t* values outside the range from -3.5 to $+3.5$ occur 0.05% of the time. Therefore the null hypothesis is extremely unlikely and we may reject the possibility that the difference in the means occurred by chance.

Historical note

The *t* distribution was derived by W. S. Gosset and published in 1908 under the pseudonym Student. His employer (The Guinness Brewing Company) would not allow him to publish under his own name.

Student's *t* and causality

If the difference between two experimental conditions does not cause a difference in the measured values, then repeated experiments will result in the standard distribution of Student's *t* values around zero, no matter how much m_a and m_b are increased.

If the difference between two experimental conditions truly causes a difference in the measured values, then repeated experiments with larger and larger values of m_a and m_b will result in distributions of Student's *t* values that deviate from zero by larger and larger amounts.

Design tip

When designing an experiment that involves random factors, consider the following for reducing random and systematic errors: (1) reduce the standard deviation of the measurements by reducing random influences as much as possible, (2) increase the sample size as much as possible to reduce the standard error of the mean, and (3) reduce the systematic error by controlling all factors that could influence the measurements but are not being studied as part of the experiment.

5.4   Least-squares fitting

Least-squares fitting is a method for finding the coefficients of an analytical curve so that the sum of the squares of the differences between the function and a data set is minimized.

5.4.1   Fitting a straight line to measured data

Let us say we have measured a quantity $f(x)$ at different values x_i, where i varies from 1 to m, and the measured values are f_i. The straight line we wish to fit to the data has the form $f(x) = a_0 + a_1 x$ (Figure 5.9).

The data provide a set of simultaneous equations:

$$f_1 = a_0 + a_1 x_1$$
$$f_2 = a_0 + a_1 x_2$$
$$\vdots$$
$$f_m = a_0 + a_1 x_m$$

Since there are many more equations than the two unknowns a_0 and a_1, it is in general impossible for all the equations to be satisfied simultaneously. Instead, we define the difference between f_i and the quantity $a_0 + a_1 x_i$ as the **residual** R_i:

$$R_i = f_i - a_0 - a_1 x_i$$

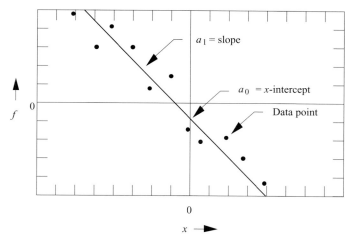

Figure 5.9 Straight-line fit to measured data (circles).

Then our task is to find the values of a_0 and a_1 that minimize the sum of squares of the residuals:

$$\sum_{i=1}^{m} R_i^2 = \text{minimum}$$

To do this, we set the first partial derivatives equal to zero:

$$\frac{\partial}{\partial a_0} \sum_{i=1}^{m} R_i^2 = 2 \sum_{i=1}^{m} R_i \frac{\partial R_i}{\partial a_0} = 0 \quad \text{and} \quad \frac{\partial}{\partial a_1} \sum_{i=1}^{m} R_i^2 = 2 \sum_{i=1}^{m} R_i \frac{\partial R_i}{\partial a_1} = 0$$

which is equivalent to:

$$\sum_{i=1}^{m} (f_i - a_0 - a_1 x_i)(-1) = 0 \quad \text{and} \quad \sum_{i=1}^{m} (f_i - a_0 - a_1 x_i)(-x_i) = 0$$

Rearranging, we have two equations in two unknowns:

$$\sum f_i = m a_0 + a_1 \sum x_i \quad \text{and} \quad \sum f_i x_i = a_0 \sum x_i + a_1 \sum x_i^2$$

which has the solutions:

$$a_0 = \frac{\sum f_i \sum x_i^2 - \sum f_i x_i \sum x_i}{m \sum x_i^2 - \left(\sum x_i\right)^2} \qquad a_1 = \frac{m \sum f_i x_i - \sum f_i \sum x_i}{m \sum x_i^2 - \left(\sum x_i\right)^2}$$

The rms deviation σ from the best-fit straight line (a measure of linearity) is given by:

$$\sigma = \sqrt{\frac{1}{m-2} \sum R_i^2}$$

5.4.2   Fitting a curve to measured data

We now generalize this treatment to handle any function with linear coefficients:

$$f(x) = \sum_{k=1}^{n} a_k g_k(x)$$

Again, let the data set be x_i, f_i, where i varies from 1 to m. The residuals are given by:

$$R_i = f_i - f(x_i) = f_i - \sum_{k=1}^{n} a_k g_k(x_i)$$

Minimizing the sum of the squares of the residuals produces the set of equations:

$$\frac{\partial}{\partial a_k} \left(\sum_{i=1}^{m} R_i{}^2 \right) = 2 \sum_{i=1}^{m} R_i \frac{\partial R_i}{\partial a_k} = 2 \sum_{i=1}^{m} R_i g_k(x_i) = 0$$

which is equivalent to:

$$\sum_{i=1}^{m} \left[f_i - \sum_{l=1}^{n} a_l g_l(x_i) \right] g_k(x_i) = 0$$

By rearranging the order of the summation, we have:

$$\sum_{l=1}^{n} a_l \sum_{i=1}^{m} g_l(x_i) g_k(x_i) = \sum_{i=1}^{m} f_i g_k(x_i)$$

Writing this equation in matrix form:

$$\begin{bmatrix} g_1(x_1) & g_1(x_2) & \cdots & g_1(x_m) \\ g_2(x_1) & g_2(x_2) & \cdots & g_2(x_m) \\ \vdots & \vdots & \ddots & \vdots \\ g_n(x_1) & g_n(x_2) & \cdots & g_n(x_m) \end{bmatrix} \begin{bmatrix} g_1(x_1) & g_1(x_1) & \cdots & g_n(x_1) \\ g_2(x_2) & g_2(x_2) & \cdots & g_n(x_2) \\ \vdots & \vdots & \ddots & \vdots \\ g_1(x_m) & g_2(x_m) & \cdots & g_n(x_m) \end{bmatrix} \begin{bmatrix} a_1 \\ a_2 \\ \vdots \\ a_n \end{bmatrix}$$

$$= \begin{bmatrix} g_1(x_1) & g_1(x_2) & \cdots & g_1(x_m) \\ g_2(x_1) & g_2(x_2) & \cdots & g_2(x_m) \\ \vdots & \vdots & \ddots & \vdots \\ g_n(x_1) & g_n(x_2) & \cdots & g_n(x_m) \end{bmatrix} \begin{bmatrix} f_1 \\ f_2 \\ \vdots \\ f_m \end{bmatrix}$$

which can be written in the more compact notation:

$$(\mathbf{g}^{\mathrm{T}}\mathbf{g})\mathbf{a} = \mathbf{g}^{\mathrm{T}}\mathbf{f}$$

where $\mathbf{g}$ is an $n \times m$ matrix, $\mathbf{g}^{\mathrm{T}}$ is its transpose, $\mathbf{a}$ is a column vector of dimension n, and $\mathbf{f}$ is a column vector of dimension m. These operations sum over the m data points and result in a set of n linear equations in n unknowns, which is usually solved by matrix inversion. In the case of least-squares fitting of a straight line (where $n = 2$), the resulting equations are readily solved by hand. For larger values of n, a matrix inversion program is usually necessary.

EXAMPLE 5.5

Use the matrix methods of this section to fit $f(x) = a_1 + a_2 x$ to the measured data f_i below:

i	1	2	3	4	5	6
x_i	0	10	20	30	40	50
f_i	87	56	35	16	-2	-13

The basis functions are $g_1(x) = 1$ and $g_2(x) = x$. The matrix equation:

$$(\mathbf{g^T\, g})\, \mathbf{a} = \mathbf{g^T\, f}$$

becomes:

$$
\begin{bmatrix} 1 & 1 & 1 & 1 & 1 & 1 \\ 0 & 10 & 20 & 30 & 40 & 50 \end{bmatrix}
\begin{bmatrix} 1 & 0 \\ 1 & 10 \\ 1 & 20 \\ 1 & 30 \\ 1 & 40 \\ 1 & 50 \end{bmatrix}
\begin{bmatrix} a_1 \\ a_2 \end{bmatrix}
=
\begin{bmatrix} 1 & 1 & 1 & 1 & 1 & 1 \\ 0 & 10 & 20 & 30 & 40 & 50 \end{bmatrix}
\begin{bmatrix} 87 \\ 56 \\ 35 \\ 16 \\ \pm 2 \\ \pm 13 \end{bmatrix}
$$

$$
\begin{bmatrix} 6 & 150 \\ 150 & 5{,}500 \end{bmatrix}
\begin{bmatrix} a_1 \\ a_2 \end{bmatrix}
=
\begin{bmatrix} 179 \\ 1{,}010 \end{bmatrix}
$$

which yields the simultaneous linear equations:

$$6a_1 + 150a_2 = 179$$
$$150a_1 + 5{,}500a_2 = 1{,}010$$

whose solutions are $a_1 = -1.98$ and $a_2 = 79.33$. So the least-squares best fit to the data is $f(x) = -1.98 + 79.33x$.

5.5 The chi-squared statistic

5.5.1 Use of chi-squared in fitting a model to data

The Student's t statistic discussed before is most frequently used to compare two measured distributions and determine the probability that the difference between them could have arisen by chance. If this probability is very small, then some factor (presumably part of the experiment) has caused the difference.

Chi-squared (χ^2) is the sum of the squares of the differences between the data and the model to be fitted to the data, weighted by the inverse of the uncertainty. It is used to compare a measured distribution with a function and determine the probability that the difference between them could have arisen by chance. If this probability is very small, then we may conclude that the function does not describe the system being measured. However, the task of science is generally to determine the equations that describe the measurable universe, and we would seek another function that does describe the measured distribution. Quite often, this is done with a family of functions $f(a_1, \ldots, a_n, x)$, where the parameters $a_1, \ldots, a_n$ are varied to minimize the χ^2 value:

$$\chi^2 = \sum_{i=1}^{m} \frac{[f_i - f(a_1, \ldots, a_n, x_i)]^2}{\sigma_i{}^2} \tag{5.5}$$

Equation (5.5) describes the difference between the function $f(a_1, \ldots, a_n, x)$ and measurements f_i made at coordinates x_i in terms of the experimental uncertainties σ_i. The parameter a describes the members of the family of functions. The assumption is that the data f_i are normally distributed with an rms uncertainty σ_i.

If the data f_i are not measurements of a continuous variable, but are the number of counts N_i recorded within intervals of x, then we have:

$$\chi^2 = \sum_{i=1}^{m} \frac{[N_i - f(a_1, \ldots, a_n, x_i)]^2}{N_i} \tag{5.6}$$

The N_i can be regarded as being normally distributed with an rms uncertainty of $\sqrt{N_i}$, provided that the $N_i \geq 30$. Equation (5.6) suffers from the problem that downward fluctuations in the N_i cause the difference $N_i - f(x_i)$ to be weighted more than equal upward fluctuations. For this reason, Equation (5.6) is considered "biased" in favor of the smaller values of N_i but is widely used because of the relative ease in taking derivatives with respect to the parameters a_i for least-squares minimization. The unbiased equation is:

$$\chi^2 = \sum_{i=1}^{m} \frac{[N_i - f(a_1, \ldots, a_n, x_i)]^2}{f(a_1, \ldots, a_n, x_i)} \tag{5.6a}$$

The number of degrees of freedom n_f is equal to the number of independent data points minus the number of parameters that are varied to minimize χ^2.

For smaller statistics, it is better to use maximum-likelihood techniques that consider the Poisson nature of the data. This leads to a more general form of χ^2 that is applicable to any measured count N_i, even zero:

$$\chi^2 = 2 \sum_{i=1}^{m} f(a_1, \ldots, a_n, x_i) - P_i$$

$$P_i = 0 \quad \text{for } N_i = 0 \tag{5.6b}$$

$$P_i = N_i \left\{ 1 + \ln[f(a_1, \ldots, a_n, x_i)/N_i] \right\} \quad \text{for } N_i > 0$$

Table 5.3(a) $P(>\chi^2)/n_f$ *probability of exceeding χ^2 per degree of freedom*

$P(>\chi^2)$	$n_f = 1$	2	3	5	7	10	15	20
1.000,00	0.000,0	0.000,0	0.000,0	0.000,0	0.000,0	0.000,0	0.000,0	0.000,0
0.900,00	0.015,8	0.105,4	0.194,8	0.322,1	0.404,7	0.486,5	0.569,8	0.622,1
0.800,00	0.064,2	0.223,2	0.335,1	0.468,5	0.546,0	0.617,9	0.687,1	0.728,9
0.700,00	0.148,5	0.356,7	0.474,6	0.600,0	0.667,3	0.726,7	0.781,4	0.813,3
0.600,00	0.275,0	0.510,9	0.623,1	0.731,1	0.784,7	0.829,6	0.868,6	0.890,4
0.500,00	0.454,9	0.693,2	0.788,7	0.870,3	0.906,5	0.934,2	0.955,9	0.966,9
0.400,00	0.708,3	0.916,3	0.982,1	1.026,4	1.040,5	1.047,3	1.048,9	1.047,6
0.300,00	1.074,2	1.204,0	1.221,6	1.212,9	1.197,6	1.178,1	1.154,8	1.138,7
0.200,00	1.642,4	1.609,5	1.547,2	1.457,9	1.400,5	1.344,2	1.287,4	1.251,9
0.100,00	2.705,5	2.302,6	2.083,8	1.847,3	1.716,7	1.598,7	1.487,1	1.420,6
0.050,00	3.841,5	2.995,8	2.604,9	2.214,1	2.009,6	1.830,7	1.666,4	1.570,5
0.020,00	5.411,9	3.912,0	3.279,1	2.677,6	2.374,6	2.116,1	1.884,0	1.751,0
0.010,00	6.634,9	4.605,2	3.781,6	3.017,3	2.639,3	2.320,9	2.038,5	1.878,3
0.005,00	7.879,4	5.298,3	4.279,4	3.349,9	2.896,8	2.518,8	2.186,8	1.999,8
0.002,00	9.549,5	6.214,6	4.931,8	3.781,5	3.228,7	2.772,2	2.375,2	2.153,6
0.001,00	10.827,6	6.907,8	5.422,1	4.103,0	3.474,6	2.958,8	2.513,2	2.265,7
0.000,50	12.115,7	7.600,9	5.910,0	4.421,1	3.716,8	3.142,0	2.647,9	2.374,9
0.000,20	13.831,1	8.517,2	6.552,0	4.837,1	4.032,4	3.379,6	2.821,9	2.515,5
0.000,10	15.136,7	9.210,4	7.035,8	5.149,0	4.268,2	3.556,4	2.950,9	2.619,3
0.000,05	16.448,1	9.903,5	7.518,2	5.458,7	4.501,8	3.731,1	3.077,9	2.721,3
0.000,02	18.189,3	10.819,8	8.154,1	5.865,4	4.807,6	3.959,1	3.243,0	2.853,7
0.000,01	19.511,4	11.513,0	8.633,9	6.171,2	5.036,9	4.129,6	3.366,2	2.952,2

For large values of N_i and f, Eqs. (5.6a) and (5.6b) are equivalent.

A general note: The least-squares and chi-squared statistics described before have the general form:

$$Q(a_1, \ldots, a_n) = \sum_{i=1}^{m} W_i R_i^2 \qquad R_i = f_i - f(a_1, \ldots, a_n, x_i)$$

where R_i are the residuals, and W_i is a weighting factor.

For simple least squares, $W_i = 1$.

For chi-squared, $W_i = \sigma_i^{-2}$.

For relative least squares, $W_i = f(a_1, \ldots, a_n, x)^{-2}$.

After computing χ^2, Tables 5.3(a) and (b) can be used to determine the probability that that value or any larger value could have arisen by chance. On the average, we expect χ^2 to be approximately equal to the number of degrees of freedom n_f, and $\chi^2 \leq n_f$ is regarded as a "good fit" of the model to the data.

Table 5.3(b) $P(>\chi^2)/n_f$ probability of exceeding χ^2 per degree of freedom

$P(>\chi^2)$	$n_f = 30$	50	70	100	150	200	300	500
1.000,00	0.000,0	0.000,0	0.000,0	0.000,0	0.000,0	0.000,0	0.000,0	0.000,0
0.900,00	0.686,6	0.753,8	0.790,4	0.823,6	0.855,2	0.874,2	0.896,9	0.919,9
0.800,00	0.778,8	0.829,0	0.855,7	0.879,5	0.901,8	0.915,0	0.930,7	0.946,4
0.700,00	0.850,3	0.886,3	0.904,9	0.921,3	0.936,4	0.945,2	0.955,6	0.965,9
0.600,00	0.914,7	0.937,3	0.948,5	0.958,1	0.966,7	0.971,6	0.977,3	0.982,7
0.500,00	0.977,9	0.986,7	0.990,5	0.993,3	0.995,6	0.996,7	0.997,8	0.998,7
0.400,00	1.043,9	1.037,8	1.033,7	1.029,5	1.025,0	1.022,2	1.018,6	1.014,8
0.300,00	1.117,7	1.094,5	1.081,3	1.069,1	1.057,2	1.049,9	1.041,2	1.032,2
0.200,00	1.208,3	1.163,3	1.138,8	1.116,7	1.095,7	1.083,0	1.068,0	1.052,8
0.100,00	1.341,9	1.263,3	1.221,8	1.185,0	1.150,5	1.130,1	1.106,0	1.081,9
0.050,00	1.459,1	1.350,1	1.293,3	1.243,4	1.197,2	1.170,0	1.138,0	1.106,3
0.020,00	1.598,7	1.452,3	1.377,0	1.311,4	1.251,2	1.215,9	1.174,7	1.134,1
0.010,00	1.696,4	1.523,1	1.434,6	1.358,1	1.288,1	1.247,2	1.199,7	1.153,0
0.005,00	1.789,1	1.589,8	1.488,8	1.401,7	1.322,4	1.276,3	1.222,8	1.170,4
0.002,00	1.905,6	1.673,1	1.556,1	1.455,8	1.364,8	1.312,2	1.251,2	1.191,8
0.001,00	1.990,1	1.733,2	1.604,5	1.494,5	1.395,1	1.337,7	1.271,4	1.206,9
0.000,50	2.072,1	1.791,2	1.651,1	1.531,7	1.424,1	1.362,1	1.290,7	1.221,3
0.000,20	2.177,1	1.865,2	1.710,3	1.578,8	1.460,8	1.392,9	1.314,9	1.239,4
0.000,10	2.254,4	1.919,4	1.753,6	1.613,2	1.487,4	1.415,3	1.332,5	1.252,5
0.000,05	2.330,2	1.972,3	1.795,8	1.646,6	1.513,3	1.437,0	1.349,5	1.265,1
0.000,02	2.428,1	2.040,4	1.850,0	1.689,4	1.546,4	1.464,7	1.371,2	1.281,2
0.000,01	2.500,8	2.090,8	1.890,0	1.721,0	1.570,7	1.485,0	1.387,1	1.293,0

EXAMPLE 5.6

Suppose we have a model function $f(x)$ that is supposed to describe a measurable function and a series of 100 measurements f_i and uncertainties σ_i, each made at a corresponding x_i. Applying Eq. (5.5), we compute $\chi^2 = 150$. What is the probability that the random fluctuations in the measurements f_i could have resulted in a $\chi^2 \geq 150$? Using Table 5.3, we look down the column for 100 degrees of freedom at find that the probability of exceeding $\chi^2 = 149.4$ is 0.001. We may conclude that the model function does not adequately describe the measured data.

EXAMPLE 5.7

Suppose we have $m = 31$ measurements a_i, and using the equations in (5.2), determine that the sample standard deviation $\sigma = 1.38$. Assuming that this σ is close to the true standard deviation σ_0, the quantity $(m - 1)\sigma^2/(\sigma_0)^2 = 30\sigma^2/(1.38)^2 = 15.75\sigma^2$ is distributed as χ^2 with 30 degrees of freedom. Using Table 5.3(b), we find that for 30 degrees of freedom, there is a 10% probability for $\chi^2 < 20.599$ and a 10% probability for $\chi^2 > 40.256$. So the 80% confidence interval for σ is $\sqrt{20.599/15.75} = 1.14$ to $\sqrt{40.256/15.75} = 1.60$.

5.5.2   Computing the probability of exceeding χ^2

The probability of exceeding χ^2 by chance is given by:

$$P(>\chi^2) = \int_{\chi^2}^{\infty} \frac{x^{(n-2)/2} e^{-x/2}}{2^{n/2}\, \Gamma(n/2)}\, dx \tag{5.7}$$

Tables 5.3(a) and (b) were computed by numerical integration of Eq. (5.7), using adaptive quadrature and inverted using Newton's method (Appendix D). Tables 5.3(a) and (b) list the values of χ^2 corresponding to the probability factors in the first column. As a general rule, $\chi^2 \leq n_f$ indicates an acceptable fit.

For $n_f > 30$, $P(>\chi^2)$ can be approximated as the Gaussian integral:

$$P(>\chi^2) = \sqrt{\frac{2}{\pi}} \int_{y}^{\infty} e^{-x^2/2}\, dx, \quad \text{where } y = \sqrt{2\chi^2} - \sqrt{2n-1}$$

An example of a poor fit would be $P < 0.1\%$ ($y > 3.09$ standard deviations). This corresponds to:

$$\sqrt{2\chi^2} - \sqrt{2n-1} > 3.09 \quad \text{and} \quad \chi^2 > 0.5\left(3.09 + \sqrt{2n-1}\right)^2$$

5.6   Solving nonlinear equations

5.6.1   Newton's method for solving f(x) = 0

This method is very effective in solving nonlinear equations of one variable. In his development of the calculus, Newton devised a method for solving $f(x) = 0$ using the first derivative $df/dx = f'(x)$ and a little geometry (Figure 5.10).

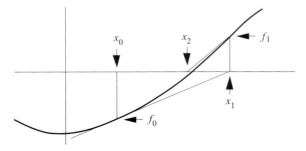

Figure 5.10 Newton's method for finding the roots of an equation.

Table 5.4 *Use of Newton's method to solve* $\sqrt[3]{10}$

i	x_i	$x_i{}^3$
0	1.00000 00000 00000 00	1.00000 00000 00000 00
1	4.00000 00000 00000 00	64.00000 00000 00000 00
2	2.87500 00000 00000 00	23.76367 18750 00000 00
3	2.31994 32892 24952 74	12.48625 23021 57114 08
4	2.16596 15551 77792 79	10.16136 92074 95129 95
5	2.15449 59251 53374 74	10.00085 27090 04352 98
6	2.15443 46917 72292 94	10.00000 00242 34792 06
7	2.15443 46900 31883 72	10.00000 00000 00000 02

If the ith approximation is x_i, the next approximation x_{i+1} is given by:

$$x_{i+1} = x_i - \frac{f(x_i)}{f'(x_i)}$$

EXAMPLE 5.8

To extract the cube root of a number a, we wish to solve the equation: $f(x) = x^3 - a = 0$. Since $f'(x) = 3x^2$, we have the iterative algorithm:

$$x_{i+1} = x_i - \frac{x_i{}^3 - a}{3x_i{}^2} = \frac{2x_i{}^3 + a}{3x_i{}^2}$$

Table 5.4 shows how this algorithm converges for $a = 10$, using the very poor initial guess $x_0 = 1$. As expected, the linear approximation is very poor far from the correct answer, and, in the first iteration, the algorithm actually jumps over the answer from 1 to 4. However, Newton's linear approximation improves rapidly as the correct answer is approached, and the last three iterations demonstrate impressively rapid convergence to 18 decimal-digit accuracy.

5.6.2   Solving *f(x) = 0* by quadratic iteration

Newton's method requires the analytical computation of df/dx. In situations where this is not convenient, quadratic iteration can be used, which only requires the computation of $f(x)$. The following steps describe the method.
1. Evaluate $f_0 = f(x_0)$, $f_- = f(x_0 - \Delta x)$, and $f_+ = f(x_0 + \Delta x)$.
2. Compute the point $x = c$, where the quadratic curve defined by $x_0 - \Delta x$, x_0, $x_0 + \Delta x$, f_-, f_0, and f_+ passes through 0.

$$a = x_0 - \Delta x f_0 / (f_0 - f_-)$$
$$b = x_0 + \Delta x f_0 / (f_0 - f_+)$$
$$c = a + f_-(b - a) / (f_- - f_+)$$

3. Assign $\Delta x = c - x_0$ and $x_0 = c$.
4. Repeat steps 2 and 3 until Δx becomes sufficiently small.

5.6.3   Numerical minimization

In fitting the parameters of an analytical model to measured data, a figure of merit Q is often defined such as in Section 5.5 on the chi-squared statistic. This function is then minimized using a computer. In most cases, Q is a nonlinear function of the parameters a_i, and analytical, closed form solutions, such as the one presented for least-squares minimization, are not possible. Usually, more brute force numerical search methods are employed.

For nonlinear equations of more than one variable, it is often necessary to use numerical minimization of a suitably defined figure of merit Q. Often such functions have contours of equal value that are rotated with respect to the independent variable axes a_i, and repeated minimization along these axes requires many iterations to reach the minimum (Figure 5.11).

There are more efficient numerical methods for multiparameter, nonquadratic function minimization, such as steepest descent, simplex, conjugate gradient, and variable metric. As described in Press et al. (1988), none of these has a clear superiority.

The steps below describe a simplified, robust version of the conjugate gradient method, which is provided in Appendix D as the function parfit.c.

1. Initialize a set of n search vectors u_i, $i = 1$ to n, to the unit vector pointing in the ith direction times the initial step size for the ith parameter. This set of search vectors will change in length but not in direction during the minimization procedure.
2. Initialize a set of n search vectors u_{i+n}, $i = 1$ to n, to the unit vector pointing in the ith direction times the initial step size for the ith parameter. This set of search vectors will change both in length and direction during the minimization procedure.
3. Use a few steps of quadratic minimization to reduce Q by moving the parameters

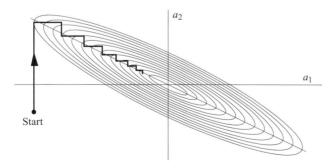

Figure 5.11 Numerical minimization of a function of two parameters by varying one parameter at a time. Unless the valley is oriented along the independent parameter axes a_i, the method requires many iterations.

a_i, $i = 1$ to n, along the search vector u_1. Readjust the length of this search vector, depending on the length of the motion vector. If the lowest value of Q is at the edge of the move, double the length of u_1. Otherwise, reduce its length by a factor of 2.

4. Repeat step 3 for all the other search vectors u_i, $i = 2, \ldots, 2n$.
5. Use a few steps of quadratic minimization to reduce Q by varying the parameters a_i, $i = 1$ to n, along the direction of advance produced in steps 3 and 4.
6. Set $u_i = u_{i+1}$, $i = n + 1$ to $2n - 1$.
7. Set u_{2n} = direction of advance produced in steps 3 and 4.
8. Repeat steps 3–7 until the change in each of the parameter values is less than some small value.

Steps 3, 4, and 5 require quadratic minimization along a line, and this is performed by the function qfit.c, which is listed in Appendix D and used by parfit.c.

Powell's theorem is that if the procedure is done n times over only the u_{n+1} to u_{2n} search vectors, they will become mutually conjugate so only n^2 line minimizations will minimize any n-parameter quadratic form. However, this procedure is unstable as the search vectors can collapse into a subspace, which prevents finding the minimum. Searching periodically by varying each parameter only (the first n line minimizations) prevents this collapse and makes the procedure more robust.

If the figure of merit Q is a chi-squared that uses the statistical uncertainty in each measurement, then the statistical uncertainty in the parameters can be estimated from the contour lines connecting equal values of χ^2 that encircle the best fit $\chi_{\min}^2$ (Figure 5.12). See Press et al. (1988) for additional information.

Appendix D provides listings of the function varfit.c, which determines the uncertainties in the fitted parameters. The steps are:

1. Find $\chi_{\min}^2$ by varying all parameters.
2. Set the parameter a_1 to its best-fit value $+s_1$, where s_1 is some step size.
3. Holding a_1 constant, vary all other parameters to minimize χ^2.
4. Determine the parabola that passes through the new value of a_1, and also passes through the best-fit value of a_1 with zero slope.

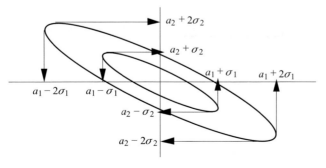

Figure 5.12 Determination of the statistical uncertainties in a two-parameter χ^2 fit as the projections of the $\chi_{\min}^2 + 1$ and best-fit $\chi_{\min}^2 + 4$ contours.

5. Determine the value of a_1 at which the parabola is equal to $\chi_{min}^2 + 1$.
6. Repeat step 3 for this new value of a_1.
7. Determine the parabola that passes through the three points closest to $\chi_{min}^2 + 1$.
8. Repeat steps 5, 6, and 7 until the value of a_1 changes by a sufficiently small amount. This value is a one-σ confidence limit.
9. Set the parameter a_1 to its best-fit value $-s_1$.
10. Repeat steps 3–8 to find the other one-σ confidence limit.
11. Repeat steps 2–10 for all the other parameters.

5.7   Monte Carlo simulation

Monte Carlo simulation is a very useful technique for estimating the probabilities of various outcomes of an experiment, given an analytical model. If the assumed model is correct, and noiseless data were available, then the data should agree perfectly with the model. In the real world, the data have a random component (or are subject to random influences) and agreement is never perfect. It is then important to ask what the chances are that: (i) the lack of agreement is caused by the randomness of the data, and (ii) the lack of agreement is caused by the inability of the assumed model to describe the actual experiment.

Figure 5.13 shows how a model with a true set of parameter values can produces a number of experimental outcomes. The actual experiment is but one of these.

Figure 5.14 shows how a model with an assumed set of parameters can generate a number of simulated experiments. The assumed model can be fit to each simulated data set and the distribution of fitted parameters resulting purely from random factors can then be studied.

For situations where each data value d_i is equal to a true value c_i plus a random component with Gaussian distribution having a known σ_i, we can use the development

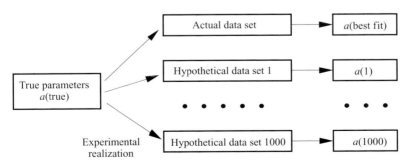

Figure 5.13 An underlying model with a set of parameter values a (true) can give rise to a number of experimental outcomes. The actual experiment is but one of these.

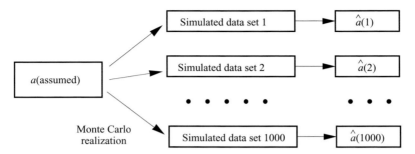

Figure 5.14 Monte Carlo simulation of experiments arising from a model with an assumed set of parameter values. The distribution of fitted parameters simulates the variations that can be expected from statistical uncertainties.

at the beginning of this chapter to synthesize data:

$$d_i = c_i + \sigma_i \sqrt{12/n} \sum_{j=ni+1}^{ni+n} (R_j - 0.5)$$

where n is chosen to generate a Gaussian distribution of sufficient accuracy and the R_j is a sequence of random numbers uniformly distributed between 0 and 1.

For situations where the experimental data must be simulated with an arbitrary random distribution $f(x)$, the following steps should be used:

1. Tabulate the cumulative integral for a series of y values.

$$g(y) = \frac{\int_{-\infty}^{y} f(x)\, dx}{\int_{-\infty}^{\infty} f(x)\, dx}$$

2. Generate a random number g_i from a distribution uniformly populated between 0 and 1.
3. Interpolate the table of step 1 to find the value y_i such that $g_i = g(y_i)$.
4. Repeat steps 2 and 3 to generate a number of simulated data values y_i. The resulting distribution will approximate $f(x)$ and simulate the randomness of a experimental data set of the same size.

Figure 5.15 shows the method where $f(x)$ is a triangular distribution. The cumulative integral $g(y)$ is zero for $y < -1$, rises quadratically to 0.5 as y increases from -1 to 0, has the same shape (but flipped) for y between 0 and $+1$, and remains at 1 for $y > 1$. The uniformly spaced horizontal arrows depict the random variables g_i and the upward arrows show the intersection with the $g(y)$ curve. The arrows have the highest density at $y = 0$, where $g(y)$ is steepest, and do not appear where $g(y)$ has zero slope.

Figure 5.16 shows the method where $f(x)$ consists of two separated triangular distributions. The cumulative integral $g(y)$ has only two regions of nonzero slope, and these each generate a triangular distribution.

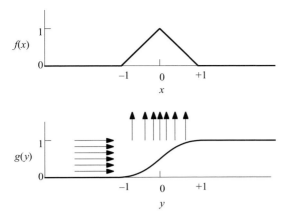

Figure 5.15 Simulation of a random variable with triangular distribution $f(x)$. The cumulative integral provides a mapping between a uniformly distributed random variable (horizontal arrows) and the desired distribution (vertical arrows).

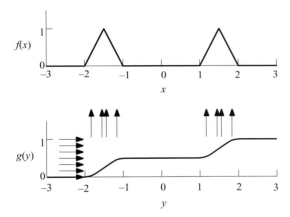

Figure 5.16 Simulation of a random variable with a double triangular distribution $f(x)$. The cumulative integral provides a mapping between a uniformly distributed random variable (horizontal arrows) and the desired distribution (vertical arrows).

5.8 Fourier transforms

Fourier transforms are used in a wide variety of signal processing applications, including:
- determining the frequency and phase content of any waveform,
- determining the Fourier series expansion of a periodic waveform,
- performing convolutions and deconvolutions by simple multiplication and division (which has many applications in signal and image processing),
- improving the signal-to-noise ratio (to the extent that the signal and noise have different frequency spectra),
- detecting a signal of unknown frequency in a large additive random noise background (SETI = Search for Extraterrestrial Intelligence).

It is interesting to note that the ear and its sensory neurons act as a Fourier transformer. Sounds entering the ear activate many frequency-sensitive "hair cells" (tiny mechanical resonators connected to sensory neurons) that span the entire range of audible frequencies. So the brain does not receive the waveform (amplitude versus time) of the sounds we hear, but the Fourier transform (amplitude versus frequency) of the waveform.

5.8.1   The integral Fourier transform with examples

The integral Fourier transform is defined by:

$$H(f) = \int_{-\infty}^{\infty} h(t)e^{-j2\pi ft}\, dt$$

and the inverse transform is defined by:

$$h(t) = \int_{-\infty}^{\infty} H(f)e^{j2\pi ft}\, df$$

One of the fundamental properties of the Fourier transform is that it transforms a time function with amplitude $h(t)$ at time t into a frequency function with amplitude $H(f)$ at frequency f. The transformed frequency function $H(f)$ contains all the information in the original time function $h(t)$, as evidenced by the ability of the inverse Fourier transform to transform $H(f)$ back into $h(t)$.

The definition of the Fourier transform can be justified by considering two mathematical facts:

1. The time integral of the product of two harmonics is zero unless they have the same frequency. If the frequencies are different the integrand oscillates equally between both signs and sums to zero. If the frequencies are the same the integrand is the square of a harmonic which sums to a positive number.
2. Any waveform $h(t)$ can be thought of as the weighted sum of harmonics of all possible frequencies (the equation for the inverse Fourier transform above).

As a result, the time integral of the product of $h(t)$ and a harmonic of frequency f (the equation for $H(f)$ above) "projects out" only that part of $h(t)$ that has frequency f.

EXAMPLE 5.9

Fourier transform of the delta function $\delta(t)$. The delta function has the properties that $\delta(t) = 0$ for $t \neq 0$ and:

$$\int_{-\infty}^{\infty} \delta(t - t_0)g(t)\, dt = g(t_0)$$

Figure 5.17 Fourier transform of a delta function.

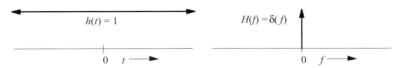

Figure 5.18 Fourier transform of a constant.

The Fourier transform is a constant for all frequencies (Figure 5.17):

$$H(f) = \int_{-\infty}^{\infty} \delta(t)e^{-j2\pi ft}\, dt = 1$$

EXAMPLE 5.10

Fourier transform of the constant $h(t) = 1$:

$$H(f) = \int_{-\infty}^{\infty} e^{-j2\pi ft}\, dt = \int_{-\infty}^{\infty} [\cos(2\pi ft) - j\sin(2\pi ft)]\, dt = \delta(f)$$

When $f \neq 0$, the integrand oscillates symmetrically about zero and has zero area. When $f = 0$, the integrand is a constant with infinite area (Figure 5.18).

Note: In this example, we have used the identities:

$$\int_{-\infty}^{\infty} \cos(2\pi ft)\, dt = \delta(f) \qquad \int_{-\infty}^{\infty} \sin(2\pi ft)\, dt = 0$$

EXAMPLE 5.11

Fourier transform of $h(t) = A\cos(2\pi f_0 t)$:

$$H(f) = A \int_{-\infty}^{\infty} \cos(2\pi f_0 t)[\cos(2\pi ft) - j\sin(2\pi ft)]\, dt$$

$$= \frac{A}{2} \int_{-\infty}^{\infty} \{\cos[2\pi(f_0 - f)t] + \cos[2\pi(f_0 + f)t]\}\, dt$$

$$= \frac{A}{2}[\delta(f_0 - f) + \delta(f_0 + f)]$$

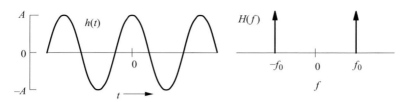

Figure 5.19 Fourier transform of $h(t) = A\cos(2\pi ft)$.

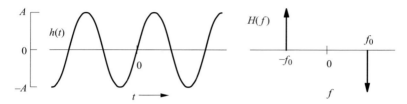

Figure 5.20 Fourier transform $H(f)$ of $h(t) = A\sin(2\pi ft)$.

The Fourier transform consists of delta functions at $\pm f_0$ (Figure 5.19).

Note: Example 5.10 ($h(t) = 1$) is a special case of Example 5.11 where $A = 1$ and $f_0 = 0$.

EXAMPLE 5.12

Fourier transform of $h(t) = A\sin(2\pi f_0 t)$:

$$H(f) = A \int_{-\infty}^{\infty} \sin(2\pi f_0 t)[\cos(2\pi ft) - j\sin(2\pi ft)]\,dt$$

$$= \frac{Aj}{2} \int_{-\infty}^{\infty} -\{\cos[2\pi(f_0 - f)t] - \cos[2\pi(f_0 + f)t]\}\,dt$$

$$= \frac{Aj}{2}[\delta(f_0 + f) - \delta(f_0 - f)]$$

The Fourier transform consists of positive delta function at $-f_0$ and a negative delta function at $+f_0$ (Figure 5.20).

Note: In Examples 5.11 and 5.12, we have used the identities:

$$e^{-j2\pi ft} = \cos(2\pi ft) - j\sin(2\pi ft)$$

$$\int_{-\infty}^{\infty} \sin(2\pi f_1 t)\cos(2\pi f_2 t)\,dt = 0, \quad \text{for all } f_1 \text{ and } f_2$$

and

$$\cos(a)\cos(b) = \frac{1}{2}[\cos(a-b) + \cos(a+b)]$$

$$\sin(a)\sin(b) = \frac{1}{2}[\cos(a-b) - \cos(a+b)]$$

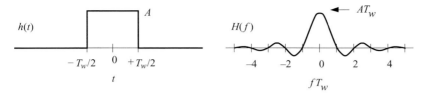

Figure 5.21 Fourier transform $H(f)$ of a rectangular pulse $h(t)$ of width T_w.

EXAMPLE 5.13

Fourier transform of a rectangular pulse of width T_w:

$$h(t) = A \quad \text{for} \quad |t| < T_w/2, \; h(t) = 0 \quad \text{for} \quad |t| > T_w/2$$

$$H(f) = A \int_{-T_w/2}^{T_w/2} [\cos(2\pi f t) - j \sin(2\pi f t)] \, dt$$

$$= \frac{A}{2\pi f} \sin(2\pi f t) \Big|_{-T_w/2}^{T_w/2} = A T_w \frac{\sin(\pi T_w f)}{\pi T_w f}$$

See Figure 5.21 for $h(t)$ and $H(f)$.

Note 1: In Example 5.9, $[h(t) = \delta(t)]$ is a special case of a rectangular function where $T_w = 0$, $A T_w = 1$, and $H(f) = \sin(\pi T_w f)/(\pi T_w f) = 1$.

Note 2: In Example 5.10, $[h(t) = 1]$ is a special case of a rectangular function where $T_w \to \infty$. In this limit, the height of the central lobe in Figure 5.21 is $A T_w \to \infty$, and the width (frequency of the first zero crossing) is $1/T_w \to 0$.

Note 3: A narrow pulse has a Fourier transform that is broad in frequency, and a wide pulse has a Fourier transform that is narrow in frequency.

Note 4: The Fourier transform of a rectangular pulse is zero at all frequencies that correspond to a whole number of cycles in time T_w.

EXAMPLE 5.14

How narrow in time must a pulse be for its frequency transform to be flat to within 3 dB from 0 Hz to a frequency f? We solve the equation:

$$\frac{\sin(\pi T_w f)}{\pi T_w f} = \frac{1}{\sqrt{2}}$$

and find a solution $T_w = 0.443/f$. For $f = 10\,\text{kHz}$, $T_w = 44.3\,\mu\text{s}$.

EXAMPLE 5.15

Fourier transform of a triangular pulse of width T_w.

$$h(t) = A(1 + 2t/T_w) \text{ for } -T_w/2 < t < 0, \; h(t) = A(1 - 2t/T_w) \text{ for } 0 < t < T_w/2:$$

$$h(t) = 0 \quad \text{for} \quad |t| > T_w/2$$

$$H(f) = 2A \int_{0}^{T_w/2} (1 - 2t/T_w) \cos(2\pi f t) \, dt = \frac{A T_w}{2} \frac{\sin^2(\pi f T_w/2)}{(\pi f T_w/2)^2}$$

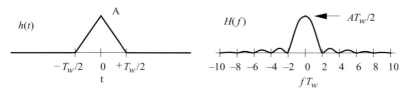

Figure 5.22 Fourier transform $H(f)$ of a triangular pulse $h(t)$ of base width T_w.

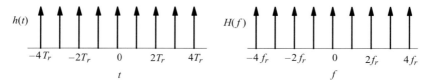

Figure 5.23 Fourier transform $H(f)$ of an infinite series of periodic delta functions $h(t)$.

See Figure 5.22 for $h(t)$ and $H(f)$.

EXAMPLE 5.16

Fourier transform of the periodic series of delta functions:

$$h(t) = \sum_{k=-\infty}^{\infty} \delta(t - kT_r)$$

$$H(f) = \int_{-\infty}^{\infty} e^{-j2\pi ft} \sum_{k=-\infty}^{\infty} \delta(t - kT_r)\, dt = \sum_{k=-\infty}^{\infty} e^{-j2\pi kf/f_r}, \quad \text{where } f_r = 1/T_r$$

When f is an integer multiple of f_r ($f = mf_r$), $H(f)$ is the infinite sum of the constant 1 and is infinite (Figure 5.23). When f is not an integer multiple of f_r, then the infinite sum is zero because its terms oscillate symmetrically about zero:

$$H(f) = f_r \sum_{k=-\infty}^{\infty} \delta(f - kf_r)$$

The Fourier transform of an infinite series of time delta functions with spacing T_r is an infinite series of frequency delta functions with spacing $f_r = 1/T_r$.

This result is important for the next section, where we use it and the Fourier convolution theorem to derive the transform of periodic time functions.

Note: Reducing I_r increases the density of delta functions in the time domain and decreases the density of delta functions in the frequency domain by the same factor. To satisfy Parseval's Theorem, the Fourier amplitude in Example 5.16 contains the factor f_r.

Parseval's Theorem

It is often useful to compare the overall normalization of a function in the time domain with its Fourier transform. Parseval's theorem relates these as follows:

$$\int_{-\infty}^{\infty} h^2(t)\, dt = \int_{-\infty}^{\infty} |H(f)|^2\, df$$

This relationship shows the equivalence of power in the time domain (voltage squared) to power in the frequency domain (Fourier amplitude squared).

5.8.2 The Fourier transform of periodic waveforms

The convolution $f(t)$ of two functions $g(t)$ and $h(t)$ is defined as:

$$f(t) = \int_{-\infty}^{\infty} g(t')h(t - t')\, dt' = g(t)^* h(t)$$

The function $h(t)$ is folded about the time axis, shifted by the time variable t, multiplied by $g(t')$, and integrated over all time.

The Fourier convolution theorem allows a convolution in the time domain to be transformed into a simple multiplication in the frequency domain. To summarize:

Fourier convolution theorem

The Fourier transform of the convolution of two functions is the product of their Fourier transforms.

We now apply this theorem to determine the Fourier transform of a series of square waves, a series of triangular waves, and the general periodic waveform.

Figure 5.24 shows a rectangular pulse of width $T_w = T_r/2$ convolved with a series of delta functions of repetition period T_r to produce a series of symmetric square waves. They are called symmetric because the waveform has the same shape when it is flipped about the time axis. To determine the Fourier transform of the convolution, we need only multiply the Fourier transform of a single rectangular pulse with the Fourier transform of the delta functions with time period T_r.

The result is a zero frequency component plus nonzero components only at odd multiples of the fundamental repetition frequency $f_r = 1/T_r$. $H(f)$ is given by:

$$H(f) = \frac{A}{2} \sum_{k=-\infty}^{\infty} \frac{\sin(\pi f/2 f_r)}{\pi f/2 f_r} \delta(f - k f_r)$$

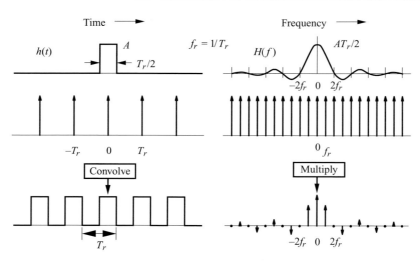

Figure 5.24 The Fourier transform of a symmetric square wave with period T_r. Left: Convolution of a single square wave of width $T_r/2$ with periodic delta functions $\sum \delta(t - mT_r)$. Right: Product of the Fourier transform of the single square wave with the periodic delta functions $\sum \delta(tf - mf_r)$, where $f_r = 1/T_r$.

which is nonzero only for $f = 0$ and odd multiples of f_r:

$$H(0) = A/2 \qquad H(f_r) = A/\pi$$
$$H(2f_r) = 0 \qquad H(3f_r) = -A/(3\pi)$$
$$H(4f_r) = 0 \qquad H(5f_r) = A/(5\pi)$$
$$\vdots \qquad\qquad \vdots$$
$$H(mf_r) = 0 \qquad H(mf_r) = A(-1)^{(m-1)/2}/(m\pi)$$
$$(m \text{ even}) \qquad (m \text{ odd})$$

Figure 5.25 shows a triangular pulse of base width T_r convolved with a series of delta functions of repetition period T_r to produce a series of symmetric triangle waves.

To determine the Fourier transform of the convolution, we need only multiply the Fourier transform of a single triangular pulse with the Fourier transform of the delta functions with time period T_r. The result is a zero frequency component plus nonzero components only at odd multiples of the fundamental repetition frequency $f_r = 1/T_r$. $H(f)$ is given by:

$$H(f) = \frac{A}{2} \sum_{k=-\infty}^{\infty} \frac{\sin^2(\pi f/2 f_r)}{(\pi f/2 f_r)^2} \delta(f - kf_r)$$

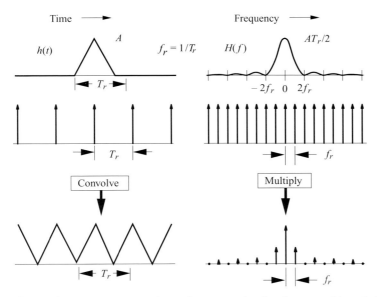

Figure 5.25 The Fourier transform of a symmetric triangle wave with period T_r. Left: Convolution of a single triangle wave of width T_r with periodic delta functions $\sum \delta(t - mT_r)$. Right: Product of the Fourier transform of the single triangle wave with the periodic delta functions $\sum \delta(tf - mf_r)$, where $f_r = 1/T_r$.

which is nonzero only for $f = 0$ and odd multiples of f_r:

$$H(0) = A/2 \qquad H(f_r) = 2A/\pi^2$$
$$H(2f_r) = 0 \qquad H(3f_r) = 2A/(3\pi)^2$$
$$H(4f_r) = 0 \qquad H(5f_r) = 2A/(5\pi)^2$$
$$\vdots \qquad\qquad \vdots$$
$$H(mf_r) = 0 \qquad H(mf_r) = 2A/(m\pi)^2$$
$$(m \text{ even}) \qquad (m \text{ odd})$$

As shown in Figure 5.26, if a time function is periodic with period P, its Fourier transform is the product of the Fourier transform of the function between 0 and P and a series of delta functions of frequency k/P. The result is nonzero only at whole-number multiples of $1/P$ and the Fourier coefficients are given by:

$$H_k = \frac{1}{P} \int\limits_0^P h(t) e^{-j2\pi kt/P} \, dt$$

The decomposition of a periodic waveform into its harmonic components is called **Fourier series analysis** and the time function $h(t)$ can be recovered by the **Fourier**

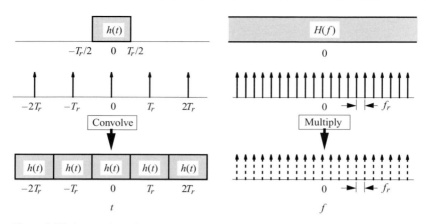

Figure 5.26 A waveform $h(t)$ that is periodic in time can be represented as the convolution of a waveform defined between 0 and T_r convolved with an infinite series of delta functions with spacing T_r. This is equivalent to multiplying the frequency content $H(f)$ with an infinite series of delta functions with spacing f_r. Thus any periodic waveform has a Fourier transform that is discrete in frequency.

series expansion:

$$h(t) = \sum_{k=-\infty}^{\infty} H_k e^{j2\pi kt/P}$$

The Fourier transform of a periodic function is nonzero only at whole-number multiples of the fundamental frequency $f_r = 1/\text{period}$. The component with frequency $k f_r$ is called the kth harmonic.

5.8.3   The Fourier transform of a periodically sampled time function

The Fourier frequency convolution theorem allows a multiplication in the time domain to be transformed into a convolution in the frequency domain.

Fourier frequency convolution theorem

The Fourier transform of the product of two functions is the convolution of their Fourier transforms.

We now apply this theorem to determine the Fourier transform of a continuous waveform that has been periodically sampled. Figure 5.27 shows the sampling of a sine wave and the multiplication of the sine wave with an infinite series of delta functions in the time domain. The corresponding Fourier transform is the convolution of the Fourier transform of the sine wave with an infinite series of delta functions in the frequency domain. Note that it is possible to recover the original sine wave by using a low-pass filter that preserves the frequency f_0 and rejects higher multiples.

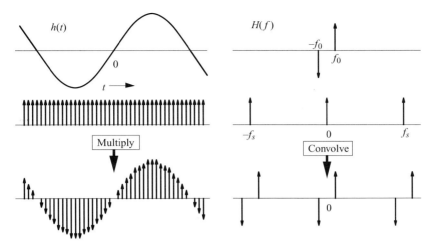

Figure 5.27 Sine wave of frequency f_0 sampled at frequency f_s. The result in the time domain is the product of the sine wave and an infinite series of sampling delta functions. The corresponding result in the frequency domain is the convolution of the Fourier transform of the sine wave and an infinite series of delta functions.

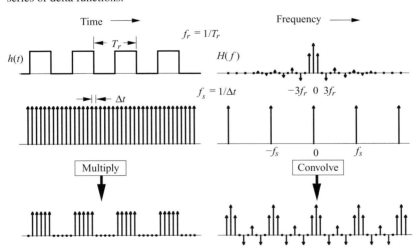

Figure 5.28 Sampling a waveform in the time domain can cause aliasing in the frequency domain. Left: Symmetric square wave with period T_r multiplied by the sampling function $\sum \delta(t - mT)$. Right: Convolution of the Fourier transform of the periodic square wave with the delta function series $\sum \delta(f - mf_s)$, where $f_s = 1/T$. Since the Fourier transform of the periodic square wave has frequency content outside of the range from $-f_s/2$ to $+f_s/2$, the convolution results in overlap of the Fourier amplitudes and aliasing.

Figure 5.28 shows the sampling of a series of rectangular waves. Because the Fourier transform of the series of square waves (derived previously using the Fourier convolution theorem) is not limited in frequency, the convolution in the frequency domain will cause a serious overlap. In particular, the lower frequencies will be contaminated by the

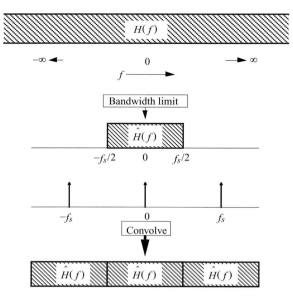

Figure 5.29 By limiting the frequency content $H(f)$ of the waveform to frequencies between $-f_s/2$ and $+f_s/2$, sampling at frequency f_s will not cause aliasing. This sampling is equivalent to convolving the bandwidth limited frequency spectrum $\hat{H}(f)$ with an infinite series of delta functions separated by frequency f_s.

addition of an infinite series of high-frequency amplitudes. This is another view of the **aliasing** phenomena described in the time domain in Chapter 3. If the waveform to be sampled is limited in frequency content to one-half the sampling frequency, then aliasing does not occur (Figure 5.29). Analog filtering (an **anti-aliasing filter**) is usually used to reject frequencies above this limit.

5.8.4   Using aliasing to advantage – the sampling oscilloscope

If an analog waveform is perfectly repetitive with a well-defined frequency f_r, then the Fourier transform is discrete in frequency and consists of an infinite series of harmonics that occur at integer multiples of the frequency f_r. If this waveform is then sampled at a frequency $f_s = f_r - d$, all harmonics will be aliased down by the frequency ratio f_r/d. It is thus possible to sample accurately repetitive waveforms with a frequency content much higher than the sampling frequency. The primary limitation is the analog bandwidth of the system input, before sampling.

In the time domain, this technique can be imagined as sampling each repetition of the waveform slightly later in time (Figure 5.30). The resulting samples are a slowed-down version of the waveform. In the frequency domain, this can be thought of as convolving an infinite series of harmonics (the kth harmonic being at frequency kf_r) with an infinite series of sampling delta functions at integer multiples of the frequency $(f_r - d)$ (Figures 5.31 and 5.32). This may also be seen in Figure 5.33 by using the periodic triangular

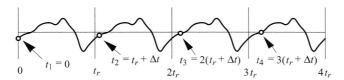

Figure 5.30 Undersampling a periodic waveform in the time domain is accomplished by sampling with a slightly longer period. Each sample is taken progressively later in phase. The result is a series of time samples with the same shape as the original waveform but at a lower frequency.

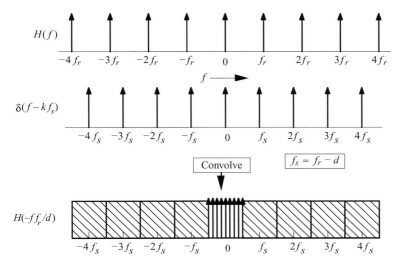

Figure 5.31 Undersampling a periodic waveform at a slightly lower frequency may be seen in the frequency domain as the convolution of an infinite series of harmonics of frequency kf_r with an infinite series of sampling delta functions at frequency if_s, where $f_s = f_r - d$. Aliasing reduces the harmonic frequencies by a factor of f_r/d.

relationship between true and observed frequencies discussed in Chapter 3. An essential requirement is that the waveform be periodic, so that the Fourier amplitude is zero between the harmonic frequencies.

If the repetition frequency of signal is f_r and the bandwidth limit is $f_{max} = n_{max} f_r$, frequency overlap is avoided if:

$$n_{max}(f_r - f_s) = f_s/2$$

solving:

$$m_{max} f_r = f_s(n_{max} + 1/2)$$

$$f_s = \frac{f_r}{1 + 1/(2n_{max})} \approx f_r[1 - 1/(2n_{max})]$$

Example: $f_{max} = 1$ GHz, $f_r = 1$ MHz, then $n_{max} = 1,000$, $f_s = 999.5$ kHz, and $d = 500$ Hz.

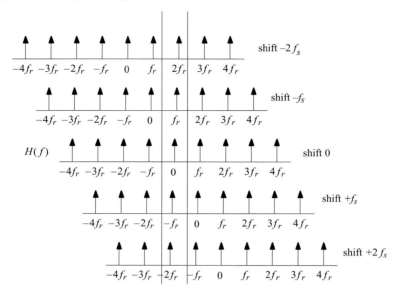

Figure 5.32 Frequency convolution process that results when a periodic waveform is sampled at a slightly lower frequency. The convolution is a series of shift and add operations.

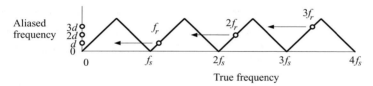

Figure 5.33 The nth harmonic at frequency nf_r is aliased to frequency nd. Since the signal is periodic with frequency f_r, only these harmonics are present.

A variant of this technique is commonly used in sampling oscilloscopes. For example, assume that the analog input bandwidth is 1 GHz, the sampling frequency is 50 MHz, and the repetition frequency is 1 MHz. During the first repetition, 50 samples will be taken every 20 ns. During the second repetition, the sampling trigger occurs 1 ns later and 50 more samples are taken. With each repetition, the sampling trigger is delayed progressively to interleave the data. After a total of 20 repetitions, 1,000 samples are taken, with spacing of 1 ns, and an effective sampling frequency of 1 GHz.

The block diagram in Figure 5.34 shows how it is possible to sample nonperiodic pulses, provided that an accurate trigger pulse can be generated. Each successive pulse is sampled at a progressively delayed time.

5.8.5   The fourier transform of a truncated time function

The Fourier transform integral of a measured signal must be truncated in time because we have no knowledge of the signal outside the time of the measurement. If the function

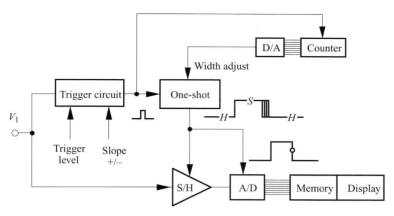

Figure 5.34 Sampling oscilloscope block diagram.

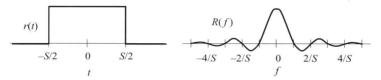

Figure 5.35 Rectangular time window and its Fourier transform.

$h(t)$ is measured between $t = -S/2$ and $+S/2$ and the Fourier transform computed between these limits, the result is the Fourier transform of $r(t)h(t)$ over all t, where the windowing function $r(t) = 1$ when t is between $-S/2$ and $S/2$, and $r(t) = 0$ otherwise:

$$\widehat{H}(f) = \int_{-\infty}^{\infty} r(t)h(t)e^{-j2\pi ft}\, dt$$

From the Fourier convolution theorem, this Fourier transform is the convolution of the true $H(f)$ and $R(f)$, the Fourier transform of $r(t)$ (Figure 5.35):

If $h(t)$ is periodic with period S, $H(f)$ is nonzero only for frequencies $f_n = n/S$. For those frequencies, $R(f_n)$ is zero except at $f_0 = 0$. Thus, the convolution does not alter $H(f)$. The problem arises for frequency components of $h(t)$ that are not integer multiples of $1/S$. The convolution causes such frequency amplitudes $H(f)$ to be spread out over a range of frequencies. This spreading is called "spectral leakage" and introduces ripples that fall off as $1/f$. Viewed in the time domain, this spectral leakage is due to the discontinuities (in value or slope) of non-periodic components at $t = -S/2$ and $t = S/2$.

A common solution to this problem is to multiply $h(t)$ by a function $c(t)$ that gradually tapers off at $t = -S/2$ and $t = S/2$ so the result has zero value and zero slope at those points. A simple and effective function is the Hann window:

$$c(t) = \frac{1}{2}[1 + \cos(2\pi t/S)]$$

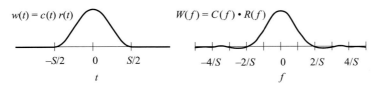

Figure 5.36 Truncated Hann (raised cosine) window and its Fourier transform.

which has the Fourier transform:

$$C(f) = \frac{1}{2}\delta(f) + \frac{1}{4}\delta(f - 1/S) + \frac{1}{4}\delta(f + 1/S)$$

The function $h(t)$ is then multiplied by the truncated Hann window $w(t) = c(t)r(t)$ whose Fourier transform $W(f)$ is given by the convolution of $C(f)$ and $R(f)$ (Figure 5.36).

$$W(f) = \frac{\sin(\pi f S)}{2\pi f} + \frac{\sin[\pi(f - 1/S)S]}{4\pi(f - 1/S)} + \frac{\sin[\pi(f + 1/S)S]}{4\pi(f + 1/S)}$$

The conclusion is that if an arbitrary waveform $h(t)$ sampled over a time period S, and $h(t)$ has any frequency components that are not periodic in S, it is essential to multiply the measurements with a tapered time window (such as the Hann) to reduce the spectral leakage. Not doing this is equivalent to convolving the true $H(f)$ with $R(f)$, which has extensive side lobes. Multiplying by the Hann window $c(t)$ is equivalent to convolving the true Fourier transform with $W(f)$. While $W(f)$ reduces greatly the contamination over large frequencies, it has larger side lobes at $\pm 1/S$ than $R(f)$. This loss in frequency resolution can be regained if it is possible to make the sampling window S larger.

5.8.6 The Fourier transform of a periodic function periodically sampled – the discrete Fourier transform

If a function $h(t)$ has periodicity S, $h(t) = h(t + S)$, and the samples are taken over this period, $S = M\Delta t$, $t_k = k\Delta t$, and $h_k = h(t_k)$. The samples can be expressed as:

$$\hat{h}(t) = \sum_{k=0}^{M-1} \delta(t - t_k)h(t)$$

The Fourier transform (normalized to one period) is given by:

$$H(f) = \int_{-\infty}^{\infty} \sum_{k=0}^{M-1} \delta(t - t_k)h(t)e^{-j2\pi ft}\,dt = \sum_{k=0}^{M-1} h(t_k)e^{-j2\pi ft_k}$$

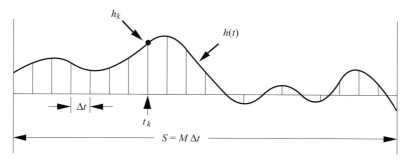

Figure 5.37 Waveform sampled at uniform time intervals Δt. The sampling frequency is $f_s = 1/\Delta t$.

From the above equation, we see that $H(f)$ is nonzero only for values of f that are integer multiples of $1/S$. Define the nth harmonic frequency as $f_n = n/S = n/(M\Delta t)$:

$$H(f_n) = H_n = \sum_{k=0}^{M-1} h_k\, e^{-j2\pi f_n k\Delta t} = \sum_{k=0}^{M-1} h_k\, e^{-j2\pi nk/M}$$

As shown in the preceding, the discrete Fourier transform (DFT) is a special case of the continuous Fourier transform where the waveform is known only at a finite number of discrete points in time. Outside the time interval where the waveform is known, it is assumed to repeat forever, both in the past and into the future. One important application of the DFT is the determination of the sine and cosine components of a periodic waveform. In many cases, these components are more useful than the shape of the waveform itself. The waveform $h(t)$ is sampled at M time intervals $t_0 = 0, t_1 = \Delta t, t_k = k\Delta t, \ldots, t_{M-1} = (M-1)\Delta t$. The sampling frequency $f_s = 1/\Delta t$. The full sampling window is $S = M\Delta t$ (Figure 5.37).

Using the notation $h_k = h(t_k)$, the DFT of h_k is defined as:

$$H_n = \sum_{k=0}^{M-1} h_k e^{-j2\pi nk/M}$$

Note: $e^{j\phi} = \cos\phi + j\sin\phi$, $e^{-j\phi} = \cos\phi - j\sin\phi$, $e^0 = e^{j2\pi} = 1$, and $e^{-j\pi} = -1$.

The significance of the DFT coefficients is that H_0 is the Fourier coefficient at frequency 0 (dc component), H_1 is the Fourier coefficient at frequency $f_1 = f_s/M$ (1 cycle per S), and H_n is the Fourier coefficient at frequency $f_n = nf_s/M$ (n cycles per S).

EXAMPLE 5.17

Perform the DFT on the waveform:

$$h_k = A + B\sin(2\pi kP/M), \quad \text{where } P \text{ is an integer}$$

$$H_0 = \sum_{k=0}^{M-1} A = MA$$

$$H_n = \sum_{k=0}^{M-1}[A + B\sin(2\pi kP/M)][\cos(2\pi kn/M) - j\sin(2\pi kn/M)], \; n > 0$$

Due to the orthogonal nature of the sine and cosine series (see the example that follows), for this h_k:

$$H_P = \sum_{k=0}^{M-1} -jB \sin^2(2\pi k P/M) = \frac{-jBM}{2}$$

$$H_{M-P} = \sum_{k=0}^{M-1} jB \sin^2(2\pi k P/M) = \frac{jBM}{2}$$

and all the other Fourier coefficients are zero. Thus, we see that the nth Fourier coefficient describes the amplitude of any sine-wave component having n complete cycles per sampling interval S.

EXAMPLE 5.18

Relate the coefficients a_j and b_j in the general expansion:

$$h_k = \sum_{m=0}^{M-1} a_m \cos(2\pi mk/M) + b_m \sin(2\pi mk/M)$$

to the complex Fourier coefficients H_n.

Firstly, we perform the discrete Fourier transform of h_k:

$$H_n = \sum_{k=0}^{M-1} \left[\sum_{m=0}^{M-1} a_m \cos(2\pi mk/M) + b_m \sin(2\pi mk/M) \right]$$
$$\times [\cos(2\pi nk/M) - j \sin(2\pi nk/M)]$$

Since:

$$\sum_{k=0}^{M-1} [\cos(2\pi m/M) \sin(2\pi nk/M)] = 0, \quad \text{for all } m \text{ and } n$$

$$\sum_{k=0}^{M-1} [\cos(2\pi mk/M) \cos(2\pi nk/M)] = 0, \quad \text{for all } m \neq n$$

$$\sum_{k=0}^{M-1} [\sin(2\pi mk/M) \sin(2\pi nk/M)] = 0, \quad \text{for all } m \neq n$$

and

$$\sum_{k=0}^{M-1} [\cos^2(2\pi nk/N)] = \sum_{k=0}^{M-1} [\sin^2(2\pi nk/N)] = M/2$$

we have for $n = 0$:

$$\boxed{H_0 = \sum_{k=0}^{M-1} a_0 = Ma_0}$$

and for $0 < n < M$:

$$
\begin{aligned}
H_n &= \sum_{k=0}^{M-1}[a_n \cos^2(2\pi nk/M) - jb_n \sin^2(2\pi nk/M)] \\
&= (M/2)(a_n - jb_n)
\end{aligned}
$$

$$
a_n = (2/M)\mathrm{Re}(H_n) \quad \text{and} \quad b_n = -(2/M)\mathrm{Im}(H_n)
$$

We see that the real part of H_n is associated with the cosine terms and the imaginary part of H_n is associated with the sine terms of the expansion.

To what do the Fourier coefficients outside the interval from 0 to $M/2$ correspond? From the definition of the DFT, the Fourier amplitude for M cycles per sampling interval S is the same as 0 cycles per S:

$$
H_M = \sum_{k=0}^{M-1} f_k\, e^{-j2\pi k} = \sum_{k=0}^{M-1} f_k = H_0
$$

Above M samples per S, all the Fourier amplitudes are equal to their lower counterparts:

$$
H_{M+n} = \sum_{k=0}^{M-1} h_k\, e^{-j2\pi k} e^{-j2\pi kn/M} = \sum_{k=0}^{M-1} h_k\, e^{-j2\pi kn/M} = H_n
$$

Between $M/2$ and M samples per S, we have the following result:

$$
H_{M-n} = \sum_{k=0}^{M-1} h_k\, e^{-j2\pi k} e^{+j2\pi kn/M} = \sum_{k=0}^{M-1} h_k\, e^{+j2\pi kn/M}
$$

If h_k is real, then $H_{M-n} = H_n{}^*$ and $H_{M/2}$ is real ($*$ denotes the complex conjugate). $H_{M/2}$ is the Fourier coefficient at frequency $M/2$ (1 cycle per $2\Delta t$ or $M/2$ cycles per M samples). This is the highest frequency that the DFT can determine. All Fourier coefficients for higher frequencies are either equal to or are complex conjugates of coefficients for lower frequencies. Thus, there are only $M/2$ *independent* Fourier coefficients.

Another manifestation of this effect is to take the Fourier transform of a sine wave with $m + M$ complete cycles per sampling interval S:

$$
h_k = \sin[2\pi k(m + M)/M]
$$

$$
H_n = \sum_{k=0}^{M-1} \sin[2\pi k(m + M)/M][\cos(2\pi kn/M) - j\sin(2\pi kn/M)]
$$

$$
H_m = -jM/2 \qquad H_{M-m} = jM/2
$$

Comparing with the results above, we see that a sine wave with $m + M$ cycles per sampling interval has the same DFT as a sine wave with m cycles per sampling interval. The appearance of frequencies above $M/2$ cycles per S as lower frequencies in the DFT is a result of inadequate sampling of the waveform.

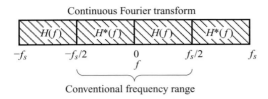

Continuous Fourier transform

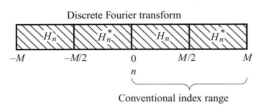

Discrete Fourier transform

Figure 5.38 Arrangement of continuous and discrete Fourier transform amplitudes for a real waveform sampled at frequency f_s.

The integral Fourier transform has a corresponding periodicity in frequency, but the frequency ranges are shifted (Figure 5.38).

In general, if the sampling frequency is inadequate, higher-frequency components of the true waveform $h(t)$ will appear as lower-frequency components in the DFT. This is called **frequency aliasing**. There is no way to correct the data after the sampling has been performed. The usual solution to this problem is to use a low-pass analog filter (**anti-aliasing filter**) that eliminates all frequencies above $f_s/2$ before sampling.

One statement of this result is the **sampling theorem**.

Sampling theorem

To be able to recover completely the continuous signal from its sampled counterpart, the sampling frequency f_s must be at least twice the highest frequency in the signal.

Each Fourier coefficient H_n is in general complex, the real part describing the cosine-like amplitude and the imaginary part describing the sine-like amplitude. The Fourier magnitude F_n is defined as:

$$F_n = \sqrt{\text{Re}(H_n)^2 + \text{Im}(H_n)^2}$$

and the phase angle θ_n is given by $\tan \theta_n = \text{Im}(H_n)/\text{Re}(H_n)$.

The inverse DFT is given by:

$$h_k = \sum_{n=0}^{M-1} \frac{H_n}{M} e^{+j2\pi nk/M}$$

Note that the same function is obtained by firstly performing the forward and then the inverse transformations given before.

What happens if we evaluate h_k outside the sampling interval S?

$$h_{M+k} = \sum_{n=0}^{M-1} \frac{H_n}{M} e^{+j2\pi n} e^{+j2\pi nk/M} = h_k$$

We see that, given a set of M Fourier coefficients, the constructed function repeats endlessly with a periodicity $S = MT$. This is analogous to the previous result that, given a set of M samples, the Fourier coefficients repeat endlessly with a periodicity $M = S/T$.

5.8.7 The fast Fourier transform

The **fast Fourier transform**, or **FFT**, is a computationally efficient method for computing the discrete Fourier transform (DFT). It was developed by John W. Tukey and James W. Cooley during the 1960s and published in 1965 as "An algorithm for the machine calculation of complex Fourier series," *Mathematics of Computation*, Vol. 19, pp. 297–301. Because they are mathematically equivalent, interpreting the results of the FFT only requires an understanding of the DFT. The computational efficiency of the FFT arises from a clever reorganization of the terms in the DFT so that numerous identical terms are computed only once. Direct computation of the DFT equation given before requires M^2 multiplications and $M(M-1)$ additions. The FFT, on the other hand, requires only $M\log_2 M$ multiplications and $2M\log_2 M$ additions.

$$H_n = \sum_{k=0}^{M-1} h_k e^{-j2\pi kn/M} = \sum_{k=0}^{M-1} h_k W^{nk} \tag{5.7}$$

where we define $W = e^{-i2\pi/M}$ and $W^0 = W^M = 1$. Writing this in matrix form and using the relationship $W^{M+nk} = W^{nk}$, we have for the example $M = 4$:

$$\begin{bmatrix} H_0 \\ H_1 \\ H_2 \\ H_3 \end{bmatrix} = \begin{bmatrix} W^0 & W^0 & W^0 & W^0 \\ W^0 & W^1 & W^2 & W^3 \\ W^0 & W^2 & W^4 & W^6 \\ W^0 & W^3 & W^6 & W^9 \end{bmatrix} \begin{bmatrix} h_0 \\ h_1 \\ h_2 \\ h_3 \end{bmatrix} = \begin{bmatrix} 1 & 1 & 1 & 1 \\ 1 & W^1 & W^2 & W^3 \\ 1 & W^2 & 1 & W^2 \\ 1 & W^3 & W^2 & W^1 \end{bmatrix} \begin{bmatrix} h_0 \\ h_1 \\ h_2 \\ h_3 \end{bmatrix}$$

The key to the efficiency of the fast Fourier transform is the factorization of the matrix, which is made possible by the interchange of certain rows:

$$\begin{bmatrix} H_0 \\ H_1 \\ H_2 \\ H_3 \end{bmatrix} = \begin{bmatrix} 1 & 1 & 1 & 1 \\ 1 & W^2 & 1 & W^2 \\ 1 & W^1 & W^2 & W^3 \\ 1 & W^3 & W^2 & W^1 \end{bmatrix} \begin{bmatrix} h_0 \\ h_1 \\ h_2 \\ h_3 \end{bmatrix}$$

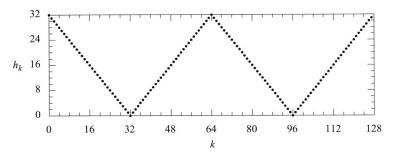

Figure 5.39 Triangle wave $h(t)$ with two cycles per 128 samples.

$$= \begin{bmatrix} 1 & 1 & 0 & 0 \\ 1 & W^2 & 0 & 0 \\ 0 & 0 & 1 & W^1 \\ 0 & 0 & 1 & W^3 \end{bmatrix} \begin{bmatrix} 1 & 0 & 1 & 0 \\ 0 & 1 & 0 & 1 \\ 1 & 0 & W^2 & 0 \\ 0 & 1 & 0 & W^2 \end{bmatrix} \begin{bmatrix} h_0 \\ h_1 \\ h_2 \\ h_3 \end{bmatrix}$$

In general, there will be $\log_2 M$ matrices with $M/2$ complex multiplications and M complex additions per matrix. Every occurrence of $W^0 = 1$ means a simple addition. Other powers of W involve multiplication by precomputed constants. As M becomes large, we have $\log_2 M$ sparse matrices. The direct method, Eq. (5.7), requires M^2 complex multiplications and $M(M-1)$ complex additions. In terms of the number of floating point operations, the FFT gains a factor of approximately $M/(2\log_2 M)$ over the direct method, which is greater than 100 for $M = 4{,}096 = 2^{12}$.

EXAMPLE 5.19

Perform the FFT of the triangle wave shown in Figure 5.39 to determine the harmonic amplitudes. Use the fft.c code given in Appendix D. Compare the original triangle wave with the sum of the constant and first harmonic, and with the sum of the constant plus the first plus the third harmonic.

From the preceding section, we learned that the complex DFT coefficients H_n could be used to recover the original waveform samples h_k:

$$h_k = \sum_{n=0}^{M-1} a_n \cos(2\pi nk/M) + b_n \sin(2\pi nk/M)$$

where:

$$a_0 = H_0/M \qquad b_0 = 0$$
$$a_n = (2/M)\text{Re}(H_n) \qquad b_n = -(2/M)\text{Im}(H_n) \qquad n > 0$$

Because the time function is symmetric, all Fourier amplitudes are real. The average value is 16, so $F_0 = 16 \times 128 = 2{,}048$. As there are two cycles in the sampling interval, the Fourier amplitude of the first harmonic is H_2, the amplitude of

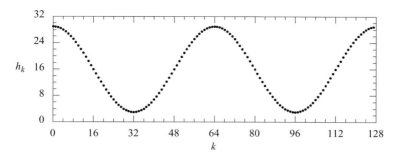

Figure 5.40 Sum of the constant term plus the first harmonic, $(h_{0,k} + h_{2,k})$.

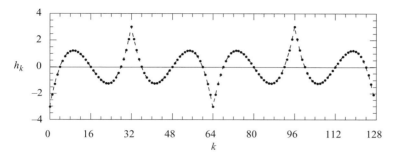

Figure 5.41 Difference between the triangle wave f_k and the constant term plus the first harmonic, $(f_k - h_{0,k} - h_{2,k})$.

the third harmonic is H_6, the amplitude of the fifth harmonic is H_{10}, etc. Performing the FFT, we find that $H_1 = 0$, $H_2 = 830.69$, $H_3 = H_4 = H_5 = 0$, $H_6 = 92.89$, $H_7 = H_8 = H_9 = 0$, $H_{10} = 33.88$, etc. Designating $h_{n,k}$ as the kth value of the nth Fourier component, and using the formulas above, we have:

$$h_{0,k} = (2{,}048/128) = 16 \qquad h_{1,k} = 0$$
$$h_{2,k} = (830.69/64)\cos(2 \times 2\pi k/128)$$
$$h_{3,k} = h_{4,k} = h_{5,k} = 0$$
$$h_{6,k} = (92.89/64)\cos(6 \times 2\pi k/128)$$
$$h_{7,k} = h_{8,k} = h_{9,k} = 0$$
$$h_{10,k} = (33.88/64)\cos(10 \times 2\pi k/128)$$

The constant plus first harmonic (Figure 5.40) is a raised cosine approximation to the triangle wave, and the difference (Figure 5.41) has 12 zero crossings. This indicates that the second harmonic is zero and anticipates the third harmonic term, which also has 12 zero crossings.

The constant plus the first and third harmonic (Figure 5.42) is a better approximation to the triangle wave, and the difference (Figure 5.43) has 20 zero crossings. This

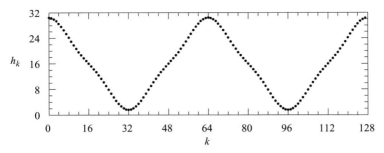

Figure 5.42 Sum of the constant term plus the first and third harmonics, $(h_{0,k} + h_{2,k} + h_{6,k})$.

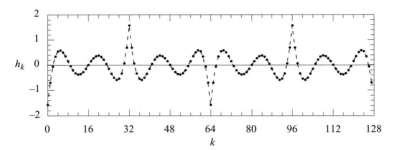

Figure 5.43 Difference between the triangle wave f_k and the constant term plus the first and third harmonics, $(f_k - h_{0,k} - h_{2,k} - k_{6,k})$.

indicates that the fourth harmonic is zero and anticipates the fifth harmonic term, which also has 20 zero crossings.

5.8.8   Use of the fast Fourier transform function and windowing

The fast Fourier transform function and its use is listed in Appendix D. It is used in Laboratory Exercises 21, 22, and 24.

Figure 5.44 shows the function $h_k = 1 + \cos(5 \times 2\pi k/128)$, which has five complete cycles in the interval from $k = 0$ to $k = 127$. Figure 5.45 shows the 128-point FFT of this function. Since the constant term is 1, the H_0 Fourier amplitude is 128 (the sum over 128 terms). Because there are five cycles per full sampling interval, the $n = 5$ and $n = M - 5 = 123$ Fourier magnitudes are nonzero. All other Fourier amplitudes are zero. The $n = 64$ coefficient corresponds to the Nyquist frequency limit.

For a general input waveform, we cannot be sure that we have sampled an integral number of its component harmonics in the truncation interval S. Figure 5.46 shows the function $h_k = 1 + \cos(5.5 \times 2\pi k/M)$ which has 5.5 cycles from $k = 0$ to $k = M - 1 = 127$.

In this case, the Fourier coefficients (Figure 5.47) are required to describe the discontinuity that occurs at the edges of the truncation interval. While h_{127} is nearly zero, the

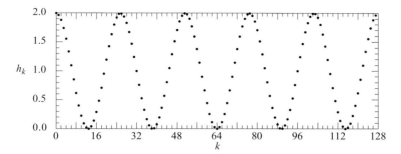

Figure 5.44 Five cycles of sine wave sampled at 128 time points. $h_k = 1 + \cos(5 \times 2\pi k/M)$. See Figure 5.45 for the Fourier transform.

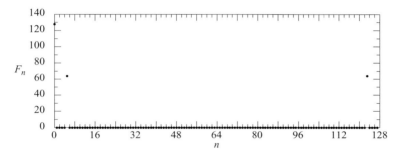

Figure 5.45 Magnitude of the Fourier coefficients of $f_k = 1 + \sin(5 \times 2\pi k/16)$. $M_0 = 128$, $M_5 = M_{59} = 64$.

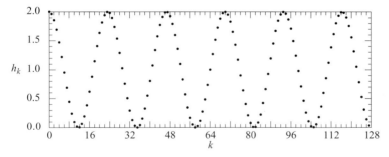

Figure 5.46 5.5 cycles of cosine wave sampled at 64 time points. $h_k = 1 + \cos(5.5 \times 2\pi k/M)$. See Figure 5.47 for the Fourier transform.

next value $h_{128} = h_0 = 2$. Since this discontinuity was not present in the non-truncated waveform, we would expect nonzero Fourier coefficients only at $H_0, H_5, H_{123}, H_6, H_{122}$. The nonzero values of all the other coefficients represent an unwanted "spectral leakage."

One solution to this problem is the use of a special "window" or truncation function that is multiplied by the sampled data to bring the result to zero gradually at the edges of the truncation interval. Many such functions have been developed, but we will use

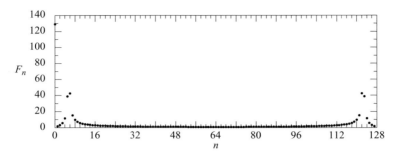

Figure 5.47 Magnitude of FFT coefficients of 5.5-cycle sine wave. Peaks are centered at $n = 5.5$ and 122.5. Note spectral leakage into other coefficients.

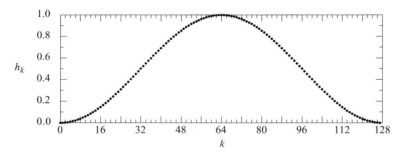

Figure 5.48 Hann window $h_k = 0.5[1 - \cos(2\pi k/M)]$.

the Hann or raised cosine window (Figure 5.48) because it is both simple and effective.

As seen in Figure 5.36, the Hann window has Fourier coefficients that are relatively small, except for H_0, H_1, and H_{127}. By the Fourier convolution theorem, multiplying the time samples of a waveform by the Hann function is equivalent to convolving the Fourier coefficients of that waveform with the Fourier coefficients of the Hann function. So the widespread spectral leakage seen in Figure 5.47 should be reduced to leakage across a single frequency coefficient.

Figure 5.49 shows the 5.5 cosine cycles (Figure 5.46) after they have been multiplied by the Hann window (Figure 5.48).

Comparing the Fourier transform of the windowed function (Figure 5.50) with that of the nonwindowed function (Figure 5.47) we see that the Hann window has greatly reduced the spectral leakage. This is consistent with the elimination of the sharp discontinuities in Figure 5.46 that transform into high-frequency amplitudes. Other effects of the Hann window are the reduction of the H_0 coefficient by a factor of 2, and the leakage into H_1.

Figure 5.51 shows five complete cycles of square wave in the interval from $k = 0$ to 127. Figure 5.52 shows the discrete Fourier transform of this function. Since the average is 0.5, the F_0 Fourier amplitude is 64. Because there are five cycles per full

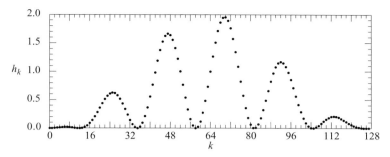

Figure 5.49 5.5 cycles of cosine multiplied by the Hann window
$h_k = 0.5[1 - \cos(2\pi k/M)][1 + \cos(5.5 \times 2\pi k/M)]$.

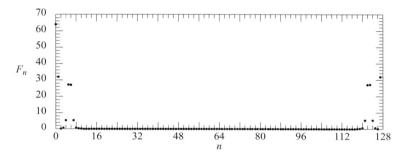

Figure 5.50 FFT of $h_k = 0.5[1 - \cos(2\pi k/M)][1 + \cos(5.5 \times 2\pi k/M)]$ (5.5-cycle cosine wave windowed with the Hann function).

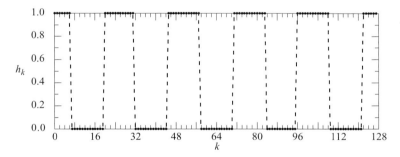

Figure 5.51 Five cycles of square wave sampled at 128 time points. The dashed line is provided to guide the eye.

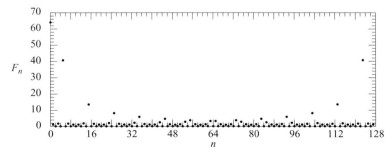

Figure 5.52 Magnitude of FFT coefficients of a five-cycle square wave. The first, third, fifth, seventh, and ninth harmonics are at $n = 5, 15, 25, 35$, and 45, respectively.

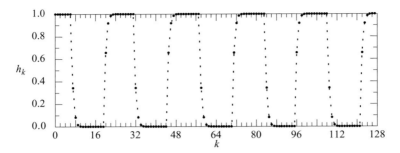

Figure 5.53 Five cycles of square wave sampled at 128 time points and filtered with a four-pole low-pass filter. The dashed line is provided to guide the eye.

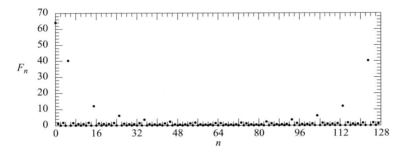

Figure 5.54 Magnitude of FFT coefficients of five cycle square wave after four-pole low-pass filtering. The first, third, fifth, seventh, and ninth harmonics are at $n = 5, 15, 25, 35$, and 45, respectively. Note reduction of aliasing artifacts.

sampling interval, all F_n values for n that are not a multiple of 5 are zero. Since the square wave is symmetric, all even harmonics $n = 10, 20, 30, 40, 50, 60$ are also zero. Since the function is real, $H_n = (H_{128-n})^*$ and $F_n = F_{128-n}$.

In Figure 5.52 we see the first, third, fifth, seventh, ninth, and eleventh harmonics of the square wave on each half of the plot. However, the thirteenth harmonic appears at Fourier indices $13 \times 5 = 65$ and $128 - 13 \times 5 = 63$ and these have "crossed over" to indices that cause them to add into lower frequencies. The fifteenth, seventeenth, etc. harmonics many be seen extending further into indices that correspond to even lower frequencies. This is a good example of how aliasing occurs in the discrete Fourier transform.

Figure 5.53 shows the five-cycle square after it has been filtered with a four-pole low-pass filter. The Fourier transform (Figure 5.54) shows a significant reduction in aliasing.

Figure 5.55 shows 5.5 cycles of square wave after filtering with a four-pole low-pass filter. The Fourier transform (Figure 5.56) shows considerable spectral leakage around the first, third, fifth, seventh, etc. harmonics. The higher harmonics have been reduced significantly by the low-pass filter and aliasing effects are minimal.

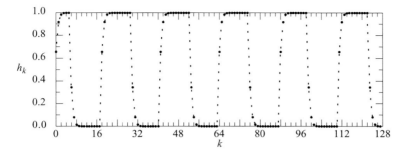

Figure 5.55 5.5 cycles of square wave sampled at 128 time points. The dashed line is provided to guide the eye.

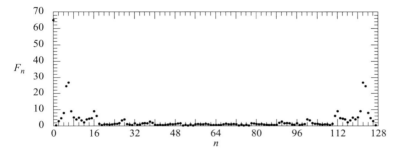

Figure 5.56 Magnitude of FFT coefficients of 5.5-cycle square wave after four-pole low-pass filtering. The first, third, fifth, and seventh harmonics are centered at $n = 5.5, 16.5, 27.5$, and 38.5, respectively. Note the spectral leakage into other Fourier coefficients.

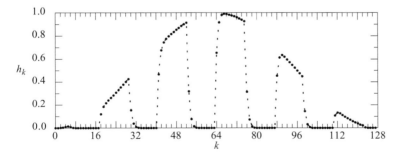

Figure 5.57 5.5 cycles of square wave after four-pole low-pass filtering and multiplication by the Hann window. The dashed line is provided to guide the eye.

Figure 5.57 shows the low-pass filtered 5.5 cycles of square wave after multiplication by the Hann window (Figure 5.48). The Fourier transform (Figure 5.58) shows a clear first harmonic at indices 5 and 6, a third harmonic at indices 16 and 17, a fifth harmonic at indices 27 and 28, a seventh harmonic at indices 38 and 39, a ninth harmonic at indices 49 and 50, and an eleventh harmonic at indices 60 and 61. Only by using both a low-pass filter (to reduce aliasing) and a Hann window (to reduce spectral leakage)

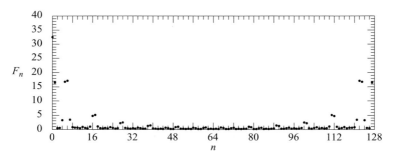

Figure 5.58 Magnitude of FFT coefficients of 5.5-cycle square wave after four-pole low-pass filtering and multiplication by the Hann window. The first, third, fifth, seventh, and ninth harmonics are centered at $n = 5.5$, 16.5, 27.5, 38.5, and 49.5, respectively. Note the reduction in spectral leakage.

can the expected frequency components be observed. The primary disadvantage of the Hann window is the appearance of side lobes. These may be seen for the first harmonic as nonzero magnitudes at indices 4 and 7.

5.8.9   Summary of sampling system design factors

The following summarizes the major considerations when designing a sampling system. As above, we assume that the samples are uniformly spaced in time with period T and sampling frequency $f_s = 1/T$, and that M samples are taken, spanning an interval $S = MT$.

Maximum signal frequency

The maximum frequency that can be reliably sampled is $f_s/2$. Higher frequencies must be removed by low-pass analog filtering (an anti-aliasing filter) before sampling, or they will appear in the $0–f_s$ frequency band.

Anti-aliasing filters

Practical low-pass filters cannot cut off with perfect sharpness. As discussed in Chapter 2, an eight-pole low-pass Butterworth filter with a corner frequency (gain $= 0.707$) at f_c will have a gain of $2^{-8} = 0.4\%$ at $2f_c$. If we regard $f < f_c$ as the pass band and $f > 2f_c$ as the stop band, the filtered signal will contain accurate amplitude information from 0 to f_c but also non-negligible components between f_c and $2f_c$. Sampling at twice this highest frequency means sampling at twice $2f_c$. To prevent frequencies above $2f_c$ from aliasing below f_c, we need to sample at $f_s > 3f_c$.

Discrete Fourier transform

The discrete Fourier transform determines the frequency components of the waveform. The first Fourier coefficient is H_0, which describes the average or 0 Hz component. The nth Fourier coefficient is H_n, which describes the component with frequency

$f_n = nf_s/M = n/(MT) = n/S$, which is n cycles per S or n cycles per M samples. For a real waveform, the highest frequency Fourier coefficient is $H_{M/2}$, which has frequency $f_{M/2} = f_s/2$.

Frequency resolution

The frequency difference between neighboring Fourier coefficients is $\Delta f = 1/S = 1/(MT)$. This is the frequency resolution in the sense that frequency components closer than Δf cannot be individually resolved. For a given sampling period T, making the frequency resolution finer requires increasing the number of samples M and correspondingly increasing the sampling time S. However, for waveforms whose frequency content is changing with time in an unknown way, a large value of S limits the speed with which those changes can be measured.

Periodic waveforms

If the waveform is perfectly periodic, with exactly m cycles in the sampling time $S = MT$, then only Fourier coefficients that are integer multiples of m will be nonzero. The lowest frequency coefficient (aside from the dc component H_0) will be the first harmonic H_m at frequency $f_m = m/S$ and the kth harmonic will be H_{km} at frequency $f_{km} = km/S$.

Nonperiodic waveforms

If the waveform is not periodic, or if a periodic waveform is not sampled for an integer number of cycles, then the last samples taken will not join smoothly with the first samples taken. This lack of continuity will generate erroneous high-frequency components (called spectral leakage) because the DFT *assumes* that the waveform has period S and will deliver Fourier coefficients that inverse transform to a periodic waveform with a discontinuity at the boundary. Multiplying the sampled valued with a window (such as the raised cosine) that smoothly joins to a zero value will eliminate the discontinuity and nearly eliminate spectral leakage. One feature of multiplying the sampled data by such windows is that it is equivalent to convolving the Fourier transform of the samples with the Fourier transform of the window. It is therefore important that the window have a minimum of high-frequency components. In the best case, the frequency resolution will be broadened so that the Fourier transform of a pure sine wave will span several Fourier coefficients. This loss in frequency resolution can be compensated by increasing $S = MT$.

5.9   Digital filters

A real-time linear digital filter has the general form:

$$y_i = A_1 x_{i-1} + A_2 x_{i-2} + \cdots + A_M x_{i-M} + B_1 y_{i-1} + \cdots + B_N y_{i-N} \tag{5.8}$$

This filter describes a transformation where each new output value y_i is a linear combination of previous input values x_{i-j} and previous output values y_{i-j}. If all the Bs are zero, the filter will have a finite impulse response (FIR). If some or all of the Bs are nonzero and sufficiently small, the filter will have an infinite impulse response (IIR) with decreasing amplitude for increasingly longer times. If some of the Bs are not sufficiently small, the filter may be unstable.

The primary advantages of the digital filter are that it is easy to change (software rather than circuit components) and permits the use of certain classes of filters that are very difficult to implement with analog components.

Note: In other treatments it is often assumed that after x_i is acquired, y_i can be computed instantaneously and the primary delay occurs between the output of y_i and the input of the next x_{i+1}. Consequently, Eq. (5.8) is frequently written as:

$$y_i = A_1 x_i + A_2 x_{i-1} + \cdots + A_M x_{i-M+1} + B_1 y_{i-1} + \cdots + B_N y_{i-N}$$

However, this assumption is not appropriate in real-time digital filtering where speed is optimized by outputing the value y_i and latching the new input value x_i *simultaneously*. The unavoidable delay in the system is the computation of the new y_i, using *previous* values of x and y.

5.9.1   The finite impulse response (FIR) filter

The finite impulse response filter is given by the general form:

$$y_i = A_1 x_{i-1} + A_2 x_{i-2} + \cdots + A_M x_{i-M}$$

The response to an input impulse:

$$x_{i<0} = 0 \qquad x_0 = 1 \qquad x_{i>0} = 0$$

is given by:

$$y_{i<1} = 0 \qquad y_1 = A_1 \qquad y_2 = A_2 \cdots y_M = A_M \qquad y_{i>M} = 0$$

Example I

$y_i = x_{i-1}$ The identity filter. The output is equal to the input, shifted in time by one cycle of the filtering procedure.

Example II

$y_i = x_{i-1} - x_{i-2}$ The linear approximation to the first derivative filter. The output is equal to the differential of the input shifted in time by 1.5 cycles.

Example III

$y_i = x_{i-1} - 2x_{i-2} + x_{i-3}$ The quadratic approximation to the second derivative filter. The output is equal to the second differential of the input shifted by two cycles.

Example IV

$y_i = 0.25x_{i-1} + 0.50x_{i-2} + 0.25x_{i-3}$ A smoothing filter. The output is smoothed and shifted by two cycles.

Example V

$y_i = (x_{i-1} + x_{i-2} + \cdots + x_{i-n+1} + x_{i-n})/n$ An averaging filter. The output is the average of the input over n samples shifted by $1 + n/2$ cycles.

5.9.2   The infinite impulse response (IIR) filter

The infinite impulse response filter is given by Eq. (5.8) where any B_i is nonzero. The outputs y_i will depend on all the inputs that have ever been received. The impulse response may exponentially decay, oscillate, or exponentially grow, depending on the coefficients B_i.

Low-pass single-pole digital filter

The digital low-pass single-pole filter has the form:

$$y_i = (1 - \alpha)x_{i-1} + \alpha y_{i-1}$$

Its response to the impulse:

$$x_{i<0} = 0 \qquad x_0 = 1 \qquad x_{i>0} = 0$$

is given by the exponential decay:

$$y_{i<1} = 0 \qquad y_1 = (1 - \alpha) \qquad y_2 = (1 - \alpha)\alpha \qquad y_i = (1 - \alpha)\alpha^{i-1}$$

This impulse response has unit area:

$$\sum_{i=-\infty}^{\infty} y_i = (1 - \alpha)(1 + \alpha + \alpha^2 + \cdots) = 1$$

Its response to the step:

$$x_{i<0} = 0 \qquad x_{i\geq0} = 1$$

is given by:

$$y_{i\leq0} = 0 \qquad y_1 = (1 - \alpha) \quad y_2 = (1 - \alpha) + \alpha(1 - \alpha) = (1 - \alpha)(1 + \alpha)$$
$$y_3 = (1 - \alpha) + \alpha(1 - \alpha)(1 + \alpha) = (1 - \alpha)(1 + \alpha + \alpha^2)$$
$$y_i = (1 - \alpha)\sum_{k=1}^{i-1}\alpha^k \qquad y_\infty = (1 - \alpha)/(1 - \alpha) = 1$$

5.9.3   Use of FIR and IIR filters to perform the DFT

The Fourier transform coefficient H_n can be viewed as the inner product of N samples and a harmonic of frequency n cycles per N samples. The most efficient method for determining the coefficients at all frequencies is the fast Fourier transform. However, if only a few coefficients need to be determined, a finite impulse response digital filter can be used to *continuously* compute H_n as follows:

$$\text{Re}(H_n) = \sum_{k=0}^{N-1} h_k \cos(2\pi nk/N)$$

$$\text{Im}(H_n) = -\sum_{k=0}^{N-1} h_k \sin(2\pi nk/N)$$

$$F_n = \sqrt{[\text{Re}(H_n)]^2 + [\text{Im}(H_n)]^2}$$

The steps are:
1. Take N samples h_k, $k = 0$ to $N - 1$.
2. Use filter to compute Hn, Fn for the desired n frequencies.
3. Acquire new sample h_N.
4. Delete h_0, shift all h_k to h_{k-1}.
5. Loop back to step 2.

The previous algorithm requires $2N$ additions and multiplications for each H_n to be computed. This computational cost can be reduced by using an infinite impulse response filter as follows:

$$H_n(t) = \sum_{k=0}^{N-1} h_k \, e^{-j2\pi nk/N}$$

$$H_n(t + \Delta t) = \sum_{k=1}^{N} h_k \, e^{-j2\pi n(k-1)/N}$$

$$= e^{j2\pi n/N} \sum_{k=1}^{N} h_k \, e^{-j2\pi nk/N} = e^{j2\pi n/N}[H_n(t) - h_0 + h_N]$$

For each H_n this algorithm requires $2N$ additions and multiplications for the first result, but subsequent values only require two additions and one complex multiplication.

The steps are:
1. Take N samples h_k, $k = 0$ to $N - 1$.
2. Use FFT to compute H_n, F_n for the desired n frequencies.
3. Acquire new sample h_N.
4. Apply complex filter to compute new values of H_n.
5. Delete h_0, shift all h_k to h_{k-1} (computational bottleneck).
6. Loop back to step 3.

The bottleneck at step 5 above can be eliminated using a cyclic pointer as follows:

$$H_n(t + \Delta t) = e^{j2\pi n/N}[H_n(t) - h_i + g]$$

The steps are:
1. Take N samples h_k, $k = 0$ to $N - 1$.
2. Use FFT to compute H_n for the desired n frequencies.
3. Set pointer $i = 0$.
4. Acquire new sample g.
5. Apply complex filter to compute new values of H_n.
6. Set $h_i = g$, $i = i + 1$.
7. If $i = N$, set $i = 0$.
8. Loop back to step 4.

5.10 Control techniques

5.10.1 Fourier control

For systems with well-defined and measurable input–output transfer functions, Fourier techniques are often useful in predicting the input waveform that will produce a desired output waveform. For example, the mechanical activator for a disk drive head has inertia that reduces the amplitude of its high-frequency response. The actual motion of the head $a(t)$ is the driving voltage waveform $b(t)$ convolved with the impulse response of the electromechanical system $c(t)$:

$$a(t) = b(t) \cdot c(t)$$

We now ask how can we *continuously* process an arbitrary waveform $a(t)$ to generate a new waveform that when convolved with $c(t)$ produces a good approximation of $a(t)$?

$$\tilde{a}(t) = [d(t) \cdot a(t)] \cdot c(t)$$

Convolution of $a(t)$ by the function $d(t)$ may be thought of as a pre-conditioning operation that compensates for the convolution with $c(t)$. By using the Fourier convolution theorem, the above convolution is equivalent to the simple multiplication of the Fourier transforms of $d(t)$, $a(t)$, and $c(t)$:

$$\mathcal{F}(\tilde{a}) = \mathcal{F}(d) \times \mathcal{F}(a) \times \mathcal{F}(c)$$

To find the compensating function $d(t)$ that will pre-process $a(t)$, we solve:

$$d(t) \approx \mathcal{F}^{-1}\left[\frac{1}{\mathcal{F}(c)}\right] = \mathcal{F}^{-1}\left[\frac{1}{C_r + jC_i}\right] = \mathcal{F}^{-1}\left[\frac{C_r - jC_i}{C_r{}^2 + C_i{}^2}\right]$$

The function $d(t)$ can be checked by verifying that $d(t) \cdot c(t)$ produces a delta function.

Laboratory Exercise 24 performs this calculation for the case where $c(t)$ is a square wave and $c(t)$ is a low-pass filter. Applications include optimal control of mechanical

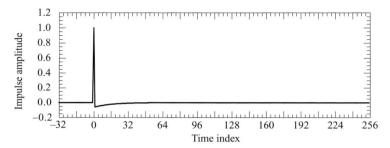

Figure 5.59 Impulse response of a single-pole high-pass filter with RC time constant = 16 time units.

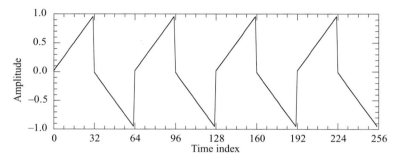

Figure 5.60 Input function that produces a four-cycle square wave when convolved with the high-pass filter response shown in Figure 5.59.

systems with well-defined inertia (such as hard disk read/write heads) and compensation for limitations in the frequency response of audio systems.

Consider the example where $a(t)$ is a square wave and $c(t)$ is a high-pass filter. The impulse response of the high-pass filter is shown in Figure 5.59 and the input function $d(t) \cdot a(t)$ is shown in Figure 5.60.

5.10.2 Analog control

Figure 5.61 shows an analog control system, which uses a differential amplifier to form an error signal, which is the difference between a sensor reading and the desired set point. This error signal is processed by a controller, which is amplified to drive an actuator that effects the desired change in the system. The typical analog controller combines the error voltage, its derivative, and its integral. Applications of analog control include airplane autopilots and "dynamic shock absorbers" for automobiles.

5.10.3 Computer-based digital control

One of the important potentials of the microcomputer is its use not merely to sense physical quantities in the real world and to perform data analysis and display, but to

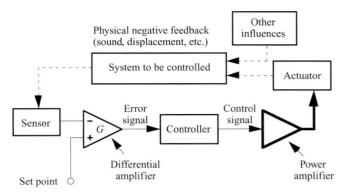

Figure 5.61 Simple analog control system, which drives an actuator with a control signal derived from the error signal, which is the difference between the set point and the sensor reading.

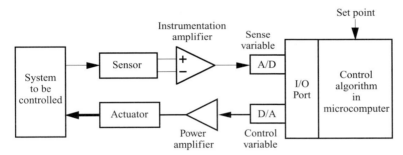

Figure 5.62 Typical microcomputer control system. The control algorithm periodically reads the sense variable and produces a control value that acts on the system to make the sense variable as close as possible to a previously entered set point.

reach out and control those quantities. In this section, we discuss some of the digital control algorithms that are used to perform that control.

The computer-based control system (Figure 5.62) can be more complex and perform better than the analog control system. The computer can input the set point, sample the sense signal, and generate a control signal that can depend on the current and previous values of the set point, the sense signal, and the control signal in ways that would be difficult to implement in an analog controller. Both the analog and computer-based control systems require a power amplifier capable of driving the actuator.

In Laboratory Exercise 25, a temperature-control system is built using a thermistor to sense temperature, a cylindrical resistor to act as an oven, and a difference amplifier and power amplifier to perform the control function. This system is used to explore the open- and closed-loop response, described in the sections that follow. In Laboratory Exercise 26, temperature control is performed by using the microcomputer to sample the sense variable, compute a control variable, and output a control voltage.

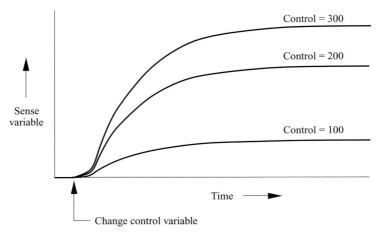

Figure 5.63 Open-loop step response as shown by a plot of the **sense** variable after an abrupt change in the **control** variable from zero to three different values.

5.10.4   Open-loop system response

Before attempting to control a system variable, it is important to understand how the **sense variable** responds in the open-loop mode to changes in the **control variable**. The measured behavior of the sense variable after an abrupt change in the control variable is called the **open-loop step response**. From this, the lag and response times of the system can be measured, which are both important time constants that play a role in the design and behavior of any control system. The **response time** is the time after an abrupt change in the control value for the sense value to make 63% $(1 - e^{-1})$ of the change to the final equilibrium value. After a period equal to five times the response time, the sensed variable should approach its asymptotic limit to within 0.7%.

Each value of the control variable will result in a different asymptotic equilibrium value of the sense variable (the open-loop response table). From this, the nonlinearities of the system response can be learned. In Figure 5.63, we show typical open-loop step responses of the sensed variable after an abrupt change in the control variable from zero to three different values. Note that the system responds after a time lag, and that the asymptotic equilibrium sense variable is not a linear function of the control variable.

5.10.5   Performance criteria for control algorithms

The **control algorithm** compares the sampled sense data to the desired sense value called the **set point**, and periodically derives the value of a **control variable** that is sent to an actuator to control the system and influence the sense data. In the control of motion, the sensor might be a digital position encoder and the actuator might be a motor. In the control of temperature, the sensor might be a thermistor or thermocouple and the actuator might be a resistor or heat pump.

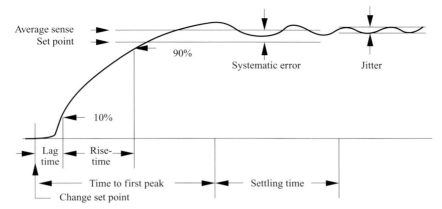

Figure 5.64 Behavior of a typical control system, where the sensed variable is shown as a function of time after an abrupt change in set point. After a lag time, the variable rapidly approaches the new set point, possibly overshoots, and possibly oscillates before settling to the sense point value. In some cases, due to inadequacies in the control algorithm, the oscillations continue indefinitely or there is a systematic error between the sensed variable and the set point that does not diminish with time.

Figure 5.64 shows the typical behavior of a control system after an abrupt change in set point. The **lag time** is the time from the change in set point to the time when the sensed variable first reaches a point 10% away from its initial value and 90% of the way from its final equilibrium value. The **risetime** is the time required for the sensed variable to change from the 10% point to the 90% point. Quite frequently, the sensed variable passes through the set-point value and overshoots before approaching from the other side.

The **time to first peak** is the time between the change in set point and the peak of the first overshoot. Possibly the sensed variable will oscillate about its average equilibrium value, and those oscillations will decay with a **settling time** that is the time required for the amplitude of the oscillations to decay by a factor of $e = 2.718,28$. It is also possible that these oscillations do not decay, but continue indefinitely.

Even long after the system has responded to the change in set point, it may exhibit jitter and systematic error. **Jitter** is the rms deviations from the mean value of the sense variable and is frequently due to noise in the control system or oscillations or "hunting" behavior of the control algorithm. **Error** in a control system is the difference between the sensed variable and the set point. The **average error**, or **accuracy**, is the difference between the set point and the time-averaged sensed variable.

5.10.6 Temperature control

One of the most commonly controlled quantities is temperature, for heating and cooling buildings or for controlling the reaction rates of chemical or biological processes. When fuel combustion or electrical resistance heating is used, the control engineer can actively control heating but must rely on heat losses for cooling. On the other hand, heat pumps

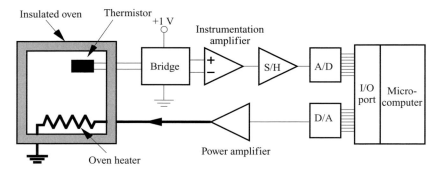

Figure 5.65 Schematic of a typical microcomputer-based temperature-control system. The thermistor and bridge provide a voltage that is converted with an A/D and read by the computer program. A control program writes a number to the D/A converter, whose output is amplified to drive an oven resistor.

or Peltier thermoelectric devices can both heat and actively cool. Figure 5.65 shows the schematic for the microcomputer-based temperature-control system explored in Laboratory Exercise 26. A thermistor is used for temperature sensing and a cylindrical ceramic resistor is used for heating. Note that almost every element in this control system has a nonlinear response:

1. the thermistor, whose resistance depends exponentially on $1/T$;
2. the bridge circuit, whose response is nonlinear when unbalanced; and
3. the ceramic resistor, whose power output is given by $P = RV^2$.

5.10.7   ON–OFF control

ON–OFF temperature control is used most often in the temperature control of building heating systems and ovens. It works by turning the heater ON when the **sense** variable is less than the **set-point** variable and turning the heater OFF when the **sense** variable is greater than the **set-point** variable. The **duty factor** is the fraction of time that the heater is ON. When the heater is turned ON, the temperature will usually rise rapidly until the **set point** is reached, the heater will be turned OFF, but the temperature may still continue to rise while the heating element comes into thermal equilibrium with the rest of the system. When the heater is turned OFF, cooling is accomplished by heat losses to the surroundings through any thermal insulation.

Representing the periodically measured sense variable as S, the set point as S', and the control variable as C, ON–OFF control can be described as the repeated application of the following algorithm:

ON–OFF control

If $S \leq S'$, then $C =$ maximum.
If $S \geq S'$, then $C =$ minimum.

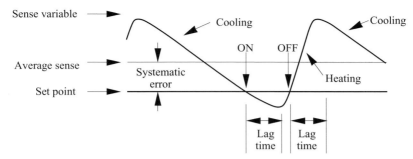

Figure 5.66 The **sense** variable as a function of time for ON–OFF control. The system passively cools until it reaches the set point when the heater is turned ON. After a time lag, the temperature rises. When the temperature exceeds the **set point**, the heater is turned OFF. The rate of heating is usually much greater then the rate of cooling, and the average temperature is above the **set point**.

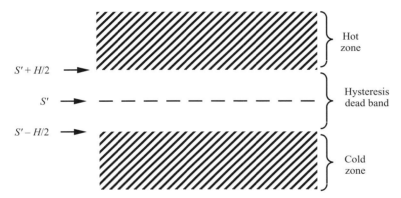

Figure 5.67 Temperature zones for ON–OFF control with a hysteresis dead band.

For effective control over a range of temperatures, the heater is chosen with sufficient capacity to overcome cooling losses at even the highest temperature. As a result, the rate of temperature rise when the heater is ON is generally much greater than the rate of temperature fall when the heater is OFF. Depending on how closely coupled the heater is to the rest of the system, there will be a lag time between switching the heater ON or OFF and any change in temperature. As seen in Figure 5.66, this results in a systematic error, where the temperature spends more time above the **set point** than below the **set point**, and the average value of the temperature can be considerably above the set point. This systematic error can be reduced by reducing the heater power in the ON state to increase the duty factor, but this reduces the maximum temperature capability of the system.

Hysteresis is used in ON–OFF control to introduce a dead band and reduce the rate at which the heater is turned on and off (Figure 5.67). A **dead band** is a region of the sensed variable where no control action takes place. While this reduces the accuracy of control, it prolongs the life of the switching hardware and heater.

Representing the periodically sampled sense variable as S, the set point as S', the hysteresis dead-band width as H, and the control variable as C, ON–OFF control with hysteresis dead band can be represented as the periodic application of the following algorithm:

ON–OFF control with hysteresis dead band

If $S \leq S' - H/2$, then $C = Z =$ maximum.
If $S \geq S' + H/2$, then $C = Z =$ minimum.
If $S' - H/2 < S < S' + H/2$, then C is unchanged.

5.10.8 Proportional control

In **proportional control**, the heater output is controlled in a fairly continuous manner, either by adjusting the voltage or by adjusting the width of a pulsed input. The sensed variable is sampled at time interval T and the open-loop system response time is T_r. The algorithm computes an **error** value as the difference between the **sense** value and the **set point**. The **control** value is changed by the quantity **error** $\times$ **gain** $\times T/T_r$ and then checked to see whether it exceeds the maximum heater input. It is necessary to keep track of small changes in the **control** variable by making it a floating-point number. It is converted to an integer only for writing to the D/A converter. Large values of **gain** result in **ringing**, or oscillations, about the set point that slowly decay, analogous to the underdamped oscillator. Large values of gain cause the **control** variable to jump between its minimum and maximum values so that the behavior is similar to the ON–OFF control.

Representing the sense variable as S (measured periodically with time interval T), the set point as S', the thermal response time as T_r, the gain as G, and the change in the control variable as ΔC, proportional control can be described as the periodic application of the following algorithm.

Proportional control

$C = C + \Delta C$
$\Delta C = (S' - S)G \, T/T_r$
If $C >$ maximum, then $C =$ maximum.
If $C <$ minimum, then $C =$ minimum.

5.10.9 PID (proportional–integral–differential) control

The behavior of the proportional control algorithm is improved substantially by the addition of terms that describe the integral of the **error** and the differential of the **error**. Note that unlike the proportional and integral terms, the difference between successive **error** terms does not depend on the set point and cannot be used by itself for control.

Representing the periodically measured sense variable as S, the set point as S', the most recent change in S as ΔS, the integral of S as $\sum S$, the proportional, integral, and differential control coefficients as D_p, D_i, and D_d, respectively, and the change in the control variable as ΔC, PID control can be described as the periodic application of the following algorithm.

PID control

$C = C + \Delta C$

$\Delta C = D_p(S - S') + D_i \sum(S - S') + D_d \Delta S$

If $C >$ maximum, then $C =$ maximum.

If $C <$ minimum, then $C =$ minimum.

5.11　Problems

5.1 A new drug has been developed for the treatment of high blood pressure (hypertension). You select a population of 18 patients suffering from hypertension. Half of the population is chosen at random to be the "control" group and these receive an inert substance (called a placebo). The other half is actually treated with the drug. The subjects do not know what group they are in. The table lists the measured systolic blood pressures before and after treatment:

	Before treatment	After treatment
Treated group		
Patient 1	160	140
Patient 2	196	180
Patient 3	158	152
Patient 4	160	135
Patient 5	194	165
Patient 6	159	123
Patient 7	181	159
Patient 8	192	185
Patient 9	216	185
Placebo group		
Patient 10	169	163
Patient 11	190	174
Patient 12	201	179
Patient 13	194	176
Patient 14	203	204
Patient 15	163	151
Patient 16	171	167
Patient 17	157	137
Patient 18	196	204

(a) Calculate the mean, the standard deviation, and the standard error of the mean for the two groups, before and after treatment (12 numbers in all).

(b) Calculate the unpaired Student's t for the change in blood pressure for each group (two numbers). Here we are only comparing the averages of the two groups, and disregarding the fact that the same patients are being used for the "before" and "after" measurements.

(c) For the unpaired Student's t, estimate the probability that the observed change in blood pressure could have arisen by chance for each group (two numbers). What number of degrees of freedom (n_f) did you use?

(d) Calculate the paired Student's t for the change in blood pressure for each group (two numbers). Here we are using the patients as their own controls, which cancels out the variation between patients and should give a more sensitive test for the effect of the drug.

(e) For the paired Student's t, estimate the probability that the observed change in blood pressure could have arisen by chance for each group (two numbers).

(f) Was the drug effective? Was the placebo effective?

5.2 Ten thousand kernels of corn have been randomly strewn over a tile floor consisting of 100 tiles.

(a) What is the average number of kernels per tile?

(b) How many tiles do you expect would have less than 80 kernels?

(c) How many would have more than 150?

5.3 Calculate chi-squared for the following data:

x_i	Model $f(x_i)$	Data value f_i	Standard deviation
0	80.0	90.0	5.0
10	50.0	55.0	2.5
20	30.0	18.5	1.5
30	20.0	18.5	1.5

(a) Estimate the probability of getting a higher chi-squared by chance.

(b) How many degrees of freedom are there?

(c) How well does the above model fit the data?

5.4 If m and n are positive integers, show that:

$$\int_{-\pi}^{\pi} \sin(mx)\sin(nx)\,dx = \int_{-\pi}^{\pi} \cos(mx)\cos(nx)\,dx \begin{cases} = 0 & \text{for } n \neq m \\ = \pi & \text{for } n = m \end{cases}$$

$$\int_{-\pi}^{\pi} \sin(mx)\cos(nx)\,dx = 0 \quad \text{for all } n, m$$

Hint: $\sin A \sin B = \frac{1}{2}[\cos(A - B)\cos(A + B)]$
$\cos A \cos B = \frac{1}{2}[\cos(A - B) + \cos(A + B)]$
$\sin A \cos B = \frac{1}{2}[\sin(A - B) + \sin(A + B)]$

5.5 You have a data-acquisition system that can sample an analog waveform, digitize it, and transfer the resulting digital number into computer memory every 10 μs.

Note: This system has no analog filtering before digitizing.

(a) What is the sampling frequency?

(b) You sample a sine wave that has a frequency $f = 25$ kHz, look at the resulting data, and observe that there are four samples per sine wave. How many samples per apparent sine wave would you expect to observe if the input frequency were 10, 50, 75, and 100 kHz?

(c) You acquire 1,024 samples of a 9,766-Hz sine wave oscillating between 0 and 2 V, and take the fast Fourier transform of the resulting data. List the location (frequency indices) of all nonzero Fourier amplitudes.

5.6 You have been using the system described in Problem 5.5 for input frequencies below 20 kHz for some time and with great success until a colleague in the next room turns on a pure 1.000-MHz (± 1 Hz) sine-wave oscillator and some of the unwanted signal gets into the analog input of your system.

(a) What would be the effect of the 1-MHz signal on your digitized data?

(b) If you took several 1,024-point data sets, where the start time was determined at random (such as by the push of a button), would the effect of the 1-MHz signal be the same for each data set? Give a reason for your answer.

(c) How could you most easily eliminate the effect of the unwanted 1-MHz signal on your digitized data?

5.7 You sample exactly 10 cycles of a 1,024-Hz square wave at a sampling frequency at 100 kHz, and take the fast Fourier transform.

(a) Over what period of time have you sampled?

(b) How many samples did you take?

(c) To what frequency does the Fourier coefficient H_1 correspond?

Hint: H_0 is the zero frequency or dc coefficient.

(d) At what Fourier coefficient do you expect the fundamental frequency to occur?

(e) At which Fourier coefficient do you expect the next nonzero harmonic to occur?

(f) At what Fourier coefficient is the highest frequency harmonic that can appear in your FFT?

5.8 You wish to develop a microcomputer-based system for monitoring the depth of liquid in a large tank by measuring the resonant frequency of the volume of air above the liquid (Figure 5.68). The tank is 10 m high and you want to determine the depth of the liquid to an accuracy of 0.1 m.

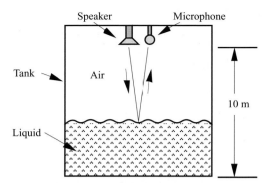

Figure 5.68 Measurement of liquid level in a tank using acoustic reflection.

- You have a speaker and microphone mounted inside the tank, at its top.
- The speed of sound in the air of the tank is $v_S = 300$ m per second.
- The fundamental resonance frequency (first harmonic) of a closed organ pipe is $0.25\, v_S/L$, where L is the length of the pipe. When the tank is empty, $L = 10$ m and the resonant frequency is 7.5 Hz. The tank is considered full when the level is 0.1 m below the speaker and microphone.
- Higher harmonics can be ignored.
- The speaker is driven by a white-noise generator, which excites all frequencies equally. Assume that the speaker and microphone have good response from 1 Hz to 10 kHz. (You have decided not to use acoustic reflection (sonar) or a sine-wave generator whose frequency can be computer controlled.)
- You decide to sample the microphone output, perform an FFT, and determine the fundamental resonant frequency by examining the Fourier amplitudes (similar to the first formant in Laboratory Exercise 22, fast Fourier transforms of the human voice, but the mouth is very large!).

(a) For a sampling interval $S = M \Delta t$ (where Δt is the time interval between samples and M is the number of samples), what is the frequency (in hertz) of the first Fourier magnitude F_1?

(b) What is the frequency corresponding to the Fourier magnitude F_n?

(c) What is the resonant frequency when the liquid is 0.1 m from the top?

(d) To satisfy the 0.1-m accuracy requirement for both nearly full and empty conditions, what are the requirements on M and Δt?

5.9 You are designing a system to sample analog data at a sampling frequency of 20 kHz in the presence of noise. An important consideration is the maximization of the signal-to-noise power ratio R, defined as:

$$R = \frac{\displaystyle\int_0^\infty S^2\, df}{\displaystyle\int_0^\infty N^2\, df}$$

where S is the signal amplitude and N is the noise amplitude. Your signal S has amplitude A from 0 Hz to 10 kHz and is zero above 10 kHz. The noise $N = 0.1\,A$ from 0 Hz to 100 kHz and is zero above 100 kHz.

(a) What is R before sampling or filtering?

(b) What is R before sampling using an ideal low-pass filter with a sharp cut-off $f_c = 10$ kHz? (Here most of the noise is excluded.)

(c) What is R after sampling but before using the filter in (b)? (Remember how high frequencies can appear in the sampled data as lower frequencies.)

(d) What is R after sampling and after using the filter in (b)?

5.10 Since an ideal low-pass filter is not commercially available, you are to design and build your own Butterworth four-pole low-pass filter for the system in Problem 5.9 above with corner frequency $f_c = 10$ kHz.

The amplitude response $G(f)$ of the filter is given by:

$$G(f) = \frac{1}{\sqrt{1 + (f/f_c)^8}}$$

(a) What is R before sampling using the four-pole filter?

Hint: Use the following table for the filter response integral F:

$$F = \int_0^{f_{max}} [G(f)^2]\,df$$

$f_{max} = 0$	$0.500\,f_c$	$1.000\,f_c$	$1.500\,f_c$	$2.000\,f_c$	$5.000\,f_c$	∞
$F = 0$	$0.500\,f_c$	$0.924\,f_c$	$1.017\,f_c$	$1.024\,f_c$	$1.025\,f_c$	$1.025\,f_c$

(b) You note that the filter of (a) has an amplitude of only 0.707 at $f_c = 10$ kHz. What should f_c be so that the filter has an amplitude accuracy of $1/2$ LSB at 10 kHz for an 8-bit A/D converter?

Hint: For small ε:

$$\frac{1}{\sqrt{1+\varepsilon}} \approx \frac{1}{1+\varepsilon/2} \approx 1 - \frac{\varepsilon}{2}$$

(c) Before sampling, what is the signal-to-noise power ratio R using this more accurate filter?

(d) After sampling, what is the signal-to-noise power ratio R using this more accurate filter?

(e) What is the signal-to-noise power ratio R after doubling the sampling frequency and using this more accurate filter?

5.11 Given the digital filter $y_i = x_{i-1} - x_{i-2}$:

(a) What is the impulse response?

Hint: An impulse has $x_0 = 1$ and $x_i = 0$ for $i \neq 0$.

(b) What is the response to the pulse $x_0 = x_1 = x_2 = 1$?

(c) To what does this digital filter most closely correspond: (i) low-pass filter, (ii) first derivative, or (iii) second derivative?

5.12 Given the following digital filter $y_i = x_{i-1} - 2x_{i-2} + x_{i-3}$:

(a) What is the impulse response?

(b) What is the response to the pulse $x_0 = x_1 = x_2 = x_3 = x_4 = x_5 = 1$? (All other $x_i = 0$.)

(c) To what does this digital filter most closely correspond: (i) low-pass filter, (ii) first derivative, or (iii) second derivative?

5.13 You want to design a temperature-control system for an electrically heated incubator for hatching chicken eggs. You are provided with a thermistor, an A/D converter, a microcomputer with parallel interface, a power relay (a device that can switch a large current ON or OFF and is controlled by a small current), and assorted electronic components.

(a) Draw a block diagram of your design for simple ON–OFF control and label all essential components.

(b) Describe the steps involved in simple ON–OFF control (or draw a flow diagram).

(c) How would (b) be modified to implement ON–OFF control with hysteresis?

(d) Do either of the techniques in (b) or (c) (theoretically) stabilize *at* the set point?

(e) How would the *hardware* be changed to implement proportional control?

5.14 Design a PID temperature control system using a thermoelectric heat pump, a thermistor, and a microcomputer with -10 to $+10$ V A/D and D/A converters. The heat pump and thermistor are in a steel box insulated with glass fibers.

(a) Sketch below a block diagram for the sensor side of the system, from the thermistor to the A/D converter.
  • Include and label all essential components.
  • Show typical voltage levels at all important points.

(b) Sketch a block diagram for the actuator side of the system, from the D/A converter to the thermoelectric heat pump.
  • Include and label all essential components.
  • Show typical voltage and current levels at all important points.

(c) List the program steps necessary for sensing and PID control.

(d) Describe the function of each of the main components of your system.

(e) What minimum and maximum temperatures do you think that this control system could achieve? Give reasons for your limits.

5.15 Show explicitly that if f_k is defined by:

$$h_k = \sum_{j=0}^{M-1} a_j \cos(2\pi jk/M) + b_j \sin(2\pi jk/M)$$

then the Fourier transform coefficient H_{M-n} is given by:

$$H_{M-n} = (M/2)(a_n + ib_n)$$

5.16 You need to test a system that samples an analog signal and performs the fast Fourier transform.

Assume:

- The sampling frequency is 32,768 Hz.
- An eight-pole Butterworth low-pass anti-aliasing filter is used with a corner frequency of 12 kHz.
- A Hann window is used to prevent spectral leakage.
- You have decided to use a 1.024-kHz symmetric square wave as a test signal and take 8,192 samples.

(a) To what frequencies (in hertz) do the first and second Fourier magnitudes (F_0 and F_1) correspond?

(b) At what Fourier index do you expect the first harmonic of the square wave to occur?

(c) At what Fourier index would you expect the next nonzero harmonic to occur?

(d) At what Fourier index do you expect the nth harmonic of the square wave to occur?

(e) What is the gain of the Butterworth filter at 16 kHz (assume dc gain $= 1$)?

(f) What is the highest frequency harmonic you would expect to see, at what Fourier index would it occur, and what would its magnitude be relative to the first harmonic?

5.17 To measure the harmonic distortion of a high-fidelity audio amplifier, you use a pure sine-wave input of exactly 100 Hz and sample the amplifier output for exactly 2 seconds at a sampling frequency of 32,768 Hz. You then take the fast Fourier transform of the digital data.

(a) To what frequencies (in hertz) do the first and second Fourier amplitudes (F_0 and F_1) correspond?

(b) What Fourier amplitude corresponds to the highest frequency that can be reliably sampled and what is that frequency?

(c) Assuming that the amplifier can amplify the 100-Hz tone perfectly with no distortion, which Fourier coefficients would be nonzero?

(d) Assuming that the amplifier introduces some distortion that causes the output to be a distorted sine wave described by both even and odd harmonics, which Fourier amplitudes would be nonzero?

Note: The even harmonics are zero only for symmetric waveforms such as the square or triangle waves you used in the laboratory exercises.

(e) If (1) the audio tone were changed to 100.25 Hz, (2) you do not multiply the data by a windowing function, and (3) the amplifier has no distortion, what would the Fourier transform look like? (Describe large, small, and zero components.)

5.18 Design a system for analyzing the harmonic content of musical instruments using the FFT. You know that the sounds will have a fundamental frequency and higher harmonics of that frequency.

The requirements are:

- Maximum frequency of interest 20 kHz (but higher frequencies may occur).
- Frequency resolution of 0.1 Hz (closest frequencies that can be clearly resolved in the FFT).
- Waveform voltage resolution $\pm 0.015\%$ of full range.
- Minimal spread of spectral leakage.

You have available the following:

- A microphone and instrumentation amplifier capable of converting music to an analog waveform with an amplitude of ± 5 V.
- A microcomputer with a counter/timer, a digital input port, and FFT program code.
- The digital input port has a "data available" status bit (input). The input port requires 1 μs to read a byte of data or the status bit. You may assume that other computer operations take a negligible amount of time.
- An external successive approximation A/D converter chip with a "start conversion" input and a "conversion complete" output. The input must be held constant during conversion.
- The counter/timer can be set up by the computer to generate external pulses with any width and any time interval.
- A 12-pole Butterworth low-pass filter with a gain $= 0.99$ at 20 kHz and 0.000,02 at 50 kHz.

Answer the following:

(a) What is the maximum allowable time period between samples?
(b) What is the minimum number of required A/D bits?
(c) What is the maximum allowable conversion time of the A/D?
(d) How long do you need to sample the waveform?
(e) What is the minimum number of samples required?
(f) Sketch your system design, showing and labeling all essential components and signal lines.
(g) List the steps (hardware and software) involved in sampling the waveform and taking the FFT.

(h) For a musical instrument with a fundamental frequency of 100 Hz, at what Fourier amplitude H_n would you expect the fundamental to occur? (Give the Fourier frequency index as n.)

(i) At what Fourier amplitude H_n would you expect the mth harmonic to occur?

5.19 In Laboratory Exercises 21 and 22, you took 512 samples, stored them in memory, called the FFT function to generate 512 complex Fourier coefficients, and computed 512 magnitude values. Design a system using digital filtering to do these same tasks continuously, rather than in "batch" mode. Assume that you have available a large number of low-cost processors suitable for digital filtering.

(a) Give the formula for your digital filter. Is it FIR or IIR?

(b) Draw a representative section of the block diagram, showing components and interconnections.

(c) How many processors are needed in all?

5.20 Design a microcomputer-based system for monitoring a 110-V, 60-Hz power line and determine the following characteristics:

• The frequency to an accuracy of 0.01 Hz.

• The amplitude of any other frequencies arising from distortions of the 60-Hz sine wave. These could arise from switching transients caused by nearby silicon-controlled rectifiers commonly used in power supplies, welding equipment, and light dimmers. See Figure 5.69 below for a typical plot of voltage versus time.

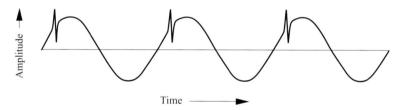

Figure 5.69 Sine wave with typical harmonic distortion caused by switching transients.

Assume the following:

• You have available an analog data-acquisition circuit with a sampling frequency of exactly 10 kHz. This circuit has a built-in digital clock, a sample-and-hold amplifier, a 12-bit A/D converter, and interface circuits allowing you to read the digitized samples with a C program.

• You are to acquire a number of samples at 10 kHz, and then perform the fast Fourier transform (FFT). The Fourier coefficients are then inspected by the program for any abnormalities and the entire procedure is repeated.

• The fundamental frequency is always between 59.9 and 60.1 Hz.

• You have decided to use a Hann window (like the one you used in Laboratory Exercises 21 and 22) to reduce spectral leakage.

• You do not know the highest frequencies that may be present in the signal.

(a) Draw a block diagram of your system. Label every essential component and signal line. (You may draw the microcomputer and each peripheral circuit or component as a separate box.)

(b) What is the minimum number of samples that you need to acquire?

(c) To what frequency does the first FFT coefficient (H_0) correspond?

(d) What Fourier coefficient corresponds to 60 Hz?

(e) If the frequency is exactly 60 Hz, and there is a distortion as shown in the figure above, what discrete Fourier coefficients would be expected to be nonzero?

(f) List the steps that the system (hardware and program) needs to do to determine the discrete Fourier coefficients.

5.21 Implement a solution to Problem 5.20 by **continuously** sampling and using digital filters continuously to compute **only** the discrete Fourier coefficients that you need. (In Problem 5.20 you were asked to acquire a number of samples **in a large batch** and then perform the FFT to compute **all** the discrete Fourier coefficients.)

(a) What is the minimum number of digital filters that you need? (Assume that each filter computes either a real or imaginary part of a discrete Fourier coefficient.)

(b) Write down a digital filter for the real part of the discrete Fourier coefficient corresponding to frequency f. (You may use either an IIR or FIR filter.)

(c) List the steps that the system (hardware and program) needs to perform to process each new data sample.

5.22 Your first assignment as an engineer with the Super Prosthetics Corp. is to design a battery-powered EMG-type control system for a motor-operated prosthetic arm. The system is to do the following:

- sense the EMG signal from three electrodes (+, −, ground) on some available muscle,
- process the raw EMG signal to produce a control signal with a frequency content of 0–2 Hz and a voltage level between 0 and 10 V,
- interface the processed signal to an A/D converter connected to a microcomputer-on-a-chip that is built into the prosthetic arm and has a suitable control program,
- interface the microcomputer D/A converter to the prosthetic motor.

Your design must work in spite of electrode and amplifier drift, and large amounts of 60-Hz interference.

Note: Low-voltage battery-powered bioinstrumentation that is clearly isolated from all other power sources does not require an isolation amplifier.

(a) Sketch a block diagram of your system. Include and label all essential components and signals.

(b) Describe or sketch the raw EMG signal. Include time and voltage scales.

(c) Describe the function of each of the main components of your system and describe or sketch the waveform that it produces.

5.23 Give the steps you would use to compute the input waveform $u(t)$ needed to produce a desired output $y(t)$ of an analog filter (Figure 5.70).

$u(t)$ → | Analog filter | → $y(t)$

Figure 5.70 Analog filter with input and output waveforms.

5.24 Draw a block diagram of a microcomputer-based system for sampling, digital storage, and playback of an audio music performance. Show all essential components and signal connections.

5.25 Given a single-pole low-pass analog filter with corner frequency $f_c = 1/(2\pi RC)$, design a digital filter that would most closely match the properties of the analog filter. Describe any differences.

5.26 After sampling a nonintegral number of periods of a periodic waveform, how would you apply windowing to reduce leakage in the FFT? Sketch a typical window and describe (simply) how it reduces leakage.

5.27 Design a simple *analog* temperature-control system using the following components:
- a thermistor;
- a thermoelectric heat pump and heat sink (one surface heats or cools, depending on input voltage polarity; the other surface is kept at room temperature by the heat sink);
- a power amplifier (single input–single output, requires ±10-V supply);
- a ± 10-V power supply;
- a steel box insulated with glass fibers;
- any components or circuits used in the laboratory exercises.

  Note: Do not use a computer or analog filtering.

  Your system should do the following:
  - keep the inside of the box at a chosen temperature,
  - allow the chosen temperature to be varied.
  (a) Sketch below a block diagram for the system.
      - Include and label all essential components and include all interconnections.
      - Show typical voltage and current levels at all important points.
  (b) Do (a) again, replacing the thermoelectric heat pump with a high-wattage, high-temperature resistor.
  (c) List the relative advantages and disadvantages of the two systems described above.

5.28 In concert halls and outdoor arenas, the sound that the audience hears is not necessarily the sound produced by the performers. The reason is that the limited response of the loudspeakers, absorption by surrounding surfaces, resonances within enclosing volumes, and feedback effects can greatly alter the frequency spectrum (Figure 5.71). One spectacular feedback effect is the loud whine that occurs at a particular frequency when the loop gain (from the recording micro-phone to the amplifier to the speakers and back through the air to the recording microphone) is greater than 1. At high volumes, it would be desirable to attenuate the signal at such resonant frequencies.

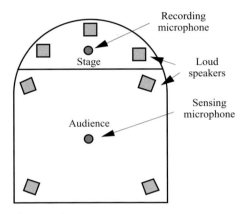

Figure 5.71 Diagram of an amphitheater as a resonant system.

Design a system for frequency filtering in a sound system so that the sound that reaches a sensing microphone in the audience has the same frequency content as the original sound received by a recording microphone on the stage. Do not worry about phase, separate stereo or quadraphonic channels, or trying to correct the sound at any other place in the audience.

Assume:
- The recording microphone and the power amplifiers that drive the loudspeakers have a flat response from 20 Hz to 25 kHz.
- The sampling microphone placed in the audience records all frequencies between 20 Hz and 25 kHz equally.
- You plan to use an eight-pole Butterworth filter in your sampling microphone circuit with a corner frequency $f_c = 20$ kHz. The frequency response of the n-pole Butterworth filter is given by:

$$G(f) = \frac{1}{\sqrt{1 + (f/f_c)^{2n}}}$$

Part 1: determining the acoustical response of the hall:
- You decide to determine the acoustical response of the hall by sending a series of sharp pulses into the power amplifiers with a repetition period of 20 Hz and recording the response with the sensing microphone.

- The pulses are so narrow that the nonzero values of their frequency spectrum are equal from 20 Hz to 30 kHz.
- You avoid windowing effects by sampling the output of the recording microphone in the audience for exactly 10 repetition periods of the 20-Hz pulses at a sampling frequency of 65,536 kHz.
- You then take the FFT to determine the frequency response of the hall

(a) How many samples will you be taking?

(b) To what frequency does the Fourier coefficient H_1 correspond?

(c) What Fourier coefficient corresponds to 20 Hz?

(d) Which Fourier coefficients are nonzero?

(e) What is the gain of the Butterworth filter at $20/\sqrt{2} = 14.14$ kHz?

   Hint: $1 + \sqrt{1 + \varepsilon} \approx 1 - \varepsilon/2$ for small ε.

(f) What is the gain of the Butterworth filter at $30/\sqrt{2} = 21.21$ kHz?

(g) What is the gain of the Butterworth filter at $40/\sqrt{2} = 28.28$ kHz?

Part 2: correcting an arbitrary waveform

Assume that between the sensing microphone on the stage and the power amplifiers that there is a filter system (called a graphic equalizer) that has 31 adjustable gains g_m that can be set by computer control (Figure 5.72). Each gain affects a narrow frequency band centered at f_m:

$$f_m = 20 \, \text{Hz}(2)^{(m-1)/3}, \quad m = 1 \text{ to } 31$$
$$f_1 = 20 \, \text{Hz}, \qquad f_{31} = 20 \, \text{kHz}$$

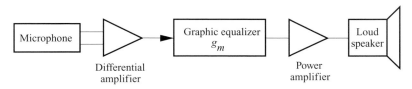

Figure 5.72 Audio recording, filtering, amplification, and loudspeaker system.

(h) Explain how would you use the response function measured in part 1 above to set the m gains of the graphic equalizer to correct an arbitrary waveform detected by the microphone on the stage. Include an equation for g_m as a function of the Fourier coefficients H_n determined in part 1.

5.29 Design a system for converting digital audio data from compact digital disk (sampled at 44 kHz) to digital audio tape (to be played at 48 kHz). Simply reading the 16-bit digital words from the compact disk and writing them to digital tape will not work because on playback all frequencies will be shifted upward by about 10%.

Note:

• The waveform produced by the D/A converter in the digital audio tape player must be an accurate replica (both in phase and frequency) of the waveform originally sampled to produce the digital disk.

• The conversion process can be performed either in batch mode or continuous mode

Do the following:

(a) Sketch the system. Show and label all essential components and signal lines.

(b) Describe how the system works.

5.30 The formula for the gain of the noninverting amplifier (Figure 2.3) is given by:

$$G_{\pm} = \frac{V_0}{V_1} = \frac{R_1 + R_2}{R_1}$$

Assume that resistors with 10% standard deviation are used with values $R_1 = 1 \text{ k}\Omega$, $R_2 = 4 \text{ k}\Omega$.

(a) What is the gain $G_{\pm}$?

(b) What is the standard deviation of $G_{\pm}$?

Hint: Use the error propagation formula.

5.31 Your job is to design an **analog altitude control system** for an airplane, using negative feedback. This system is supposed to keep the airplane at a nearly constant altitude (height above sea level), despite updrafts, downdrafts, and changes in engine speed.

The vertical acceleration is measured using a **piezoelectric transducer** connected to a mass. When the unit it accelerated, the force causes charges to separate. The sensitivity is 10 pC/g ($g = 10 \text{ m/s}^2$, the acceleration of gravity), and the capacitance is 1 nF.

Integrating this signal (Figure 5.73) gives the vertical velocity. Integrating again gives the altitude.

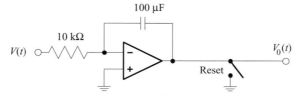

Figure 5.73 Circuit for performing the first integral of $V(t)$. An input voltage of 1 V will produce an output V_0 that increases at 1 V/s. A switch can set V_0 to zero at any time t_0.

Set your overall circuit gain so that if the system is reset and subjected to a 0.1 g acceleration for 1 s, the second integral output is 50 mV.

The altitude of the airplane is changed by a **stepping motor** that adjusts the angle of the trailing horizontal tail surfaces (the elevators). If the elevators are angled down, the tail of the airplane is forced up, and the airplane dives. If the elevators are angled up, the tail of the airplane is forced down, and the airplane climbs.

(a) Sketch your system design. Include the sensors, actuators, and any other necessary electronics (but keep it simple). Include and label all essential components and interconnections.

(b) For this item, assume that the stepping motor has been disconnected (open-loop condition), the integrators have been reset at $t = 0$ s, that the system is subjected to a $0.1g$ downward acceleration from $t = 0$ to 1 s, and that the acceleration is zero after $t = 1$ s. Plot voltage versus time from $t = 0$ to 2 s for the (i) accelerometer output, (ii) first integral output, and (iii) second integral output.

(c) For this item, assume that the stepping motor has been reconnected (closed-loop condition). Describe how the various components of the system function when the airplane encounters a downdraft.

5.32 You are given a "black box" electronic circuit with one input V_{in} and one output V_{out}. You are to measure $|V_{out}/V_{in}|$ as a function of frequency from 100 to 100,000 Hz in 100-Hz steps (there are a total of 1,000 frequency values as follows: $100, 200, \ldots, 99{,}900, 100{,}000$ Hz). Rather than setting the sine-wave generator to 1,000 different frequencies and measuring $|V_{out}/V_{in}|$ directly, you decide to input a periodic series of 1-μs wide pulses, sample the output, take the FFT of the sampled values, and compute the magnitude F_n of each complex Fourier coefficient H_n. You know that a periodic series of 1-μs wide pulses at a repetition frequency f_r has a fundamental at f_r plus harmonic multiples of f_r, and all Fourier magnitudes are nearly equal from the fundamental to 100 kHz (*Note*: the amplitude drops to 0.707 at 443 kHz). One of your design goals is to minimize the number of samples to minimize: (i) the amount of memory needed, and (ii) the size of the FFT to be computed.

(a) Draw a block diagram of all components and essential interconnections. Label all components, control lines, and data lines.

(b) What is your pulse repetition rate?

(c) What is your sampling frequency?

(d) How long is your sampling window?

(e) How many samples will you take?

(f) What frequency does the Fourier coefficient H_n correspond to?

(g) For an arbitrary linear, time-invariant black box, what Fourier magnitudes F_n are sure to be zero (or very small)?

(h) If the black box contains a single-stage low-pass filter with a corner frequency of 1 kHz, describe the magnitude ratios F_n/F_1 that you would expect.

5.33 Imagine that many years ago, a spacecraft was sent to measure the magnetic fields in the great void between the Sun and the nearest star. Every 100 seconds the measurements are digitized and then phase and amplitude encoded (like the 56-kbaud modem that connects your computer to the internet) to produce a one-second-long analog-like signal with a frequency content between 1 and 3,000 Hz. Because the spacecraft is far from Earth and has limited battery power, the data signal is weaker than the background noise from the rest of the universe (Figure 5.74).

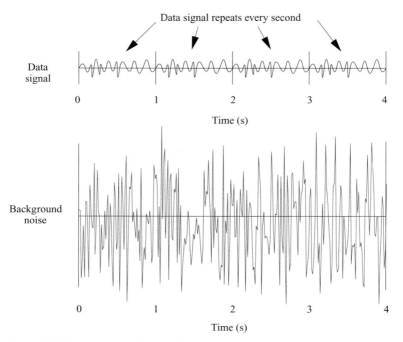

Figure 5.74 Periodic data signal and white background noise.

To be able to detect the weak data signal, you use three techniques:

  (i) the signal is band-pass filtered before transmission,
 (ii) the spacecraft sends the same one-second-long signal a 100 times with a period of exactly one second,
(iii) you use your knowledge of the FFT of a periodic signal to separate the background noise further from the data signal.

To do this, you perform the following steps to the signal received at the Earth:

1. low-pass filter the signal (weak data signal plus background noise);
2. sample the filtered signal for exactly 100 s at 10,486 Hz ($2^{20} = 1,048,576$ samples), (assume that the time it takes for the signal to reach the Earth is known at all times so that sampling always begins at the exact start of the first one-second-long signal);
3. take the FFT;

4. subtract as much of the background noise as possible;
5. recover one cycle of the data signal;
6. demodulate to transform the one-second modulated analog signal into the original digital signal (assume that you have a modem that does this).

Note: If $a(t) = b(t) + c(t)$, then $\text{FFT}(a) = \text{FFT}(b) + \text{FFT}(c)$

Do the following:

(a) Describe (or sketch) the Fourier magnitudes F_n from the FFT in step 3 as a function of the frequency index n.
(b) To what frequency does the Fourier magnitude F_n correspond?
(c) Design a Butterworth low-pass anti-aliasing filter that has a gain >0.99 for frequencies below 3,000 Hz and a gain <0.001 for all frequencies that could alias below 3,000 Hz.
(d) Explain whether a Hann window would improve the recovered waveform.
(e) Describe in detail how a computer program would implement steps 2–6. (*Note*: There are over seven program steps.)

5.34 Show that if the function $h(t)$ has the following properties:

1. periodic with period P: $h(t) = h(t + P)$, and
2. the first half of each period has the same shape but opposite sign of the second half of each period: $h(t) = -h(t + P/2)$,

then the Fourier transform:

$$H(f) = \int_{-\infty}^{\infty} h(t)e^{-j2\pi ft} \, dt$$

has all even harmonics equal to zero, i.e:

$$H(n/P) = 0, \qquad n = \text{even}$$

5.35 Design an analog control system using PID control for an elevator in a building (Figure 5.75).

• The elevator: (1) hangs from a cable in the elevator shaft, (2) can move up and down on a set of vertical rails from the basement to the fifth floor, and (3) carries a maximum of 10 passengers.
• The cable is wound around a large drum of radius R, and a large electric motor is provided to turn the drum. The distance between floors is $D = 4\pi R = 4$ m (to an accuracy of 1 mm).
• The other end of the cable comes off the drum and is connected to a counter-weight which can also move up and down on a set of rails. The counterweight has mass W equal to the mass of the elevator M plus the mass of five passengers (assume all passengers have the same mass p).

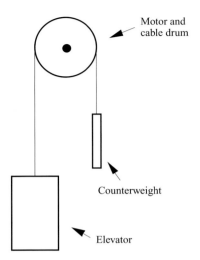

Figure 5.75 Elevator with motor-driven pulley and counterweight.

- In the elevator car and on each floor are a set of buttons for the six possible floors: 0 (the basement), 1, 2, 3, 4, and 5.
- There is a floor selector circuit that produces an output that corresponds to the last button pushed: -10 V for floor 0 (the basement), -6 V for floor 1, etc., up to $+10$ V for floor 5. When a new floor is selected, the elevator door closes, the elevator moves to the desired floor, and the door opens.
- The elevator lifting motor has a power-converter circuit that converts a -10- to $+10$-V input signal V_{in} (input impedance 10 kΩ) to a high current -600- to $+600$-V potential V_m that powers the motor. $V_m = 60\, V_{in}$.
- The motor provides the force needed to hold the elevator at a selected floor by counteracting the imbalance between the mass of the elevator plus n passengers $(M + np)$ and the mass of the counterweight $(W = M + 5p)$.
- To hold the elevator when empty $V_m = -300$ V; with five passengers $V_m = 0$ V; with ten passengers $V_m = +300$ V.
- The motor also provides the force needed to accelerate the elevator (plus counterweight) from one floor to another.
- You have a 10-kΩ helical resistor (10 turns).
- You also have a difference amplifier with a 1-MΩ input impedance on both the $+$ and $-$ inputs, an output impedance of 100 Ω, and a gain G that can be set by you.

(a) Sketch your circuit design, including the floor selector circuit, the motor power converter, the elevator motor, the difference amplifier, the PID control circuit, and any other components that may be necessary.

(b) For $G = 100$ and 1,000, tabulate the voltages at important points in your circuit after the elevator has five passengers, the second floor has been selected, and the elevator has come to rest.

(c) For $G = 100$ and 1,000, tabulate the voltages at important points in your circuit after five more passengers enter the elevator (total $= 10$) while the floor selector circuit is still set to the second floor. Describe any motion of the elevator relative to (b) above.

(d) For $G = 100$ and 1,000, tabulate the voltages at important points in your circuit after all passengers get off the elevator while the floor selector circuit is still set to the second floor. Describe any motion of the elevator relative to (c) above.

(e) For $G = 1,000$, plot the voltages at important points in your circuit from time $t = 0$ to 100 s during the following events:

$t = 0$ s	the elevator is at rest with five passengers at floor 2
$t = 10$ s	five more passengers instantaneously jump on
$t = 20$ s	the fourth floor is selected
$t = 30$ s	the elevator comes to rest
$t = 40$ s	all passengers instantaneously jump off
$t = 50$ s	the basement is selected
$t = 70$ s	the elevator comes to rest

Assume that from rest, the elevator takes 1 s to accelerate to its limiting speed, and 1 s to decelerate to rest.

5.36 Design a circuit for controlling a street light. The street light is to be turned on at night and turned off during the day. Components available are:
- *pin* photodiode;
- street light (200 V, 10 A);
- electromechanical relay switch (an input of 10 V and 1 A is needed to close the switch, and an output of 200 V and 10 A is needed to light the street light);
- additional components as needed, but keep it simple.

(a) Sketch your design, showing and labeling all essential components and inter-connections.

(b) List voltages at key points in the circuit at night.

(c) List voltages at key points in the circuit during the day.

(d) Describe three important common-sense considerations for mounting the *pin* photodiode.

5.37 The convolution $a(t)$ of two functions $b(t)$ and $c(t)$ is defined as:

$$a(t) = b(t) \cdot c(t) = \int\limits_{-\infty}^{+\infty} b(t)c(t - t')\,dt'$$

Also define a unit rectangle waveform $r(t)$:

$$r(t) = 1 \quad \text{for } \pm 0.5 < t < +0.5, \quad r(t) = 0 \text{ otherwise}$$

(a) Sketch the convolution $a(t)$ of $b(t) = r(t)$ and $c(t) = \delta(t - 1)$.

(b) Sketch the convolution $a(t)$ of $b(t) = r(t)$ and $c(t) = \sum_{k=-\infty}^{+\infty} \delta(t - 2k)$

(c) Compute the convolution $a(t)$ of $b(t) = r(t)$ and $c(t) = r(t)$. The result should be a familiar function shown in the Course Reader and encountered in Laboratory Exercise 21.

(d) Using the Fourier transform of $r(t)$ (Example 5.13) and the Fourier convolution theorem, compute the Fourier transform of $a(t) = r(t) \cdot r(t)$.

5.38 You have been asked to help design a Doppler ultrasound system for measuring the speed of approaching vehicles on a highway. The system sends a continuous tone of 100-kHz sound waves in a well-defined direction and there is a receiver alongside that receives the Doppler-shifted echo. Your part in the project is to design the sampling and signal processing hardware and software, starting from the echo receiver.

- The Doppler-shifted frequency is given by $f' = f\left(\frac{(1+v/v_s)}{(1-v/v_s)}\right)$ where v is the speed of the approaching vehicle and v_s is the speed of sound in air (assume 300 m/s).
- To simplify and speed your calculations, use the approximation $f' \approx f(1 + 2v/v_s)$.
- Assume that the echo receiver signal is the sum of 0.1-V p–p echo and an unavoidable 10-V p–p primary 100-kHz tone that leaks into the echo receiver.
- The echo circuit has wide-band amplification with white noise, so you decide to use a low-pass eight-pole Butterworth anti-aliasing filter that effectively accepts frequencies below f_1 and rejects frequencies above $2f_1$, where f_1 is a frequency of your choosing.
- Your system samples at frequency f_s, takes M samples (where M is a power of 2), performs the FFT, and must be able to determine the speed of an approaching vehicle between 3 and 60 m/s to an accuracy of ±0.3 m/s.

Answer the following:

(a) What are the echo frequencies for vehicle speeds of 3, 30 (67 mph), 30.3, and 60 m/s (134 mph)?

(b) How long must your sampling window be to distinguish 30 m/s from 30.3 m/s clearly?

(c) How can you reduce the spectral leakage from the 10-V p–p 100-kHz primary onto the 0.1-V p–p echo frequency?

(d) Considering the maximum signal frequency (corresponds to 60 m/s) and the white noise in the echo receiver circuit, what value of f_1 does your low-pass filter require?

(e) Considering the value of f_1 from (d) above, and that the filter rejects frequencies above $2f_1$, what is the minimum sampling frequency that prevents the aliasing of white noise between f_1 and $2f_1$ into frequencies below f_1?

(f) How many samples will you take for each measurement of vehicle speed?

(g) Sketch all FFT magnitudes versus frequency index for a vehicle speed of 30 m/s. You will need to use a vertical axis labeled in powers of 10. Provide an additional label for the horizontal axis in hertz. Assume that the white noise is 10% of the Fourier magnitude of the echo signal.

5.39 You sample exactly five cycles of a 15-Hz square wave (after anti-aliasing filtering) and compute the FFT. The magnitude of your FFT coefficients are plotted in Figure 5.76. Explain the nonzero values at $n = 5, 15, 20, 25, 35, 45, 55, 73, 83, 93, 103, 108, 113,$ and 123. (You do not need to explain the amplitudes, just why they are nonzero.)

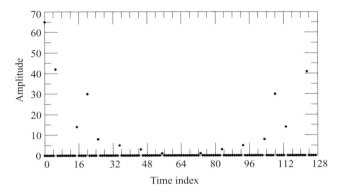

Figure 5.76 FFT of data taken in the laboratory.

5.40 You are given a linear, time-invariant system that acts as a single-stage low-pass filter plus a damped oscillator so that a step change at the input results in output oscillations that exponentially decay with time. The impulse response $c(t)$ is the sum of a decaying exponential and a decaying harmonic (Figure 5.77):

$$c(t) = e^{-t/\tau} + 2e^{-t/\tau}\cos(2\pi f_0 t), \quad \text{where } f_0 = 100 \text{ Hz} \quad \text{and} \quad \tau = 1 \text{ s}$$

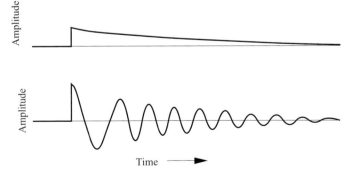

Figure 5.77 Impulse response $c(t) =$ the sum of a decaying exponential and a decaying cosine wave.

(a) Derive the equation of the Fourier transform of the impulse response (explain your reasoning).

(b) Sketch the Fourier transform (magnitudes only) of the impulse response.

(c) How would you compute the input that would make a square-wave output? *Note*: The Fourier transform of the decaying exponential:

$$h(t) = 0 \quad \text{for } t < 0; \qquad h(t) = Ae^{-t/\tau} \quad \text{for } t \geq 0$$

$$H(f) = \frac{A}{\sqrt{1 + 4\pi^2 f^2 \tau^2}}$$

5.12 Additional reading

Michael Andrews, *Programming Microprocessor Interfaces for Control and Instrumentation*, Prentice Hall, Englewood Cliffs, NJ, 1982.

Karl J. Åström and Björn Wittenmark, *Computer Controlled Systems*, Prentice-Hall, Englewood Cliffs, NJ, 1984.

E. Oran Brigham, *The Fast Fourier Transform and its Applications*, Prentice Hall, Englewood Cliffs, NJ, 1988.

Robert W. Hornbeck, *Numerical Methods*, Prentice Hall, Englewood Cliffs, NJ, 1975.

Benjamin C. Kuo, *Digital Control Systems*, Holt, Rinehart, and Winston, New York, 1980.

Benjamin C. Kuo, *Automatic Control Systems*, Prentice Hall, Englewood Cliffs, NJ, 1987.

Lyman Ott and William Mendenhall, *Understanding Statistics*, Duxbury Press, Boston, MA, 1985.

Charles L. Phillips and Royce D. Harbor, *Feedback Control Systems*, Prentice Hall, Englewood Cliffs, NJ, 1988.

Charles L. Phillips and H. Troy Nagle, Jr., *Digital Control System and Design*, Prentice Hall, Englewood Cliffs, NJ, 1984.

William H. Press, Brian P. Flannery, Saul A. Teukolsky, and William T. Vetterling, *Numerical Recipes in C*, Cambridge University Press, New York, NY, 1988.

C. B. Rorabaugh, *Digital Filter Designers Handbook*, McGraw-Hill, New York, NY, 1993.

George W. Snedecor, *Statistical Methods*, Iowa State University Press, Ames, IA, 1965.

Samuel D. Stearns and Ruth A. David, *Signal Processing Algorithms*, Prentice Hall, Englewood Cliffs, NJ, 1988.

Laboratory Exercise 20

Analog ↔ digital conversion and least-squares fitting

Purpose

To digitize a set of analog voltages with an A/D converter, to use a D/A converter to recover the voltages, and to use the least-squares technique to determine the accuracy and linearity of the procedure.

Equipment

- IBM-compatible Pentium microcomputer with Windows NT operating system and Microsoft Visual C++ compiler
- Printer (shared with other laboratory stations)
- Data Translation DT3010 interfacing board
- ±12-V power supplies
- Two 10-μF, 25-V electrolytic capacitors (put between power and ground at circuit board binding posts)
- One 10-kΩ trimpot
- Digital multimeter
- Superstrip circuit board

Background

1.   Analog-to-digital (A/D) converter characteristics

The purpose of the A/D converter is to transform an analog voltage into a binary number. For an ideal linear A/D, the digital representation n corresponding to the analog voltage

V is given by:

$$n = \left[\frac{V - V_{\min}}{\Delta V} + \frac{1}{2} \right]_{\text{INTEGER}}$$

$V_{\min} + \Delta V/2$ is the measured transition voltage $V_{0,1}$ at which n switches between 0 and 1. ΔV is the average difference between neighboring transition voltages:

$$\Delta V = \frac{V_{2^N-2,2^N-1} - V_{0,1}}{2^N - 2}$$

For an N-bit A/D, the maximum transition voltage is $V_{2^N-2,2^N-1}$.

2.   Digital-to-analog (D/A) converter characteristics

The purpose of the D/A converter is to convert a binary number to an analog output voltage. For an ideal linear D/A, the output voltage V_n corresponding to the digital input n is given by:

$$V_n = V_{\min} + n\Delta V, \quad \text{where } \Delta V = \frac{V_{\max} - V_{\min}}{2^N - 1}$$

where $V_{\min} = V_0$ is the lowest measured output voltage and $V_{\max} = V_{2^N-1}$ is the highest measured output voltage. This may be verified by evaluating the equation for V_n at $n = 0$ and $n = 2^N - 1$.

3.   Relationship between A/D and D/A conversion

If the values of $V_{\min}$ and ΔV are the same for the A/D and D/A converters, and they have a linear response, then the two conversion processes do not introduce any systematic errors. The A/D output n is produced only by input voltages in the range between the transition voltages:

$$V_{n-1,n} = V_{\min} + (n - 0.5)\Delta V \quad \text{and} \quad V_{n,n+1} = V_{\min} + (n + 0.5)\Delta V$$

The average input value in this range is $V_{\min} + n\Delta V$, which is the same value given by the D/A equation for V_n. As a result, when an arbitrary input voltage is digitized by an A/D and when the digital value is converted back by a D/A, the analog values may differ by as much as $\Delta V/2$, but the average value of the difference will be zero.

4.   Two-parameter least-squares fit

The model is $y = a + bx_i$, where x_i are measured values, $i = 1$ to m, and a and b are unknown. As shown in Chapter 5, the least-squares best-fit coefficients are given by:

$$a = \frac{st - rq}{ms - r^2} \quad \text{and} \quad b = \frac{mq - rt}{ms - r^2}$$

where

$$r = \sum x_i, \quad s = \sum x_i^2, \quad q = \sum x_i y_i, \quad \text{and} \quad t = \sum y_i$$

The residuals are $R_i = a + bx_i - y_i$, where y_i is the measured analog output value corresponding to the analog input x_i. The rms deviation between the data the best-fit straight line (a measure of linearity) is given by:

$$\text{rms} = \sqrt{\frac{1}{m} \sum R_i^2}$$

Additional reading

Section 3.2 Digital-to-analog converter circuits
Section 3.3 Analog-to-digital converter circuits
Section 5.4 Least-squares fitting
Appendix E DT3010 analog input and analog output

Procedure

1. Circuit

Construct a voltage divider using a 20-turn 10-kΩ trimpot connected between -11 and $+11$ V. Connect the wiper to the digital multimeter and the positive input of the data-acquisition circuit (channel 0^+). Connect the negative input (channel 0^-) and all digital and analog grounds of the data-acquisition board to your external power-supply ground.

2. Program

Write a program that does the following:
1. Asks the user to: (i) connect the multimeter to the trimpot, (ii) set an analog input voltage with the trimpot, (iii) enter the multimeter value into the keyboard, and (iv) press return.
2. Digitizes the voltage with the A/D converter.
3. Stores the digital number and sends it to the D/A.
4. Asks the user to: (i) connect the multimeter to the analog output, (ii) enter the multimeter value into the keyboard, and (iii) press return.

5. For each step, stores three numbers: the analog input, the converted digital equivalent, and the analog output.

3. Data

Run the program for about 40 analog voltage levels spaced over the full range from -10 to $+10$ V. Uniform spacing is not necessary; it is better if the input voltage values are randomly chosen.

Laboratory report

1. Setup

Draw a simple block diagram of your experimental setup.

2. Data summary and analysis

2.1 Do a two-parameter least-squares fit of a and b of the model $y = a + bx$ to your data, where x is the analog input to the A/D and y is the analog output from the D/A. Print your results in a table with the following headings:

Index	Input	Measured output	Best linear fit	Residual $R_i = a + bx_i - y_i$	$100 R_i / x_i$
i	x_i	y_i	$a + bx_i$	(mV)	(%)

2.2 Do a two-parameter least-squares fit to the A/D converter using the model:

$$n = a + bV_i$$

2.3 Do a two-parameter least-squares fit to the D/A converter using the model:

$$V = a + bn_i$$

3. Discussion and conclusions

3.1 Discuss how the principles of this laboratory exercise relate to the recording and playback used in compact digital disks and digital audio tapes.

3.2 Discuss how a 12-bit D/A converter can be used to test an eight-bit A/D converter entirely under program control.

3.3 From your analysis, determine whether the A/D or the D/A converter contributed more to the departures from linearity of the entire system. Did some of their nonlinearities cancel?

4.   Questions

4.1 How did the A/D and D/A V_{min} values compare?

4.2 How does the least-squares parameter a of the entire system relate to the A/D and D/A V_{min} values?

4.3 How well did the best least-squares fit values of a and b agree with your expectations?

4.4 Why is it best to choose random voltages in testing A/D converters?

5.   Program and laboratory data sheets

5.1 Include printouts of your program code, data, and output.

5.2 Include your handwritten data sheets (or a copy), which should consist of a log of the procedures you used, any special circumstances, and the measurements you recorded manually.

Laboratory Exercise 21

Fast Fourier transforms of sampled data

Purpose

To sample sine, square, and triangle waves, and to compute and display their fast Fourier transforms. To observe spectral leakage and the effect of windowing. To observe aliasing and the effect of a low-pass anti-aliasing filter. To compare the observed harmonic amplitudes of the square and triangle waves with expected values.

Equipment

- IBM-compatible Pentium microcomputer with Windows NT operating system and Microsoft Visual C++ compiler
- Printer (shared with other laboratory stations)
- Data Translation DT3010 interfacing board
- Wave generator
- Oscilloscope
- ± 12-V power supplies
- Superstrip circuit breadboard
- Two 10-μF, 25-V electrolytic capacitors (put between power and ground at circuit board binding posts)
- Four 0.1-μF, CK-05 bypass capacitors (put between power and ground at all integrated circuits)
- Two LF356 op amps for anti-aliasing filter
- Four 6.8-kΩ resistors
- Two 20-kΩ trimpots
- One 560-pF capacitor
- One 1,000-pF capacitor
- Two 1,200-pF capacitors

- One 1,800-pF capacitor
- One 4,700-pF capacitor
- One 5,600-pF capacitor
- One 15,000-pF capacitor

Background

1.   Use of the fft.c function

Your program should include the fft.c function in the project. Your code should include the following statements:

```
void fft(double [], double[], int, int);
double xr[1024], xi[1024];
int nu, ie;
. . .
fft(xr, xi, nu, ie);
. . .
```

If ie = -1, fft.c performs the forward Fourier transform $f_k \rightarrow F_n$.
If ie = $+1$, fft.c performs the reverse Fourier transform $F_n \rightarrow f_k$.

The number of points N is given by $N = 2^{nu}$. For $N = 1024$, nu $= 10$.

To perform the discrete Fourier transform of a real time series, set xr[k] $= f_k$, and xi[k] $= 0$. The fft.c function will return with the complex Fourier amplitudes in the same xr and xi arrays. Note that the xr frequency amplitudes correspond to the cosine-like terms and the xi frequency amplitudes correspond to the sine-like terms.

The ratio of the real and imaginary parts of the Fourier coefficients depends on the point (phase) of the waveform at which you start sampling. Since the wave generator and the computer program are not synchronized, the phase angle ϕ will vary randomly from run to run and is not too important. However, the magnitude F_n of the Fourier coefficient H_n does not depend on ϕ:

$$\tan \phi = \mathrm{Im}(H_n)/\mathrm{Re}(H_n) \qquad F_n = \sqrt{\mathrm{Re}(H_n)^2 + \mathrm{Im}(H_n)^2}$$

Additional reading

Section 5.8 Fourier transforms
E. Orhan Brigham, *The Fast Fourier Transform and its Applications*, Prentice Hall, Englewood Cliffs, NJ, 1988.

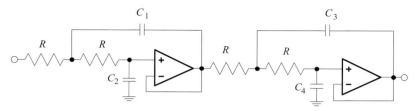

Laboratory Figure 21.1 Butterworth four-pole low-pass filter used to block frequencies above one-half the sampling frequency.

Procedure

1.   Anti-aliasing filter

Build the Butterworth four-pole low-pass filter shown in Laboratory Figure 21.1 with a 3-dB corner frequency at 5 kHz. This filter will be used to block frequencies above one-half the sampling frequency. See Laboratory Figures 4.1 and 4.2 for op-amp pin assignments and external connections. See Section 2.6 and Table 2.3 to see how the filter component values were derived.

Test the filter using your sine-wave generator. The filter should reduce frequencies at and above 10 kHz (the Nyquist limit of the data-acquisition circuit), reduce the amplitude to 0.707 at 7 kHz (the corner frequency of the filter), and preserve the amplitude (gain = 1) at frequencies below 7 kHz.

For $f_c = 7$ kHz and $R = 6.8$ kΩ, we want:

$C_1 = 6,750$ pF (5,600 pF and $\approx$1,200 pF in parallel)
$C_2 = 5,750$ pF (4,700 pF and $\approx$1,000 pF in parallel)
$C_3 = 16,300$ pF (15,000 pF and $\approx$1,200 pF in parallel)
$C_4 = 2,390$ pF (1,800 pF and $\approx$560 pF in parallel)

Common capacitors, however, are seldom more accurate than 20%, and 5% accuracy will require measuring various parallel combinations with a capacitance meter. Higher accuracy is not required for this anti-aliasing filter. See Appendix H for standard capacitor numerical codes.

Save this filter circuit for Laboratory Exercise 22 (Fast Fourier transforms of the human voice) and Laboratory Exercise 24 (Digital control using Fourier deconvolution).

2.   Computer sampling using the analog input port

Connect the input of the filter to the wave generator and connect the output of the filter to the positive input (channel 0^+, DT3010 line 1) of the data-acquisition card (Laboratory

Laboratory Figure 21.2 Equipment setup.

Figure 21.2). Connect the negative input (channel 0^-, DT3010 line 2) and all digital and analog grounds (DT3010 lines 50, 57, 81, 82, 83, 106) of the data-acquisition card to your external power-supply ground.

3.   Program

Write a C program to do the following:

1. Sample 1024 amplitude values at about 20 kHz and store them in an int array.
2. Subtract 2,048 from each value to produce the data array f_k, ($k = 0$ to 1023). (This is necessary because the analog input device converts input voltages from -10 to $+10$ V to numbers from 0 to 4,095.)
3. Write the 1024 values f_k to a disk file for plotting.
4. At the option of the user, multiply the data by the Hann (or raised cosine) window:

$$h_k = 0.5\,[1 - \cos(2\pi k/1024)]$$

and write the 1024 new values to a disk file for plotting.
5. Perform the FFT to compute 1024 complex Fourier coefficients H_n. Compute arrays of magnitude values:

$$F_n = \sqrt{\mathrm{Re}(H_n)^2 + \mathrm{Im}(H_n)^2}$$

and magnitude squared values.
6. Write the magnitude values to a disk file for plotting.

The data-acquisition loop should look like the one you used in Laboratory Exercise 10 to demonstrate aliasing.

4.   FFT of periodic waveforms

4.1 Wave generator and data-acquisition setup. Set your wave generator at about 2 kHz and adjust the amplitude and offset knobs to produce a sine wave that oscillates between about -5 and $+5$ V. Take 1024 samples at your maximum rate (approximately 20 kHz). Display the plot file on the screen. Note the number of cycles and adjust the frequency of the wave generator so that about 6.5 cycles are sampled. Other values such as 5.5 or 7.5 cycles are suitable, but for the purposes of this laboratory exercise, it is important that you do not sample an integral number of cycles.

4.2 FFT of a noninteger number of sine waves. Take 1024 samples, subtract 2,048 from each, **but do not multiply the data by the Hann window**. Use the FFT function to get 1024 real and imaginary components of the frequency amplitudes. Display the magnitude values on the screen using the plot program and print the plot on the printer. Print the real, imaginary, and magnitude values.

Note: If the lower magnitudes F_n do not equal their upper counterparts F_{1024-n}, your program has an error.

4.3 FFT of a noninteger number of Hann-windowed sine waves. Take 1024 samples, subtract 2,048 from each, **and multiply the data by the Hann window**. Take the FFT, print the plot of the magnitude values, and print the real, imaginary, magnitude, and magnitude squared values.

4.4 FFT of a noninteger number of Hann-windowed square waves. Without changing the frequency or amplitude, switch the wave generator from sine-wave to square-wave mode. As in 4.3 above, take 1024 samples, subtract 2,048 from each, and multiply the data by the Hann window. Take the FFT, print the plot of the magnitude values, and print the real, imaginary, and magnitude values.

4.5 FFT of a noninteger number of Hann-windowed triangle waves. Without changing the frequency or amplitude, switch the wave generator from square-wave to triangle-wave mode. As in 4.3 above, take 1024 samples, subtract 2,048 from each, and multiply the data by the Hann window. Take the FFT, print the plot of the magnitude values, and print the real, imaginary, and magnitude values.

4.6 FFT of a noninteger number of Hann-windowed square waves without using the anti-aliasing filter. Bypass the anti-aliasing filter and repeat 4.4 above. Take the FFT, print the plot of the magnitude values, and print the real, imaginary, and magnitude values.

Laboratory report

1.   Setup

Draw a simple block diagram of your experimental setup.

2.   Data summary and analysis

For the nonwindowed sine wave, note the spread of magnitude values near the fundamental frequency. To determine the combined magnitude value for the fundamental frequency, add the neighboring magnitude squared values and take the square root of the sum. Do the same for the windowed sine wave, noting differences in the extent of the spread.

For the square and triangle waves, note the peaks in the magnitude squared values, and identify the corresponding harmonic numbers. The nth harmonic has a frequency n times the primary frequency of the square or triangle wave. Due to the effects of windowing, each harmonic will span several adjacent Fourier coefficients.

For each harmonic up to the ninth, tabulate: (1) the frequency, (2) the harmonic number, (3) the combined magnitude, (4) the ratio of the combined magnitude of the first harmonic to the combined magnitude of each of the higher harmonics, and (5) the expected ratio of the first to each of the higher harmonics. To determine the combined magnitude value for each harmonic, add the neighboring magnitude squared values and take the square root of the sum. (Note that the magnitude values for the square wave should decrease with increasing frequency as f^{-1} and the magnitude values for the triangle wave should decrease as f^{-2}.)

3. Discussion and conclusions

3.1 Discuss procedure section 1 (construction and testing of the anti-aliasing filter).

3.2 Discuss procedure section 4.1 (adjustment of wave generator for half-integer cycles).

3.3 Discuss procedure section 4.2 (FFT of a noninteger number of sine waves).

3.4 Discuss procedure section 4.3 (FFT of a noninteger number of Hann-windowed sine waves). Compare with procedure section 4.2, when the Hann window was not used.

3.5 Discuss procedure section 4.4 (FFT of a noninteger number of Hann-windowed square waves).

3.6 Discuss procedure section 4.5 (FFT of a noninteger number of Hann-windowed triangle waves).

3.7 Discuss procedure section 4.6 (FFT of a noninteger number of Hann-windowed square waves, but without the anti-aliasing filter). Compare with procedure section 4.4, when the anti-aliasing filter was used.

3.8 Explain why the Fourier amplitudes that you measured occurred at the Fourier frequency indexes that they did.

3.9 Discuss the limitations of the spectrum-analysis technique used in this laboratory exercise.

4. Questions

4.1 How would you expect the last 64 values of the magnitude (i.e. $F_{513} \rightarrow F_{1023}$) to compare with the first 64 values after the dc term (i.e. $F_1 \rightarrow F_{512}$)? Give an explicit relationship.

4.2 For the FFT of the sine wave, did you observe any nonzero amplitudes at integral multiples of the fundamental frequency? What could cause such nonzero Fourier coefficients?

4.3 Since the data-acquisition loop was not synchronized with the output of the wave generator, would you expect the real and imaginary components of the Fourier amplitude to be the same if the exercise was repeated?

4.4 What benefit did the Hann window have for sampling and Fourier transforming a noninteger number of cycles?

4.5 What benefit did the anti-aliasing filter have for sampling and Fourier transforming the square wave?

5.   Program and laboratory data sheets

5.1 Include printouts of your program code, and printed output (plots of magnitude values and printout of numerical values of frequency index, real and imaginary Fourier coefficients, and magnitude) from procedure sections 4.2–4.6.

5.2 Include your handwritten data sheets (or a copy), which should consist of a log of the procedures you used, any special circumstances, and the measurements you recorded manually.

Laboratory Exercise 22

Fast Fourier transforms of the human voice

Purpose

To sample periodically the audio waveform produced by the human voice when uttering various vowel sounds and to perform a fast Fourier transform (FFT). To investigate which features in the FFT depend on the identity of the speaker and which features depend on the vowel that is spoken.

Equipment

- IBM-compatible Pentium microcomputer with Windows NT operating system and Microsoft Visual C++ compiler
- Printer (shared with other laboratory stations)
- Data Translation DT3010 interfacing board
- Oscilloscope
- +5- and ±12-V power supplies
- Superstrip circuit breadboard
- Three 10-μF, 25-V electrolytic capacitors (put between power and ground at circuit board binding posts)
- Six 0.1-μF, CK-05 bypass capacitors (put between power and ground at all integrated circuits)
- Four LF356 op amps (one for the microphone amplifier, two for the anti-aliasing filter, one for the power amplifier)
- LM12 80-W power op-amp circuit on heat sink
- Small loudspeaker
- Microphone
- One 1-kΩ resistor (microphone amplifier)
- Four 6.8-kΩ resistors (Butterworth filter)
- One 10-kΩ resistor (microphone amplifier)

- Two 100-kΩ resistors (microphone circuit)
- Three 20-kΩ trimpots (op-amp offset adjust)
- One 560-pF capacitor (Butterworth filter)
- One 1,000-pF capacitor (Butterworth filter)
- Two 1,200-pF capacitors (Butterworth filter)
- One 1,800-pF capacitor (Butterworth filter)
- One 4,700-pF capacitor (Butterworth filter)
- One 5,600-pF capacitor (Butterworth filter)
- One 15,000-pF capacitor (Butterworth filter)
- Two 1-μF electrolytic capacitors (microphone circuit)
- AD625 or LH0036 instrumentation amplifier for microphone
- For AD625 instrumentation amplifier chip, use resistors below:
  - One 25-kΩ trimpot (offset adjust)
  - Two 20-kΩ resistors (R_F)
  - One 3.9-kΩ resistor (R_G for gain of 10)
- For LH0036 instrumentation amplifier chip, use resistors below:
  - One 100-kΩ trimpot (offset adjust)
  - One 5.1-kΩ resistor (R_G for gain of 10)
  - One 3.3-kΩ resistor (offset adjust)
  - One 33-kΩ resistor (offset adjust)

Background

Voiced sounds are produced by forcing air through the **glottis** (the opening between the vocal chords) with the tension of the vocal chords adjusted so that they vibrate to produce quasi-periodic pulses of air that excite the vocal tract. When you "hold your breath" you use these muscles to close the glottis completely. The **vocal tract** consists of the **nasal pharynx** (the connection between the nasal passages and the back of the mouth), the **oral pharynx** (the connection between the back of the mouth and the esophagus), and the mouth. The jaw, lips, and tongue can be moved to change the acoustical properties of the vocal tract. In addition, by lowering the soft palate, the nasal tract can be used to produce nasal sounds. See Laboratory Figure 22.1 for an illustration of these structures.

Vowels are voiced sounds produced with a fixed vocal tract. English examples are *a* as in father, *e* as in see, *o* as in go, *u* as in up.

Diphthongs are gliding voiced sounds that begin with one vowel and end with another. English examples are "bay," "boy," "you."

Nasals are voiced sounds produced by total constriction of the vocal tract and the lowering of the soft plate so that the air passes through the nasal cavity and the sound is radiated at the nostrils. The mouth still plays an important role as a resonant cavity.

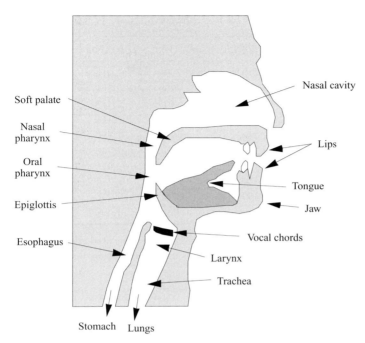

Laboratory Figure 22.1 Structures in the head that are involved in speech. Air from the lungs is forced between the vocal chords to produce quasi-periodic pulses of air. The vocal tract (pharynx and mouth) acts as a resonant cavity and the jaw, lips, tongue, and soft palate can be moved to change the sounds produced. The epiglottis is lowered for swallowing food. The glottis (space between the vocal chords) is closed during breath holding.

English examples are *m* (vocal tract constricted at the lips) and *n* (vocal tract constricted just back of the teeth). Nasals have a concentration of low-frequency energy and the mid-range shows no prominent peaks.

Unvoiced fricatives are produced by exciting the vocal tract by a steady flow of air that becomes turbulent in the region of a constriction in the vocal tract. The vocal chords do not vibrate. Examples are *f* (constriction near lips), *s* (constriction near middle of vocal tract), and *sh* (constriction near back of vocal tract). In these cases, the waveform is nonperiodic and similar to white noise. There are also many other identified speech components, such as semivowels (such as *w*, *l*, and *r*), voiced fricatives (such as *v* and *z*), voiced stops (such as *b*, *d*, and *g*), unvoiced stops (such as *p*, *t*, and *k*), and affricatives (such as *j* and *h*).

In this laboratory exercise, we will concentrate on the vowels, because they have characteristic frequency spectra. The resonant frequencies associated with each vowel are called "formants." For example, for the average male speaker, the *o* in hot should have its first formant f_1 at 730 Hz, the second f_2 at 1,090 Hz, and the third f_3 at 2,440 Hz. See Laboratory Table 22.1 and Laboratory Figure 22.2 for additional examples.

Laboratory Table 22.1 *Average formant frequencies for the vowels*

IPA symbol	Typical word	f_1 (Hz)	f_2 (Hz)	f_3 (Hz)	f_2/f_1	f_3/f_1
i	beet	270	2,290	3,010	8.5	11.1
I	bit	390	1,990	2,550	5.1	6.5
ϵ	bet	530	1,840	2,480	3.8	4.7
æ	bat	660	1,720	2,410	2.6	3.7
Λ	but	640	1,190	2,390	2.3	4.6
a	hot	730	1,090	2,440	1.5	3.3
OW	bought	570	840	2,410	1.5	4.2
U	foot	440	1,020	2,240	2.3	5.1
u	boot	300	870	2,240	2.9	7.5
ER	bird	490	1,350	1,690	2.8	3.4

Source: Gordon E. Peterson and Harold L. Barney, "Control methods used in a study of the vowels," *J. Accoust. Soc. Am.*, Vol. 24 (1952): 175–184. Reprinted with permission of AT&T Corp., NJ.

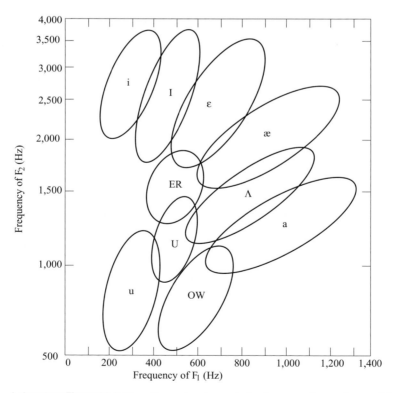

Laboratory Figure 22.2 Two-dimensional correlation between first and second formant frequencies. *Source:* Gordon E. Peterson and Harold L. Barney, "Control methods used in a study of the vowels," *J. Accoust. Soc. Am.*, Vol. 24 (1952): 175–184. Reprinted with permission of AT&T Corp., NJ.

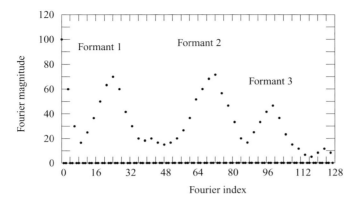

Laboratory Figure 22.3 FFT of a typical vowel sound. In this example, three cycles were sampled and as a result only every third Fourier magnitude is nonzero. The formants are caused by resonances in the vocal tract and define the vowel sound. Different vowels may be spoken by varying the shape of the vocal tract.

A simple model for human speech vowels is an impulse generator (vocal chords) followed by a time-varying filter (mouth and tongue). The regular impulses are characteristic of the speaker and produce a FFT with nonzero Fourier amplitudes over a wide range of frequencies. The mouth and tongue transmit these harmonics to the outside world with an intensity pattern that is characteristic of the vowel being spoken. By doing an FFT of a steady vowel sound, you should be able to identify two or three primary resonances (called formants) that are characteristic of that particular vowel (Laboratory Figure 22.3). Note that the actual frequencies of the formants depend on the pitch of the speaker's voice and the ratios of the formant frequencies are more indicative of the vowel spoken.

Additional reading

R. W. Broderson, P. J. Hurst, and D. J. Allstot, "Switched-capacitor applications in speech processing," *IEEE Symp. Circ. Sys.*, Vol. 3 (1980): 732–737.

Gordon E. Peterson and Harold L. Barney, "Control methods used in a study of the vowels," *J. Accoust. Soc. Am.*, Vol. 24 (1952): 175–184.

L. R. Rabiner and R. W. Schafer, *Digital Processing of Speech Signals*, Prentice Hall, Englewood Cliffs, NJ, 1978.

Procedure

1.   Microphone pre-amplifier

Build the instrumentation amplifier shown in Laboratory Exercise 5. For the AD625, $R_F = 20\ \text{k}\Omega$ and R_G should be $3.9\ \text{k}\Omega$ for a gain of about 10. For the LH0036, R_G

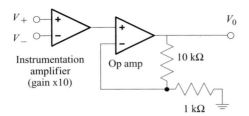

Laboratory Figure 22.4 Microphone amplifier circuit.

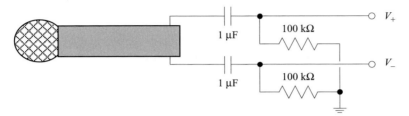

Laboratory Figure 22.5 Dynamic microphone and circuit for connection to instrumentation amplifier.

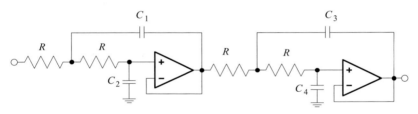

Laboratory Figure 22.6 Butterworth four-pole anti-aliasing filter.

should be 5.1 kΩ for a gain of about 10. Follow the instrumentation amplifier with a noninverting amplifier with a gain of about 10 (Laboratory Figure 22.4). Build the microphone circuit as shown in Laboratory Figure 22.5 and connect it to the instrumentation amplifier inputs. The microphone amplifier output should be in the -5- to $+5$-V range for the anti-aliasing filter, which has unity gain.

2. Anti-aliasing filter

Build the Butterworth four-pole filter shown in Laboratory Figure 22.6 with a 3-dB corner frequency at 7 kHz. See Laboratory Figures 4.1 and 4.2 for op-amp pin assignments and external connections. Connect 0.1-μF capacitors between pin 4 (-12 V) and ground and between pin 7 ($+12$ V) and ground. See Section 2.6 and Table 2.3 for filter component values.

Test the filter using your sine-wave generator. The filter should attenuate frequencies at and above 10 kHz (the Nyquist limit of the data-acquisition circuit), reduce the amplitude to 0.707 at 7 kHz (the corner frequency of the filter), and preserve the amplitude (gain = 1) at frequencies below 5 kHz.

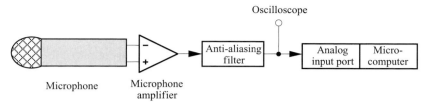

Oscilloscope

Microphone   Microphone amplifier   Anti-aliasing filter   Analog input port   Micro-computer

Laboratory Figure 22.7 Equipment setup for sampling vowel sounds.

For $f_c = 7$ kHz and $R = 6.8$ kΩ, we want

$C_1 = 6{,}750$ pF (5,600 pF and $\approx 1{,}200$ pF in parallel)
$C_2 = 5{,}750$ pF (4,700 pF and $\approx 1{,}000$ pF in parallel)
$C_3 = 16{,}300$ pF (15,000 pF and $\approx 1{,}200$ pF in parallel)
$C_4 = 2{,}390$ pF (1,800 pF and ≈ 560 pF in parallel)

Common capacitors, however, are seldom more accurate than 20%, and 5% accuracy will require measuring various parallel combinations with a capacitance meter. Higher accuracy is not required for this anti-aliasing filter. See Appendix H for standard capacitor numerical codes.

Connect the filter circuit input to the output of the instrumentation amplifier and check that the filter output does not exceed the -5- to $+5$-V range when speaking into the microphone in a moderate voice.

3.   Data-acquisition and playback circuit

Connect the output of the Butterworth low-pass filter to the data-acquisition card. Connect the negative input (channel 0^-) and all digital and analog grounds of the data-acquisition card to your external power-supply ground (Laboratory Figure 22.7). Connect the analog output to the current amplifier and speaker for playback.

4.   Program

For each vowel sound, use the "Standard Analog I/O loop" to sample and store 4,096 values. Window the data as you did in Laboratory Exercise 21 (Fast Fourier transforms of sampled data). Use a 100-Hz sine wave to measure your sampling rate f_s (it should be above 10-kHz).

Use the FFT function to transform the 4,096 data values into 4,096 real and imaginary components of the frequency coefficients. Compute the array of magnitude values:

$$F_n = \sqrt{\mathrm{Re}(H_n)^2 + \mathrm{Im}(H_n)^2}$$

and write them to a file for printing and plotting.

5.   Determination of sampling frequency

Set your wave generator for a 1-kHz sine wave and sample 4,096 values using the standard I/O loop from Laboratory Exercise 10. Below the bottom of the loop, write the samples to a file. Observe the file and determine how many samples cover 10 cycles. One-tenth of this number is the sampling frequency in kilohertz. If this number is significantly less than 20 kHz, check your I/O loop for unnecessary steps.

6.   FFT of vowel sounds

Using the microphone, anti-aliasing filters, and computer, sample 4,096 values for two different vowel sounds chosen from Laboratory Table 22.1, as spoken by both laboratory partners (four data sets). Observe the filter output on the oscilloscope. The vowel sounds should appear as a periodic signal. Write these values to a file and print the plots.

For each vowel sound, perform the FFT, compute the Fourier magnitude array, write it as a file to disk, and print the plot. As there are individual differences between speakers, do not be too concerned if your data do not agree closely with Laboratory Table 22.1.

For the two speakers, expand the plots (first 100 Fourier magnitudes) that best show the individual harmonics of the vocal chord vibrations and print the plots.

7.   Sampling and playback of human speech

7.1 Power amplifier. Connect the analog output to the LM12 power op-amp circuit. Connect the output of the LM12 to the small speaker (Laboratory Figure 22.8). The circuit is designed to protect the speaker by blocking dc and limiting the voltage to ± 1 V ac. To avoid feedback, keep the LM12 output and speaker wires as far as possible from the LM12 input wires.

7.2 Program. Modify your program to operate in two modes:

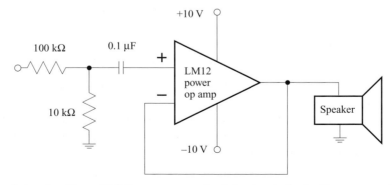

Laboratory Figure 22.8 Power op-amp circuit for the playback of human speech.

Mode 1: Use the "Standard Analog I/O loop" to sample about 10 s of human speech. By using "calloc" and pointer arithmetic (see Appendix C), it is possible to acquire and store the 200k or so values required.

Mode 2: Use the "Standard Analog I/O loop" to output the 10 s of speech on the speaker.

7.3 Sampling and playback. Use the system to sample and playback some of your favorite phrases. Record your observations on fidelity and clarity.

Laboratory report

1.   Setup

Draw a simple block diagram of your experimental setup.

2.   Data summary and analysis

2.1 Plot the time samples of the four data sets (procedure section 6) to compare the time waveforms of different vowels spoken by the same speaker and the same vowels spoken by different speakers.

2.2 Plot the four sets of Fourier magnitudes (procedure section 6) to compare the frequency content of different vowels spoken by the same speaker and the same vowels spoken by different speakers. Include a plot of the first 100 Fourier magnitudes so that the fundamental harmonics of the vocal chords can be seen and measured.

2.3 The vocal chords produce a periodic oscillation that will appear in your Fourier magnitudes as a low-frequency spike (the fundamental) plus all higher harmonics. This is similar to what you observed in Laboratory Exercise 21 for the square wave, except that both even and odd harmonics will be present. From the sampling frequency f_s, the number of samples M, and the index of the lowest harmonic n_1, estimate the fundamental frequency (first harmonic) of the vocal chords for each of your two speakers as $(f_s n_1 / M)$.

2.4 For each of the four Fourier magnitude plots (two vowels spoken by two speakers) label the first, second, and third harmonics. The kth harmonic will occur at or near Fourier index $n_k = k n_1$. Due to the effects of windowing, each harmonic will span several adjacent Fourier coefficients.

2.5 You should see that the magnitude of the harmonics vary with frequency. Even though all harmonics of the vocal chords are present, some frequency bands are enhanced by the position of the mouth and tongue, which form a resonant cavity. The lowest such band is the first formant and appears at frequency f_1. In your four plots, identify the three most prominent formants and label their frequencies f_1, f_2, and f_3.

Compare these frequencies and the ratios f_2/f_1 and f_3/f_1 with those in Laboratory Table 22.1.

3.   Discussion and conclusions

3.1 Discuss your measurement of sampling frequency (procedure section 5) and the accuracy of the measurement.

3.2 Examine the four Fourier coefficient data sets (procedure section 6). Discuss which features of the Fourier magnitude plot are characteristic of the vowel sound and which are characteristic of the speaker.

3.3 Discuss procedure section 7. Consider sound fidelity in terms of frequency and amplitude accuracy, and higher harmonics produced by the D/A. Consider the digital storage of one hour of high-fidelity stereo music.

4.   Questions

4.1 To what frequency (in hertz) did the Fourier coefficient $H_{M/2}$ correspond?

4.2 How well did the ratio of formant frequencies f_2/f_1 and f_3/f_1 compare with Laboratory Table 22.1?

4.3 For the two speakers, what were the frequency differences (in hertz) between neighboring vocal chord harmonics?

4.4 How would you design a computer program to determine which vowel was spoken, independent of speaker? Show your answer in a list of steps and comment on any major problems.

5.   Program and laboratory data sheets

5.1 Include printouts of your program code and output of FFT magnitude values.

5.2 Include your handwritten data sheets (or a copy), which should consist of a log of the procedures you used, any special circumstances, and the measurements you recorded manually.

Laboratory Exercise 23

Digital filtering

Purpose

To filter sine waves with analog and digital low-pass filters and to compare their properties as a function of frequency.

Equipment

- IBM-compatible Pentium microcomputer with Windows NT operating system and Microsoft Visual C++ compiler
- Printer (shared with other laboratory stations)
- Data Translation DT3010 interfacing board
- Oscilloscope
- ± 12-V power supplies
- Two 10-μF, 25-V electrolytic capacitors (put between power and ground at circuit board binding posts)
- Two 0.1-μF, CK-05 capacitors (put between power and ground at integrated circuit)
- Superstrip circuit board
- One LF356 op-amp integrated circuit
- Sine-wave generator
- One 51-kΩ resistor (low-pass filter)
- One 0.1-μF, CK-05 capacitor (low-pass filter)
- One 20-kΩ trimpot (op-amp offset adjust)

Background

1.   Analog low-pass single-pole filter

The analog low-pass single-pole filter shown in Laboratory Figure 23.1 has a closed-loop gain G and phase shift ϕ given as follows:

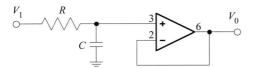

Laboratory Figure 23.1 Single-pole low-pass analog filter. See Laboratory Figure 4.3 for external components.

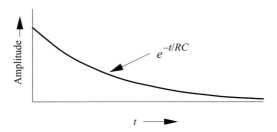

Laboratory Figure 23.2 Impulse response of a low-pass single-pole analog filter.

$$|G| = \frac{1}{\sqrt{1 + (f/f_c)^2}}, \qquad \tan(\phi) = -f/f_c \qquad (23.1)$$

where f is the frequency and $f_c = (2\pi RC)^{-1}$ is the corner frequency. For low frequencies, $|G| = 1$. At $f = f_c$, $|G| = 1/\sqrt{2}$. At very high frequencies, $|G| = f_c/f$. For $f \ll f_c$, the phase shift is $0°$. At $f = f_c$, the output is phase shifted by $45°$. For $f \gg f_c$, the output is phase shifted by $90°$.

This analog filter has an impulse response $e^{-t/RC}$, as shown in Laboratory Figure 23.2. The impulse basically puts a charge on the output of capacitor C, which decays with time constant RC.

2. Digital low-pass single-pole filter

The digital low-pass single-pole filter has the form:

$$y_i = (1 - \alpha)x_{i-1} + \alpha y_{i-1}$$

Its response to the impulse:

$$x_{i<0} = 0, \qquad x_0 = 1, \qquad x_{i>0} = 0$$

is given by the exponential decay:

$$y_{i<1} = 0, \qquad y_1 = (1 - \alpha), \qquad y_2 = (1 - \alpha)\alpha, \qquad y_i = (1 - \alpha)\alpha^{i-1}$$

This impulse response has unit area:

$$\sum_{i=-\infty}^{\infty} y_i = (1 - \alpha)(1 + \alpha + \alpha^2 + \cdots) = 1$$

Since the operations of reading the analog input port, performing the filtering, and writing the result to the analog output port take a combined time Δt, the digital filter has an additional delay relative to the analog filter. The digital filter approaches the analog filter in this respect as $\Delta t \rightarrow 0$.

Additional reading

Samuel D. Stearns and Ruth A. David, *Signal Processing Algorithms*, Prentice Hall, Englewood Cliffs, NJ, 1988.
Section 5.9 Digital filters

Procedure

1.   Analog filtering

Build the low-pass single-pole analog filter shown in Laboratory Figure 23.1 with $R = 51\,\text{k}\Omega$ and $C = 0.1\,\mu\text{F}$. The corner frequency is at $f_c = (2\pi RC)^{-1} = 31\,\text{Hz}$. See Laboratory Figure 4.2 for op-amp external connections. Using the setup shown in Laboratory Figure 23.3, measure the response of the analog filter at 10, 30, 100, 300, 1,000, and 3,000 Hz. Record both $V_{\text{out}}/V_{\text{in}}$ and the phase shift.

2.   Digital filtering

Using the setup shown in Laboratory Figure 23.4, sample the sine-wave generator with the IBM data-acquisition circuit at the frequencies listed earlier. Since the corner frequency is 31 Hz, well-below one-half the sampling frequency, you will not need the anti-aliasing filter used in Laboratory Exercise 22. Adjust the sine wave for a peak-to-peak amplitude of about 10 V and an average amplitude of zero. If time permits, filter the output as you did in Laboratory Exercise 10.

Write a program to sample the sine wave, perform the digital filtering:

$$y_i = (1 - \alpha)x_{i-1} + \alpha y_{i-1} \tag{23.2}$$

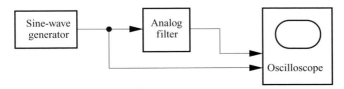

Laboratory Figure 23.3 Setup for analog filtering of sine waves.

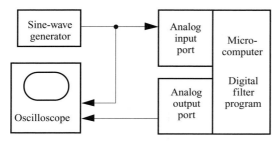

Laboratory Figure 23.4 Setup for digital filtering of sine waves.

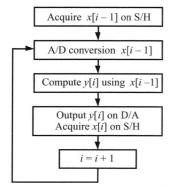

Laboratory Figure 23.5 Flow chart for digital-filtering program.

and write the result to the D/A converter as rapidly as possible. See Laboratory Figure 23.5 for the flow chart. In Equation (23.2), x_{i-1} is the most recent value read by the A/D, y_{i-1} is the previous output value, and $0 < \alpha < 1$ is a filter-design parameter.

Using a 10-Hz input sine wave, observe the steps in the output waveform and estimate your filter process time Δt.

The impulse response of this low-pass, single-pole digital filter is given by $y_i = (1 - \alpha)\alpha^{i-1}$, which we wish to equate to the impulse response of the analog filter:

$$y_i = (1 - \alpha)\alpha^{i-1} = (1 - \alpha)e^{-t/RC}$$
$$t = (i - 1)\Delta t, \qquad \alpha = e^{-\Delta t/RC}$$

Use floating-point multiplication in Eq. (23.2). Integer arithmetic is more difficult and actually slower on microcomputers with floating-point processors.

When you begin filtering, the values of y_{i-1} will be inaccurate, particularly if you choose $y_0 = 0$. As the filter is implemented many times, the successive values of y_i "forget" the initial erroneous value used for y_0, with a time constant of RC. It is thus necessary to discard the first five $RC/\Delta t = -5/\ln \alpha$ samples.

Use these data to record the filter gain and phase shift as a function of frequency as you did in procedure section 1.

Laboratory report

1.   Setup

Draw a simple block diagram of your experimental setup.

2.   Data summary and analysis

2.1 Tabulate and plot the observed values of V_{out}/V_{in} for both filters and the ideal values (Eq. (23.1)) as a function of frequency.

2.2 Tabulate and plot the observed values of phase for both filters and the ideal values (Eq. (23.1)) as a function of frequency.

2.3 Tabulate and plot the difference in phase shifts $\Delta\phi$ and the time shift $\Delta t = \Delta\phi/(2\pi f)$ between the two filters as a function of frequency.

3.   Discussion and conclusions

3.1 Discuss the differences between your analog and digital filters. Consider implementation, accuracy, and gain and phase versus frequency.

3.2 Discuss situations where an analog filter would be preferable and where a digital filter would be preferable.

3.3 Discuss how the filter process time Δt affected the phase shift of the digital filter.

3.4 Discuss how the digital filter would function at frequencies $f > 1/(2\Delta t)$.

4.   Questions

4.1 What value of α did you use for your low-pass digital filter?

4.2 What value of α would you use in this exercise if your sampling frequency were 100 kHz?

4.3 Compute the impulse responses for the following filters:

$$y_i = x_{i-1} + 0.5\ y_{i-1} + 2\ y_{i-2}$$

$$y_i = x_{i-1} + 0.9\ y_{i-4}$$

$$y_i = x_{i-1} - y_{i-1}$$

Hint: Use $x_0 = 1$ $x_{i\neq0} = 0$

4.4 At what frequency did the digital filter have a phase shift of 90°? At this frequency, what was the phase shift of the analog filter?

4.5 Was the difference in time shift Δt between the two filters independent of frequency? What did you expect?

5.   Program and laboratory data sheets

5.1 Include printouts of your program code and output.

5.2 Include your handwritten data sheets (or a copy), which should consist of a log of the procedures you used, any special circumstances, and the measurements you recorded manually.

Laboratory Exercise 24

Process compensation using Fourier deconvolution
and digital filtering

Purpose

To sample the impulse response $c(t)$ of an analog circuit (a single-pole low-pass filter), and use FFT techniques to compute a corresponding digital deconvolution filter b. If a waveform $a(t)$ is firstly digitally filtered with b and then sent through the analog circuit, the output is a close approximation to the original $a(t)$. The digital filter b is computed as the inverse FFT of $(1/\text{FFT}(c(t)))$.

Equipment

- IBM PC with Data Translation 3010 board
- Printer (shared with other laboratory stations)
- Superstrip circuit board
- ± 12-V power supply
- Two 10-μF, 25-V electrolytic capacitors (put between power and ground at circuit board binding posts)
- One LF 356 op amp
- Two 0.1-μF, CK-05 capacitors (put one between the LF356 pin 4 and ground, and the other between pin 7 and ground)
- One 20-kΩ trimpot (to adjust the op-amp output offset if needed)
- One 0.1-μF, CK-05 capacitor (low-pass filter)
- One 1-kΩ resistor (low-pass filter)
- One 100-kΩ resistor (low-pass filter)
- One 51-kΩ resistor (low-pass filter)

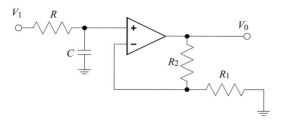

Laboratory Figure 24.1 Single-pole low-pass analog filter. See Laboratory Figure 4.2 for external components. Connect 0.1-μF capacitors between pin 4 (-12 V) and ground and between pin 7 ($+12$ V) and ground.

Background

1.   Analog low-pass single-pole filter

As described in Chapter 2 and Laboratory Exercise 23, the analog low-pass single-pole filter shown in Laboratory Figure 24.1 has a closed-loop gain G:

$$|G| = \frac{(R_1 + R_2)/R_1}{\sqrt{1 + (f/f_c)^2}}$$

where f is the frequency and $f_c = (2\pi RC)^{-1}$ is the corner frequency. For low frequencies, $|G| = (R_1 + R_2)/R_1$. At $f = f_c$, $|G| = (R_1 + R_2)/(\sqrt{2}R_1)$. At very high frequencies, $|G| = (f_c/f)(R_1 + R_2)/R_1$.

After an abrupt change in V_1, the voltage on capacitor C changes with an exponential time constant RC to equal the new value of V_1 asymptotically.

A voltage impulse at V_1 of amplitude ΔV and duration ΔT (assumed to be much shorter than RC) puts a charge $\Delta V \Delta T/R$ on the output of capacitor C, which decays with time constant RC. The voltage transient at the input of the op amp is thus given by:

$$V(t) = \frac{\Delta V \Delta T}{RC} e^{-t/RC}$$

and plotted in Laboratory Figure 24.2.

2.   The Fourier convolution theorem applied to digital control

The control problem we address here is determining the digital filter $b(t)$ necessary to pre-process the signal so that after it passes through the low-pass filter, the output is similar to the original $a(t)$. In other words, we want to find the $b(t)$ that compensates

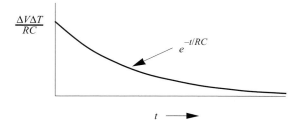

Laboratory Figure 24.2 Response of a low-pass single-pole analog filter to an impulse of amplitude ΔV and duration $\Delta T \ll RC$.

for what the low-pass filter does to the signal:

$$\tilde{a}(t) = [a(t) \cdot b(t)] \cdot c(t)$$

where $\cdot$ is the convolution operator.

The Fourier convolution theorem states that:

$$\mathcal{F}(a) = \mathcal{F}(a) \times \mathcal{F}(b) \times \mathcal{F}(c)$$

where $\times$ is a simple multiplication and $\mathcal{F}(a)$, $\mathcal{F}(b)$, and $\mathcal{F}(c)$ are the Fourier transforms of $a(t)$, $b(t)$, and $c(t)$, respectively.

From simple division, we have:

$$\mathcal{F}(b) = 1/\mathcal{F}(c)$$

which means that $b(t)$ can be determined from:

$$b(t) = \mathcal{F}^{-1}[1/\mathcal{F}(c)]$$

So the steps are:
1. Measure the impulse response $c(t)$.
2. Take the Fourier transform of $c(t) = C_r + jC_i$.
3. Compute one divided by the Fourier transform of $c(t)$:

$$1/\mathcal{F}(c) = \frac{1}{C_r + jC_i} = \frac{C_r - jC_i}{C_r^2 + C_i^2}$$

4. Compute a finite impulse response filter as the inverse Fourier transform:

$$b(t) = \mathcal{F}^{-1}[1/\mathcal{F}(c)] = \mathcal{F}^{-1}\left[\frac{C_r - jC_i}{C_r^2 + C_i^2}\right]$$

where the filter coefficients are given by:

$$b_k = \mathcal{F}^{-1}\left[\frac{C_{rk} - jC_{ik}}{C_{rk}^2 + C_{ik}^2}\right]$$

and the filter equation is:

$$y_{n+1} = \frac{1}{B} \sum_{k=0}^{N} x_{n-k} b_k, \quad B = \sum_{k=0}^{N} bk$$

Now you can continuously sample $a(t)$, continuously filter with digital filter $b(t)$, and send the output to the analog circuit. The circuit output should be a delayed version of $a(t)$.

In this laboratory exercise, $c(t)$ is the output of the analog filter after it has been driven with a short pulse produced by the analog output port.

Note: In the general case, if the magnitude of $\mathcal{F}(c)$ is too small, the division $1/\mathcal{F}(c)$ may result in numerical errors and the digital filter $b(t)$ may produce unreliable results. Stated in another way, it is difficult to produce frequency components in the output of a system if those frequency components are poorly transmitted by the system.

Additional reading

E. Oran Brigham, *The Fast Fourier Transform and its Applications*, Prentice Hall, Englewood Cliffs, NJ, 1988.

Procedure

1.   The low-pass single-pole analog filter

Build the low-pass single-pole analog filter shown in Laboratory Figure 24.1 with $R = 51\,\mathrm{k}\Omega$ and $C = 0.1\,\mu\mathrm{F}$. For dc gain $= 1$, set $R_1 = \infty$ and $R_2 = 0\,\Omega$. For dc gain $= 100$, set $R_1 = 1\,\mathrm{k}\Omega$ and $R_2 = 100\,\mathrm{k}\Omega$. The RC exponential decay time will be about 5.1 ms and the corner frequency is $f_c = (2\pi RC)^{-1} \approx 31$ Hz.

2.   Measurement of the impulse response of the low-pass filter

As shown in Laboratory Figure 24.3, connect the analog output port to the input of the low-pass filter and to one channel of the oscilloscope (dc coupling). Connect the output of the low-pass filter to the analog input port and to another channel of the oscilloscope (dc coupling). Use the first channel to trigger the oscilloscope.

Set the low-pass filter gain to 20 by using $R_1 = 1\,\mathrm{k}\Omega$ and $R_2 = 20\,\mathrm{k}\Omega$.

Write and run a program that does the following:

1. Load the first 1,022 elements of array nb (nb[0] to nb[1021]) with 2,048, nb[1022] with 4,095 and nb[1023] with 2,048. Send the array to the analog output port in a tight

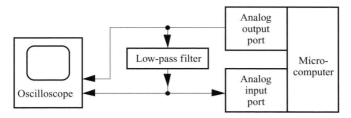

Laboratory Figure 24.3 Setup for sending a computer-generated impulse through a low-pass filter and recording the output waveform.

loop. This will output 0 V for about 100 ms, output a 10-V pulse about 100 μs wide, and then set the output back to 0 V.

The output of the filter circuit should be:

$$V(t) = \frac{\Delta V \Delta T G}{RC} e^{-t/RC} = \frac{(10\ \text{V})(100\ \mu\text{s})(20)}{5.1\ \text{ms}} e^{-t/5.1\ \text{ms}}$$
$$= (3.9\ \text{V}) e^{-t/5.1\ \text{ms}}$$

2. Immediately start an analog input loop and store 1,024 analog input port values in array nc. Include a dummy digital filter loop and an analog output step in the loop. These are not used at this stage, but it will ensure that the impulse response is sampled at the same rate that will be used for filtering input in procedure section 4. The values in array nc are the impulse response of the filter, which should exponentially decay to zero with a time constant of 5.1 ms.

3. Transform the nc A/D values to an array of voltage values cr using:

cr = (nc - 2048) / 204.8

Set all elements of the ci array to 0.0. Write the array cr to a file for subsequent plotting.

4. Perform $\mathcal{F}(c)$ using the fft.c function. The cr and ci arrays will then contain the complex Fourier coefficients C_r and C_i. Compute the magnitude array cm = SQRT(cr^2 + ci^2) and write it to a file for subsequent plotting.

5. Compute $b = \mathcal{F}^{-1}[1/\mathcal{F}(c)]$ using the fft.c function. Check that bi is small relative to br (ideally, b is real and all elements of the bi array should be 0). Write the array br to a file for subsequent plotting.

6. Loop over the elements of the br array and determine brlim, the largest value of fabs(br[i]). Transform br from voltage values to scaled D/A output values by using:

nb[i] = 2047.5 (br[i]/brlim + 1)

This transformation guarantees that the nb array contains only values from 0 to 4,095. Save this array for procedure section 4.

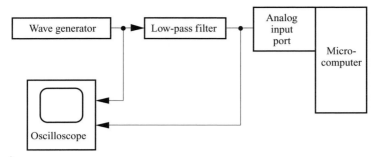

Laboratory Figure 24.4 Setup for sending a waveform through a low-pass filter and recording the output.

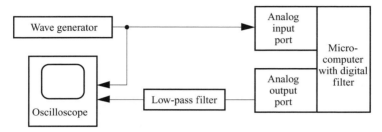

Laboratory Figure 24.5 Setup for sending a waveform through a digital deconvolution filter and then through a low-pass filter and recording the output.

3.   Response of low-pass filter to square-wave input

Set the wave generator for about 50-Hz square wave and send it through the low-pass analog filter as shown in Laboratory Figure 24.4. Record the output.

4.   Response of combined digital deconvolution filter and low-pass analog filter to square-wave input

Using the same wave-generator settings as in procedure section 3 above, connect the low-pass filter to the analog output port as shown in Laboratory Figure 24.5.

Set the low-pass analog filter gain to 1 by using $R_1 = \infty$ and $R_2 = 0 \ \Omega$.

Write a program to perform the following steps:

1. Read an input value x_i from the analog input port.
2. Compute the next output value y_{n+1} using the following digital filter:

$$y_{n+1} = \sum_{k=0}^{N} x_{n-k} b_k$$

where b_k is the kth element of the br[] array and N is chosen to provide accurate filtering without slowing the process too much.

3. Output the value y_{i+1} using the analog output port.
4. Loop back to step 1 in a tight loop.
   Plot the outputs of the wave generator, the D/A converter, and the low-pass filter.

Laboratory report

1.   Setup

Draw simple block diagrams of your experimental setups.

2.   Data summary and analysis

2.1 From the plots produced in procedure section 3, determine the period of the square wave and the RC time constant of the filter.

2.2 From the impulse sent to the low-pass filter and the expected filter response (see background section 1), compute the expected amplitude and shape of the low-pass filter impulse response.

2.3 Include the plots from procedure section 4 and compare the outputs of the wave generator and the low-pass filter.

3.   Discussion and conclusions

3.1 Compare the observed low-pass filter output from procedure section 3 with the expected RC filter response. Discuss agreement and reasons for any observed differences.

3.2 Compare the observed low-pass filter impulse response from procedure section 3 with the expected RC filter response. Discuss agreement and reasons for any observed differences.

3.3 Compare the input square wave with the observed response in procedure section 4. How well was the digital deconvolution filter able to compensate for the low-pass analog filter? Discuss agreement and reasons for any observed differences.

3.4 Discuss how the control method used in this laboratory exercise could be improved, its limitations, and its general applicability.

4.   Questions

4.1 How would the approximate shape you observed in procedure section 2 (filter output with 50-Hz square wave input) be expected from the frequency response of the filter, the frequency content of the square wave, and what you learned in Laboratory Exercise 21 (FFT transforms of sampled data)? (*Hint*: Use the Fourier convolution theorem.)

4.2 Why was the filter gain set to a high value (20 rather than 1) when recording the impulse response?

4.3 What problems would arise in this laboratory exercise if the Butterworth low-pass filter had 16 poles?

4.4 Why is this method limited to linear, time-invariant systems?

5.   Program and laboratory data sheets

5.1 Include printouts of your program code and output.

5.2 Include your handwritten data sheets (or a copy), which should consist of a log of the procedures you used, any special circumstances, and the measurements you recorded manually.

Laboratory Exercise 25

Analog temperature control using a resistive heater

Purpose

To sense the temperature of a small oven with a thermistor bridge circuit and instrumentation amplifier. To generate an error signal as the difference between the sense signal and a set point. To amplify the error signal and use a resistor to implement analog ON–OFF and proportional temperature control.

Equipment

- Superstrip circuit board
- +5-V power supply (for thermistor bridge)
- ±12-V power supplies (for instrumentation amplifiers)
- 2-A, +12-V power supply (high current to drive oven resistor)
- four 10-μF, 25-V electrolytic capacitors (put between power and ground at circuit board binding posts)
- one dial thermometer (temperature range -10 to $+110$ °C, with 4-mm diameter metal stem, 13 cm long)
- three ceramic 5-Ω, 12-W resistors (to be connected in series, mechanically and electrically)
- one precision thermistor (Omega type YSI 44004 [1207]–2,252 Ω at 25 °C)
- LM12 power op-amp circuit on heat sink
- Two 2.4-kΩ resistors (thermistor bridge)
- One 20-kΩ trimpot (thermistor bridge)
- One 100-Ω resistor (for power amplifier)
- Two AD625 or LH0036 instrumentation amplifiers
- For AD625 instrumentation amplifiers, use resistors below:
  - Two 20-kΩ trimpots (offset adjust)
  - Four 20-kΩ resistors(R_F)

- One 390-Ω resistor (R_G for gain of 100)
- Two 3.9-kΩ resistors (R_G for gain of 10)
- For LH0036 instrumentation amplifiers, use resistors below:
- Two 100-kΩ trimpots (offset adjust)
- Two 3.3-kΩ resistors (offset adjust)
- Two 33-kΩ resistors (offset adjust)
- One 510-Ω resistor (R_G for gain of 100)
- Two 5.1-kΩ resistors (R_G for gain of 10)

Background

1. Control modes

Three 12-W, 5-Ω ceramic resistors connected in series will act as a small cylindrical oven. The temperature inside the oven will be sensed by a thermistor and a dial thermometer. The thermistor will be one element of a bridge circuit with one variable resistor. This resistor is adjusted so the bridge has a small positive output voltage at room temperature and about 50 mV at 40 °C. An instrumentation amplifier will be used to amplify this signal by about 10 to provide a **sense signal** in the range from 0 to 0.5 V. A second instrumentation amplifier with a gain of 10 will be used to compare the sense signal with a **set point** and generate an **error signal** in the range from 0 to 5 V. The error signal will be amplified by a LM12 power op amp to drive the oven resistor.

Open-loop step response
An error signal is abruptly applied and the temperature response measured. Use this mode when measuring the open-loop step response and the relationship between the error signal and the sense signal.

Mode 1 (ON–OFF)
The difference amplifier gain is made large. If the error signal is negative (too hot), the heating resistor is shut off. If the error signal is even slightly positive (too cold), the heating resistor is set to maximum power.

Mode 2 (proportional)
The difference amplifier gain is set to a moderate value. As before, if the error signal is negative (too hot), the heating resistor is shut off. However, the heating resistor voltage is proportional to the error voltage and the system will reach an equilibrium temperature that changes very slightly, even if the ambient temperate changes.

2.   Performance criteria

In this laboratory exercise we will begin with the oven at some thermal equilibrium temperature T_0, abruptly change the set point S, and observe the change in temperature T with time t. The oven temperature will first change toward the desired temperature, possibly overshoot, and possibly oscillate. There are several criteria used to evaluate the performance of control algorithms, as described in Chapter 5. These are: lag time, risetime, time to first peak, settling time, jitter, and accuracy.

Additional reading

Section 5.10 Control techniques

Procedure

1.   Circuit

The overall schematic of the analog control system is shown in Laboratory Figure 25.1. Set up the thermistor bridge as shown in Figure 4.22 with $V_b = 1$ V, $R_T = 2.5$ kΩ (room temperature), $R_3 = R_4 = 2.4$ kΩ, and $R_1 = 20$ kΩ (variable).

Amplify the differential bridge output with the first instrumentation amplifier, using external connections as shown in Laboratory Exercise 5. Amplify the difference between the sense signal and the set point with the second instrumentation amplifier (used here as a difference amplifier). For the AD625, use $R_F = 20$ kΩ and $R_G = 3.9$ kΩ for a gain of 10. For the LH0036, use $R_G = 5.1$ kΩ for a gain of 10.

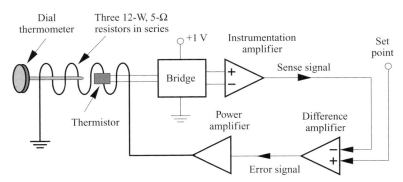

Laboratory Figure 25.1 Block diagram of the temperature control system. See Laboratory Figure 12.1 for the thermistor bridge circuit and Laboratory Figure 5.2 for the instrumentation amplifier circuit.

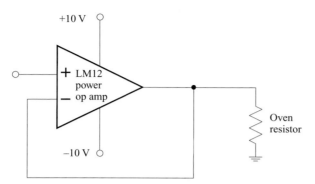

Laboratory Figure 25.2 LM12 power op-amp amplifier circuit.

Power the two instrumentation amplifier circuits with the low-current ±12-V power supplies. Connect the error signal to the input of the LM12 power op-amp circuit (Laboratory Figure 25.2). The LM12 power op-amp circuit is mounted on a heat sink. The three 12-W ceramic load resistors are to be connected in series to act an oven. *You must use the +12-V, 2-A supply for the LM12 circuit.*

2.   Open-loop response

1. Set the error signal to produce an initial LM12 output of 2 V. Record the temperature in the oven (dial thermometer), the sense value, and the resistor voltage at the start, at 30 and 60 s, and then every minute until equilibrium is approached (approximately 15 minutes). Record the time required to heat up from 10 to 90% of the temperature difference between the initial and final values.
2. Repeat for LM12 outputs of 4 and 6 V. The result relates the resistor voltage to the equilibrium dial thermometer temperature and sense value.
3. Set the LM12 output to 2 V and record the time required to cool down from 90 to 10% of the temperature difference.

3.   ON–OFF closed-loop response

Set the difference amplifier gain of 100 and a set point near the mid-range of the sense values recorded in procedure section 2. With this gain, a small error signal will produce a maximum power amplifier output. Connect the sense value, set point, and error voltage as shown in Laboratory Figure 25.1. Record these values, as well as the LM12 output every minute for 10 minutes.

4.   Proportional closed-loop step response

Set the difference amplifier gain to about 10 and a set point near the mid-range of the sense values recorded in procedure section 2. Connect the sense value, set point, and error voltage as shown in Laboratory Figure 25.1. Record these values, as well as the LM12 output every minute for 10 minutes.

Laboratory report

1.   Setup

Include a simple block diagram of your experimental setup.

2.   Data summary and analysis

2.1 Plot open-loop sense and dial thermometer temperature versus control LM12 voltage at equilibrium (data of procedure section 2).

2.2 Compute and plot the open-loop step response in terms of lag time, 10–90% risetime, and jitter (data of procedure section 2).

2.3 For the closed-loop step responses of the two control algorithms used (procedure sections 3 and 4), tabulate lag time, risetime, time to first peak, settling time, jitter, and systematic error.

3.   Discussion and conclusions

3.1 Discuss procedure sections 2, 3, and 4.

3.2 Describe how the control strategy you would use for a system with a long response time and a large inertia differs from the strategy you would use for a system with a short response time and a small inertia.

4.   Questions

4.1 Which of the two control algorithms you used had the least jitter? The least systematic error? To what temperature errors in degrees celsius do these correspond?

4.2 After an abrupt change in set point, which control algorithm (ON–OFF or proportional) reached 90% first (sum of lag time and risetime)?

5.   Laboratory data sheets

Include your handwritten data sheets (or a copy), which should consist of a log of the procedures you used, any special circumstances, and the measurements you recorded manually.

Laboratory Exercise 26

Temperature control using the computer and a resistive heater

Purpose

To sense the temperature of a small oven with a thermistor bridge circuit and instrumentation amplifier, to digitize the analog signal, and to write a C program to implement and compare several algorithms for the digital control of temperature.

Equipment

- IBM-compatible Pentium microcomputer with Windows NT operating system and Microsoft Visual C++ compiler
- Printer (shared with other laboratory stations)
- Data Translation DT3010 interfacing board
- Parallel port ribbon cable
- Superstrip circuit board
- ±12-V power supplies
- 2-A, +12-V power supply
- +5-V power supply (for thermistor bridge)
- Three 10-μF, 25-V electrolytic capacitors (put between power and ground at circuit board binding posts)
- One dial thermometer (temperature range −10 to +110 °C, with 4-mm diameter metal stem, 13 cm long)
- Three ceramic 5-Ω, 12-W resistors (to be connected in series, mechanically and electrically)
- One precision thermistor (Omega type YSI 44004 [1207] – 2,252 Ω at 25 °C)
- LM12 power op-amp circuit on heat sink
- One 20-kΩ trimpot (thermistor bridge)
- One 100-Ω resistor (power amplifier circuit)
- Two 2.4-kΩ resistors (thermistor bridge)

- AD625 or LH0036 instrumentation amplifier
- For the AD625 instrumentation amplifier, use resistors below:
  - One 20-kΩ trimpot (offset adjust)
  - Two 20-kΩ resistors (R_F)
  - One 3.9-kΩ resistor (R_G for gain of 10)
- For the LH0036 instrumentation amplifier, use resistors below:
  - One 100-kΩ trimpot (offset adjust)
  - One 3.3-kΩ resistor (offset adjust)
  - One 33-kΩ resistor (offset adjust)
  - One 5.1-kΩ resistor (R_G for gain of 10)

Background

1. Control modes

Three 12-W, 5-Ω ceramic resistors connected in series will act as a small cylindrical oven. The temperature inside the oven will be sensed by a thermistor and a dial thermometer. The thermistor will be one element of a bridge circuit with one variable resistor. This resistor is adjusted so the bridge has a small positive output voltage at room temperature and about 50 mV at 40 °C. An instrumentation amplifier will be used to amplify this signal by about 100 to provide 0–5 V for the analog input port. You will write a program that digitizes this **sense signal**, compares it with a previously entered **set point**, and generates the difference as an **error signal**. You will program one of several control algorithms to generate a number that is sent out via the digital-to-analog converter. The resulting analog signal is amplified by a LM12 power amplifier and used to drive the ceramic resistors. Note that almost every element in this control system has a nonlinear response: (i) the thermistor, whose resistance depends exponentially on $1/T$; (ii) the bridge circuit; and (iii) the ceramic resistor, whose power output is given by $P = RV^2$.

Write a program that provides one manual and two automatic control modes. The program firstly asks for a sampling interval Δt (s), then every time that interval passes, read and display: (i) the A/D value, (ii) the error signal, and (iii) the D/A control signal.

Mode 0 (manual)

The program asks for a control signal (2,048–4,095) and sends this value to the D/A. At each multiple of the sampling time Δt you will be able to see the effect on the A/D sense signal. Use this mode when measuring the open-loop step response and the relationship between the control signal and the sense signal.

Mode 1 (ON–OFF)

The program asks for a set point (2,048–4,095). At every time interval Δt, the sense signal is sampled and an error signal is computed as sense signal minus set point. The control variable is then determined as follows:

If error signal <0, control D/A is set to 4,095 ($+10$ V).

If error signal >0, control D/A is set to 2,048 (0 V).

Mode 2 (proportional)

The program asks for a set point (2,048–4,095) and a gain value. At every time interval Δt, the sense signal is sampled, an error signal computed, and control D/A is changed by the quantity ($-$gain $\times$ error). A typical value for gain is $\eta(\Delta t / T_r)$, where η is the ratio of the change in control signal to the change in sense signal, and T_r is the thermal response time. Large values of gain are equivalent to ON–OFF control and result in oscillations. Small values of gain result in very slow response. This is very analogous to the underdamped and overdamped harmonic oscillator except that thermal systems generally have large phase lags. Have your program calculate "control $= (-$gain $\times$ error)" using "float" variables for accuracy and convert it to an integer for output to the D/A.

Mode 3 (proportional-derivative)

As above, but the change in the control variable has an additional term proportional to the change in sense.

2.   Performance criteria

In this laboratory exercise we will begin with the oven at some thermal equilibrium temperature T_0, abruptly change the set point S, and observe the change in temperature T with time t. The oven temperature will first change toward the desired temperature, possibly overshoot, and possibly oscillate. There are several criteria used to evaluate the performance of control algorithms, as described in Chapter 5. These are: lag time, risetime, time to first peak, settling time, jitter, and accuracy.

Additional reading

Section 5.10 Control techniques
Section 5.10.6 Temperature control

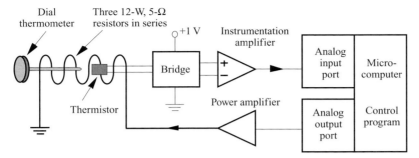

Laboratory Figure 26.1 Block diagram of the temperature control system. See Laboratory Figure 12.1 for the thermistor bridge circuit and Laboratory Figure 5.2 for the instrumentation amplifier circuit.

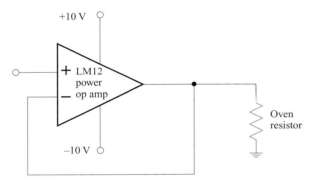

Laboratory Figure 26.2 LM12 power op-amp circuit.

Procedure

1.   Circuit

The overall schematic of the control system is shown in Laboratory Figure 26.1. Set up the thermistor bridge as shown in Figure 4.14 with $V_b = 1$ V, $R_T = 2.5$ kΩ (room temperature), $R_3 = R_4 = 2.4$ kΩ, and $R_1 = 20$ kΩ (variable). Amplify V_0 with your instrumentation amplifier circuit shown in Laboratory Exercise 5. Adjust the gain as needed to provide a suitable voltage swing for the A/D converter. Connect the instrumentation amplifier output to the data-acquisition circuit. For the AD625 and a gain of 10, use $R_F = 20$ kΩ and $R_G = 3.9$ kΩ. For the LH0036 and a gain of 10, use $R_G = 5.1$ kΩ.

Power all instrumentation amplifier circuits with the low-current ± 12-V power supplies. Connect the output of the D/A to the input of the op amp and the output of the op amp to the LM12 power op amp (Laboratory Figure 26.2). The three 12-W ceramic load resistors are to be connected in series to act an oven. *You must use the* +12-V, 2-A *supply for the LM12 circuit.*

2. Program

Write a program to implement mode 0 (manual), mode 1 (ON–OFF), and mode 2 (proportional) *or* mode 3 (proportional-derivative). At the start of your program, ask the user for a value Δt for the sampling interval. Read the timer (as you did in Laboratory Exercises 2 and 3) and sample the A/D value every Δt seconds. Mode 0 requires entering a D/A control value, modes 1 to 3 require entering a set point, and mode 3 requires entering a proportional gain value.

3. Open-loop response table

1. In manual mode 0, set the control variable to produce a D/A output of 2 V. Wait for the temperature to stabilize, and measure and record the temperature in the oven with your dial thermometer. Also record the A/D input voltage, the A/D output number, the resistor voltage, and compute the resistor power.
2. Repeat for D/A outputs of 3, 4, 5, and 6 V. The result is a table that relates the D/A input (and resistor power) to the dial thermometer temperature and A/D output. Ideally, the gain of the thermistor bridge amplifier should be chosen so that (at temperature equilibrium) a low value of D/A input corresponds to a low value of A/D output, and an D/A input of about 3,500 corresponds to an A/D output of about 3,500. Do not approach 2,048 or 4,095 too closely to avoid saturation.
3. In mode 0, set the D/A output to 2 V and record the time required to cool down from 90 to 10% of the temperature difference. (This is the same temperature excursion measured as the risetime.)

4. Open-loop step response

In manual mode 0, abruptly change the D/A output from 2 to 5 V, and record the open-loop step response. To do this, have your program write the time and the A/D input number at the end of each 5-s time interval to a file for later printing.

5. ON–OFF closed-loop response

In mode 0 with a D/A output voltage value that corresponds to an A/D value of about 2,500, wait for the temperature to stabilize. Enter mode 1 and abruptly change the A/D set point to some value between 3,000 and 4,000. Record the closed-loop step response. To do this, have your program write the time, the A/D input number, the error value, and the D/A output number at the end of each 5-s time interval to a file for later printing.

6.   Proportional closed-loop step response

In mode 0 with a D/A value that corresponds to an A/D value of about 2,500, wait for the temperature to stabilize. Enter mode 2 and abruptly change the A/D set point to some value between 3,000 and 4,000. A typical value for the gain is $\eta(\Delta t / T_r)$, where η is the ratio of the change in control signal to the change in sense signal (determined from procedure step 3), Δt is the sampling time interval, and T_r is the open-loop response time. Have your program write the time, the A/D input number, the error value, and the D/A output number at the end of each 5-s time interval to a file for later printing.

7.   Proportional-derivative closed-loop step response

As for mode 2, with the addition of a derivative coefficient.

Laboratory report

1.   Setup

Include a simple block diagram of your experimental setup.

2.   Data summary and analysis

2.1 Tabulate the open-loop response (equilibrium sense and dial thermometer temperature versus control D/A number from procedure section 3).

2.2 Plot the open-loop response data (A/D input versus time from procedure section 4). Tabulate lag time, risetime, and jitter.

2.3 For the closed-loop step responses of the two control algorithms used (procedure sections 5, and 6 or 7), tabulate lag time, risetime, time to first peak, settling time, jitter, and systematic error.

3.   Discussion and conclusions

3.1 Discuss procedure sections 3, 4, 5, and 6 or 7.

3.2 Describe how the control strategy you would use for a system with a long response time and a large inertia differs from the strategy you would use for a system with a short response time and a small inertia.

4.   Questions

4.1 Which of the control algorithms you used had the least jitter? The least systematic error? To what temperature errors in degrees celsius do these correspond?

4.2 After an abrupt change in set point, which control algorithm reached 90% first (sum of lag time and risetime)?

5.   Program and laboratory data sheets

5.1 Include printouts of your program code and output.

5.2 Include your handwritten data sheets (or a copy), which should consist of a log of the procedures you used, any special circumstances, and the measurements you recorded manually.

Laboratory Exercise 27

Temperature control using the computer and a thermoelectric heat pump

Purpose

To sense the temperature of a small breaker of water with a thermistor bridge circuit and instrumentation amplifier. To write a C program to read the analog temperature signal and compute a control signal that is sent to a power amplifier and thermoelectric heat pump. To implement and compare several algorithms for the digital control of temperature.

Equipment

- IBM-compatible Pentium microcomputer with Windows NT operating system and Microsoft Visual C++ compiler
- Printer (shared with other laboratory stations)
- Data Translation DT3010 interfacing board
- Parallel port ribbon cable
- Superstrip circuit board
- ±12-V power supplies (low current for instrumentation amplifier)
- +5-V power supply (for thermistor bridge)
- ±12-V power supplies (>5 A for power amplifier)
- Three 10-μF, 25-V electrolytic capacitors (put between power and ground at circuit board binding posts)
- One dial thermometer (temperature range -10 to $+110\,°C$, with 4-mm diameter metal stem, 13 cm long)
- One precision thermistor (Omega type YSI44004 [1207] – 2,252 Ω at 25 °C)
- Cambion No. 801-3959-01 thermoelectric device (maximum current 10 A at 5 V)
- Thick aluminum plate (heat sink)
- Small beaker, approx. 10 ml

- Heat sink compound: zinc oxide in silicone paste
- Small 5 mil Mylar sheet
- LF356 op amp (for power amplifier circuit)
- LM12 power op-amp circuit mounted on heat sink
- One 20-kΩ trimpot (thermistor bridge)
- Two 2.4-kΩ resistors (thermistor bridge)
- AD625 or LH0036 instrumentation amplifier
- For the AD625 instrumentation amplifier, use resistors below:
  - One 20-kΩ trimpot (offset adjust)
  - Two 20-kΩ resistors (R_F)
  - One 3.9-kΩ resistor (R_G for gain of 10)
- For the LH0036 instrumentation amplifier, use resistors below:
  - One 100-kΩ trimpot (offset adjust)
  - One 3.3-kΩ resistor (offset adjust)
  - One 33-kΩ resistor (offset adjust)
  - One 5.1-kΩ resistor (R_G for gain of 10)

Background

1. Control modes

A thermoelectric heat pump thermally mounted to a heat sink will be used to control the temperature of a small beaker of water. The temperature of the water in the beaker will be sensed by a thermistor and a dial thermometer. The thermistor will be one element of a bridge circuit with one variable resistor. This resistor is adjusted so the bridge has a small positive output voltage at room temperature and about 50 mV at 40 °C. An instrumentation amplifier will be used to amplify this signal by about 100 to provide 0–5 V for the analog input port. You will write a program that digitizes this **sense signal**, compares it with a previously entered **set point**, and generates the difference as an **error signal**. You will program one of several control algorithms to generate a number that is sent out via the digital-to-analog converter. The resulting analog signal is amplified by a LM12 power op amp and used to drive the Peltier heat pump. Note that almost every element in this control system has a nonlinear response: (i) the thermistor, whose resistance depends exponentially on $1/T$; (ii) the bridge circuit; (iii) the LM12 amplifier, which has an offset; and (iv) the ceramic resistor, whose power output is given by $P = RV^2$.

Write a program that provides one manual and two automatic control modes. The program first asks for a sampling interval Δt (s), then every time that interval passes, read and display: (i) the A/D value, (ii) the error signal, and (iii) the D/A control signal.

Mode 0 (manual)

The program asks for a control signal (0–4,095) and sends this value to the D/A. At each multiple of the sampling time Δt you will be able to see the effect on the A/D sense signal. Use this mode when measuring the open-loop step response and the relationship between the control signal and the sense signal.

Mode 1 (ON–OFF)

The program asks for a set point (0–4,095). At every time interval Δt, the sense signal is sampled and an error signal is computed as sense signal minus set point. The control variable is then determined as follows:

If error signal <0, control D/A is set to 4,095 ($+10$ V for max. heating).
If error signal >0, control D/A is set to 0 (-10 V for max. cooling).

Mode 2 (proportional)

The program asks for a set point (0–4,095) and a gain value. At every time interval Δt, the sense signal is sampled, an error signal computed, and control D/A is changed by the quantity ($-$gain $\times$ error). A typical value for gain is $\eta(\Delta t / T_r)$, where η is the ratio of the change in control signal to the change in sense signal, and T_r is the thermal response time. Large values of gain are equivalent to ON–OFF control and result in oscillations. Small values of gain result in very slow response. This is very analogous to the underdamped and overdamped harmonic oscillator except that thermal systems generally have large phase lags. Have your program calculate "control = ($-$gain $\times$ error)" using "float" variables for accuracy and convert it to an integer for output to the D/A.

Mode 3 (proportional-derivative)

As above, but the change in the control variable has an additional term proportional to the change in sense.

2.   Performance criteria

In this laboratory exercise we will begin with the oven at some thermal equilibrium temperature T_0, abruptly change the set point S, and observe the change in temperature T with time t. The oven temperature will firstly change toward the desired temperature, possibly overshoot, and possibly oscillate. There are several criteria used to evaluate the performance of control algorithms, as described in Chapter 5. These are: lag time, risetime, time to first peak, settling time, jitter, and accuracy.

Additional reading

Section 5.10 Control techniques
Section 5.10.6 Temperature control

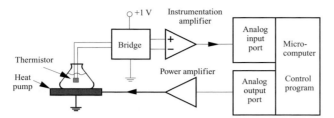

Laboratory Figure 27.1 Block diagram of the temperature control system. See Laboratory Figure 12.1 for the thermistor bridge circuit and Laboratory Exercise 5 for the instrumentation amplifier circuit.

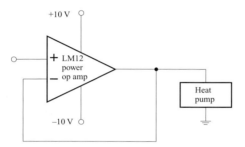

Laboratory Figure 27.2 LM12 power op-amp circuit for driving the thermoelectric heat pump.

Procedure

1.   Circuit

The overall schematic of the control system is shown in Laboratory Figure 27.1. Set up the thermistor bridge as shown in Figure 4.14 with $V_b = 1$ V, $R_T = 2.5$ kΩ (room temperature), $R_3 = R_4 = 2.4$ kΩ, and $R_1 = 20$ kΩ (variable). Amplify V_0 with your instrumentation amplifier circuit shown in Laboratory Exercise 5. Adjust the gain as needed to provide a suitable voltage swing for the A/D converter. Connect the instrumentation amplifier output to the data-acquisition circuit. For the AD625 and a gain of 10, use $R_F = 20$ kΩ and $R_G = 3.9$ kΩ. For the LH0036 and a gain of 10, use $R_G = 5.1$ kΩ.

Power all instrumentation amplifier circuits with the low-current ± 12-V power supplies. Connect the output of the D/A to the input of the op amp and the output of the op amp to the LM12 power amplifier (Laboratory Figure 27.2). The three 12-W ceramic load resistors are to be connected in series to act as an oven. *You must use the ± 12-V, 5-A supply for the LM12 circuit.*

2.　Program

Write a program to implement mode 0 (manual), mode 1 (ON–OFF), and mode 2 (proportional) *or* mode 3 (proportional-derivative). At the start of your program, ask the user for a value Δt for the sampling interval. Read the timer (as you did in Laboratory Exercises 2 and 3) and sample the A/D value every Δt seconds. Mode 0 requires entering a D/A control value, modes 1–3 require entering a set point, and mode 3 requires entering a proportional gain value.

3.　Open-loop response table

1. In manual mode 0, set the control variable to produce a D/A output of 2 V. Wait for the temperature to stabilize, and measure and record the temperature in the oven with your dial thermometer. Also record the A/D input voltage, the A/D output number, the resistor voltage, and compute the resistor power.
2. Repeat for D/A outputs of −4, −2, +2, and +4 V. The result is a table that relates the D/A input (and resistor power) to the dial thermometer temperature and A/D output. Ideally, the gain of the thermistor bridge amplifier should be chosen so that (at temperature equilibrium) a low value of D/A input corresponds to a low value of A/D output, and a D/A input of about 3,500 corresponds to an A/D output of about 3,500. Do not approach 0 or 4,095 too closely to avoid saturation.
3. In mode 0, set the D/A output to 2 V and record the time required to cool down from 90 to 10% of the temperature difference. (This is the same temperature excursion measured as the risetime.)

4.　Open-loop step response

In manual mode 0, abruptly change the D/A output from 2 to 5 V, and record the open-loop step response. To do this, have your program write the time and the A/D input number at the end of each 5-s time interval to a file for later printing.

5.　ON–OFF closed-loop response

In mode 0 with a D/A output voltage value that corresponds to an A/D value of about 1,000, wait for the temperature to stabilize. Enter mode 1 and abruptly change the A/D set point to some value around 3,000. Record the closed-loop step response. To do this, have your program write the time, the A/D input number, the error value, and the D/A output number at the end of each 5-s time interval to a file for later printing.

6.   Proportional closed-loop step response

In mode 0 with a D/A value that corresponds to an A/D value of about 1,000, wait for the temperature to stabilize. Enter mode 2 and abruptly change the A/D set point to some value around 3,000. A typical value for the gain is $\eta(\Delta t/T_r)$, where η is the ratio of the change in control signal to the change in sense signal (determined from procedure step 3), Δt is the sampling time interval, and T_r is the open-loop response time. Have your program write the time, the A/D input number, the error value, and the D/A output number at the end of each 5-s time interval to a file for later printing.

7.   Proportional-derivative closed-loop step response

As for mode 2, with the addition of a derivative coefficient.

Laboratory report

1.   Setup

Include a simple block diagram of your experimental setup.

2.   Data summary and analysis

2.1 Tabulate the open-loop response (equilibrium sense and dial thermometer temperature versus control D/A number from procedure section 3).

2.2 Plot the open-loop response data (A/D input versus time from procedure section 4). Tabulate lag time, risetime, and jitter.

2.3 For the closed-loop step responses of the two control algorithms used (procedure sections 5, and 6 or 7), tabulate lag time, risetime, time to first peak, settling time, jitter, and systematic error.

3.   Discussion and conclusions

3.1 Discuss procedure sections 3, 4, 5, and 6 or 7.

3.2 Describe how the control strategy you would use for a system with a long response time and a large inertia differs from the strategy you would use for a system with a short response time and a small inertia.

4.   Questions

4.1 Which of the control algorithms you used had the least jitter? The least systematic error? To what temperature errors in degrees celsius do these correspond?

4.2 After an abrupt change in set point, which control algorithm reached 90% first (sum of lag time and risetime)?

5.   Program and laboratory data sheets

5.1 Include printouts of your program code and output.

5.2 Include your handwritten data sheets (or a copy), which should consist of a log of the procedures you used, any special circumstances, and the measurements you recorded manually.

Appendix A: Grounding and shielding

A.1 Introduction

Noise in electronic circuits includes three basic categories:
1. noise received with the original signal and indistinguishable from it,
2. intrinsic noise in the circuit (Johnson, shot, etc.), and
3. interference noise generated by components in the circuit or picked up from outside the circuit.

This appendix considers only the last category, which is the only form of noise that can be influenced by choices of wiring and shielding. The sections below discuss interference noise from common impedance paths and from capacitive coupling, and list general rules to follow.

A.2 Interference noise due to common impedance

Whenever two circuit elements share a common current path, the impedance of that path can couple signals between them. This most commonly occurs when several circuits use a single conductor to return current back to their power supplies. This is shown in Figure A.1, where analog and digital circuits are connected to their power supplies through conductors having impedance Z. The $+5$- and ±15-V current supply wires do not share a common impedance and are relatively free from interference noise. On the other hand, the current return path (usually referred to as "ground") has a common impedance for all circuits and power supplies. A current transient in a digital circuit will briefly shift the "ground" reference for the analog circuits. This problem will generally not be visible using a voltage meter because the effect is brief. To see such problems, look at a "ground" wire near an analog circuit with an oscilloscope set for extreme vertical gain, and trigger the oscilloscope on transient pulses on a $+5$-V supply conductor near a digital circuit.

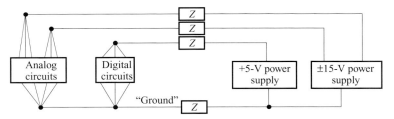

Figure A.1 Connection of circuit current return paths from all circuits to all power supplies by a single conductor. The common impedance couples interference noise from the digital circuits to the analog circuit ground reference.

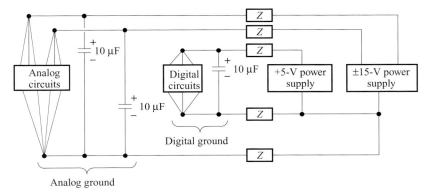

Figure A.2 Schematic showing separate digital and analog ground return conductors, and the use of capacitors to provide current during brief peak loads. Pulses on the digital ground return interfere with the analog ground much less than in Figure A.1, because these conductors do not share a common impedance.

This problem is solved by keeping the digital and analog grounds separate at the circuits and connecting them at the power supplies (Figure A.2). In this way, there is no common impedance on the current return conductors.

A.3 Interference noise due to capacitive coupling

A capacitor exists between any two conductors, providing that the space between them is filled with a dielectric insulator (e.g. epoxy, plastic film, air, vacuum). Figure A.3 shows two circuits with separate power supplies that are coupled by the stray capacitance between their signal conductors. In the case where the input impedance of the receiving circuit is largely resistive with resistance R, the coupled noise signal V_1 is related to the noise signal V_n at the source by the equation:

$$V_1 = \frac{V_n R}{R + 1/(2j\omega C_S)} = \frac{-2V_n\omega R C_S}{j - 2\omega R C_S}$$

$$\frac{|V_1|}{|V_n|} = \frac{2\omega R C_S}{\sqrt{1 + (2\omega R C_S)^2}} = \frac{f/f_0}{\sqrt{1 + (f/f_0)^2}}$$

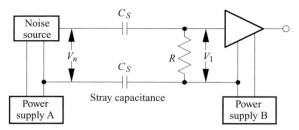

Figure A.3 Connection of circuit current return paths from all circuits to all power supplies by a single conductor. The common impedance couples interference noise from the digital circuits to the analog circuit ground reference.

where $f_0 = 1/(\pi RC_S)$, the RC corner frequency and $f = 2\pi\omega$, the frequency of the noise source.

For a typical case $R = 1\,k\Omega$, $C_S = 1\,pF$, and $f = 1\,MHz$, $|V_1|/|V_n| = 0.012$. This problem is worse at higher frequencies. For example, the 10-ns risetimes and 5-V swings of TTL logic can couple 100-MHz transients of many millivolts through stray capacitances as low as 0.001 pF.

Capacitive coupling is essential in wireless communications and permits radio and television transmission between the transmitting tower and the receiver. However, electronic circuits can broadcast unwanted signals to communication receivers, and there are strict rules limiting radio emissions from high-frequency digital circuits such as computers.

One of the most effective means of blocking capacitive interference is by placing a conductive shield around each circuit (the principle of the "Faraday cage"). Since there is a stray capacitance from this shield to every conductor within, its effectiveness is reduced if it is also used to carry return currents to the power supply.

A.4   General rules to follow

- Think where the currents flow: all the current that enters your circuit from the power supplies must somehow return to those same power supplies. Since every conductor has a finite dc resistance and ac impedance, the conductors that carry that current may be called "ground" but may have dc shifts and voltage spikes.
- Keep digital and analog ground returns separate: conductors that serve as the "digital ground" can carry large voltage spikes as the digital circuits change logic state, and these grounds serve as very poor analog circuit grounds. Remember that analog circuits have no "noise immunity."
- Printed circuit boards and well-designed breadboards have a "ground plane" that is best used as electromagnetic shielding from outside electrical interference. Do not use this plane as a conductor to return current to the power supply.

- Both the power-supply and the ground-return leads should be sufficiently heavy wire to reduce both dc and ac voltage drops to acceptable levels.
- Put electrolytic capacitors (in the 1- to 100-μF range) between the power-supply conductors and their ground-return conductors at the point where they connect to the circuit board. Their purpose is to stabilize the supply voltages from external electrical interference (60 Hz to 1 MHz), and brief (microsecond to millisecond) drops in supply voltage due to momentary current demands by both external circuits and internal circuit components. These capacitors are not useful in protecting against the 1- to 10-ns voltage spikes caused by switching in digital circuits.
- Put Mylar capacitors (in the 0.1-μF range) across the voltage supplies of all digital integrated-circuit chips that demand very brief surges of current and across all analog chips that have to be protected from such surges.
- Never blindly trust your power supply and ground connections; look at them with an oscilloscope while the circuit is running.
- An op amp is usually shown with only three signal terminals, the two differential inputs and an output that is referenced to "ground." In reality, however, one or both of the power-supply terminals acts as a fourth output conductor, which is expected to be a low impedance at all frequencies within the amplifier bandwidth. For example, some amplifiers such as the 741 have excellent low-frequency power-supply rejection, but are limited in their ability to compensate for variations in the negative supply that are faster than the closed-loop bandwidth. Thus, other high-speed switching circuits that draw current pulses from the negative supply can cause fluctuations in the output of such op amps.

Appendix B: Experimental uncertainties

B.1   Multimeter accuracy

The following is the typical accuracy for a student-type digital multimeter, information often provided on the back of the meter. The shorthand notation $(A + B)$ means that the uncertainty of the reading is guaranteed to be less than $\pm(A\%$ of reading $+ B$ least significant digits):

    Max counts 2,000 (e.g. 2.000 V or 200.0 mV or 2,000 Ω)

<200 V	dc $(0.1 + 1)$ 45 Hz to 10 kHz $(0.5 + 2)$
<200 mA	dc $(0.3 + 1)$ 45 Hz to 10 kHz $(1.0 + 2)$
<200 kΩ	$(0.2 + 1)$
>2 MΩ	$(0.5 + 1)$

    For example, for a reading of 1.453 V dc, the uncertainty is less than $\pm(1.453 \times 0.1\% + 0.001)$ V $= \pm0.0025$ V. For a reading of 1.453 V ac, the uncertainty is less than $\pm(1.453 \times 0.5\% + 0.002) = \pm0.009$ V.

    Note that in the latter case, the **accuracy** (adherence to the accepted standard) is ±0.009 V, but the **precision** (ability to detect small changes) is one least significant digit, or ±0.001 V.

    A triangular distribution with full width $\pm W$ has a standard deviation $\sigma = 0.408W$. So as a rule of thumb, the multimeter uncertainty can be thought of as representing approximately 2.5 standard deviations.

B.2   Propagation of random error

Given the function $f(a_1, a_2, \ldots, a_n)$, where the variables $a_1, a_2, \ldots, a_n$ are independently random and have standard deviations σ_{ai}, then standard deviations can be combined as follows:

$$\sigma_f{}^2 = \left(\frac{\partial f}{\partial a_1}\right)^2 \sigma_{a1}{}^2 + \left(\frac{\partial f}{\partial a_2}\right)^2 \sigma_{a2}{}^2 + \cdots + \left(\frac{\partial f}{\partial a_n}\right)^2 \sigma_{an}{}^2$$

EXAMPLE B.1

Adding two random variables

$$f = R_1 + R_2 \qquad \partial f / \partial R_1 = \partial f / \partial R_2 = 1 \qquad \sigma_f{}^2 = \sigma_{R_1}{}^2 + \sigma_{R_2}{}^2$$

General rule

When adding (or subtracting) random numbers, the variance (σ^2) of the result is the sum of the variances.

EXAMPLE B.2

Multiplying two random variables

$$f = R_1 R_2 \qquad \partial f / \partial R_1 = R_2 \qquad \partial f / \partial R_2 = R_1$$

$$\sigma_f{}^2 = R_2{}^2 \sigma_{R_1}{}^2 + R_1{}^2 \sigma_{R_2}{}^2$$

$$\frac{\sigma_f{}^2}{f^2} = \frac{R_2{}^2 \sigma_{R_1}{}^2 + R_1{}^2 \sigma_{R_2}{}^2}{R_1{}^2 R_2{}^2} = \frac{\sigma_{R_1}{}^2}{R_1{}^2} + \frac{\sigma_{R_2}{}^2}{R_2{}^2}$$

General rule

When multiplying (or dividing) random numbers, the fractional variance of the result is the sum of the fractional variances.

Appendix C: C programming tips

C.1   Declare all variables

C.1.1   Declaring an array

To declare an integer array a of dimension 10, put the following statement at the top of your program:

int a[10];

   In C, subscripts start from 0, so that valid elements are a[0], a[1], ... , a[9]. Writing into a[10] may do serious damage to other data!

Note the square brackets and the ";" after every statement.

C.1.2   Overflow warning

The following will overflow if b*b has more than 16 bits:

long a;
int b;
a=b*b;

   The solution is to make both a and b long.

C.1.3   Variables used in C programs

Table 1.2 lists the variables that are used in C programs.

C.2   Arithmetic statements

```
a=b+c;      /* addition */
a=b−c;      /* subtraction */
```

```
a=b*c;        /* multiplication */
a=b/c;        /* division */
a=b%c;        /* b modulo c (b and c both "int") */
```

Note: There is no low-level exponentiation operator such as a**b or aˆb.

```
a=1459;        /* a = decimal 1459 */
b=0127;        /* b = octal 127 (note leading zero) */
c=0xA3FC;      /* c = hexadecimal A3FC (note leading
                  "zero ex") */
```

C.3   Conditional tests

```
if (expression)
    statement 1
else
    statement 2
```

- "expression" uses conditional operators, such as a < b or a != b or a == b or a <= b (see Section C.4).
- "statement" can be a single statement line ending with ";" or many statement lines, each ending in ";", enclosed between "{" and "}".

C.4   Conditional operators

These operators are used with conditional tests; see Section C.3.

```
= =   equal to
! =   not equal to
<     less than
< =   less than or equal to
>     greater than
> =   greater than or equal to
& &   logical AND
||    logical OR
```

C.5   Indexed looping

```
for (expression 1; expression 2; expression 3)
    statement;
```

is equivalent to:

```
expression 1;
while(expression 2) {
    statement;
    expression 3;
    }
```

Usually, expressions 1 and 3 are assignments and expression 2 is relational, e.g:

```
for (i=0; i<100 ; ++i)
    expression;
```

will loop 100 times with i taking on the values $0, 1, 2, 3, \ldots, 99$.

C.6   Bitwise logical operators

```
&   AND
|   inclusive-OR
^   exclusive-OR
≪   left shift
≫   right shift
~   complement
```

Common uses of these operators are in selectively zeroing (masking) portions of a word and in combining two words without having to deal with 2's complement arithmetic.

For example, to zero the 16 most significant bits of the 32-bit word a and to retain the 16 least significant bits, use:

```
a = a & 0x00FF;
```

To combine the two 16-bit words a and b into a single 32-bit word c, where a contains the most significant 16 bits and b contains the least significant 16 bits, use:

```
c = a ≪ 16 | b;
```

C.7   Increment and decrement operators

++i increments i by 1 and then uses its value as the value in the expression
i++ uses i as the value in the expression, then increments i by 1
--i decrements i by 1 and then uses its value as the value in the expression
i-- uses i as the value in the expression, then decrements i by 1

C.8 The printf **statement**

The first part of the argument is in quotes (" ") and may contain text to be printed and special format commands, always beginning with % or /. The second part of the argument is a list of quantities (numbers or text data) to be printed according to the % commands (Table C.1). Each such quantity in the list is associated with the corresponding % command.

Example of printf use: printf("index=%d data=%10.3f",a,b);

Summary of "\" commands (Brian W. Kernighhan and Dennis M. Ritchie, *The C Programming Language,* Prentice Hall, Englewood Cliffs, NJ, 1988, p. 193):

\a   audible alert

\n   new line (carriage return and linefeed)

\r   carriage return (no linefeed)

\b   backspace (to make ϕ, θ, etc.)

\t   horizontal tab (note use in Section C.10)

\nnn   ASCII character corresponding to nnn (octal)

C.9 Defining your own functions

```
main()
{
double fun(double, double);      /* declare function prototype */
double x, y, z;                  /* define variables */
. . .                            /* your main program goes here */
z = fun(x, y);                   /* call to function */
. . .                            /* more of your program */
}                                /* end of main program */
```

Table C.1 *Examples of % format commands*

Format	Variable Type	Output Printed As
%5d	"int"	decimal, field width = 5
%10ld	"long"	decimal, field = 10
%10.3f	"float"	decimal, field = 10, 3 places after "."
%15.8lf	"double"	decimal, field = 15, 8 places after "."
%4x	"int"	hexadecimal (4 bits, 0 to F), field = 4
%10lx	"long"	hexadecimal, field =10
%c	"char"	single ASCII character

```
double fun(double x, double y)      /* note: no final semicolon */
{
double u, v;                        /* declare local variables */
...                                 /* body of code to compute fun(x, y) */
return(u);                          /* return local variable value */
}
```

Note 1: The function name "fun" can be any name of your choosing.

Note 2: If the function code appears before main(), then the initial type declaration of the function is not required.

C.10   "Including" your own functions

It is a common practice to maintain a "library" of functions that can be used by many different programs. In this way, only one copy of each function need be maintained. The code for each function is in a file by itself and that file is "included" in the main program code to be compiled as though it were all in one file. For example, the file "fun.c" might look like:

```
double fun(double x, double y)      /* note: no final semicolon */
{
double u, v;                        /* declare local variables */
...                                 /* body of code to compute fun(x,y) */
return(u);                          /* return local variable value */
}
```

And the main program would look like:

```
#include "fun.c"
main()
{
double fun(x, y);      /* declare function prototype */
double x, y, z;        /* define variables */
...                    /* your main program goes here */
z = fun(x, y);         /* call to function */
...                    /* more of your program */
}                      /* end of main program */
```

Note: Since the included file precedes its first call, a separate type declaration in the main program code is not necessary.

C.11   Opening and writing to files of arbitrary name

This very important feature is to be used for most of the data recording in systems where several microcomputers share a printer. It allows you to save your data on disk and then print it later, whenever the printer is free.

At the top of your program, put the statements:

```
char    out_file_name[16];
FILE    *out_file;
```

Later in your program, ask the user for a filename, read it, and open the file:

```
printf("Output file name:\t")
scanf("%s", out_file_name);
out_file = fopen(out_file_name, "w");
```

To write to the newly created file:

```
fprintf(out_file, "This file is %s", out_file_name);
```

(or anything else you want to write).

C.12   Using library functions

To use the standard I/O library, put "#include <stdio.h>" before "main()." This is necessary for all programs that use printf, scanf, fprintf, fscanf, etc.

To use the math function library, put "#include <math.h>" before "main()." All the math functions listed below have type "double" arguments and return type "double."

cos(x)	cosine of x, x in radians
sin(x)	sine of x, x in radians
tan(x)	tan of x, x in radians
atan(x)	arc tangent of x in radians
atan2(y, x)	arc tangent of y/x (range spans 2π)
exp(x)	exponential, base e
log(x)	base e (natural) logarithm
log10(x)	base 10 logarithm
pow(x, y)	power function x^y
sqrt(x)	square root
fabs(x)	absolute value (use abs(i) for type "int")

C.13   Allocating large storage arrays

Many C compilers for microcomputers do not allow arrays to be declared that are larger than 32 kbytes. This limitation can be overcome by using the ANSI C library function calloc to allocate storage during program execution. The following example allocates an array of 100,000 doubles (800,000 bytes):

At the top of your program, define a pointer to the array:

double *a;

The storage allocation is performed using the code:

a = (double *) calloc(100000, sizeof(double));

In use, the code *(a + i) stands for the ith value of the array, where $i = 0$ to 99,999.

Note 1: The value returned by calloc is a 32-bit pointer, and its arguments are of type size-t, which has the same size as a pointer.

Use the ANSI C function free to deallocate storage.

Note 2: Some C compilers (for example, Borland Turbo C++ for the IBM PC) require non-ANSI functions (farcalloc) and pointer declarations ("huge") to allocate large arrays.

C.14   General format rules for C programs

- Every single-line statement must end in ";".
- Multi-line statements must be inclosed between "{"and"}".
- Every variable must be declared (char, int, long, float, double).
- Comments must be inclosed between /* and */.

Appendix D: Numerical methods and C functions

D.1   Introduction

This appendix describes some numerical methods useful in microcomputer-based data analysis. The fast Fourier transform function performs a discrete Fourier transform of a waveform sampled at periodic time intervals. The PARFIT function permits non-linear fitting by varying parameters to minimize a goodness-of-fit quantity (such as chi-squared). The VARFIT function varies the parameters about their best-fit values to determine their statistical uncertainties. The ADSINT function performs numerical integration using adaptive quadrature and dynamically divides the integration interval into smaller steps to evaluate the integral with the desired accuracy and the minimum number of integrand evaluations. An example is given for computing the probability of exceeding a particular value of Student's t. Several methods for function inversion are given using both Newton's method and quadratic approximation. The ADSINT function and Newton's method were used to compute the chi-squared and Student's t probability tables in Chapter 5. The RAMDOME function provides floating point numbers randomly distributed between 0 and 1 with 2^{31} different values, a period of 2^{61} numbers, and with a low degree of correlation.

D.2   Fast Fourier transform

The fast Fourier transform (FFT) is discussed in Section 5.8. This algorithm is basically a way to compute the discrete Fourier transform rapidly.

The C code listing (Section D. 2.1) was adapted by the author from the FORTRAN code in E. Orhan Brigham, *The Fast Fourier Transform*, © 1974, p. 164, by permission of Prentice-Hall, Inc., Englewood Cliffs, N.J. This C function is used with the statement fft(xr, xi, nu, ie);, where xr and xi are the real and imaginary parts, respectively, of a coefficient array having 2^{nu} elements. If ie < 0, the forward transform is performed (time samples into frequency coefficients), and if ie ≥ 0, the reverse transform is performed (frequency coefficients into time samples). The function returns the result of the transformation in the arrays xr and xi. Note that raising 2 to an integer power and bit reversal are performed using shifts rather than multiplication and division by 2. The code allocates local storage as needed so that the limit to the size of the array to be transformed is limited only by available memory.

D.2.1   C program code

```c
#include <stdio.h>
#include <math.h>
#include <stdlib.h>
int ibitr(int, int);
void fft(double [], double [], int, int);
void fft(double xr[], double xi[], int nu, int ie)
{
static int n, n1, n2, nu1, p, k1n2, i, k, l;
static double arg, c, s, tr, ti;
double *stab, *ctab;
#define TWOPI 6.2831853 /* 2 PI */
#define PITWO 1.5707963 /* PI/2 */
/* allocate storage for sine and cosine tables */
n=1<<nu;
stab = (double *) calloc(n, sizeof(double));
ctab = (double *) calloc(n, sizeof(double));
if ( (stab == NULL) || (ctab == NULL) )
    {
    printf("Can't allocate fft.c data storage - exit\n");
    exit(1);
    }
n2=n/2;
nu1=nu-1;
for (i=0; i<n; ++i)
    {
    arg=TWOPI*i/n;
    stab[i]=sin(arg);
    ctab[i]=sin(arg+PITWO);
    }
k=0;
for (l=0; l<nu; ++l)
    {
    while (k<n)
```

```
{
for(i=0; i<n2; ++i)
    {
    n1=1<<nu1;
    p=ibitr(k/n1, nu);
    s=stab[p];
    c=ctab[p];
    if(ie>0) s=-s;
    k1n2=k+n2;
    tr = xr[k1n2]*c + xi[k1n2]*s;
    ti = xi[k1n2]*c - xr[k1n2]*s;
    xr[k1n2] = xr[k] - tr;
    xi[k1n2] = xi[k] - ti;
    xr[k] = xr[k] + tr;
    xi[k] = xi[k] + ti;
    k=k+1;
    }    /*end i loop */
k=k+n2;
}        /*end while */
k=0;
nu1=nu1-1;
n2=n2/2;
}        /*end l loop */
for (k=0; k<n; ++k)
{
i=ibitr(k, nu);
if(i>k)
    {
    tr=xr[k];
    ti=xi[k];
    xr[k]=xr[i];
    xi[k]=xi[i];
    xr[i]=tr;
    xi[i]=ti;
    }
}
if (ie>0)
    {
    for(i=0; i<n; ++i)
        {
        xr[i]=xr[i]/n;
```

```
        xi[i]=xi[i]/n;
        }
    }
free(stab);
free(ctab);
return;
}
int ibitr(int j, int nu)
{
int rm, lm, i, bitr;
rm=1;
lm=1<<(nu-1);
bitr=0;
for (i=0; i<nu; ++i)
    {
    if ((j&rm)!=0) bitr=bitr|lm;
    rm=rm<<1;
    lm=lm>>1;
    }
return (bitr);
}
```

D.3   Minimization function PARFIT

The PARFIT function is based on Powell's conjugate gradient method, with modifications by the author to improve its performance. A user-defined function, called fun(m,n,x), computes a goodness-of-fit for any set of parameters x. PARFIT calls fun repeatedly and seeks the first minimum downhill from the starting point. The primary advantages are its robustness, small size, and its ability to handle nonstationary correlations (i.e., it is able to turn corners in its pursuit of the minimum).

The function is called using the statement:

parfit(m,n,x,e,conv,maxit,nprint);

where the initial arguments of the function are in the array x of dimension n. The initial step size is in the array e. If any e[i] is < 0, then the corresponding parameter x[i] is never varied. The parameter maxit is the maximum number of iterations. The parameter nprint controls printing to the standard output device. If nprint > 0, then PARFIT prints output for every iteration that is an integral multiple of nprint. If nprint ≤ 0, output is suppressed. The message "MINIMUM NOT FOUND" should not be suppressed. The function returns when the change in the x vector is less than the vector e*conv or when the iteration counter exceeds maxit.

D.3.1   Flow chart

1. Initialize a set of n search vectors u_i, $i = 1$ to n, to the unit vector pointing in the ith direction times the initial step size for the ith parameter. This set of search vectors will change in length but not in direction during the minimization procedure.
2. Initialize a set of n search vectors u_{i+n}, $i = 1$ to n, to the unit vector pointing in the ith direction times the initial step size for the ith parameter. This set of search vectors will change both in length and direction during the minimization procedure.
3. Use a few steps of quadratic minimization to reduce Q by moving the parameters a_i, $i = 1$ to n, along the search vector u_1. Readjust the length of this search vector, depending on the length of the motion vector. If the lowest value of Q is at the edge of the move, double the length of u_1. Otherwise, reduce its length by a factor of 2.
4. Repeat step 3 for all the other search vectors u_i, $i = 2, \ldots, 2n$.
5. Use a few steps of quadratic minimization to reduce Q by varying the parameters a_i, $i = 1$ to n, along the direction of advance produced in steps 3 and 4.
6. Set $u_i = u_{i+1}$, $i = n + 1$ to $2n - 1$.
7. Set $u_{2n} =$ direction of advance produced in steps 3 and 4.
8. Repeat steps 3 to 7 until the change in each of the parameter values is less than some small value.

   Steps 3, 4, and 5 require quadratic minimization along a line, and this is performed by the function qfit.c, which is listed in Appendix D.3.2 and used by parfit.c.

   Powell's theorem states that if the procedure is done n times over only the u_{n+1} to u_{2n} search vectors, they will become mutually conjugate so only n^2 line minimizations will minimize any n-parameter quadratic form. However, this procedure is unstable as the search vectors can collapse into a subspace, which prevents finding the minimum. Searching periodically by varying each parameter only (the first n-line minimizations) prevents this collapse and makes the procedure more robust.

D.3.2   C program code

```
#include <stdio.h>
#include <math.h>
#include <stdlib.h>
extern FILE *outfile;
extern void fun(int, int, double[]);
extern void parfit(int, int, double [], double [], double, int, int);
extern void varfit(int, int, double [], double [], double [], double [],
```

```
double, int, int);
extern double ncalls;
void qfit(int, int, double *, double *, double *, double, int);
double *x2, *c, *xi, *y;
void parfit(int m, int n, double x[], double e[], double conv, int maxit,
int nprint)
{
static int ipow, npow, ipar, ifconv, iit, ifprint, nfc;
double *base, *c, *abse, *d, *xp, *advance;
double fp;
/*    parfit - function to find a minimum of a multi-parameter function
            using Powell's quadratic minimization
        x[i] is the argument array
        n is the array size
        e[i] the initial step size array
            (if any e[i] is zero then x[i] is never varied)
        c[i] is the change in x[i]
        d[i] is the dynamic step size
        maxit is the maximum number of interations allowed by the user
        parfit exits when the change in all x[i] is less than | e[i]|*conv
        If nprint is negative or zero- no output
        If nprint is positive, stream outputs result for every
            iteration that is an integral multiple of nprint
        The message "minimum not found" cannot be supressed
*/
fun(m, n, x);
if (maxit == 0) return;
base = (double *) calloc(12*(n + 1), sizeof(double));
xi = (double *) calloc( (2*n*n + n), sizeof(double));
if (( base == NULL) || (xi == NULL) )
    {
    fprintf(outfile, "Can't allocate storage for parfit- exit\n" );
    printf( "Can't allocate storage for parfit - exit\n" );
    exit(1);
    }
/* set pointers for parfit function arrays */
xp = base;
abse = base + (n + 1);
d = base + 2 * (n + 1);
advance = base + 3 * (n + 1); /* double length vector */
/* set pointers for qfit function arrays */
```

```
x2 = base + 5 * (n + 1);
c = base + 6 * (n + 1);
y = base + 7 * (n + 1);
ncalls=1.;
for (ipar = 0; ipar < n; ipar++) abse[ipar] = fabs(e[ipar]);
for (ipar = 0; ipar <= 2 * n; ipar++) advance[ipar] = 0.;
/* set up xi array- first half varies each parameter
      separately, second half contains Powell's
      vectors of motion history */
npow = 0;
for (ipow = 0; ipow < n; ipow++)
    {
    if ( e[ipow] == 0.) continue;
    for (ipar = 0; ipar < n; ipar++)
        {
        if (ipar == ipow) xi[ipar + n* npow] = abse[ipar];
        else xi[ipar + n * npow] = 0.;
        }
    npow++;
    }
for (ipow = 0; ipow < n; ipow++)
    {
    if ( e[ipow] == 0.) continue;
    for (ipar = 0; ipar < n; ipar++)
        {
        if (ipar == ipow) xi[ipar + n* npow] = abse[ipar];
        else xi[ipar + n * npow] = 0.;
        }
    npow++;
    }
npow /= 2.;
iit=0;
ifconv=0;
while ( (ifconv == 0) && (iit < maxit)) /* main iteration loop */
    {
    ifprint=0;
    if ( (nprint!=0) && (iit%nprint==0) ) ifprint=1;
    if ( (nprint!=0) && (iit ==1) ) ifprint =1;
    if (ifprint!=0)
        {
        printf(" Iteration %5d    %7.0f function calls
```

```
            fun =%20.9e\n", iit,ncalls,x[n]);
        fprintf(outfile,
            "Iteration %5d    %7.0f function calls
                fun =%20.9e\n", iit,ncalls,x[n]);
        for(ipar=0; ipar<n; ipar++)
            {
            printf("%14.6f", x[ipar]);
            fprintf(outfile, "%14.6f", x[ipar]);
            if(ipar%5 == 4)    fprintf(outfile,"\n");
            }
        printf("\n");
        fprintf(outfile, "\n");
        }
    for (ipar = 0; ipar < n; ipar++) xp[ipar] = x[ipar];
    /* loop over parameter vectors plus Powell's vectors */
    for (ipow = 0; ipow < 2*npow; ipow++)
        {
        fp = x[n];
        for (ipar = 0; ipar < n; ipar++)
            {
            d[ipar] = xi[ipar + n * ipow];
            if (xi[ipar + n * ipow] == 0.) continue;
            } /* end ipar loop */
        /* line minimization call */
        qfit(m, n, x, abse, d, conv, 5);
        for (ipar = 0; ipar < n; ipar++)
            {
            xi[ipar + n * ipow] = d[ipar];
            if (xi[ipar + n * ipow] == 0.) continue;
            }
            advance[ipow] += fp - x[n];
        } /* end ipow loop */
    fp = x[n];
    /* minimize along advance vector for this iteration */
    for (ipar = 0; ipar < n; ipar++)
        d[ipar] = 0.5 * (x[ipar] - xp[ipar]);
    qfit(m, n, x, abse, d, conv, 5);
    advance[2 * npow] += fp - x[n];
    /* move down Powell's advance history vectors */
    for (ipow = npow; ipow < (2 * npow - 1); ipow++)
        {
```

```
        for (ipar = 0; ipar < n; ipar++)
            xi[ipar + n * ipow] = xi[ipar + n * (ipow + 1)];
    }
    /* put this iteration's advance into last history vector
        and test for convergence
        quit when dynamic step size is less than e[i]*conv*/
    ifconv=1;
    for (ipar = 0; ipar < n; ipar++)
        {
        xi[ipar + n * (2*npow - 1)] = d[ipar];
        if ( (abse[ipar] > 0) &&
            ( fabs(xp[ipar] - x[ipar]) > abse[ipar] *conv ) )
            ifconv=0;
        }
    iit++;
    ifprint=0;
    /* print results if we have converged */
    if ( (nprint!=0) && (ifconv!=0) ) ifprint=1.;
    /* print results if we have run out of iterations */
    if ( (nprint!=0) && (iit==maxit) ) ifprint=1;
    if ( ifprint!=0 )
        {
        printf("Iteration %5d   %7.0f function calls
            fun =%20.9e\n", iit,ncalls,x[n]);
        fprintf(outfile,
            "Iteration %5d   %7.0f function calls   fun =%20.9e\n",
            iit,ncalls,x[n]);
        for(ipar=0; ipar<n; ipar++)
            {
            printf(" %14.6f ", x[ipar]);
            fprintf(outfile, " %14.6f ", x[ipar]);
            if(ipar%5 == 4) fprintf(outfile,"\n");
            }
        printf("\n");
        fprintf(outfile, "\n");
        fprintf(outfile, "advance Δ fun amounts\n");
        for (ipow = 0; ipow <= 2 * npow; ipow++)
            fprintf(outfile, "ipow = %d   advance %f\n",
                ipow, advance[ipow]);
        } /* end ifprint case */
    } /*   end iteration loop */
```

```
if (ifconv == 0)
        fprintf(outfile, "\nminimum not found    \n");
if ( (nprint!=0) && (ifconv!=0) )
        fprintf(outfile, "minimum found\n");
free(xi);
free(base);
return;
} /* end of parfit function */
/*    function qfit- performs a one parameter recursive subspace
minimization */
void qfit(int m, int n, double x[], double xstep[], double d[], double
conv, int itmax)
{
int i, iter, ifconv;
double s1, s2, s3, s4, f1, f2, f3, f4;
/*

    d[] defines search direction
    s1, s2, s3 positions (maintained in sequence)
    f1, f2, f3   fun values (maintained so that f1 and f3 bracket f2)
    on entrance x[i] and x[n] are assumed valid
    on exit x[i] and x[n] are valid
    phase 1 brackets the minimum f1>f2<f3
    (s1, s2, s3 kept in sequence)
    phase 2 does a quadratic estimate of the minimum position
    xstep[] * conv is the convergence array. qfit quits when one
        iteration moves x[] along a vector that would fit in the
        xstep[]*conv box
*/
/* save x, f at entry */
for (i = 0; i < n; i++) x2[i] = x[i];
s2 = 0.;
f2 = x[n];
/* take first step and compute x1[] */
s1 = -1.;
for (i = 0; i < n; i++) x[i] = x2[i] + s1 * d[i];
fun(m, n, x);
f1 = x[n];
/* if f1 is lower, swap f1 and f2, s1 and s2 */
if (f1 < f2)
    {
    s1 = 0.;   s2 = -1.;
```

```
        f3 = f1;   f1 = f2;   f2 = f3;
        }
/* compute s3 */
s3 = 2. * s2 - s1;
for (i=0; i<n; i++) x[i] = x2[i] + s3 * d[i];
fun(m, n, x);
f3 = x[n];
iter=0;
while (1)
    {/* enter with s1, s2, s3- compute s4- new s1, s2, s3
        at end of loop - s1, s2, s3 kept in sequence
        but min value can be f1, or f2, or f3*/
    for (i = 0; i < n; i++)
        y[i] = x[i];   /* y[] stores the values of x[] at the start
                        of the while loop */
    if ( (s2 - s1)/(s3 - s2) <0)
        {
        printf("s1, s2, s3 = %f %f %f not in sequence\n",
            s1, s2, s3);
        fprintf(outfile, "s1, s2, s3 = %f %f %f not in sequence\n",
            s1, s2, s3);
        }
    if ( (f1 < f2) && (f1 < f3) ) /* f1 is lowest value */
        s4 = 2. * s1 - s3;
    else if ( (f3 <= f1) && (f3 <= f2) ) /* f3 is lowest value */
        s4 = 2. * s3 - s1;
    else /* f2 is lowest value – only use quadratic interpolation
        formula if f2 is lowest, and s1-s2 and s2-s3 intervals
        differ by <10x */
        {
        if (fabs(s1-s2) > 10.*fabs(s2-s3)) s4 = 2. * s2 - s3;
        else if (fabs(s2-s3) > 10.*fabs(s1-s2)) s4 = 2. * s2 - s1;
        else
            {
        s4 = s2 - 0.5 *
            ( (s2-s1)*(s2-s1)*(f2-f3) - (s2-s3)*(s2-s3)*(f2-f1) ) /
            ( (s2-s1)*(f2-f3) - (s2-s3)*(f2-f1) );
        /* If s4 has become exactly equal to s2, jiggle it a bit
            in the direction of s3 */
        if (s2 == s4)
            s4 += 0.01 * (s3-s1);
```

```
            }
        }
    for (i=0; i<n; i++) x[i] = x2[i] + s4 * d[i];
    ifconv = 1;
    for (i = 0; i < n; i++)
        {
        if (xstep[i] == 0) continue;
        if ( fabs(x[i] - y[i]) > conv * xstep[i]) ifconv = 0;
        }
    if (iter >= itmax) break;
    if (ifconv == 1) break;
    fun(m, n, x);
    f4 = x[n];
    /* keep f4 and reestablish s1, s2, s3 sequence */
    if ( (s4 - s1)/(s1 -s2) > 0 )
        { /* s4, s1, s2 are in sequence- eliminate f3 */
        if ( ((f3<f1) && (f3<f2) && (f3<f4) )
            {
            printf("error- eliminating lowest value\n");
            fprintf(outfile, "error- eliminating lowest value\n");
            }
        s3 = s2; s2 = s1; s1 = s4;
        f3 = f2; f2 = f1; f1 = f4;
        }
    else if ( (s1 - s4)/(s4 -s2) > 0 )
        { /* s1, s4, s2 are in sequence- eliminate f3 */
        if ( ((f3<f1) && (f3<f2) && (f3<f4) )
            {
            printf("error- eliminating lowest value\n");
            fprintf(outfile, "error- eliminating lowest value\n");
            }
        s3 = s2; s2 = s4;
        f3 = f2; f2 = f4;
        }
    else if ( (s2 - s4)/(s4 -s3) > 0 )
        { /* s2, s4, s3 are in sequence- eliminate f1 */
        if ( ((f1<f2) && (f1<f3) && (f1<f4) )
            {
            printf("error- eliminating lowest value\n");
            fprintf(outfile, "error- eliminating lowest value\n");
            }
```

```
            s1 = s2;  s2 = s4;
            f1 = f2;  f2 = f4;
            }
        else if ( (s2 - s3)/(s3 -s4) > 0 )
            { /* s2, s3, s4 are in sequence- eliminate f1 */
            if ( (f1<f2) && (f1<f3) && (f1<f4) )
                {
                printf("error- eliminating lowest value\n");
                fprintf(outfile, "error- eliminating lowest value\n");
                }
            s1 = s2;  s2 = s3;  s3 = s4;
            f1 = f2;  f2 = f3;  f3 = f4;
            }
        else
            { /* s1, s4, s2 sequence error */
            printf("error- sequencing failure after f4\n");
            fprintf(outfile, "error- sequencing failure after f4\n");
            }
        iter++;
        } /* end of interation loop */
/* set x equal to best fit of s1, s2, s3 */
if ( (f1 < f2) && (f1 < f3) ) {s2 = s1;  f2 = f1; }
if ( (f3 < f2) && (f3 < f1) ) {s2 = s3;  f2 = f3; }
    for (i=0; i < n; i++) x[i] = x2[i] + s2 * d[i];
x[n] = f2;
s4 = fabs(s2);
if (s4 < 2.)
    {
    if (s4 < 0.01) s4 = 0.01;
    for (i=0; i < n; i++) d[i] *= s4;
    }
return;
} /* end qfit function */
```

D.4   The uncertainty estimation function VARFIT

The VARFIT function was developed by the author to determine parameter uncertainties on a small computer. It uses PARFIT to find the largest and smallest values of each parameter that give a best-fit χ^2 equal to the overall best-fit $\chi_{min}^2 + 1$. It calls PARFIT repeatedly to perform the necessary multi-parameter minimizations.

D.4.1 Flow chart

1. Find $\chi_{min}^2 + 1$ by varying all parameters.
2. Set the parameter a_1 to its best fit value $+s_1$, where s_1 is some step size.
3. Holding a_1 constant, vary all other parameters to minimize χ^2.
4. Determine the parabola that passes through the new value of a_1, and also passes through the best-fit value of a_1 with zero slope.
5. Determine the value of a_1 at which the parabola is equal to $\chi_{min}^2 + 1$.
6. Repeat step 3 for this new value of a_1.
7. Determine the parabola that passes through the three points closest to $\chi_{min}^2 + 1$.
8. Repeat steps 5, 6 and 7 until the value of a_1 changes by a sufficiently small amount. This value is a one-σ confidence limit.
9. Set the parameter a_1 to its best-fit value $-s_1$.
10. Repeat steps 3 to 8 to find the other one-σ confidence limit.
11. Repeat steps 2 through 10 for all the other parameters.

D.4.2 C program code

```
#include <stdio.h>
#include <math.h>
#include <time.h>
#include <stdlib.h>
#define MAXIT 20
extern FILE *outfile;
extern void fun(int, int, double[]);
extern void parfit(int, int, double[], double[], double, int, int);
extern void varfit(int, int, double[], double[], double[], double[],
double, int, int);
extern double chisq;
extern char comments[];
void varfit(int m, int npar, double par[],
    double step[], double em[], double ep[],
    double conv, int maxit, int nprint)
{
static double f0, f1, f2, f3, f4, f5, fx, dir, dx;
static double xx2, xx3, xx4, ff2, ff3, ff4;
static double g1, g2, g3, g4, g5, g6[2];
static double fp[6], fpmin, ax, temp, sqrtf32;
static int i, ipar, jpar, istep, iistep, iprint;
static int idir, nout, iout, saved[8];
static int imin, nsave, iflo, ifhi, ig6, ifnew;
```

```
char s[80];
char apm[5];
double *base;
double *x1;
double *x2;
double *x3;
double *x4;
double *x5;
double *x;
double *e;
double *plus1m;
double *plus1p;
apm[0] = '-';
apm[1] = '+';
if (maxit == 0)
    return;
base = (double *) calloc(9*(npar + 1), sizeof(double));
if (base == NULL)
    {
    fprintf(outfile, "Can't allocate storage for varfit- exit\n");
    printf("Can't allocate storage for varfit - exit\n");
    exit(1);
    }
x1 = base;
x2 = base + (npar + 1);
x3 = base + 2 * (npar + 1);
x4 = base + 3 * (npar + 1);
x5 = base + 4 * (npar + 1);
x = base + 5 * (npar + 1);
e = base + 6 * (npar + 1);
plus1m = base + 7 * (npar + 1);
plus1p = base + 8 * (npar + 1);
/* Make local copies of the parameter and step arrays */
for (ipar=0; ipar<npar; ipar++)
    {
    x[ipar] = par[ipar];
    e[ipar] = step[ipar];
    }
x[npar] = 0.;
/* call parfit to verify minimum- save fun value and parameters */
parfit(m, npar, x, e, conv, maxit, nprint);
```

```
f1 = x[npar];       /* f1 is the minimum varying all parameters */
chisq = x[npar];
for(ipar=0; ipar<npar; ipar++) x1[ipar] = x[ipar];
/* Loop over parameters to find errors - skip if step <= 0
      (big loop to very near end of file) */
for(ipar=0; ipar<npar; ipar++)
      {
      if (step[ipar] == 0)
       continue;       /* Ignore this parameter if step size is zero */
      if (em[ipar] == 0 && ep[ipar] == 0)
            continue;        /* No uncertainties requested
                              for this parameter */
      /* step this parameter first backward and then forward */
      for(idir=0; idir<2; idir++)
            /* Loop over directions */
            {
            if ( idir == 0 && em[ipar] == 0 )
                  continue;   /* In negative direction, with no
                              negative uncertainty */
            if ( idir == 1 && ep[ipar] == 0 )
                  continue;      /* In positive direction, with no
                              positive uncertainty */
            e[ipar] = 0;      /* Makes parfit() hold x[ipar] constant */
            printf("\n\n**** PARAMETER %d%c ****%s\n",
                  ipar, apm[idir], comments);
            fprintf(outfile, "\n**** PARAMETER %d%c ****%s\n",
                  ipar, apm[idir], comments);
            for (i=0; i<npar; i++)   /* set all xs to best fit values */
                  {
                  x[i] = x1[i];
                  x2[i] = x1[i];
                  }
            f2=f1;
            fp[2] = f2 - (f1+1.);
            printf("x2 %12.6f   f2 %12.6f   fp2 %f\n",
                  x2[ipar], f2, fp[2]);
            fprintf(outfile,"x2 %12.6f   f2 %12.6f   fp2 %f\n",
                  x2[ipar], f2, fp[2]);
            if(idir==0) /* step in negative direction */
                  x[ipar] = x1[ipar] + em[ipar];
            if(idir==1) /* step in positive direction */
```

```
        x[ipar] = x1[ipar] + ep[ipar];
    /* vary all other xs */
    parfit(m, npar, x, e, conv, maxit, nprint);
    f3 = x[npar];
    chisq = x[npar];
    /* Now, f1 (=f2) is the minimum varying all parameters;
            f3       is the minimum at x[ipar] varying all parameters
    */
    if (f1-f3 > 1.E-4)
        {
        fprintf(outfile, "*** lower minimum found by
            varfit (#1)***\n");
        printf("*** lower minimum found by varfit (#1)***\n");
        fprintf(outfile, "chisq= %f\n", chisq);
        printf("chisq= %f\n", chisq);
        for(jpar=0; jpar<npar; jpar++)
            {
            printf ("%5d   %20.13f\n", jpar, x[jpar]);
            fprintf (outfile, "%5d   %20.13f\n", jpar, x[jpar]);
            par[jpar] = x[jpar];     /* Update input
                                        array of parameters */

            }
        par[npar] = x[npar];
        m = 1;
        fun(m, npar, x);
        return;
        }
    for (i=0; i<npar; i++) x3[i] = x[i];
    fp[3] = f3 - (f1+1.);   /* The goal is to make this
        quantity zero */
    printf("x3 %12.6f   f3 %12.6f   fp3 %f\n",
        x3[ipar], f3, fp[3]);
    fprintf(outfile,"x3 %12.6f   f3 %12.6f   fp3 %f\n",
        x3[ipar], f3, fp[3]);
    if (fp[3] > 0) ax = 0.0; else ax = 0.1;
    /* use quadratic interpolation to get a third point,
        assuming that the slope is zero at best fit point x1, f1 */
    sqrtf32 = sqrt(fabs(f3 - f2));
    if (sqrtf32 < 0.5) sqrtf32 = 0.5;
    for (i=0; i<npar; i++)
        {
```

```
        if ( (idir==0) & (em[i]==0) & (step[i]==0) ) continue;
        if ( (idir==1) & (ep[i]==0) & (step[i]==0) ) continue;
        x[i] = x2[i] + (x3[i] - x2[i] )/ sqrtf32;
        /* if f3 is not above f2+1, overstep a bit */
        if (ax > 0) x[i] += ax * ( x[i] - x3[i] );
        }
/* vary all other xs */
parfit(m, npar, x, e, conv, maxit, nprint);
f4 = x[npar];
/* f4 is the value after the initial quadratic interpolation
or extrapolation */
chisq = x[npar];
if (f1-f4 > 1.E-4)
    {
    fprintf(outfile, "*** lower minimum found by
        varfit (#2)***\n");
    printf("*** lower minimum found by varfit (#2)***\n");
    fprintf(outfile, "chisq= %f\n", chisq);
    printf("chisq= %f\n", chisq);
    for(jpar=0; jpar<npar; jpar++)
        {
        printf ("%5d   %20.13f\n", jpar, x[jpar]);
        fprintf (outfile, "%5d   %20.13f\n", jpar, x[jpar]);
        par[jpar] = x[jpar];      /* Update input array
                                    of parameters */
        }
    par[npar] = x[npar];
    m = 1;
    fun(m, npar, x);
    return;
    }
for (i=0; i<npar; i++) x4[i] = x[i];
fp[4] = f4 -(f1+1.);
printf("x4 %12.6f   f4 %12.6f   fp4 %f\n",
    x4[ipar], f4, fp[4]);
fprintf(outfile,"x4 %12.6f   f4 %12.6f   fp4 %f\n",
    x4[ipar], f4, fp[4]);
/* At this point x2 is the best fit, x3 is one step, and x4 is
a new estimate obtained by quadratic interpolation */
ifnew = 0;
for (istep=1; istep<MAXIT; istep++)
```

```
/* Loop to bracket f1+1 */
{
if (fp[3] > 0.) break;
fp[4] = f4 - (f1+1.);
if (fp[4] > 0.) break; /*skip if f1+1 is bracketed */
for (i=0; i<npar; i++)
    /*step in same direction to bracket f1+1 */
    {
    if(step[i]<=0) continue;
    /* first step an amount equal to the last step,
    then progressively larger multiples,
    but always limited by step[i] */
    dx = istep * ( x4[i] - x3[i] ) + 0.001 * ( x4[i] - x2[i] );
    if (fabs(dx) > step[i] )
        dx *= step[i]/fabs(dx);
    x[i] = x4[i] + dx;
    }
ifnew = 1;
/* vary all other xs */
parfit(m, npar, x, e, conv, maxit, nprint);
fx = x[npar];
chisq = x[npar];
if (f1 - fx > 1.E-4)
    {
    fprintf(outfile,
        "*** lower minimum found by varfit (#3)***\n" );
    printf("*** lower minimum found by varfit (#3)***\n" );
    fprintf(outfile, "chisq= %f\n", chisq);
    printf("chisq= %f\n", chisq);
    for(jpar=0; jpar<npar; jpar++)
        {
        printf ("%5d   %20.13f\n", jpar, x[jpar]);
        fprintf (outfile, "%5d   %20.13f\n", jpar, x[jpar]);
        par[jpar] = x[jpar];
        /* Update input array of parameters */
        }
    par[npar] = x[npar];
    m = 1;
    fun(m, npar, x);
    return;
    }
```

```
           for (i=0; i<npar; i++)
               {
               x2[i] = x3[i];
               x3[i] = x4[i];
               x4[i] = x[i];
               }
           f2 = f3;
           f3 = f4;
           f4 = fx;
           fp[4] = f4 - (f1+1.);
           printf("x4 %12.6f   f4 %12.6f   fp4 %f\n",
               x4[ipar], f4, fp[4]);
            fprintf(outfile,"x4 %12.6f   f4 %12.6f   fp4 %f\n",
               x4[ipar], f4, fp[4]);
           } /* end loop to bracket f1+1 */
    for (istep=1; istep<MAXIT; istep++)
        { /* iterate to find a better solution to f(x) = f1+1.
            by quadratic interpolation */
        printf("\nparameter %d%c begin varfit iteration %d\n",
            ipar, apm[idir], istep);
        fprintf(outfile,
            "\nparameter %d%c begin varfit iteration %d\n",
            ipar, apm[idir], istep);
        if ( fabs(fp[4]) < 5E-7)
            { /* if previous estimates are good enough- skip */
            f5 = f4;
            for (i = 0; i<npar; i++) x5[i] = x4[i];
            break;
            }
        fp[2] = f2 -(f1+1.);
        fp[3] = f3 -(f1+1.);
        fp[4] = f4 - (f1+1.);
        if ( (ifnew==1) || (istep>1) )
        {
        printf("x2 %12.6f   f2 %12.6f   fp2 %f\n", x2[ipar], f2, fp[2]);
        fprintf(outfile,"x2 %12.6f   f2 %12.6f   fp2 %f\n", x2[ipar], f2, fp[2]);
        printf("x3 %12.6f   f3 %12.6f   fp3 %f\n", x3[ipar], f3, fp[3]);
        fprintf(outfile,"x3 %12.6f   f3 %12.6f   fp3 %f\n", x3[ipar], f3, fp[3]);
        printf("x4 %12.6f   f4 %12.6f   fp4 %f\n", (x4[ipar]), f4, fp[4]);
        fprintf(outfile,"x4 %12.6f   f4 %12.6f   fp4 %f\n", x4[ipar], f4, fp[4]);
```

```
}
/* set x[ipar] to the closed form solution of
     f(x) = f1+1, where f(x) is a polynomial passing through
     the three known points f2(x2), f3(x3), f4(x4)
     asume that solution is bracketed by known points */
ff2 = f2 -f4;
ff3 = f3 - f4;
xx2 = x2[ipar] - x4[ipar];
xx3 = x3[ipar] - x4[ipar];
g1 = ff3 * xx2 * xx2 - ff2 * xx3 * xx3;
g2 = ff2 * xx3 - ff3 * xx2;
g3 = xx2 * xx3 * (xx2 - xx3);
g4 = f1+1. - f4;
g5 = g1 * g1 + 4. * g2 * g3 * g4;
if (g5 < 0)
    {
     printf("g5 < 0 error: i %d   istep %d\n", i, istep);
     printf("x2[ipar] %f   f2 %f \n",x2[ipar], f2);
     printf("x3[ipar] %f   f3 %f \n",x3[ipar], f3);
     printf("x4[ipar] %f   f4 %f \n",x4[ipar], f4);
     fprintf(outfile, "g5 < 0 error: i %d   istep %d\n",
         i, istep);
     fprintf(outfile, "x2[ipar] %f   f2 %f \n",
         x2[ipar], f2);
     fprintf(outfile, "x3[ipar] %f   f3 %f \n",
         x3[ipar], f3);
     fprintf(outfile, "x4[ipar] %f   f4 %f \n",
         x4[ipar], f4);
     g5 = 0.;
    } /* end if g5<0 */
g5 = sqrt(g5);
/* compute both roots */
g6[0] = (-g1 - g5) / (2 * g2);
g6[1] = (-g1 + g5) / (2 * g2);
for (ig6 = 0; ig6<2; ig6++)
    {
    x[ipar] = x4[ipar] + g6[ig6];
    iflo = 0;   ifhi = 0;
    if ( x[ipar] < x2[ipar] ) iflo = 1; else ifhi = 1;
    if ( x[ipar] < x3[ipar] ) iflo = 1; else ifhi = 1;
```

```
                    if ( x[ipar] < x4[ipar] ) iflo = 1; else ifhi = 1;
                    if ( (iflo==1) && (ifhi==1) ) break;
                    if (ig6==1)
                    {
                    printf( "non bracket error: ipar %d   istep %d\n",
                        ipar, istep);
                    fprintf(outfile,
                        "non bracket error: ipar %d   istep %d\n",
                        par, istep);
                    printf( "x2[ipar] %f f2    %f \n",x2[ipar], f2);
                    printf( "x3[ipar] %f f3    %f \n",x3[ipar], f3);
                    printf( "x4[ipar] %f f4    %f \n",x4[ipar], f4);
                    fprintf(outfile, "x2[ipar] %f f2    %f \n", x2[ipar], f2);
                    fprintf(outfile, "x3[ipar] %f f3    %f \n", x3[ipar], f3);
                    fprintf(outfile, "x4[ipar] %f f4    %f \n", x4[ipar], f4);
                    } /* end if (ig6==1) */
                    } /* end loop ig6 = 0, 1 */
            temp = ( x[ipar] - x4[ipar] ) / ( x2[ipar] - x4[ipar] );
            for (i=0; i<npar; i++)
                {
                if(step[i]<=0) continue;
                if (i != ipar) x[i] = x4[i] + ( x2[i] - x4[i] ) * temp;
                }
            /* use parfit to find f(x) for the new point */
            /* vary all other xs */
            parfit(m, npar, x, e, conv, maxit, nprint);
            f5 = x[npar];
            chisq = x[npar];
            if (f1-f5 > 1.E-4)
                {
                fprintf(outfile,
                    "*** lower minimum found by varfit (#4)***\n");
                printf( "*** lower minimum found by varfit (#4)***\n");
                fprintf(outfile, "chisq= %f\n", chisq);
                printf( "chisq= %f\n", chisq);
                for(jpar=0; jpar<npar; jpar++)
                    {
                    printf ("%5d    %20.13f\n", jpar, x[jpar]);
                    fprintf (outfile, "%5d    %20.13f\n", jpar, x[jpar]);
                    par[jpar] = x[jpar];
```

```
                        /* Update input array of parameters */
                        }
                    par[npar] = x[npar];
                    m = 1;
                    fun(m, npar, x);
                    return;
                    }
            for (i=0; i<npar; i++) x5[i] = x[i];
            fp[5] = f5 - (f1+1.);
            printf("x5 %12.6f   f5 %12.6f   fp5 %f\n",
                x5[ipar], f5, fp[5]);
            fprintf(outfile,
                    "\nx5 %12.6f   f5 %12.6f   fp5 %f\n",
                x5[ipar], f5, fp[5]);
            if (fabs(fp[5]) < 5E-7) break;
            /* eliminate outlying point, but save last estimate and
            any solo points on one side of (f1+1) */
            fp[5] = f5 - (f1+1.);
            /* start with fp2, fp3, fp4, fp5 not saved */
            for (i=2; i<6; i++) saved[i] = 0;
             nsave = 0;
            fpmin = 1E20;
            imin = 0;
            for (i=2; i<6; i++)
                { /* save closest fp above f1+1 */
                if ( (fp[i] > 0.) && (fp[i] < fpmin) )
                    {fpmin = fp[i]; imin = i; }
                }
            if (imin >0)
                {
                saved[imin] = 1;
                nsave++;
                }
            fprintf(outfile,
                "\nsave closest above f1+1: imin    %d
                    fpmin %f \n", imin, fp[imin]);
            printf("\nsave closest above f1+1: imin    %d
                    fpmin %f \n", imin, fp[imin]);
                fpmin = -1E20;
                imin = 0;
```

```
for (i=2; i<6; i++)
    { /* save closest fp below f1+1 */
    if ( (fp[i] < 0.) && (fp[i] > fpmin) )
        {fpmin = fp[i]; imin = i; }
    }
if (imin >0)
    {
    saved[imin] = 1;
    nsave++;
    }
fprintf(outfile,
    "save closest below f1+1: imin  %d fpmin %f \n",
    imin, fp[imin]);
printf("save closest below f1+1: imin  %d
    fpmin %f \n", imin, fp[imin]);
 /* save fp5 if not already saved */
if (saved[5] == 0)
    {
    saved[5] = 1;
    nsave++;
    }
while (nsave < 3)
    /* if only 1 or 2 saved, save closest to f1+1 */
    {
    fpmin = 1E20;
    imin =0;
    for (i = 2; i <5; i++)
        {
        if ( (saved[i] == 0) && (fabs(fp[i]) < fpmin) )
            {fpmin = fabs(fp[i]); imin = i; }
        }
    if (imin == 0) break;
    if (imin > 0)
        {
        saved[imin] = 1;
        nsave++;
        }
    fprintf(outfile,
        "save closest to f1+1: imin   %d   fpmin %f \n",
        imin, fp[imin]);
```

```
        printf("save closest to f1+1: imin   %d   fpmin %f \n",
            imin, fp[imin]);
        }
    iout = 0;
    for (i=2; i<5; i++) if(saved[i] == 0) iout = i;
    if (iout == 0)
        {
        printf ("error- can't find point to be eliminated\n");
        fprintf (outfile,
            "error- can't find point to be eliminated\n");
        }
    if (iout==2)
        {
        f2 = f5;
        for (i=0; i<npar; i++) x2[i] = x5[i];
        }
    if (iout==3)
        {
        f3 = f5;
        for (i=0; i<npar; i++) x3[i] = x5[i];
        }
    if (iout==4)
        {
        f4 = f5;
        for (i=0; i<npar; i++) x4[i] = x5[i];
        }
        }/* end loop over interpolation steps */
    if(idir==0)
        {
        plus1m[ipar] = f5 - f1;
        em[ipar] = x5[ipar] - x1[ipar];
        if (fabs( plus1m[ipar] - 1.) > 1E-6)
            {
            fprintf(outfile,"*** par %d%c   dx %f at
                chisq + %f\n", ipar, apm[idir], em[ipar],
                plus1m[ipar] );
            em[ipar] = 9999.;
            }
        }
    if(idir==1)
```

```
                         {
                         plus1p[ipar] = f5 - f1;
                         ep[ipar] = x5[ipar] - x1[ipar];
                         if (fabs( plus1p[ipar] - 1.) > 1E-6)
                             {
                             fprintf(outfile, "*** par %d%c   dx %f at
                                 chisq + %f\n", ipar, apm[idir], ep[ipar],
                                 plus1p[ipar] );
                             ep[ipar] = 9999.;
                             }
                         }
                     printf("par %d %f uncertainty    %f at chisq + %f\n",
                         ipar, x1[ipar], x5[ipar] - x1[ipar], f5-f1);
                     fprintf(outfile,"par %d %f    uncertainty %f at chisq + %f\n",
                         ipar, x1[ipar], x5[ipar] - x1[ipar], f5-f1);
                     iprint = 0;
                     for (i = 26; i < npar; i++) if (x5[i] != 0)
                         {
                         printf("%2d %11.6f", i , x5[i] );
                         fprintf(outfile, "%2d %11.6f", i , x5[i] );
                         iprint++;
                         if ( (iprint%3) == 2) {printf("\n"); fprintf(outfile, "\n"); }
                         }
                     } /* end loop over directions */
                 e[ipar] = step[ipar];
                 x[ipar] = x1[ipar];
                 } /* end loop over parameters */
         free(base);
         return;
         } /* end varfit function */
```

D.5   Numerical evaluation of functions defined by integrals

A number of methods have been devised for the numerical evaluation of integrals, including the trapezoidal rule and Simpson's rule, that integrate the function tabulated on a uniform spacing, and more sophisticated methods, such as Gaussian quadrature, that use a carefully chosen spacing (corresponding to zeros of a particular polynomial) and special weighting.

The disadvantage of Simpson's rule is that a large number of function calls are necessary to cover the whole range of integration with sufficient density, even if significant contributions come from only a small part of the range. This is a particular problem for integrals that have an infinite integration limit and do not permit a change of variable that simultaneously avoids infinities in both the limits of integration and the integrand. The disadvantage of Gaussian quadrature methods is that the points of function evaluation must be chosen before the integral (and its accuracy) can be evaluated.

The method that follows uses adaptive quadrature, where the points of function tabulation and the accuracy are determined as the method proceeds. Without using any pre-determined set of function points, the method develops a fine spacing, where the quadratic approximation results in the greatest error, and a relatively coarse spacing, where the contributions to the error are small. The method can even be used for functions with infinite limits of integration without having to find a suitable change of variable.

D.5.1   Flow chart

The algorithm to evaluate the integral of a function $F(x)$ numerically to an accuracy e by using adaptive quadrature, given the limits of integration a and b is:

1. Initialize the first segment by dividing the interval from a to b into four equal parts and by evaluating the integrand at the five boundary points. Compute the error for this segment and declare this segment as having the largest error.
2. Move all data beyond the segment with the maximum error four places to create space for a new segment.
3. Divide the segment with the largest error into two equal segments, using the old integrand values and evaluating only the new values needed.
4. Compute and store the errors for the two new segments.
5. Add the errors for all segments thus far created and determine the segment with the largest error. If the sum is greater than the desired error and if the number of segments is less than the limit, loop back to step 2.
6. Exit the function, returning the value of the integral, the estimated upper limit on the error, and the number of segments created.

D.5.2   C program code

```
/* adaptive quadrature integration function */
double adsint(double a,double b,int m,double errmax,double
*errptr,int *nsegptr)
/*
*    nmax is the maximum number of segments calculated
*    integral taken over a,b interval
*    if positive, errmax is maximum error of the integral
*    if negative, - errmax is the maximum fractional error of the integral
```

```
*    user must define a function igrand(x,n) that evaluates the
*       integrand of type n
*/
{
int n,i,iseg,nseg,isegm;
#define NMAX 201       /* absolute maximum number of segments*/
#define NMAX4 801
#define NSEGMAX 199
double x[NMAX4], f[NMAX4], s[NMAX], e[NMAX];
double h,dx,xtemp,emax,s1,s2,sum,err;
if(a==b)
     {
     *errptr=0;
     *nsegptr=0;
     return(0);
     }
/* initialize first segment */
nseg=1;
iseg=0;
h=b-a;
dx=h/4.;
for(i=0;i<5;++i)
     {
     x[i]=a+i*dx;
     xtemp=x[i];
     f[i]=igrand(xtemp,m);
     }
s1=(f[0]+4.*f[2]+f[4])*h/6.;
s2=(f[0]+4.*f[1]+2.*f[2]+4.*f[3]+f[4])*h/12.;
err=fabs(s1-s2)/15.;
e[iseg]=err;
s[iseg]=s2;
isegm=0;
/* loop until error summed over all segments is less than errmax
     or until maximum number of segments is reached */
while(1)
     {
     /* move down all segments above isegm */
     n=nseg*4+4;
     x[n]=x[n-4];
     f[n]=f[n-4];
```

```
for (iseg=nseg;iseg>(isegm+1);–iseg)
    {
    s[iseg]=s[iseg-1];
    e[iseg]=e[iseg-1];
    for (i=0;i<4;++i)
        {
        n=iseg*4+i;
        x[n]=x[n-4];
        f[n]=f[n-4];
        }
    }
nseg=nseg+1;
/* expand segment isegm */
n=isegm*4;
x[n+6]=x[n+3];
x[n+4]=x[n+2];
x[n+2]=x[n+1];
f[n+6]=f[n+3];
f[n+4]=f[n+2];
f[n+2]=f[n+1];
x[n+1]=(x[n]+x[n+2])/2.;
x[n+3]=(x[n+2]+x[n+4])/2.;
x[n+5]=(x[n+4]+x[n+6])/2.;
x[n+7]=(x[n+6]+x[n+8])/2.;
xtemp=x[n+1];
f[n+1]=igrand(xtemp,m);
xtemp=x[n+3];
f[n+3]=igrand(xtemp,m);
xtemp=x[n+5];
f[n+5]=igrand(xtemp,m);
xtemp=x[n+7];
f[n+7]=igrand(xtemp,m);
/* determine integrals and errors for the two new segments */
for(iseg=isegm;iseg<isegm+2;++iseg)
    {
    n=iseg*4;
    h=x[n+4]-x[n];
    s1=(f[n]+4.*f[n+2]+f[n+4])*h/6.;
    s2=(f[n]+4.*f[n+1]+2.*f[n+2]+4.*f[n+3]+f[n+4])*h/12.;
    e[iseg]=fabs(s1-s2)/15.;
    s[iseg]=s2;
```

```
        }
    /* loop to find total error and the index isegm corresponding
    to maximum error*/
    emax=-999.;
    err=0.;
    sum=0.;
    for(iseg=0;iseg<nseg;++iseg)
        {
        err=err+e[iseg];
        sum=sum+s[iseg];
        n=iseg*4;
        if(e[iseg]>emax)
            {
            emax=e[iseg];
            isegm=iseg;
            }
        }
        if(nseg>=NSEGMAX) break;
        if(errmax>=0 && err<errmax) break;
        if(errmax<0 && (err/sum)<-errmax) break;
    }   /* end main while loop */
/* final statistics */
sum=0;
err=0;
for(iseg=0;iseg<nseg;++iseg)
    {
    sum=sum+s[iseg];
    err=err+e[iseg];
    }
*errptr=err;
*nsegptr=nseg;
return(sum);
}
```

D.5.3 Example of a program to compute the probability of exceeding a particular value of Student's *t* (of either sign)

```
#include<stdio.h>
#include<math.h>
double gamman2(int);
double adsint(double,double,int,double,double *,int *);
```

```
double igrand(double,int);
double igrand(double x,int m)
{
double a;
if (m==0)
    {
    a=exp(-x*x/2);
    }
else
    {
    a=1.+x*x/m;
    a=exp(-log(a)*(m+1.)/2.);
    }
return(a);
}
main()    /* program for computing P(>| Student's t|)*/
{
double a,b,an,q,gamma1,gamma2,err,errmax,ng,st,ans;
int nseg,ndeg,np,n,n1,i;
double q1,q2,q3;
while(1)
{
printf(" Prob of exceeding | t |\ n");
printf(" enter Student's t (<0 quits):");
scanf(" %lf",&st);
if (st<0) break;
printf("enter ndeg (0 means inf., <0 quits):");
scanf("%d",&ndeg);
if (ndeg<0) break;
if(st<2.)
    {
    a=0;
    b=st;
    }
    else
    {
    a=st;
    b=1.e15;
    }
printf("a = %12.4lf  b = %15.6le  ndeg= %5d\n",a,b,ndeg);
/* maximum ndeg is between 3500 and 4000*/
```

```
if(ndeg==0) q = sqrt(2./PI);
if(ndeg>0 && ndeg<3000)
    {
    gamma1=gamman2(ndeg);
    gamma2=gamman2(ndeg+1);
    q=2.*gamma2/(gamma1*sqrt(PI*ndeg));
    }
if(ndeg>=3000)
    {
    q1=ndeg;
    q2=(q1+1.)/q1;
    q3=pow(q2,q1/2.);
    q=2.*q3/sqrt(2.*E*PI);
    }
printf("ndeg =%5d   q=%15.9lf\n",ndeg,q);
errmax=-1.e-8;   /* specify fractional error */
ans=q*adsint(a,b,ndeg,errmax,&err,&nseg);
if (st<2.) ans=1.-ans;
printf("probability of exceeding = %20.10le\n",ans);
printf("error=%20.10le   %d segments\n\n",q*err,nseg);
}
}
```

D.5.4   Function to compute the gamma function of integral or half-integral arguments

```
double gamman2(int n)
{
double ng,n1,n2,gamma;
int np;
ng=n/2;
np=n%2;
if (np==1)
    {
    gamma = sqrt(PI);
    n1=0.5;
    }
    else
    {
    gamma=1;
    n1=1;
    }
```

```
for (n2=n1;n2<ng;++n2) gamma=gamma*n2;
return(gamma);
}
```

D.6   Function inversion using Newton's method

Frequently, it is necessary to invert a function $F(x)$ that does not have a sufficiently simple form for analytic inversion, but can be differentiated. In this case, we approximate the value x_n that yields a function value $F(x_n)$ that is very close to a given F_n by iterating over the algorithm that follows. This method is particularly applicable to functions that are defined by integrals (e.g. Gaussian error function, probability of exceeding Student's t or χ^2, gamma function), because the magnitude of $F'(x)$ is simply the integrand. See the previous section for the adaptive quadrature method of numerical evaluation of such functions.

D.6.1   Flow chart

The algorithm to find the value x_n such that $|F(x_n) - F_n| < e$, given a function $F(x)$, its derivative function $F'(x)$, the desired value F_n, an initial guess x_0, and the maximum error e follows.

    Loop over steps 1 to 5:
1. Evaluate $F_0 = F(x_0)$.
2. If $|F_0 - F_n| < e$, then quit loop.
3. Evaluate $F_1 = F'(x_0)$.
4. Use linear extrapolation to find a better value for x_0:

$$x_0 = x_0 + \frac{F_n - F_0}{F_1}$$

5. If x_0 is out of legal range, set x_0 to the edge of the range.

D.7   Function inversion using quadratic approximation

Frequently, it is necessary to invert a function $F(x)$ that does not have a sufficiently simple form for analytic inversion. Furthermore, it may be difficult or impossible to take the first derivative and use Newton's method. In these cases, it is more convenient to use numerical methods that evaluate $F(x)$ at suitably chosen points to find the value $x = x_n$ such that $F(x_n)$ is nearly equal to a given value F_n. The method shown uses a first guess and a step size to evaluate the function about the first guess, do a quadratic approximation to find a better value of x_n, and iterate until $F(x_n)$ is sufficiently close to F_n.

D.7.1 Flow chart

The algorithm to find the value x_n such that $|F(x_n) - F_n| < e$, given a function $F(x)$, the desired value F_n, an initial guess x_0, an initial step size Δx, and the maximum error e follows.

Loop over steps 1 to 6:

1. Evaluate $F_0 = F(x_0)$.
2. If $|F_0 - F_n| < e$, then quit loop.
3. Compute $x_- = x_0 - \Delta x$ and $x_+ = x_0 + \Delta x$.
   If x_- and/or x_+ are outside the legal range, adjust their values to the edge of the range.
4. Evaluate $F_- = F(x_-)$, and $F_+ = F(x_+)$.
5. Compute the point $x = c$, where the quadratic curve defined by x_-, x_0, x_+, F_-, F_0, and F_+ passes through F_n.
   $$a = x_0 + (x_- - x_0)(F_0 - F_n)/(F_0 - F_-)$$
   $$b = x_0 + (x_+ - x_0)(F_0 - F_n)/(F_0 - F_+)$$
   $$c = a + (b - a)(F_- - F_n)/(F_- - F_+)$$
6. Compute the distance between x_0 and the new estimate c, which is to be used as the next step size.
   If $x_- \geq c \geq x_+$, then $\Delta x = c - x_0$ and $x_0 = c$.
   If $c < x_-$, then $x_0 = x_-$ and $\Delta x = 2\Delta x$.
   If $c > x_+$, then $x_0 = x_+$ and $\Delta x = 2\Delta x$.

D.8 Random number generator

The function randome.c listed below computes pseudo-random numbers in the range from 0 to 1 using one of the methods proposed by Pierre L'Ecuyer in "Efficient and portable combined random number generators," *Communications of the ACM*, Vol. 31, pp 742–749, 774, 1988.

D.8.1 Previous methods

The most commonly employed method is the Lehmer linear congruential generator (LCG): $s(i + 1) = (A^*s(i) + C)$ MOD M. This can be thought of as a roulette wheel with M consecutive numbers which is spun an amount $(A^*s(i) + C)$. When $C = 0$, this becomes the multiplicative linear congruential generator (MLCG), which has the maximum possible period $P = M - 1$ if M is prime and A is a primitive element modulo M (i.e. A^N MOD $M = 1$ is true only for $N = M - 1$). By using a technique described by Linus Schrage in "A more portable Fortran random number generator," *ACM Trans. Mathematical Sofware*, Vol. 5, No. 2, pp 132–138, 1979, it is possible in 31-bit arithmetic to perform the MLCG when M is as large as $2^{31} - 1$ and A is as large as 1.4×2^{15}. However, the MLCG has the well-known shortcoming that if successive numbers are plotted in a multi-dimensional space, the points will lie only on a set of parallel hyperplanes. Such correlations are unacceptable in Monte Carlo computations where several random numbers are needed to define each event. One solution (the method of Bayes and Durham as described by Donald Knuth in Chapter 3 of *The Art of Computer Programming*, Addison-Wesley, Reading, MA, 1981), is the use of a large (typically 97 element) "state table" which is randomly accessed to scramble and break up these correlations. An added advantage is that the period is increased to a very large number. These techniques are described in the book *Numerical Recipes* by C. Williams, H. Press, Brian P. Flannery, Saul A. Teukolsky, and William Vetterling, Cambridge University Press, 1988, and are the basis for many commonly used random number generators. Of special note is the ANSI C standard generator rand() which has only 32k values, a period of 32k, and is aptly described in William H. Press, Brian P. Flannery, Saul A. Teukolsky, and William A. Vetterling, *Numerical Recipes in C*, Cambridge University Press, 1988, as being worthless.

D.8.2   L'Ecuyer's method

The method combines two 31-bit MLCGs to generate $\approx 2.1 \times 10^9$ distinct random 31-bit numbers with a period of $2.305,84 \times 10^{18} (\approx 2^{61})$ and no detectable correlations.
- By combining the outputs of two MLCG generators with different and suitably chosen M and A values, the period is increased to 2^{61} and the correlations are reduced to an undetectable level.
- The state of the generator can be uniquely described by two 31-bit seed numbers rather than by a seed number and a large state table.

The author thanks Orin Dahl of LBNL for bringing L'Ecuyer's method to his attention. L'Ecuyer's method is also used at CERN for Monte Carlo computations in high energy physics.

D.8.3   C program use

Use:

long s1, s2;

double ran;

randome (&s1, &s2, &ran);

The function uses the seeds s1 and s2 to compute two new seeds, which are combined to form the random number ran.

Seed numbers (input and output)

&s1 and &s2 are pointers to two seed numbers that uniquely represent the state of the generator.

s1 is between 1 and 2,147,483,562 (2^{31}–86) inclusive.

s2 is between 1 and 2,147,483,398 (2^{31}–250) inclusive.

Random number (output)

ran is a pseudo-random number equal to n/2,147,483,563, where
n is a whole number between 1 and 2,147,483,562 inclusive.

Multiple runs

• For multiple debugging runs, where the same sequence of random numbers is desired, the seeds can be started at the same values (e.g. s1 = 1 and s2 = 1).

• For multiple production runs, where independent sequences of random numbers are desired, it is suggested that the two seeds be written to a file at the end of each run and then read at the start of the next run to continue the sequence.

Note: If the generator is started for example with s1 = 1 and s2 = 1, then 2.3×10^{18} numbers are available before the sequence repeats. By cycling only one of the generators before starting (e.g. starting with s1 = 1 and s2 = 40,692), a different sequence of 2.3×10^{18} numbers is available. This is how the full theoretical period of 2^{61} can be realized.

D.8.4   C program code

```
randome(long *s1ptr, long *s2ptr, double *ranptr) {
if ( (*s1ptr = 40014*(*s1ptr%53668) - 12211*(*s1ptr/53668) ) < 0)
    *s1ptr += 2147483563;
if ( (*s2ptr = 40692*(*s2ptr%52774) - 3791*(*s2ptr/52774) ) < 0)
    *s2ptr += 2147483399;
if ( (*ranptr = *s1ptr - *s2ptr) < 1.) *ranptr += 2147483562.;
*ranptr = *ranptr/2147483563.;
return; }
```

Appendix E: Summary of Data Translation DT3010 PCI plug-in card

E.1   Introduction

The Data Translation, Inc. DT3010 PCI plug-in card has the following functions:
- Two 8-bit binary ports that can be configured for either input or output. There is no handshaking capability.
- Four differential analog input channels, using a four-input multiplexer (MUX) and a single 12-bit A/D converter. The input ranges are -10 to $+10$ V, or 0 to $+10$ V.
- Two independent analog output channels, using two 12-bit D/A converters. The output ranges are -10 to $+10$ V, or 0 to $+10$ V.

See Table E.1 for pin assignments.

To use the DT3010 software, your program should start out like this:

```
#include <windows.h>
#include <stdio.h>
#include "DAboard.h"
int main
{
    unsigned int val;
    InitAll();          /* necessary to initialize the DT3010 data
acquisition board */
<your program, including binary and analog I/O>
}
```

The "DAboard.c" and "InitAll.c" code files must be included in the compiling of the project.

E.2   Parallel output

The DT3010 has a binary I/O device with two 8-bit parallel ports that can be configured for either input or output. In this book we will be using lines 97 to 104 for output. A

Table E.1 *Pin assignments for the DT3010 on the DT740 screw terminal panel*

Pin	Signal name	Pin	Signal name
1	Analog input ch 0+	9	
2	Analog input ch 0−	10	
3	Analog input ch 1+	11	
4	Analog input ch 1−	12	
5		13	
6		14	
7		15	
8		16	
17		25	
18		26	
19		27	
20		28	
21		29	
22		30	
23		31	
24		32	
33		41	Analog output ch 0+
34		42	Analog output ch 0−
35		43	Analog output ch 1+
36		44	Analog output ch 1−
37		45	
38		46	
39		47	
40		48	
49	+5 V reference (do not use!)	57	Digital ground
50	Analog ground	58	Ch 0 clock in
51		59	Ch 0 clock out
52		60	Ch 0 external gate in
53		61	Ch 1 clock in
54		62	Ch 1 clock out
55		63	Ch 1 external gate in
56		64	
65		73	
66		74	
67		75	
68		76	
69		77	External A/D trigger (TTL only)
70		78	
71		79	
72		80	

Table E.1 *Pin assignments for the DT3010 on the DT740 screw terminal panel (cont.)*

81	Digital ground	89	Binary input bit 0 (LSB)
82	Digital ground	90	Binary input bit 1
83	Digital ground	91	Binary input bit 2
84		92	Binary input bit 3
85		93	Binary input bit 4
86		94	Binary input bit 5
87		95	Binary input bit 6
88		96	Binary input bit 7 (MSB)
97	Binary output bit 0 (LSB)	105	
98	Binary output bit 1	106	Analog ground
99	Binary output bit 2	107	Analog trigger
100	Binary output bit 3	108	
101	Binary output bit 4	109	
102	Binary output bit 5	110	
103	Binary output bit 6	111	
104	Binary output bit 7(MSB)	112	

binary one produces a TTL high voltage (about 4 V) and a binary zero produces a TTL low voltage (about 0.5 V). The C code function for binary output is

```
olDaPutSingleValue(hDout, val, 0, 1.0);
```

In most cases, handshaking is not needed for binary output, and the above function can be used by the program to assert the binary number "val" on the external output lines. A binary one produces a TTL high voltage (about 4 V) and a binary zero produces a TTL low voltage (about 0.5 V). The voltages remain on the output lines until the program writes a new binary number.

Warning. When a new binary number is written and more than one bit changes, there will be a brief time (about 5 ns) when erroneous values will occur.

If such occasional erroneous values can be tolerated and when the external circuit only needs to read the current value of a binary number, handshaking is not required.

However, if: (1) occasional erroneous output values cannot be tolerated, or (2) the computer is asked to generate a binary number in response to a request from an external circuit, or (3) the computer must transmit a series of numbers faithfully to an external circuit, the following handshaking steps are essential:
1. The external circuit signals "ready for data."
2. The computer generates the binary number, asserts it on the output lines, and after the voltages have settled, signals "data available."
3. The external circuit reads the output lines and removes the "ready for data" signal.
4. The computer removes the "data available" signal.

This handshaking can be accomplished by devoting one of the input bits to "ready for data" and one of the output bits to "output data available." The handshaking steps 1–4 are repeated for each new output number.

E.3   Parallel input

The DT3010 has a binary I/O device with two 8-bit parallel ports that can be configured for either input or output. In this course we will be using lines 89–96 for input. A TTL high-input voltage (3–5 V) reads as a binary one and a low-input voltage (0–0.5 V) reads as a binary zero. The C code function for binary input is:

```
olDaGetSingleValue(hDin, &val, 0, 1.0);
```

Warning. When the external circuit changes the data and more than one bit changes, there will be a brief time (about 5 ns) when erroneous values will occur.

If such occasional erroneous values can be tolerated and when the computer only needs to read the current value on the input lines, handshaking is not required.

However, if: (1) occasional erroneous output values cannot be tolerated, or (2) the external circuit is asked to generate a binary number in response to a request from the program, or (3) the external circuit must transmit a series of numbers faithfully to the computer, the following handshaking steps are essential:

1. The program signals "ready for data."
2. The external circuit generates the binary number, asserts it on the output lines, and after the voltages have settled, signals "data available."
3. The program reads the output lines and removes the "ready for data" signal.
4. The external circuit removes the "data available" signal.

This handshaking can be accomplished by devoting one of the input bits to "ready for data" and one of the output bits to "output data available." The handshaking steps 1–4 are repeated for each new input number.

E.4   Analog output

The C code function for analog output is:

```
olDaPutSingleValue(hDa, val, channel, 1.0);
```

Usually, the analog output port is used to generate a constant voltage or a voltage waveform and handshaking is not required.

E.5   Analog input

The C code function for analog input is:

olDaGetSingleValue(hAd, &val, channel, 1.0);

If the program only needs to record the current value of an analog voltage, hand-shaking is not required.

Handshaking is used when the program acquires analog data from an external circuit that has an unpredictable response time. It allows the program to control the external circuit and to wait until the data are valid before reading them.

E.6   Using the DT3010 board with the Microsoft visual C++ compiler

See Laboratory Exercise 1.

Appendix F: Using the digital oscilloscope to record waveforms

F.1   Introduction

The instructions below describe how to use the HP VEE panel driver for the HP 54600 digital oscilloscope to record and print waveforms.

F.2   Capturing the waveform

- Open HP VEE from the shortcut on the desktop.
- Open the "Instrument Manager" from the "I/O" pull-down menu.
- Under "HP-IB7" select "Digital Scope (hp54600b@707)" (just select it, don't double click on it).
- Click on the "Panel Driver" button.
- A panel driver will appear that allows you to control the oscilloscope.
- Select the timebase and channel sensitivity on the oscilloscope to get a good waveform.
- Press the "Stop" button on the oscilloscope to capture the waveform on the oscilloscope screen.
- To transfer the digitized waveform from the oscilloscope into the panel driver memory, click once on the black window in the panel driver window.

F.3   Printing the waveform

- Right click on the panel driver and select "Add Terminal: Data Output."
- Select "WF_Ch1 (waveform)" if you want to print the waveform on channel 1.
  If you are sampling both channels 1 and 2, also select "WF_Ch2 (waveform)."
- One or two output labels (depending on the number of channels selected) will appear on the blue part of the panel driver located on its right end.

- Select "Waveform (time)" from the "Display" pull-down menu.
- Place the display in the HP VEE window.
- The default setting allows one input trace to go into the display. To add a second waveform, right click on the display window and select "Add Terminal: Data Input."
- Connect the output channel of the panel driver to the desired input channel of the display by clicking the black dots of the desired channel ports.
- Once everything is connected, run the HP VEE program by clicking on the "Run" button (looks like an arrow). The waveform in the panel driver should show up in the display window.
- Scale the displays by clicking on "Auto Scale."
- To print the waveforms, right click on the black window of the display window and select "plot."

Appendix G: Electrical hazards and safety

G.1   Introduction

An intact dry 1 cm^2 patch of skin has a typical electrical resistance of about 100 kΩ, which is primarily provided by epithelium, the horny outermost layer of skin. Under normal circumstances, the skin provides considerable protection against brief contact with electrical potentials even as high as 120 V. However, if the skin is wet or perspiring, the electrical resistance is greatly reduced and dangerous currents can be conducted. If the skin has a cut, or if Ag(AgCl) electrodes with electrode paste are used, the resistance can decrease to below 1 kΩ. Under these conditions, dangerously high currents can be produced by potentials as low as 12 V. See Table G.1 for the physiological effects of various current levels. With a skin resistance of 100 kΩ, 500 V would be required to produce a current of 5 mA, but with a skin resistance of 1 kΩ, only 5 V will produce the same current. At the other extreme, high-voltage generators with potentials of 100 kV are used in science museums to show safely the effect of static electricity on people whose heads are endowed with large quantities of hair. The important lesson here is that the primary factor in electrical hazards is the current, not the voltage, passing directly through the heart. This is of particular concern to designers of equipment that uses electrodes placed near or on the surface of the heart, such as coronary catheters, or during surgery.

The 99.5 percentile threshold rms current from arm to arm required for perception and for involuntary muscular contraction is shown in Figure G.1.

In summary:

- At 60 Hz, 99.5% of the population will not perceive currents <400 μA across the arms.
- A 60 Hz, 99.5% of the population will be able to "let go" if the current across the arms is <10 mA.
- These threshold currents are minimum at 60 Hz, 2–3 times higher at 0 Hz, and 5–10 times higher at 10 kHz.

Table G.1 *Physiological effects of 60-Hz currents from a 1-s contact with the body extremities**,†

Current	Effect
500 μA	Threshold of perception
5 mA	Unpleasant, but accepted safe level
>20 mA	Pain, fatigue, possible physical injury, inability to "let go" due to involuntary muscular contraction
>75 mA	Possible respiratory paralysis and loss of normal rhythm of heart muscle contraction (ventricular fibrillation) resulting in death
>1 A	Sustained contraction of the heart (which may revert to normal rhythm if the duration is sufficiently short)
>10 A	Severe burns and physical injury

*The danger is greatly reduced if the current cannot pass through the chest.

†When electrodes are attached directly to the surface of the heart, ventricular fibrillation can occur with currents as low as 20 μA.

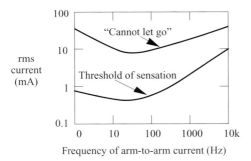

Figure G.1 Current thresholds for the 99.5 percentile of men and women.

G.2   Electrical power

To understand the nature of electrical hazards, it is first necessary to understand the nature of the systems that supply electrical power. Electrical power stations typically produce 200 mW to 2 GW at 4 kV and a frequency of 60 Hz. This is transformed to higher voltages such as 110 kV to over 500 kV for long-distance transmission. Regional substations transform this voltage down to 4–12 kV for distribution along individual streets. Service transformers mounted on wooden poles transform this down to 220 V for connection to individual buildings (Figure G.2). A third output conductor is connected to the center tap of the output transformer (Figure G.3).

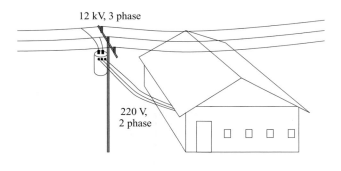

12 kV, 3 phase

220 V,
2 phase

Figure G.2 Electrical power is distributed along streets at a typical voltage of "12-kV" three phase. Any two of these conductors will provide two phase power at 12 kV rms.

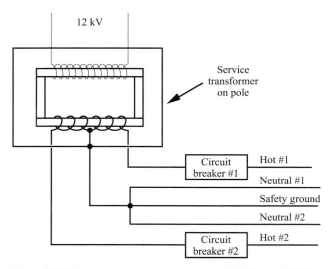

12 kV

Service
transformer
on pole

Circuit
breaker #1   Hot #1

Neutral #1

Safety ground

Neutral #2

Circuit   Hot #2
breaker #2

Figure G.3 The service transformer converts 12 kV into 220-V two phase and provides a center tap conductor that is used for current return (neutral) and to provide a safety ground.

Note: Each of the "12-kV" three phase conductors actually has a voltage $V_n = 10$ kV $\sin(\omega t + \phi_n)$ with respect to ground, where $\phi_n = n2\pi/3$ and $n = 1, 2, 3$. Any two of these conductors have a voltage difference 17 kV $\sin(\omega t + \phi')$ between them, which has an rms value of 12 kV. Each of the 220-V two phase conductors has a voltage $V_n = 156$ V $\sin(\omega t + n\pi)$ with respect to ground, where $n = 1, 2$.

Within the building the center tap is split into neutral and ground conductors. For equipment requiring 110 V, one hot and one neutral conductor is used. For equipment requiring 220 V, both hot conductors are used. Circuit breakers are used to interrupt the current on individual circuits if the current exceeds preset safety limits, typically 10–100 A. In normal use, the current supplied by a hot conductor is returned by the corresponding neutral conductor. The safety ground is connected to equipment cases and does not normally carry current.

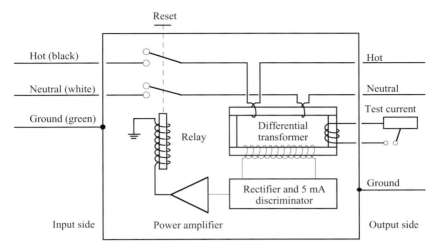

Figure G.4 Ground fault interrupter (GFI) circuit. A differential transformer, rectifier, and discriminator is used to determine when the difference in current between the hot and neutral conductors exceeds the safety limit of 5 mA and interrupt both conductors. The contacts are held open until a reset button is pushed. In addition, a test button is provided that sends an unbalanced current through a transformer winding.

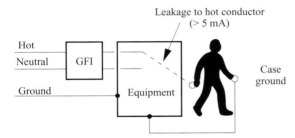

Figure G.5 Situation where accidental contact is made with a hot conductor and case ground of the same piece of equipment. The circuit is interrupted by the GFI circuit if the current exceeds 5 mA.

G.3   The ground fault interrupter circuit

The ground fault interrupter (GFI) circuit uses a differential transformer and amplifier to detect any difference in current between the 120-V "hot" (black) conductor and the "neutral" (white) conductor (Figure G.4). If this difference exceeds 5 mA, the current in both conductors is interrupted by a pair of circuit breakers. The GFI circuit protects against accidental contact with a hot conductor and either the case ground of the same piece of equipment (Figure G.5) or the neutral or case ground of another piece of equipment (Figure G.6).

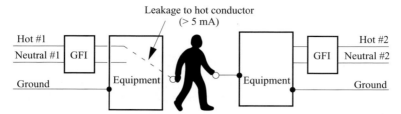

Figure G.6 Situation where accidental contact is made with hot conductor No. 1 and either case ground or the neutral conductor of another piece of equipment. The circuit is interrupted by the GFI circuit No. 1 if the current exceeds 5 mA.

G.4   The isolation transformer

If electrical equipment is not connected directly to the power line, but connected through a transformer, the equipment is isolated because the transformer output conductors cannot carry any appreciable current to the power line conductors. Only by coming in contact with both isolation transformer output conductors will an electrical shock be received. Coming in contact with only one of the transformer output conductors and any other single conductor (power line, water pipe, etc.) will not cause a shock.

G.5   Typical accident scenarios

1. Due to a fault in electrical insulation, the metal case of a piece of equipment is energized. The victim touches the faulty equipment with one hand and a properly grounded piece of equipment with the other hand.
2. The victim touches live conductors in equipment whose case has been removed for repair. Note that one of these conductors may be at "ground" potential, but should be considered "live" if it can return current to its source.
3. Although the equipment is disconnected from power, the victim touches a charged high-voltage capacitor after the case is opened for repair.

G.6   Methods of accident prevention

1. Connect all external metal parts to a common ground that does not normally carry current. This is the familiar green "U ground" or "house ground" in electrical receptacles. It is expected that people will be touching all combinations of equipment cases and water pipes so it is important that all of these are at the same potential.

2. Double insulate all dangerous voltages. This is used for power tools, but is not effective in the presence of water and is not used for medical equipment due to the danger of a saline solution spill.

3. When designing high-voltage (>1 kV) systems, especially those that require large capacitors, provide two bleeder resistors in parallel that automatically eliminate dangerous potentials a short time after the equipment is turned off. (This is more reliable than a single bleeder resistor which may burn out after several years without being noticed.) When servicing such equipment, firmly attach a grounding strap to all dangerous potentials.

4. Use GFI circuits (described before), which interrupt the current in the typical case of accidental contact with a hot conductor and ground.

5. Use an isolation transformer, as described above.

6. Isolation of patient-connected parts by a low-capacitance transformer or opto-isolator (LED and photodiode) or by radio telemetry.

7. Use current-limiting circuits in all conductive paths connected to the patient.

Appendix H: Standard resistor and capacitor values

H.1 Standard resistor values and color codes

Carbon resistors are the most frequently used in printed circuits and come in a limited selection of values. The tables below list those values that resistor manufacturers commonly provide. In this book, all circuits used in the laboratory exercises specify resistor values taken from this list.

For carbon resistors with tolerances of 2% or larger, the first three colors are used to determine the value of the resistor. The first two colors (Table H.1) indicate the first two digits and the third color (Table H.2) indicates a multiplier. The fourth color (Table H.3) indicates the resistor tolerance. Values of commonly available 2, 5, and 10% resistors and their first three color codes are listed in Table H.4.

For metal film resistors with tolerances of 1% or smaller, the first four colors are used to determine the value of the resistor. The first three colors indicate the first three digits and the fourth color indicates a multiplier. The fifth color indicates the resistor tolerance.

Silver is used to indicate a multiplier of 0.01. For resistor values below $10\,\Omega$, the values corresponding to 11, 13, 16, 20, 24, and 30 are not available.

Tolerances of 1% or better are available in metal film resistors.

H.2 Standard capacitor values and codes

Capacitors are manufactured in a number of types and are available in a limited selection of values. The tables below list values that capacitor manufacturers commonly provide. In this book, all circuits used in the laboratory exercises specify capacitor values taken from this list.

Capacitors with values below $1\,\mu F$ are marked with numbers similar to the color codes used for carbon resistors. The first two digits indicate a value in picofarads and

Table H.1 *First two colors of the resistor color code*

Color	Black	Brown	Red	Orange	Yellow	Green	Blue	Violet	Gray	White
Digit	0	1	2	3	4	5	6	7	8	9

Table H.2 *Third color of the resistor color code*

Color	Gold	Black	Brown	Red	Orange	Yellow	Green	Blue	Violet
Multiplier	0.1	1	10	100	1,000	10 k	100 k	1 M	10 M

Table H.3 *Fourth color of the resistor color code*

Color	None	Silver	Gold	Red	Brown	Green	Blue	Violet	Gray
Tolerance (%)	20	10	5	2	1	0.5	0.25	0.10	0.05

Table H.4 *Standard 2, 5, and 10% resistor values and their color codes*

First two colors		Third color						
		Gold	Black	Brown	Red(k)	Orange(k)	Yellow(k)	Green(M)
Brown	black	1.0	10	100	1.0	10	100	1.0
Brown	brown		11	110	1.1	11	110	1.1
Brown	red	1.2	12	120	1.2	12	120	1.2
Brown	orange		13	130	1.3	13	130	1.3
Brown	green	1.5	15	150	1.5	15	150	1.5
Brown	blue		16	160	1.6	16	160	1.6
Brown	gray	1.8	18	180	1.8	18	180	1.8
Red	black		20	200	2.0	20	200	2.0
Red	red	2.2	22	220	2.2	22	220	2.2
Red	yellow		24	240	2.4	24	240	2.4
Red	violet	2.7	27	270	2.7	27	270	2.7
Orange	black		30	300	3.0	30	300	3.0
Orange	orange	3.3	33	330	3.3	33	330	3.3
Orange	blue	3.6	36	360	3.6	36	360	3.6
Orange	white	3.9	39	390	3.9	39	390	3.9
Yellow	orange	4.3	43	430	4.3	43	430	4.3
Yellow	violet	4.7	47	470	4.7	47	470	4.7
Green	brown	5.1	51	510	5.1	51	510	5.1
Green	blue	5.6	56	560	5.6	56	560	5.6
Blue	red	6.2	62	620	6.2	62	620	6.2
Blue	gray	6.8	68	680	6.8	68	680	6.8
Violet	green	7.5	75	750	7.5	75	750	7.5
Gray	red	8.2	82	820	8.2	82	820	8.2
White	brown	9.1	91	910	9.1	91	910	9.1

Table H.5 *Standard ceramic disk and plastic/metal foil capacitor values and their corresponding three-digit codes*

First two code numbers	Third code number				
	0(pF)	1(pF)	2(pF)	3(μF)	4(μF)
10	10	100	1,000	0.01	0.1
12	12	120	1,200	0.012	0.12
15	15	150	1,500	0.015	0.15
18	18	180	1,800	0.018	0.18
22	22	220	2,200	0.022	0.22
27	27	270	2,700	0.027	0.27
33	33	330	3,300	0.033	0.33
39	39	390	3,900	0.039	0.39
47	47	470	4,700	0.047	0.47
56	56	560	5,600	0.056	0.56
68	68	680	6,800	0.068	0.68
82	82	820	8,200	0.082	0.82

the third digit indicates a power-of-ten multiplier. For example, a capacitor marked "104" is 100,000 pF or 0.1 μF.

Ceramic disk capacitors come in values ranging from 10 to 100,000 pF (0.1 μF) and voltage ratings from 25 to 500 V. The first two digits indicate a value (in picofarads) and the third digit indicates a power-of-ten multiplier. See Table H.5 for standard values and corresponding three-digit codes.

Plastic and metal foil capacitors (Mylar, polypropylene, polyester) come in values ranging from 1,000 pF (0.001 μF) to 10 μF and voltage ratings from 50 to 500 V. Capacitors with values below 1,000,000 pF (1 μF) are marked with the first two digits indicating a value (in picofarads) and the third digit indicating a power-of-ten multiplier. Capacitors with values 1 μF and above are marked directly in microfarads.

Tantalum electrolytic capacitors are designed for use in printed circuits and typically have values ranging from 1 to 330 μF and have voltage ratings typically from 6 to 50 V. One lead is commonly marked + and must be used in the correct polarity or the capacitor may explode. Standard capacitance values are a subset of the values shown in Table H.4 (..., 12, 15, 22, 33, 47, 68, 100 μF, ...).

Aluminum electrolytic capacitors are used when large values of capacitance are required, such as in power supplies. While the smaller sizes can be mounted in printed circuits, the larger units consist of aluminum cans that require separate mechanical attachment. They have values ranging from 1 to 100,000 μF and have voltage ratings typically from 6 to 50 V. One lead is commonly marked + and must be used in the correct polarity or the capacitor may explode. Standard capacitance values are a subset of the values shown in Table H.4 (..., 100, 150, 220, 330, 470, 680, 1,000 μF, ...).

Appendix I: ASCII character codes

I.1 ASCII character set codes

Table I.1 lists the control codes and printable characters commonly used by computers and their decimal, octal, and hexadecimal equivalents. The letters ASCII stand for the American Standard Code for Information Interchange. Table I.2 lists the two and three character control codes and their meaning.

Of particular importance are the following:

NUL   Ignored in printing, used in C programming to indicate the end of a character string

BEL   Used to produce an audible "beep"

HT   Horizontal tab, used to line up columns of text – HT is also used (rather than US) to delimit fields in spread sheet and data base programs

LF   Line feed, which moves the cursor down but not left

FF   Form feed, used to go to the top of the next page

CR   Carriage return, which technically only moves to the beginning of the current line (not the beginning of the next line). Consequently, most systems translate the single CR character into a LF followed by CR. CR is also used (rather than RS) to delimit records in spread sheet and data base programs

SP   the "space" character

On the modern keyboard, Hex 1C (file separator) is mapped to the right arrow, Hex 1D (group separator) is mapped to the left arrow, Hex 1E (record separator) is mapped to the up arrow, and Hex 1F (unit separator) is mapped to the down arrow.

Table I.1 *ASCII character codes*

Decimal	Octal	Hexadecimal	Character	Decimal	Octal	Hexadecimal	character
0	000	00	NUL	44	054	2C	,
1	001	01	SOH	45	055	2D	−
2	002	02	STX	46	056	2E	.
3	003	03	ETX	47	057	2F	/
4	004	04	EOT	48	060	30	0
5	005	05	ENQ	49	061	31	1
6	006	06	ACK	50	062	32	2
7	007	07	BEL	51	063	33	3
8	010	08	BS	52	064	34	4
9	011	09	HT	53	065	35	5
10	012	0A	LF	54	066	36	6
11	013	0B	VT	55	067	37	7
12	014	0C	FF	56	070	38	8
13	015	0D	CR	57	071	39	9
14	016	0E	SO	58	072	3A	:
15	017	0F	SI	59	073	3B	;
16	020	10	DLE	60	074	3C	<
17	021	11	DC1	61	075	3D	=
18	022	12	DC2	62	076	3E	>
19	023	13	DC3	63	077	3F	?
20	024	14	DC4	64	100	40	@
21	025	15	NAK	65	101	41	A
22	026	16	SYN	66	102	42	B
23	027	17	ETB	67	103	43	C
24	030	18	CAN	68	104	44	D
25	031	19	EM	69	105	45	E
26	032	1A	SUB	70	106	46	F
27	033	1B	ESC	71	107	47	G
28	034	1C	FS→	72	110	48	H
29	035	1D	GS←	73	111	49	I
30	036	1E	RS↑	74	112	4A	J
31	037	1F	US↓	75	113	4B	K
32	040	20	SP	76	114	4C	L
33	041	21	!	77	115	4D	M
34	042	22	"	78	116	4E	N
35	043	23	#	79	117	4F	O
36	044	24	$	80	120	50	P
37	045	25	%	81	121	51	Q
38	046	26	&	82	122	52	R
39	047	27	'	83	123	53	S
40	050	28	(	84	124	54	T
41	051	29	)	85	125	55	U
42	052	2A	*	86	126	56	V
43	053	2B	+	87	127	57	W

Table I.1 *ASCII character codes (cont.)*

Decimal	Octal	Hexadecimal	Character	Decimal	Octal	Hexadecimal	character
88	130	58	X	108	154	6C	l
89	131	59	Y	109	155	6D	m
90	132	5A	Z	110	156	6E	n
91	133	5B	[	111	157	6F	o
92	134	5C		112	160	70	p
93	135	5D	]	113	161	71	q
94	136	5E	∧	114	162	72	r
95	137	5F		115	163	73	s
96	140	60	`	116	164	74	t
97	141	61	a	117	165	75	u
98	142	62	b	118	166	76	v
99	143	63	c	119	167	77	w
100	144	64	d	120	170	78	x
101	145	65	e	121	171	79	y
102	146	66	f	122	172	7A	z
103	147	67	g	123	173	7B	{
104	150	68	h	124	174	7C		
105	151	69	i	125	175	7D	}
106	152	6A	j	126	176	7E	~
107	153	6B	k	127	177	7F	DEL

Table I.2 *ASCII control codes*

Hexadecimal	Code	Definition	Hexadecimal	Code	Definition
00	NUL	Null	11	DC1	Device control 1
01	SOH	Start of heading	12	DC2	Device control 2
02	STX	Start of text	13	DC3	Device control 3
03	ETX	End of text	14	DC4	Device control 4
04	EOT	End of tape	15	NAK	Negative acknowledge
05	ENQ	Enquiry	16	SYN	Synchronize
06	ACK	Acknowledge	17	ETB	End of transmitted block
07	BEL	Bell	18	CAN	Cancel
08	BS	Backspace	19	EM	End of medium
09	HT	Horizontal tab	20	SP	Space
0A	LF	Line feed	1A	SUB	Substitute
0B	VT	Vertical tab	1B	ESC	Escape
0C	FF	Form feed	1C	FS	File separator
0D	CR	Carriage return	1D	GS	Group separator
0E	SO	Shift out	1E	RS	Record separator
0F	SI	Shift in	1F	US	Unit separator
10	DLE	Data link escape	7F	DEL	Delete

Glossary

A/D converter: circuit for producing a digital representation of the voltage level at its analog input.

Absolute accuracy: agreement between measured values and *ideal* values (e.g. output voltage of a D/A converter or transition voltages of an A/D converter).

Absolute accuracy, analog-to-digital converter: agreement between the transition voltages and their ideal values, before adjustment for zero offset and gain error.

Absolute accuracy, digital-to-analog converter: agreement between the actual output voltages and their ideal values, before adjustments for zero offset and gain error.

Absolute digital encoder: device that produces an absolute digital output that depends on a physical quantity, such as position or angle.

Absolute zero: temperature corresponding to the lowest possible thermal energy ($-273.16\,^{\circ}\text{C}$ or $-459.69\,^{\circ}\text{F}$).

Accelerator: add-on circuit for a microcomputer that increases microprocessor speed by providing a higher speed processor.

Accuracy (absolute) of a sensor: agreement between measured values and the accepted standard (see also *precision*).

Accuracy, control system: difference between the set point and the sense signal averaged over time.

Accuracy, sensor: agreement between the response curve (relationship between physical input and electronic output) and the ideal curve.

Acquisition time (sample-and-hold): time from the hold-to-sample edge of the control pulse to the time when the sample-and-hold amplifier has reached its final value (within a specified error).

Action potential: transient membrane potential changes caused by the motion of sodium and potassium ions that occur when a nerve or muscle fiber fires. It is usually initiated at one end of the cell by a neurotransmitter chemical that makes the cell membrane temporarily permeable to sodium ions, which rush into the cell to increase its interior potential from its normal -85 to $+20$ mV. This change in potential also makes the adjacent membrane temporarily permeable to sodium ions, which causes the process to propagate down the length of the cell. The membrane potential is restored by an outflow of potassium ions, and the ion concentration is restored by a process called the sodium–potassium pump.

Actuator: device that converts electrical energy into physical energy.

Adaptive quadrature integration: method for numerical integration that minimizes the number of function values needed to achieve a given accuracy by identifying and selectively subdividing the interval that contributes the largest error (see Appendix D).

Address bus: a common set of lines used to communicate addresses to many listeners, where each listener has a unique address.

Address decoder: circuit for generating a select signal when a pre-determined address appears.

Ag(AgCl) electrode: composite electrode consisting of intermixed Ag metal and AgCl salt (which is conductive). The electrode can convert Cl^- ions in solution into electrons in a wire by combining the Cl^- ions with Ag metal to produce AgCl salt and electrons in the Ag metal. The electrode can convert electrons into Cl^- ions in solution by decomposing AgCl into Ag metal and Cl^-.

Aliasing, frequency: when a waveform is periodically sampled, frequency components above one-half the sampling frequency appear as erroneous lower frequency components.

Alpha particle: bare helium nucleus (two protons plus two neutrons) emitted naturally by many heavy isotopes and can be artificially produced by ionizing and accelerating helium gas.

Alumel: aluminum–nickel alloy used in thermocouples.

Amplifier: circuit for converting a signal (the input) into an output that is an accurate replica amplified in voltage, current, or both.

Amplifier, current: amplifier circuit whose output voltage is proportional to the input current.

Amplifier, differential: amplifier circuit whose output voltage is an amplified waveform of the voltage difference between the two inputs and insensitive to the average voltage of the two inputs.

Amplifier, instrumentation: amplifier circuit that has: (1) output voltage proportional to the voltage difference between the two inputs; (2) very high input impedance; (3) low output impedance; and (4) controllable, constant gain over a large bandwidth.

Amplifier, isolation: differential amplifier circuit able to withstand high voltages from dc to 60 Hz between the input and output.

Amplifier, power: amplifier able to deliver an electrical output greater than 1 W. This requires discrete power transistors.

Amplifier, summing: amplifier circuit whose output voltage is an amplified version of the sum of the voltages of two or more inputs.

Amplitude demodulation: process of removing a higher frequency sine wave from its product with a lower frequency signal. In AM (amplitude modulation) radio reception, the high-frequency sine-wave carrier is removed using a rectifier and a low-pass filter.

Amplitude modulation: multiplication of two waveforms. In AM (amplitude modulation) radio transmission, one waveform is the audio signal and the other is a sine wave of higher frequency used as a carrier wave.

Analog: characterizing a signal or circuit that can take on an essentially infinite number of different values within a finite range.

Analog delay (sample-and-hold): the delay between the analog input and the analog output in sample mode.

Analog filter: analog circuit for selectively suppressing certain frequencies while retaining others.

Analog interface: circuit that can: (1) convert an analog input voltage into digital data, (2) convert digital data into an analog output voltage, (3) be addressed by a computer program for reading and writing digital data, and (4) be addressed by a computer program to set external control lines and read status bits.

Analog multiplexer: circuit with two or more analog inputs and one analog output that can be connected to any one of the inputs by digital control.

Analog position sensor: a sensor that converts position into an analog electrical signal.

Analog-to-digital converter: circuit for converting an analog voltage into a binary number with a unique relationship (e.g. linear, logarithmic).

Angle (conversion factors): 1 radian (rad) $= 180°/\pi = 57.2958°$. 1 mrad $= 0.057,295,8° = 3.437,75$ arc min $= 206.265$ arc sec.

Angle sensor: device that produces an electrical system that depends on angular position.

Angle sensor (resistor): arc-shaped resistor and a sliding contact mounted to a shaft. If the ends of the resistor are at different fixed voltages, the voltage of the sliding contact depends on the shaft angle.

Anti-aliasing filter: analog filter designed to eliminate high frequencies above the Nyquist sampling limit (one-half the sampling frequency) that would otherwise appear as erroneous lower frequencies in the sampled data.

Aperture delay (sample-and-hold amplifier): the time between the edge of the sample-to-hold command and the time when the input was equal to the held value. Equal to the control delay plus one-half the aperture time minus the analog delay.

Aperture jitter (sample-and-hold amplifier): the rms variation in aperture delay caused by noise in the control signal and switching circuit.

Aperture time (sample-and-hold amplifier): the time required for the switch to open and characterizes only the response time of the switch. (The time between the edge of the sample-to-hold command and the time when the switch begins to open is the control delay.) The value that is held at the output is determined by the average input value during the aperture time.

Apex (of the heart): portion of the heart closest to the feet.

ASCII character code: internal 8-bit code used by computer programs to describe the entire alphanumeric character set, including nonprinting characters such as spaces and horizontal tabs.

Assert (data): generate analog or digital signal potentials on signal lines. This does not insure that the signal has been read by any other circuit (see *handshaking*).

Asynchronous: characterizing a process that does not occur at predictable times.

Atria: initial pumping chambers of the heart. The right atrium receives blood from the venous circulation and pumps it into the right ventricle. The left atrium receives blood from the lungs and pumps it into the left ventricle.

Atrial systole: period of active atrial contraction between the P wave and the closing of the AV valves.

AV (atrioventricular) node: a cluster of cells leading from the lower portion of the right atrium to the ventricular septum that conducts the depolarization wave very slowly (70 ms) to allow the atrial systole to reach completion before the ventricular systole begins.

AV (atrioventricular) valves: the tricuspid valve between the right atrium and the right ventricle; and the mitral valve between the left atrium and the left ventricle.

Average (sample mean): sum of the measured values divided by their number.

Average error of a control system: difference between the set point and the average sense value, which corresponds to the physical quantity being controlled.

Average step size (of an A/D converter): average difference between all neighboring transition voltages. For an N-bit A/D converter, can be computed as the last transition voltage $V(2^N - 2, 2^N - 1)$ minus the first transition voltage $V(0, 1)$, divided by $2^N - 2$.

Average step size (D/A): the average value of the steps between output voltages: $V_n - V_{n-1}$.

Avogadro's number: 6.0228×10^{26} molecules/mole.

Band-pass filter: filter that passes frequencies only between two limits (a minimum and a maximum).

Bandwidth: for a circuit with a voltage gain $G(f)$ as a function of frequency f, and a gain G_0 in the passband, the bandwidth $\Delta f = \int G^2(f)\,df/G_0^2$. For a low-pass one-pole filter, the exponential time constant $\tau = RC$, the corner frequency f_c, and the bandwidth Δf are related by: $f_c = 1/(2\pi\tau)$ and $\Delta f = (\pi/2)f_c = 1.571 f_c$.

Base (of the heart): portion of the heart closest to the head, where the valves and great vessels are located.

Baseline restoration: the process of suppressing slow changes in a variable that would cause it to drift out of range. Usually accomplished by using a high-pass filter.

Beer's law: the transmission of light through a colored solution is inversely proportional to the exponential of "the concentration of the solution × the optical path length."

Bessel filter: a filter with a phase shift that is proportional to the frequency. This introduces a fixed time delay to all frequency components that are passed and transmits pulses with minimum distortion.

Beta of a thermistor: the thermistor resistance R at temperature T is given by $R(T) = R(T_0)\exp(\beta(1/T - 1/T_0))$, where $R(T_0)$ is the resistance at the reference temperature T_0.

Beta particle: a moving electron, usually produced by radioactive decay where a neutron is converted into a proton plus an electron plus an anti-electron neutrino.

Bias: (1) deviation from a desired value in a particular direction, or (2) fixed applied voltage.

Bimetal switch: switch made from a bimetallic strip (materials with different thermal expansion bonded together) and a moveable contact point. Differential thermal expansion causes the bimetallic strip to open or close the switch at a specific temperature that can be adjusted by moving the contact point.

Binary numbers: a representation where the rightmost bit has the value of 1 and each bit has a value twice the value of the bit to the right.

Binned data: a representation where the data are grouped into bands of value and the number in each band is plotted against the central value of the band.

Binomial distribution: distribution of the probability of a specified number of successes in a given number of independent trials, in each of which the probability of success is the same.

Blood pressure: pressure waveform in the arterial circulation. Clinically, the maximum (systolic) and minimum (diastolic) pressures measured with a pressure cuff on the upper arm and a sound sensor on the brachial artery.

BNC connector: twist-lock connector commonly used at the ends of a coaxial cable.

Bode plot: plot of amplitude versus frequency for an amplifier or filter circuit.

Boltzmann's constant: constant that relates the temperature with the mean statistical energy. The value is $k = 8.617,09 \times 10^{-5}$ eV/K $= 1.380,6 \times 10^{-23}$ J/K $= 1.380,6 \times 10^{-16}$ erg/K.

Bridge circuit: a network of four resistors that has zero output only when the resistors are balanced.

Brush position encoder: senses angular position by using conductive brushes and circular disks with conductive patterns.

Buffer amplifier: amplifier placed between a source and a load to preserve the voltage of a waveform. The buffer amplifier increases the available current and is able to drive a load that would otherwise reduce the voltage of the source.

Bundle branches: conduct the depolarization wave from the bundle of His around the inner (endocardial) surface of the ventricles.

Bus: one or more conductors used in common by multiple senders or by multiple receivers. Circuits are needed to insure that only one sender at a time asserts data on a bus and that only the intended receiver accepts data from the bus. The bus is used in the microprocessor for selective addressing (via the address bus) and for selective data transfer (via the data bus). It can also be used to allow many circuits to intercommunicate while using the minimum number of conductors.

Butterworth filter: a filter that is maximally flat in its passband.

Byte I/O: transfer of data in 8-bit units.

Calender–van Dusen equation: equation describing the resistance of a thermistor as a quadratic polynomial of the temperature T: $R(T) = R(0)(1 + AT + BT^2)$.

Calibration: (1) The process of using standard values to adjust an instrument so that subsequent readings correspond to their true values. (2) The process of measuring deviations from a set of standard values so that subsequent readings can be converted to their true values.

Calorie: the amount of heat needed to raise 1 g of water 1 °C at 15 °C (not to be confused with the kg calorie used in nutrition).

Cardiac cycle: one beat of the heart.

CD–ROM (compact disk – read only memory): an optical disk that contains information in the form of small pits that were produced during manufacturing. It can be read by the compact disk technology developed for the audio entertainment industry.

CD–R (compact disk – write once read many): an optical disk that can be written once by a laser during use and the written portions read thereafter as a CD–ROM.

CD–RW (compact disk – write many read many): an optical disk that can be written multiple times by a laser and read as a CD–ROM.

Celsius temperature scale: temperature scale defined with 0 °C at the freezing point of water and 100 °C at the boiling point of water at standard pressure.

Charge of electron: 1 electron charge $= 1.602{,}2 \times 10^{-19}$ C.

Charge transfer error (sample-and-hold): output offset error in the held value caused by charge dumped onto the holding capacitor from the switching circuit. Ideally, this should be a fixed offset, independent of input voltage level.

Chebyshev filter: a filter that maximizes sharpness of the frequency roll-off, but introduces ripples in the passband. The filter parameters are determined by numerical optimization.

Chi-squared: sum of the squares of data value − model function/by one standard deviation of the data value. A best fit of the model to the data is achieved by varying the model parameters to minimize chi-squared.

Chip (integrated circuit): silicon crystal processed by doping and deposition to produce complete electrical circuits. Some chips contain over 1 million transistors.

Chromel: chromium–nickel alloy used in thermocouples.

Circulatory system: the system for pumping and circulating blood to the lungs and organs. It consists of the heart, arteries, arterioles, capillaries, venules, and veins.

Clock: [noun] a circuit for producing periodic pulses, usually used to synchronize a process or keep track of elapsed time.

Clock: [verb] to store data in a register at the instant that a control pulse (called the clock pulse) has a logic transition. (As a verb, synonymous with latch and strobe.)

Closed-loop gain: gain of an amplifier circuit when the negative feedback loop is complete.

CMRR (common-mode rejection ratio): ratio of differential gain to common-mode gain.

Colors (visible): the electromagnetic spectrum visible to the human eye.

Violet	420 nm
Blue	450 nm
Green	550 nm (peak response of the eye)
Yellow	580 nm
Orange	600 nm
Red	700 nm

Note that the wavelength range for visible light is 400–800 nm.

Common-mode gain: the gain of a differential amplifier for signals present at both inputs. Expressed as the change in output voltage divided by the change in the common input voltage.

Common-mode rejection (CMR): $20 \log_{10}(\text{CMRR})$, where CMRR is the common-mode rejection ratio, or the ratio of differential gain to common-mode gain.

Common-mode rejection ratio (CMRR): ratio of differential gain to common-mode gain.

Comparator circuit: circuit with two analog inputs and one digital input. The output state is determined from the relative value of the two inputs. Usually consists of a differential amplifier with very high gain and voltage limited outputs.

Compiler: software that converts a higher level language such as C or FORTRAN into the executable instruction codes of a computer processor.

Complex impedance: the complex constant Z that describes the relationship between voltage $V(t)$ and current $I(t)$ in a linear system at frequency $f = 2\pi\omega$. If the excitation voltage is $V(t) = V_0\sin(\omega t)$, and the resulting current is $I(t) = I_0 \sin(\omega t + \phi)$, $Z = R + jX$, where $(V_0/I_0)^2 = R^2 + X^2$ and $\tan(\phi) = X/R$.

Constantan: copper–nickel alloy used in thermocouples.

Contact bounce: mechanical rebound of contacts causing multiple pulsing.

Control delay (sample-and-hold): the time between the edge of the sample-to-hold command and the time when the switch begins to open.

Control register: a circuit that takes its input from program write statements and transforms those bits (1 or 0) into logic voltage levels (TRUE or FALSE) on external output lines.

Control system: a system that senses the current and possibly past values of a physical quantity to be controlled, and drives an actuator so that the measured values correspond to a pre-determined set point. (See also *ON–OFF control*, *proportional control*, and *PID control*.)

Conversion time (analog-to-digital converter): time required to convert an analog input to a digital output.

Convolution of two functions: integral of the product of two functions, where the argument of one of the functions has been flipped in sign and shifted. The convolution is a function of the amount of shift.

Corner frequency (of a Butterworth filter): frequency at which the amplitude drops by 3 dB (a factor of $0.707 = 2^{-1/2}$).

Counter/timer circuit: when used as a counter, can accumulate the number of digital pulses at its input, and this number can then be read by a program statement. When used as a timer, can accumulate the number of internal or external clock pulses, and this number can then be read by a program statement.

Cross-talk: unwanted signal coupled from one conductor or circuit to another.

Current: 1 ampere (A) $= 1$ C/s $= 6.241 \times 10^{18}$ electrons/s.

Current amplifier: amplifier circuit whose output voltage is proportional to the input current.

Curve fitting: process of determining the parameters of a function so that the function best describes a set of data.

D-type flip-flop: a digital circuit whose output is set equal to its input by a logic command pulse. At all other times, the output does not depend on the input.

D/A converter: circuit for producing an analog voltage level that corresponds to an input digital number.

D/A glitch: brief erroneous spike that occurs in the output of a D/A converter when two or more input bits change at slightly different times.

Damped harmonic oscillator: an oscillator where the second derivative of the displacement depends on the displacement (harmonic term) and on the first derivative of the displacement (damping term).

Dark current (of a photodiode): current that passes though a biased photodiode in the absence of light.

Darlington amplifier: current-amplifying circuit consisting of two transistors as cascaded emitter followers.

Data: any information to be processed.

Data bus (for connecting two or more parallel outputs): hardware circuit where a number of parallel outputs can be selectively connected to a common set of parallel signal lines called a bus. Only one parallel output can be asserted on the bus at any one time.

Dead band, control system: the range of sense values for which the control variable is zero. This means that the quantity is not controlled when it is in the dead band.

Debouncing: process for removing secondary pulses that result from mechanical bouncing of an electrical switch.

Decade: a factor of ten, such as in frequency or amplitude.

Decimal: most commonly used base-10 numerical representation of binary numbers.

Decrement: decrease a number, usually by 1.

Deglitcher (for digital-to-analog converter): circuit for blocking erroneous D/A output values that result when a new input word is presented and two or more switches change state at slightly different times.

Degrees of freedom: in fitting a model function to a set of data values, the number of degrees of freedom is equal to the number of data values minus the number of parameters varied in the fit.

Delta function: $\delta(t_0)$ is nonzero only at $t = t_0$, and has unit integral.

Density of Hg: 13.6 g cm^{-3}.

Depolarization sequence of heart: time sequence of electrochemical events that cause the pumping muscles of the heart to depolarize in a specific order. Mechanical contraction occurs after depolarization.

DFT (Discrete Fourier transform): the DFT of the time series h_k is given by

$$H_n = \sum_{k=0}^{m-1} h_k \exp(-j2\pi nk/M)$$

Dial thermometer: a spiral of two metals with different thermal expansion coefficients fixed at the one end and connected to a needle at the other end. Changes in temperature cause the needle to rotate and the temperature can be read by the numbers printed on a dial mounted behind the needle.

Diamagnetic: materials in which an applied magnetic field is slightly reduced due to the alteration of electron orbits. Such materials move from a stronger magnetic field to a weaker magnetic field.

Diastole: when used alone, means ventricular diastole, the period of ventricular relaxation between the closing of the semilunar valves and the closing of the AV valves.

Diastolic pressure: lowest pressure in the circulatory system, when the heart has ended its resting phase and is about to beat again (see also *systolic pressure*).

Dichrotic notch: brief drop in arterial pressure due to backflow associated with the closing of the semilunar valves.

Differential amplifier: amplifier circuit whose output voltage is an amplified waveform of the voltage difference between the two inputs and insensitive to the average voltage of the two inputs.

Differential gain: the gain of a differential amplifier, expressed as the change in output voltage divided by the change in the difference in input voltages.

Differential linearity error (of an A/D converter): difference between the spacing of neighboring transition voltages and their average spacing.

Differential linearity error (of a D/A converter): the difference between the output step sizes and the average step size.

Differentiating filter: a high-pass filter that detects the time rate of change of a waveform.

Diffusion emf: the electromotive force that drives all free charge carriers to diffuse and fill the space available.

Digital angle encoder: a disk containing a digital code pattern that is sensed to produce a digital representation of angle.

Digital counters/timers: circuit for counting the number of clock pulses, and for allowing an external circuit to read the number accumulated. When used as a counter, the clock pulses are generated by an external circuit. When used as a timer, the clock pulses are generated by an internal oscillator of known frequency.

Digital encoder: sensor for measuring angular or linear position by detecting marks on one or more digital strips.

Digital filter: filter whose digital output is a linear combination of previous digital input and output values. The finite impulse response (FIR) filter does not depend on previous output values. The infinite impulse response (IIR) depends on previous output values.

Digital interface: circuit for transferring digital information between external lines and a microcomputer. It consists of data buffers for the temporary storage of data, and control and status registers for handshaking.

Digital-to-analog converter: circuit for converting a digital number into a corresponding analog voltage.

Diode: electronic two-terminal device that conducts current only when the applied voltage exceeds a certain value.

Diphthong: in human speech, gliding sounds that begin with one vowel and end with another.

Discrete Fourier transform (see also *fast Fourier transform*): the discrete Fourier transform of the time series h_k is given by

$$H_n = \sum_{k=0}^{M-1} h_k \exp(-j2\pi nk/M).$$

It determines the amplitude of the frequency components of a periodically sampled waveform.

Disk memory: rotating magnetic surface for storing large amounts of digital information. Unlike random access memory, information is retained when power is lost.

Dissipation constant: ratio of a change in internal power dissipation to a change in temperature (W/°C).

Droop (sample-and-hold): the drift in output voltage in hold mode due to charge leakage from the hold capacitor through the switch, the amplifier, or the capacitor itself.

Dual element sensor: an assembly with two independent sensing elements used for differential sensing applications.

Dual slope (or integrating) analog-to-digital converter: circuit that charges a capacitor for a fixed amount of time with a current that depends on input voltage, and then discharges the capacitor at a fixed current. The time required for discharge is measured by counting digital clock pulses to determine the converted number.

Duplex, full: serial digital transmission using separate conductors for send and receive.

Duplex, half: serial digital transmission where send and receive transmissions take place over the same conductor.

Duty factor: fraction of time that a system is operating or in an active state.

Dynode: surface in a photomultiplier tube coated with a special semiconductor. When energetic electrons (50–200 eV) strike a dynode, many electrons are released. A series of ten dynodes is used to produce typical electron gains of 10^7.

ECL: emitter-coupled logic.

Editor: software that allows the opening, viewing, changing, and storage of computer files.

Einthoven triangle: formed by the three points of the body used to record the standard electrocardiogram: the right arm, the left arm, and the left leg. Lead I is the potential between the left arm and the right arm, lead II is the potential between the left leg and the right arm, and lead III is the potential between the left leg and the left arm.

Electric charge (conversion factors):

1 electron charge $= 1.602,2 \times 10^{-19}$ C.

1 coulomb (C) $= 6.241 \times 10^{18}$ electron charges.

1 faraday = 1 mole of electron charge = 96,487 C.

Electrocardiogram (ECG): signals detected at the surface of the skin due to the depolarization of the heart muscles.

Electrode offset potential: voltage resulting from electrochemical reactions in the electrodes.

Electromagnetic isolation amplifier: isolation amplifier that uses high-frequency coupling (typically 100 kHz to 1 MHz) to isolate the input and output stages.

Electromagnetic pickup: electrical signals received from another circuit due to voltage fluctuations on common conductors or capacitive coupling.

Electromyogram (EMG): electrical signals detected at the surface of the skin caused by skeletal muscle depolarization.

Electron mobility: the ratio of electron velocity to electric field strength.

Electronic ice point (for a thermocouple): absolute temperature sensor (usually a thermistor or solid-state temperature sensor) used to determine the temperature of the sensing junction of a thermocouple. This information is used to correct the thermocouple output to the value it would have if the reference junction were at $0\,^{\circ}$C.

Electronic transducer: a transducer where the input or the output is electrical in nature (voltage, current, or resistance).

Electrooculogram (EOG): electrical signals detected on the forehead due to the currents in the retina. Differential amplification produces a signal that depends on the orientation of the eyes.

End-point voltage, A/D converter: lowest and highest voltages at which the lowest and highest bit transitions occur. Outside of this range, the output number does not depend on the input voltage.

End-point voltage, D/A converter: the highest and lowest voltages produced by any input number.

Energy (conversion factors):

1 electron volt (eV) $= 1.6021 \times 10^{-12}$ erg $= 1.6021 \times 10^{-19}$ J

1 eV/molecule $= 1.6021 \times 10^{-12}$ ergs/molecule

$\qquad\qquad = 23.060$ kcal/mole

1 erg $= 1$ dyn cm $= 1$ g cm^2/s$^2 = 6.241 \times 10^{11}$ eV

1 joule(J) = 1 newton meter = 1 kg m^2/s^2 = 1 V C
1 joule(J) = 10^7 erg = 6.241 × 10^{18} eV
1 kWh = 3.600 × 10^6 J = 3,409.54 Btu = 8.591,8 × 10^5 cal
1 calorie (g) (cal) = 4.190,0 J = 3.974 × 10^3 Btu
$\qquad\qquad$ = 1.163,9 × 10^{-6} kWh
$\qquad\qquad$ = 6.946,8 × 10^{-17} erg = 4.336,1 × 10^{-5} eV
1 ton TNT = 4.2 × 10^{16} erg

Euler's identity: $e^{j\phi}\cos(\phi) + j\sin(\phi)$, where $j = \sqrt{-1}$.

Exclusive-OR: logic process that produces an output of 0 for inputs of (0, 0) or (1, 1) and an output of 1 for inputs of (0, 1) or (1, 0).

Fahrenheit temperature scale: temperature scale defined with 32 °F at the freezing point of water and 212 °F at the boiling point of water at standard pressure.

Fast Fourier transform (FFT): efficient numerical method for computing the discrete Fourier transform.

Feedthrough (sample-and-hold): fraction of the input signal that appears at the output in hold mode, caused primarily by the capacitance of the open switch.

Ferromagnetic: materials in which an applied magnetic field is increased due to the cooperative action of magnetically oriented groups of molecules.

File: named unit of information on computer systems, typically a block of data, a report, source code for a program, or an executable program.

Filter, anti-aliasing: analog filter used to block input frequencies greater than one-half the sampling frequency.

Filter, differentiating: a high-pass filter that detects the time rate of change of a waveform.

Filter, integrating: a low-pass filter that time averages a waveform.

Flash A/D converter: A/D converter consisting of $2^N - 1$ comparators and address logic. One input of each comparator is connected in common to the voltage to be digitized, the other input of each comparator is connected to one value of an ascending series of reference voltages. The address of the comparator whose inputs are nearly equal is the digital output.

Flip-flop (edge-triggered): a digital circuit whose output is set equal to its input by an edge of a logic command pulse. At all other times, the output does not depend on the input.

Force (unit conversion):
1 dyne (dyn) = 1 g cm/s^2
1 newton (N) = 1 kg m/s^2 = 10^5 dyn = 0.224,7 lb (at surface of Earth)

Force transducer: device that changes its electrical characteristics as a function of force. Usually consists of strain gauges bonded to an elastic element.

Formants (of musical instruments): resonant frequencies due to shape and construction.

Formants (of vowels): resonant frequency bands of the human vocal tract.

Fourier convolution theorem: the Fourier transform of the convolution of two functions is the simple product of the Fourier transforms of the two functions.

Fourier frequency convolution theorem: the Fourier transform of the simple product of two functions is the convolution of the Fourier transforms of the two functions.

Fourier series expansion (of a periodic function): representation of an arbitrary real periodic waveform as a sum of harmonic functions:

$$h(t) = \sum_{k=-\infty}^{\infty} H_k \times \exp(j2\pi kt/P).$$

The expansion coefficients H_k are given by:

$$H_k = (1/P) \int_0^P h(t) \exp(-j2\pi kt/P) \, dt.$$

Frequency aliasing: when a waveform is periodically sampled, frequency components above one-half the sampling frequency appear as erroneous lower frequency components.

Frequency modulation (of a carrier): process of varying the frequency of a sine-wave carrier in proportion to the amplitude of another waveform. In FM (frequency modulation) radio transmission, the second waveform is the audio signal to be transmitted.

Frequency response: the output versus input relationship as a function of frequency.

Frequency scaling: process of multiplying all frequencies by a common factor. If the inverse Fourier transform of $H(f)$ is $h(t)$, then the inverse Fourier transform of $H(kf)$ is $(1/|k|)h(t/k)$.

Frequency shift theorem: process of shifting all frequencies by a constant frequency. If the inverse Fourier transform of $H(f)$ is $h(t)$, then the inverse Fourier transform of $H(f - f_0)$ is $\exp(j2\pi t f_0)h(t)$.

Full-duplex: serial digital transmission using separate conductors for send and receive.

Full-wave rectifier: circuit whose output voltage is equal to the absolute value of the input voltage.

Fundamental (first harmonic): harmonic (sinusoidal) component of a periodic signal that has the same period as the signal.

Gain–bandwidth product: product of the gain and the bandwidth of an amplifier. This is a constant for the operational amplifier, since its gain is inversely proportional to frequency.

Gain, common mode: the gain of a differential amplifier for signals present at both inputs. Expressed as output voltage divided by average input voltage.

Gain, differential: the gain of a differential amplifier, expressed as output voltage divided by the difference in input voltages.

Gain error (analog-to-digital or digital-to-analog converters): Difference between the measured and ideal output versus input slope.

Gamma ray: energetic ($>100 \, \text{keV}$) electromagnetic radiation, usually produced during the decay of radioactive isotopes and present in cosmic rays.

Gauge factor (strain gauge): the ratio G_S of the fractional change in resistance $\Delta R/R$ to the fractional change in length $\Delta L/L$. $\Delta R/R = G_S(\Delta L/L)$.

Gaussian distribution of error: a random distribution $G(x)$ with mean μ and standard deviation σ, given by:

$$G(x) = \exp[-(x - \mu)^2/(2\sigma^2]/\sqrt{2\pi\sigma^2}.$$

Glitch (of a D/A converter): brief erroneous spike that occurs in the output of a D/A converter when two or more input bits change at slightly different times.

Glottis: opening between the vocal chords.

GPIB (IEEE 488) interface: general purpose interfacing bus developed by the Hewlett Packard Corporation in 1965 and standardized by the IEEE. The GPIB cable has 24 conductors: 8 data lines, 8 control lines, and 8 ground lines.

Gravitational force: $F = GM_1M_2/R^2$, where $G = 6.67 \times 10^{-11}\,\text{N}\,\text{m}^2/\text{kg}^2$
$$= 6.67 \times 10^{-8}\,\text{dyn}\,\text{cm}^2/\text{g}^2.$$

Gray code: binary code with the property that only one bit changes from one number to the next.

Ground: conductor used to return current back to the power supply.

Ground fault interrupter: circuit used to detect a current imbalance between the hot and neutral conductors and to interrupt the circuit when the imbalance exceeds 5 mA. The GFI is used to protect personnel in the event of accidental contact with a hot conductor and ground.

Half-duplex: serial digital transmission where send and receive transmissions take place over the same conductor.

Half-flash A/D converter: circuit that uses a flash converter to determine the most significant half of the bits, a D/A converter to produce a corresponding voltage, and a difference amplifier and flash converter to determine the least significant half of the bits.

Half-wave rectifier: circuit that passes positive input voltages and blocks negative input voltages.

Handshaking: communication procedure used to ensure that both sender and receiver are ready before the transaction and that the data have been successfully transferred. Typical signals are "ready to send," "ready to receive," "data available," "data taken."

Hann window: a function that is used to multiply a truncated time domain series to eliminate the discontinuities in amplitude and slope at the ends of the truncation interval. The Hann function is given by $h_k = 0.5[1 - \cos(2\pi k/M)]$, where M is the number of elements in the series.

Harmonic (of a periodic signal): component of the signal that has an exact whole number multiple of the fundamental frequency of the signal itself. The complex amplitudes of the harmonic components can be determined from the Fourier transform of the signal.

Harmonic (signal): a sinusoidal wave having a pure frequency f.
$$V(t) = A\cos(2\pi f t + \theta) = B\cos(2\pi f t) + C\sin(2\pi f t)$$

$$B = A\cos(\theta) \quad C = -A\sin(\theta)$$
$$A = \sqrt{B^2 + C^2} \quad \tan(\theta) = -C/B$$

Harmonic distortion: variations in shape from the ideal harmonic form. Can be detected by nonzero Fourier amplitudes at >2 multiples of the fundamental frequency.

Harmonic oscillator (damped): an oscillator where the second derivative of the displacement depends on the displacement (harmonic term) and on the first derivative of the displacement (damping term).

Heart murmur: abnormal sounds produced by the blood passing through deformed cardiac valves.

Heart sounds: sounds produced by the heart, primarily due to closing of the heart valves. The first heart sound is produced by the closing of the AV valves and the second heart sound is produced by the closing of the semilunar valves.

Heat sink: usually a metal structure with fins for convective cooling used to transfer heat from a circuit element such as a power transistor or thermoelectric heat pump.

Hexadecimal: numerical representation of binary numbers with a base of 16 (4 bits). Differs from decimal representation in that 10, 11, 12, 13, 14, and 15 are represented by A, B, C, D, E, and F, respectively.

High-pass filter: filter that blocks low frequencies.

Hysteresis: dependence of the output on previous history. Occurs naturally in magnetic and mechanical systems (backlash).

Ice point: see *electronic ice point*.

IEEE-488 (GPIB) interface: general purpose interfacing bus developed by the Hewlett Packard Corporation in 1965 and standardized by the IEEE. The GPIB cable has 24 conductors: 8 data lines, 8 control lines, and 8 ground lines.

Impulse response: output response of a system when given an impulse delta function impulse at its input.

Incandescent: producing light by thermal agitation of electrons. Visible incandescent light is produced by objects hotter than $800\,°C$.

Inclusive-OR: logic process that produces an output of 0 when the inputs are (0, 0) and an output of 1 otherwise.

Increment: increase a number, usually by 1.

Incremental encoder (angular or linear): sensor that uses a single row of marks that can be sensed and counted to determine relative position.

Inductance: 1 henry (H) = 1 V s/A/turn.

Infrared radiation: electromagnetic radiation between 800 nm (red light) and 10^6 nm. Form of radiation used in making noncontact temperature measurements.

Inner product: the inner product P of two vectors a_i and b_i is given by $P = \sum a_i b_i$. The inner product of two functions $g(x)$ and $h(x)$ is given by $P = \int g(x) h(x) \, dx$. Two vectors or functions are orthogonal if their inner product is zero.

Input impedance: ratio of input current to input voltage. May be a complex function of frequency.

Input offset voltage: the input offset voltage of an amplifier is the output offset voltage when the input(s) are grounded divided by the amplifier gain. Note that the input offset voltage cannot be measured directly at the input.

Instrumentation amplifier: amplifier circuit that has: (1) output voltage proportional to the voltage difference between the two inputs, (2) very high input impedance, (3) low output impedance, and (4) constant gain over a large bandwidth.

Integral Fourier transform: the Fourier transform $H(f)$ of the function $h(t)$ is given by

$$H(f) = \int_{-\infty}^{\infty} h(t) \exp(-j2\pi f t)\, dt.$$

Integrated circuit chip: silicon crystal processed by doping and deposition to produce complete electrical circuits. Some chips contain over 1 million transistors.

Integrating (or dual slope) analog-to-digital converter: circuit that charges a capacitor for a fixed amount of time with a current that depends on input voltage, and then discharges the capacitor at a fixed current. The time required for discharge is measured by counting digital clock pulses to determine the converted number.

Integrating filter: a low-pass filter that time averages a waveform.

Interrupt: signal processed by a microcomputer that stops the current execution sequence, saves the current memory location and the state of the registers, and branches to a memory location specific to that interrupt.

Interrupt vector: series of memory locations to which a microprocessor branches when processing interrupts.

Ionic potential: potential produced by ions in solution.

Isolation amplifier: differential amplifier circuit able to withstand high voltages from dc to 60 Hz between the input and output.

Jitter (control system): deviation (rms) of the sense variable about its mean.

Johnson noise: random voltage generated in a resistor due to the thermal agitation of electrons within it. Produces 129-μV rms at room temperature across a 1-MΩ resistor within a bandwidth of 1 MHz.

Junction (thermocouple): point in a thermocouple where two dissimilar metals are joined.

Kelvin temperature scale: temperature scale defined with 0 K at absolute zero (the absence of all thermal agitation) and 273.16 K at the triple point of water.

Korotkoff sounds: sounds heard at the brachial artery as the pressure of a cuff on the upper arm is changed.

Lag time (control system): time after a step change in the set point that the sense variable reaches 10% of its final change.

Lambert–Beer law: the transmission of light through a colored solution is inversely proportional to the exponential of the concentration of the solution × the optical path length.

Laser: device that produces coherent, monochromatic, monodirectional light by stimulated emission.

Latch: [noun] a circuit that transfers data from input lines to internal storage when a "latch" pulse occurs. (May also have tri-state output buffers that transfer data from internal storage to output lines when a "select" level is asserted.)

Latch: [verb] the act of transferring digital data from the input to internal storage by using a digital control "latch" pulse. (As a verb, synonymous with clock and strobe.)

Least significant bit (LSB): bit in a binary number that represents "one," (has the least value of all the bits).

Least-squares fitting: process of determining the parameters of a function so that the function best describes a set of data in the least-squares sense (i.e. minimum sum of the squares of the differences between the function and the data).

LED (light emitting diode): a diode made of a high bandgap semiconductor so that electrons promoted by a forward voltage can emit visible light.

Length (conversion factors):

1 angstrom (Å) $= 10^{-8}$ cm $= 10^{-10}$ m $= 0.1$ nm

1 cm $= 0.032,808$ ft $= 0.393,70$ in $= 0.109,36$ yd

1 in $= 2.54$ cm (exactly) $= 25,400$ μm

1 ft $= 30.48$ cm (exactly)

1 km $= 10^5$ cm $= 0.621,4$ mi $= 3,280.8$ ft

1 m $= 1.094$ yd $= 39.37$ in $= 3.280,8$ ft

1 mile $= 1.609,34$ km $= 5,280$ ft (exactly) $= 1,760$ yd $= 8$ furlongs

1 light ns $= 29.979$ cm $= 0.983,6$ ft

Earth radius (equatorial) $= 6.378,2 \times 10^8$ cm

Solar radius $= 6.960 \times 10^{10}$ cm

1 astronomical unit (AU) $= 1.496 \times 10^{13}$ cm $= 9.296 \times 10^7$ mile

1 light year (LY) $= 9.460,6 \times 10^{17}$ cm $= 0.306,6$ pc $= 63,280$ AU

1 parsec (pc) $= 3.262$ LY $= 206,265$ AU $= 3.086 \times 10^{18}$ cm

Light: 1 Candella $= 1/638$ watts per steradian at 540×10^{12} Hz (555.17 nm).

Line driver circuit: a circuit with differential low impedance outputs designed to drive signals over long wires.

Linearity error (A/D or D/A converter): deviation between measured response and a straight line between the end points (or best-fit line).

Linearity error (general): difference between the measured output versus input response and the ideal linear behavior.

Liquid crystal thermometer: device consisting of a graded liquid crystal and a temperature scale. At a specific temperature only a limited length of the liquid crystal is dark.

Loudspeaker: actuator for converting an electronic waveform into an acoustic waveform (sound).

Low-pass filter: a filter that blocks high frequencies.

Luminescent: producing electromagnetic radiation as electrons drop from an electronic excited state to a lower excited state. Characterized by a relatively cool temperature and a narrow range of emission wavelengths.

Magnitude of a complex number: the square root of the sum of the squares of the real and imaginary components.

Maintainability: the ability to keep a system in repair and available for use.

Mass (conversion factors):

$1\,g = 1\,cm^3\,water\,(4\,°C) = 0.352,74\,oz\,(avdp)$

$1\,kg = 1\,liter\,water\,(4\,°C) = 2.204,6\,lb\,(avdp) = 35.274\,oz\,(avdp)$

$1\,lb(avdp) = 453.592\,g = 16\,oz\,(avdp)$

$1\,lb(apoth\,or\,Troy) = 373.42\,g = 12\,oz\,(apoth\,or\,Troy)$

Avoirdupois (pronounced av'ur du poyz) is the most commonly used system of weights in the United States and Great Britain. Apothecary is used for drugs and Troy is used for precious metals and gems.

Measuring junction: the thermocouple junction used to measure an unknown temperature. The other junction (the reference junction) is kept at a known temperature.

Membrane potential: ionic potential inside a cellular membrane relative to a distant point in the ionic media.

Metal–foil strain gauge: metal trace deposited on a thin Mylar substrate. The fractional change in resistance $\Delta R/R$ is related to strain ($\Delta L/L$) by the gauge factor G_S: $\Delta R/R = G_S(\Delta L/L)$.

Microcomputer: system consisting of a microprocessor, and memory for storing instructions and data. Practical considerations also usually require a keyboard, pointing device, video display, printer, magnetic and optical disks, and associated interface circuits.

Microphone: sensor for converting an acoustic waveform into an electronic waveform.

Microprocessor: integrated circuit able to read instructions and data from memory, and execute those instructions. These instructions include transfer of data from memory to registers, arithmetic operations, and conditional branching.

Missed codes (analog-to-digital converter): output codes that cannot be produced by any analog input voltage. Usually caused by inaccuracies in internal resistors.

Modem (modulator–demodulator): circuit for converting logic levels into two tones (modulation) for transmission over commercial telephone lines and for converting those tones back into logic levels (demodulation).

Most significant bit (MSB): bit in a binary number that represents the highest value of all the bits. For an N-bit number, the MSB has the value 2^{N-1}.

Motor unit (of a muscle): a motor nerve and the muscle cells that it enervates.

Nasal sounds: in human speech, sounds produced by raising the soft palate.

Negative feedback: process of inverting a fraction of the output and adding it to the input. This tends to stabilize the input versus output response.

Neuron: a nerve cell.

Neutron: uncharged, unstable nuclear particle with about the same mass as the proton. It is present in all nuclei except hydrogen, and when isolated, decays to produce an electron, a proton, and an electron anti-neutrino. It is produced in nuclear reactors and is present in cosmic rays.

Newton's method: method for the numerical solution of equations that uses the analytical first derivative.

Nibble: Four bits of binary data, or one-half of a byte. One nibble can be represented by a single hexadecimal number.

Noise: any component of the output that would be interpreted as a signal but does not depend on the quantity being sensed. This includes thermal noise in resistors, shot noise in amplifier elements, and external electrical interference.

Noise factor (of an amplifier): the ratio of the measured output noise to the ideal output noise, considering the input impedance as a source of Johnson noise and the gain of the amplifier.

Notch filter: filter that passes all frequencies, except for a narrow band.

Null hypothesis: the assumption that the factor being tested does not affect the data. If the computed probability of the null hypothesis is very low (say below 0.1%), then the null hypothesis can be rejected.

Null modem cable: an RS-232 interfacing cable that can be used to connect two local circuits together so that: (1) modems and phone connections are not needed, and (2) the handshaking lines are tied so that each circuit thinks that the other circuit is always ready to accept data.

Number of degrees of freedom: in fitting a model function to a set of data values, the number of degrees of freedom is equal to the number of data values minus the number of parameters varied in the fit.

Numerical integration: method of performing an integration by evaluating the integrand at specific points and combining the values. Simpson's method and the trapezoidal rule use evenly spaced points and simple weighted summation. Other methods are more efficient and more complicated.

Nyquist frequency: the maximum frequency that a sampling system can sample and recover. Equal to one-half the sampling frequency.

Nyquist theorem: to recover a waveform from its sampled values, the highest frequency present must greater than or equal to one-half the sampling frequency.

Octal: numerical representation of binary numbers with a base of 8 (3 bits).

Octave: a factor of two, such as in frequency.

Offset voltage (input and output): output offset voltage is the output voltage when the input is zero. The input offset voltage is the output offset voltage divided by the gain.

ON–OFF control: control algorithm that drives the actuator to its extreme limits, depending on the sign of the error signal (set point minus sense signal). Special case of proportional control with infinite gain.

ON–OFF temperature control: when sense is below set-point temperature, turn heater on full. When sense is above set-point temperature, turn heater off.

Open-loop gain (op amp): the gain of the operational amplifier as a function of frequency without feedback or output loading.

Operational amplifier (ideal): device with differential inputs V_- and V_+, and an output V_0. The output is given by $V_0 = A(V_+ - V_-)$, where A is infinite at all frequencies. No current flows into either input (infinite input impedance) and the output impedance is zero.

Operational amplifier (realistic): differs from the ideal operational amplifier in that: (1) the gain A is finite and drops at high frequency, (2) the output impedance is nonzero, (3) the input impedance is not infinite and different for the two inputs, and (4) input and output leakage currents produce an output offset voltage that depends on external resistance paths and temperature.

Optical encoder (angle or position): position sensor that uses strips of patterns that can be optically sensed to determine position.

Optical isolation amplifier: isolation amplifier that uses a time-varying light beam to isolate the input and output stages.

OR (exclusive and inclusive): the exclusive-OR of two inputs is zero if they are the same and one if they are different. The inclusive-OR of two inputs is zero only if both are zero, otherwise it is one.

Orthogonal functions: two functions are orthogonal if their inner product is zero (see also *inner product*).

Output impedance: effective series impedance of a circuit output.

Output offset voltage: output voltage when the input voltage is zero.

P-wave: electrical signal produced by the depolarization of the atria of the heart.

Parallel input port: circuit that stores digital data after being latched by an external circuit and transfers the data to memory under computer control.

Parallel output port: circuit for storing digital data by computer control and allowing it to be read by an external circuit.

Paramagnetic: materials in which an applied magnetic field is slightly increased due to the alteration of electron orbits. Such materials move from a weaker magnetic field to a stronger magnetic field.

Parseval's relation: relationship between the time integral of a function and the frequency integral of its Fourier transform.

Passband (of a filter): the range of frequencies that are passed unattenuated by a filter.

Peltier effect: transfer of heat energy between two thermocouple junctions by means of a current flow between them.

Peltier emf: diffusion potential produced when materials with different electron mobilities are placed in electrical contact. This emf acts to drive electrons from the material with the higher electron mobility to the material with the lower electron mobility.

Peltier thermoelectric heat pump: device for using electrons to pump heat. Usually consists of a number of semiconductor pairs with alternating higher and lower electron mobilities connected in series.

Pharynx: the air passage at the back of the mouth from the nose to the esophagus.

Phase of Fourier coefficient: for a complex Fourier coefficient F, the phase ϕ is given by $\tan(\phi) = \text{Im}(F)/\text{Re}(F)$.

Phase shift: the phase shift between the output voltage and the input voltage.

Phonocardiogram: sounds produced by the heart and detectable at the surface of the chest.

Photocathode: special semiconductor material placed at the entrance window of a photomultiplier tube to convert photons to electrons in a vacuum by the photoelectric effect.

Photoconductive current: current produced by electron–hole pairs in a biased semiconducting photodetector.

Photoconductive mode (of a photodiode): mode of operation where the diode is reverse biased at a fixed voltage and the current is proportional to the light intensity received.

Photodiode: a diode that is constructed to permit incident light to produce electron–hole pairs to be separated in the junction. The resulting current is proportional to the intensity of the incident light.

Photomultiplier tube: a vacuum tube containing a photocathode for converting light into photoelectrons, a series of dynode plates for amplifying the electrons, and an anode for collecting the electron pulse. Capable of detecting single photons with a quantum efficiency of about 20%.

Photon: fundamental quantum of electromagnetic radiation, with zero rest mass, one unit of angular momentum, and an energy E and wavelength λ related by $E = (1{,}241 \text{ eV/nm})\lambda$.

Photon energy (conversion factor): $E = hc/\lambda$, where $hc = 1{,}240$ eV nm.

Photovoltaic mode (of a photodiode): no external bias – diode is a source of current and/or voltage when light shines on it.

Photovoltaic potential: voltage produced by electron–hole pairs in an unbiased semiconducting photodetector.

PID control: control algorithm that uses a linear combination of the error signal P (the difference between the set point and the sense signal), its time integral I, and its time derivative D to generate a control signal.

Piezoelectric effect: property of some insulating crystals to become mechanically strained when an electric field is applied and to generate an electric field when a mechanical strain is applied.

pin **photodiode:** an intrinsic silicon crystal with p-type and n-type contacts. Light shining into the intrinsic layer makes electron–hole pairs which are collected as a current at the contacts.

Pink noise: random noise with an equal amount of noise power per octave (factor of 2) in frequency (see also *white noise*).

Planck's constant: $h = 6.625,4 \times 10^{-27}$ erg s.

Platinum resistance thermometer: consists of a fine platinum wire or deposit on an insulating substrate with an accurately known temperature versus resistance relationship.

Poisson statistics: probability $P(n|x)$ of n occurrences given the average probability x: $P(n|x) = x^n e^{-n}/n!$

Poisson's ratio: minus the ratio of the transverse strain to the longitudinal strain.

Polling: process where a computer program repeatedly checks if a process has been completed or that a device needs data.

Position sensor: a sensor whose output depends on position.

Positive feedback (in comparators): used to stabilize a comparator in its current output state and reduce the chance that input noise will reverse the output state.

Positron: anti-electron, positive in charge, produced in accelerators and by beta decay, where a proton is converted into a neutron plus a positron plus an electron neutrino.

Potentiometer: a resistor whose value depends on a shaft angle or position.

Power (unit conversion factors):
$1\,W = 1\,V\,A = 0.238\,cal/s = 107\,erg/s$
$1\,horsepower(hp) = 746\,watts(W) = 178.298\,cal/s$
$\quad = 2{,}547.2\,Btu/h = 550.22\,ft\,lb/s$

Power amplifier: amplifier able to drive an actuator, such as a loudspeaker, motor, or resistance heater. This usually requires discrete power transistors.

Power-supply sensitivity: ratio of the percent change in output voltage to a percent change in power-supply voltage.

Precision: ability of an instrument to detect small changes in the measured quantity reliably. Implies the ability to measure the same value under repeated identical conditions.

Precision of a sensor: ability to detect small changes in the measured quantity and measure the same value under repeated identical conditions (see also *accuracy*).

Pressure: Force per unit area
$1\,atmosphere\,(atm) = 760\,mm\,Hg = 29.921,3\,in\,Hg$
$\quad = 1.013,25 \times 10^6\,dyn/cm^2 = 14.696,0\,lb/in^2$
$1\,bar = 106\,dyn/cm^2$
$1\,dyn\,cm^{-2} = 9.869,23 \times 10^{-7}\,atm = 7.500,6 \times 10^{-4}\,mm\,Hg$
$\quad = 1.450,38 \times 10^{-5}\,lb/in^2$

Proportional control: control algorithm that amplifies the error signal (the difference between the set point and the sense signal) to generate a control signal.

Purkinje network: conducts the depolarization wave through the ventricular wall from the inner (endocardial) to the outer (epicardial) surfaces.

QRS-complex: electrical signals produced by the depolarization of the ventricles and the repolarization of the atria.

Quadratic equation:

$$ax^2 + bx + c = 0, \quad x = \left(-b \pm \sqrt{b^2 - 4ac}\right)/2a.$$

Quadrature addition: combining numbers by taking the square root of the sum of their squares.

Quantizing error (analog-to-digital and digital-to-analog converters): inability to convert (A/D) or produce (D/A) an arbitrary voltage due to the discrete nature of the digital information. The quantizing error of an N-bit converter is $\pm(V_{max} - V_{min})/2^{N+1}$, where V_{min} to V_{max} is the analog range of the converter.

R–2R ladder (of a D/A converter): resistor network consisting only of resistors of value R and $2R$ that permits N switches to control a binary sequence of currents $I_i = I_0 2^i$. Used in integrated circuit D/A converters to generate a current proportional to a binary number.

Rakine temperature scale: temperature scale defined with 0 R at absolute zero (the absence of all thermal agitation) and 459.67 R at the triple point of water.

Random-access memory (RAM): semiconductor memory that can be written and read by a computer program. Information is lost when power is lost.

Rectifier circuit, full wave: circuit whose output voltage is the absolute value of the input voltage.

Rectifier circuit, half wave: circuit that passes positive input voltages and blocks negative input voltages.

Reference junction (of a thermocouple): the thermocouple junction which is held at a known temperature. The other junction (the measuring or sensing junction) is used to measure an unknown temperature.

Refractory metal thermocouple: thermocouple with melting point above $3,600\,°C$, most commonly made from tungsten and tungsten–rhenium alloy. At such high temperatures, can only be used in an inert or vacuum environment.

Refractory period (of a nerve or muscle): period of time after an action potential or mechanical contraction when an additional stimulus produces little or no response.

Register: circuit used to sample and store digital data when given a control pulse (called a clock, latch, or strobe pulse). Specific examples are the edge-triggered flip-flop and the transparent latch.

Relative accuracy error (A/D converter): agreement between measured transition voltages $V_{n,n+1}$ plotted as a function of n and a straight line passing through the first and last transition voltages.

Relative accuracy error (D/A converter): agreement between the measured output voltages V_n and a straight line passing through the lowest and highest output voltages.

Relative digital position encoder: sensor that uses a single row of marks that can be sensed and counted to determine relative position.

Reliability: the ability of a system to be available for use.

Repeatability: the ability of an instrument to measure the same value under repeated identical conditions.

Residual: difference between a model function and the measured data.

Resistance ratio characteristic: the ratio of thermistor resistance at 25 °C to the resistance at 125 °C.

Resistive strain element: element whose resistance depends on strain.

Resistor color codes: see Appendix H.

Resolution (analog-to-digital and digital-to-analog converters): inability to convert (A/D) or produce (D/A) an arbitrary voltage due to the discrete nature of the digital information. The resolution of an N-bit converter is $\pm(V_{max} - V_{min})/2^{N+1}$, where V_{min} to V_{max} is the analog range of the converter.

Response curve of a sensor: (1) output as a function of input, or (2) change in output with time after a step change in input.

Response time of a sensor (exponential time constant τ): Time required for a sensor to record $1 - e^{-1} = 63.2\%$ of a step change in the measured quantity. After a time t, the sensor is within $e^{-t/\tau}$ of the final value. After five time constants, the sensor is within 0.67% of the final value. The 10–90% risetime takes $(\ln(0.9) - \ln(0.1))\tau = 2.197\tau$.

Risetime: the time for the output of a circuit to change from 10 to 90% of its final value after a step input. For a low-pass one-pole filter, the exponential time constant $\tau = RC$, the corner frequency f_c, the bandwidth Δf, and the risetime t_r are related by: $f_c = 1/(2\pi\tau)$, $\Delta f = (\pi/2)f_c = 1.571 f_c$, and $t_r = (\ln(0.9) - \ln(0.1))$, $\tau = 2.197\tau$ (see also *slew rate*).

RMS noise: root mean square of a signal about its ideal value.

RS-232 interface: serial, asynchronous interfacing standard with separate conductors for receive and transmit. Can be used with commercial telephone systems.

RTD: resistance temperature detector. Material (usually platinum) whose resistance is used to measure temperature.

SA (sinoatrial) node: cluster of highly conductive cells in the back wall of the right atrium where the cardiac depolarization wave is initiated.

Saccadic eye motion: rapid motion of the eyes that occurs when the target is moving faster than normal eye control will allow.

Sample mean: sum of the measured values divided by their number.

Sample-and-hold acquisition time: time from the hold-to-sample edge to the time when the output has reached its final value within a specified error.

Sample-and-hold amplifier: circuit that either samples or holds, depending on a logic input. The output is equal to its input in sample mode and held at a fixed value in hold mode. The held value is equal to the input value at the sample-to-hold edge.

Sample-and-hold droop rate: the drift in output voltage in hold mode due to charge leakage from the hold capacitor through the switch, the amplifier, or the capacitor itself.

Sample-to-hold charge transfer error: output offset error in the held value caused by charge dumped onto the holding capacitor from the switching circuit. Ideally, this should be a fixed offset, independent of input voltage level.

Sample-to-hold offset: shift in output level at the sample-to-hold transition after charge transfer has been accounted for.

Sampling theorem (Nyquist): to recover a waveform from its sampled values, the highest frequency present must be less than or equal to one-half the sampling frequency.

Schmitt trigger: a comparator circuit with a large amount of hysteresis (caused by positive feedback) so that the output state is relatively unaffected by noise at the input.

SCSI: small computer standard interface commonly used for magnetic "hard" disks and CD–ROM drives.

Seebeck effect: when two dissimilar metals are joined, at the two junctions held at different temperatures, a current will flow in the circuit.

Seebeck emf: the sum of two Thompson emfs and two Peltier emfs produced when two different metals are joined at their ends and the junctions are at different temperatures.

Select input: an input (used in many integrated circuits) that must be asserted before the circuit can function.

Self-heating of a thermistor: tendency of a thermistor in a resistance bridge to be heated by the bridge current through it.

Semilunar valves: the aortic valve between the left ventricle and the aorta; and the pulmonary valve between the right ventricle and the pulmonary artery.

Sense signal: in a control system, the signal (derived from a sensor) that represents the physical quantity to be controlled.

Sensing junction (of a thermocouple): the thermocouple junction used to measure an unknown temperature. The other junction (the reference junction) is kept at a known temperature.

Sensitivity: the minimum change in a measured quantity that an instrument can detect. Also the ratio of the change in output to a change in input.

Sensitivity of a sensor: change in output per unit change in physical quantity being sensed.

Sensor: transducer whose electrical properties depend on a physical quantity, and can be used (directly or with a bridge, etc.) to produce an electrical signal that is related to that quantity.

Septum (of the heart): muscular wall between the left and right ventricles. Contains the bundle of His where ventricular depolarization originates.

Serial I/O port: circuit for the asynchronous exchange of data with peripheral devices, such as keyboards and video display devices.

Set point: in a control system, the desired value at which the sense signal should be maintained.

Settling time (amplifier or D/A converter or S/H): time required for the output to remain within a specified error band of its final value after a step input.

Seven-segment decoder: circuit for converting a binary word into its visual equivalent in the form of lit segments. The number 8 is represented with all seven segments lit, and the number 0 has only the outer six segments lit.

Shot noise: noise that arises from the discrete nature of the charge carriers (electrons). All electrical currents have shot noise because the number of carriers per unit time is random (see *Poisson statistics*).

Sigma–delta A/D converter: a system that rapidly samples an analog voltage and uses a one-bit D/A in feedback to produce a string of ones and zeros. The pattern of ones and zeros is processed with digital filters to determine a digital representation of the analog voltage with considerable accuracy.

Sign extension: the process of extending the most significant bit to the left to convert a 2's complement number to a number with a larger number of bits. If the initial number has its most significant bit equal to one (i.e. a negative number), the leftmost portion of the new word is filled with ones.

Sinter: to fuse particles together by heating.

Slew rate: the rate at which a circuit can change its output voltage. Expressed in volts per microsecond and limited by the maximum output current and the output capacitance (see also *risetime*).

Small-signal rectifier: op-amp circuit able to take the absolute value of both small and large input voltage waveforms.

Specific heat: the thermal energy necessary to cause a unit change in temperature. Expressed in joules per degrees celsius or calories per degrees celsius.

Spectral leakage (as seen in the FFT of a periodic signal): when a frequency component of a signal does not have a whole number of cycles in the sampling window, the discontinuity at the edges of the window generates frequency components not present in the original signal. This appears as leakage into the adjoining Fourier coefficients and can be described by convolving the true Fourier transform with the Fourier transform of a rectangular sampling window.

Sphygmomanometer: device with a pressure gauge and inflatable cuff used on the upper arm to measure blood pressure.

Spontaneous emission: process where an electron in an excited state spontaneously returns to the ground state with the emission of a photon.

Stability: the ability to maintain the same response and noise level, despite the effects of time and usage.

Standard deviation: the standard deviation of m measurements a_i is given by

$$\sigma_a = \sqrt{\frac{1}{m-1}\sum_{i=1}^{m} R_i^2} = \sqrt{\frac{1}{m-1}\sum_{i=1}^{m}(a_i - \bar{a})^2}.$$

Standard error of the mean: the standard error (standard deviation) of the mean is given by

$$\sigma_{\bar{a}} = \sqrt{\frac{1}{m^2} \sum_{i=1}^{m} \sigma_{ai}{}^2} = \sqrt{\frac{\sigma_a{}^2}{m}}.$$

Status register (general): transforms logic voltage levels on external lines (TRUE or FALSE) to bits (1 or 0) that can be read by the program.

Status register (parallel input port): set by the input port whenever new data are strobed onto the input registers. Read by the microcomputer program to detect the availability of new data. Reset when program reads the data.

Step response: output response of a system when the input abruptly changes from zero to a standard value.

Stethoscope: a device for collecting heart sounds from the surface of the chest and directing them into the ear.

Stimulated emission: the process where the electromagnetic field of a photon stimulates an electron in an excited state to emit a photon coherent with the first. The laser is based on this process.

Strain: fractional change in length $\Delta L / L$.

Strain gauge (resistive): element whose fractional change in resistance $\Delta R / R$ is a known function of the strain $\Delta L / L$.

Stress: force per unit area.

Strobe: to store data in a register at the instant that a control pulse (called the strobe pulse) has a logic transition. (As a verb, synonymous with latch and clock.)

Student's *t* test: statistical test for comparing the means of two sets of measurements and determining the probability that the difference could have occurred by chance.

Successive approximation A/D converter: A/D converter that determines each output bit in sequence by using an D/A converter as follows: (1) the most significant undetermined bit of the D/A input is set to one and all lesser bits are set to zero; (2) the bit is determined to be zero if the analog input is less than the D/A output, otherwise it remains one. Steps 1 and 2 are repeated for each bit until all are determined.

Summing amplifier: amplifier circuit whose output voltage is an amplified version of the sum of the voltages of two or more inputs.

Switch, bimetallic: see bimetal switch.

Switch debouncing: process for removing secondary pulses that result from mechanical bouncing of an electrical switch.

Synchronous communication: transfer of data between two circuits using a common clock.

Systole: when used alone, means ventricular systole, the period of active ventricular contraction between the closing of the AV valves and the closing of the semilunar valves (see also *atrial systole*).

Systolic pressure: maximum systemic pressure that occurs immediately after ventricular contraction, when the aortic valve opens (see also *diastolic pressure*).

T-wave: electrical signal produced by the repolarization of the left and right ventricle muscles of the heart.

Temperature coefficient: change of a physical quantity (e.g. resistance, delay, dark current, output offset voltage) per unit change in temperature.

Temperature sensor: sensor whose electrical output depends on temperature.

Terminal, video display: consists of a display device (usually a cathode ray tube), a serial interface circuit, buffer memory, and a keyboard.

Thermal equilibrium: condition where heat flow into each component of the system is equal to heat loss to the other components or to the surrounding medium. As a result, the temperature of each component does not change with time. Note that each component may be at a different temperature.

Thermistor: a sintered semiconducting material that exhibits a large drop in electrical resistance with increasing temperature.

Thermistor bridge: circuit used to convert the resistance of a thermistor to a differential voltage.

Thermocouple: a circuit formed by joining two dissimilar materials. The open circuit output voltage (Seebeck emf) is a function of the temperature difference between the two junctions. One junction (the reference junction) is kept at a known temperature and the other (the sensing junction) is used to measure the unknown temperature.

Thermoelectric heat pump: device that converts electrical energy into a temperature difference by pumping heat energy.

Thermometer: a transducer (not necessarily electronic) that can transform temperature into a more readily observable signal.

Thermopile: Arrangement of thermocouples where the Seebeck emfs add to produce a larger open circuit voltage.

Thompson emf: diffusion potential produced when a conductor has a temperature gradient along its length. This emf acts to drive electrons from the warmer end to the cooler end.

Time scaling: process of multiplying all times by a common factor. If the Fourier transform of $h(t)$ is $H(f)$, then the Fourier transform of $h(kt)$ is $(1/|k|)F(f/k)$.

Time shift theorem: process of shifting all frequencies by a constant frequency. If the Fourier transform of $h(t)$ is $H(f)$, then the Fourier transform of $h(t - t_0)$ is $\exp(-j2\pi f t_0)H(f)$.

Toggle: [verb] to switch from one binary logic state to the other.

Tracking A/D converter: circuit that tracks an analog input by incrementing and decrementing a digital number. The number is constantly converted with a D/A converter and compared with the A/D analog input. The number is incremented if the D/A output is less than the analog input, and decrements the number if the D/A output is greater than the analog input.

Transducer (electronic): transducer where the input is electrical (i.e. an actuator), the output is electrical (i.e. a sensor), or both are electrical.

Transducer (general): device for converting one form of energy into another.

Transition voltages (of an A/D converter): Specific input voltages at which the output switches between one output number and the next.

Transparent mode latch: digital circuit whose output is equal to its input when the transparent mode is selected by a logic gate input. When not in the transparent mode, the output is held at its last transparent mode value and does not depend on the input.

Triple point: for a specific substance, the unique temperature and pressure at which the gas, liquid, and solid phases simultaneously exist.

Tri-state buffer: digital circuit with input and output lines, and a select line. When the select line is in one logic state, output = input. Otherwise the output is in a high-impedance state and neither drives nor loads any circuit connected to it. Used to permit several outputs to be connected in a common bus.

TTL: transistor–transistor logic, characterized by a pair of transistors in series between power-supply conductors at 0 and 5 V. Only one transistor conducts at a time, and this establishes whether the output is a logic high or low. During logic switching, high currents briefly flow.

2's complement representation: system where numbers are sign inverted by a 2's complement transformation.

2's complement transformation: transformation where each bit is reversed (complemented) and a one is added. Used in "2's complement arithmetic" to reverse the sign of a number so that an adder can perform a subtraction.

Type I and II errors: a Type I error is committed if the null hypothesis is rejected when it is true. A Type II error is committed if the null hypothesis is accepted when it is false. The Type II error is more acceptable and often the mark of a careful researcher who might rather wait to see the results of a more sensitive experiment than run the risk of making a false claim.

UART (universal asynchronous receiver/transmitter): serial RS232 interface circuit found in most older computers.

Unity-gain amplifier: a buffer amplifier used to preserve voltage. The buffer amplifier increases the available current and is able to drive a load that would otherwise reduce the voltage of the source.

Unity-gain bandwidth: the bandwidth of an amplifier when set for unity gain.

Variance: the square of the standard deviation

Velocity (conversion factors):

    1 mile/h = 1.609,3 km/h = 1.466,7 ft/s = 88 ft/min = 0.447,04 m/s

    1 ft/s = 1.097,3 km/h = 0.304,8 m/s = 0.681,82 mile/h

    1 m/s = 3.280,8 ft/s = 2.236,9 mile/h

Velocity of light in vacuum: $c = 2.997,924,58 \times 10^8$ m/s.

Ventricles: final pumping chambers of the heart. The right ventricle receives blood from the right atrium and pumps it to the lungs where it picks up oxygen. The left ventricle receives blood from the left atrium and pumps it to the rest of the body through the systemic arteries.

Video display terminal: consists of a display device (usually a cathode ray tube), a serial interface circuit, buffer memory, and a keyboard.

Virtual ground: the ground that exists at the negative input of an op amp, due to negative feedback and the grounding of the positive input.

Virtual short rule: if an op amp is in a negative feedback circuit, its output is not saturated, and if the open-loop gain is high, then negative feedback acts to keep the positive and negative terminals at the same potential.

VME interface: public domain standard which allows asynchronous communication among up to 21 processors and many more devices.

Vocal tract: consists of the mouth and pharynx.

Voiced sounds: sounds in human speech produced by forcing air through the glottis (the opening between the vocal chords).

Voltage follower: circuit whose output voltage is equal to the input voltage.

Voltage gain: the output voltage divided by the input voltage.

Vowel: voiced sounds produced with a fixed vocal tract.

Waveform recovery: recovery of an analog waveform from its digitized samples.

Waveform sampling: the process of sampling, digitizing, and storing a waveform.

Wheatstone bridge: a network of four resistors and a voltage source connected such that if all four resistors are matched, the output is zero.

White noise: random noise with an equal amount of noise power in each frequency interval (see also *pink noise*).

Wired-OR: circuit technique for connecting open collector logic outputs together. For negative logic (low = 1), this produces a logic OR function, since any low output can pull the result low. For positive logic (high = 1), this produces a logic AND function, since all outputs must be high for the result to be high.

Word: in microcomputer usage, the 16- or 32-bit entity that the microprocessor reads from memory or writes to memory with a single command.

WORM (write once read many) disk: an optical disk that can be written once during use and the written portions read thereafter as a CD–ROM.

X-ray: electromagnetic radiation (typically above 1 keV) produced when a charged particle (usually an electron) is accelerated, or when it drops from one orbital to one of lower energy.

Youngs's modulus: ratio of stress (force per unit area) to strain (fractional change in length).

Zero-offset error (analog-to-digital converter): difference between the lowest transition voltage $V_{0,1}$ and its ideal value.

Zero-offset error (digital-to-analog converter): difference between the lowest voltage that can be produced and its ideal value.

Index